Sustainability Solutions

An Interdisciplinary and Transdisciplinary Nexus

Sustainability Solutions addresses a wide range of global challenges with applied solutions in the interdisciplinary and transdisciplinary sustainability nexus. Titles in the series present a strong sustainable development focus, and topics which connect environment, economic and social applications are encouraged. Topics could include environmental impacts, vulnerability, risk assessment and reduction, human-environmental systems (e.g. food, water, health) and innovative solutions to climate change. Volumes in the series present novel solutions to societal challenges using innovative application, for example demonstrating a solutions-based approach using machine learning, artificial intelligence, decision support systems, advance geospatial technology. Case studies and shorter report volumes which present knowledge transfer for policymaking and governance are also encouraged.

Keywords:

Interdisciplinary environmental solutions, transdisciplinary knowledge transfer, geospatial technologies, natural resource management, decision support system, artificial intelligence, deep learning, smart technology, environmental science, economic, social and spatial, human-environment interactions, sustainable development

Avijit Roy · Margubur Rahaman · Pradip Chouhan ·
Farooq Ahmed
Editors

Environment and Public Health

Insights Towards Theory, Evidences and Sustainable Solutions

Springer

Editors
Avijit Roy
Department of Geography
Malda College
Malda, West Bengal, India

Pradip Chouhan
Department of Geography
University of Gour Banga
Malda, West Bengal, India

Margubur Rahaman
Department of Migration and Urban Studies
International Institute for Population Sciences
Mumbai, Maharashtra, India

Farooq Ahmed
Department of Anthropology
The Islamia University of Bahawalpur
Bahawalpur, Pakistan

ISSN 3059-278X ISSN 3059-2798 (electronic)
Sustainability Solutions
ISBN 978-3-031-99769-3 ISBN 978-3-031-99770-9 (eBook)
https://doi.org/10.1007/978-3-031-99770-9

This Springer imprint is published by the registered company Springer Nature Switzerland AG
The registered company address is: Gewerbestrasse 11, 6330 Cham, Switzerland

Foreword by Prof. S. B. Singh

In the contemporary era of complex global health challenges, understanding the nexus between environmental factors and public health outcomes is both timely and imperative. The need for comprehensive, multidisciplinary scholarship becomes indispensable. The edited volume *Environment and Public Health: Insights Towards Theory, Evidences and Sustainable Solutions* is a timely and well-curated contribution to this essential discourse.

What sets this volume apart is its ability to weave together theoretical insights, empirical evidence and practical interventions across diverse geographic and socio-economic contexts. The editors—Dr. Avijit Roy, Dr. Margubur Rahaman, Prof. Pradip Chouhan and Dr. Farooq Ahmed—have brought together an impressive group of contributors from South Asia and beyond. Their expertise spans fields as varied as medical anthropology, geospatial sciences, health geography and sustainable development. The structure of the book into five thematic parts ensures a comprehensive exploration—from historical epidemiology to cutting-edge artificial intelligence applications in environment and health monitoring.

This book serves both as a foundational text for students and a valuable reference for policymakers, researchers, and practitioners working at the nexus of environment and health. I commend the editors and authors for this scholarly achievement and believe this work will significantly contribute to shaping responsive, resilient and sustainable health systems globally.

Prof. S. B. Singh
(Retd.) Department of Geography
Banaras Hindu University
Varanasi, Uttar Pradesh, India

Foreword by Prof. Praveen Kumar Pathak

The inextricable link between environmental conditions and public health outcomes has become increasingly evident in the face of global climate change, rapid urbanization and emerging health threats. In this context, the edited volume *Environment and Public Health: Insights Towards Theory, Evidences and Sustainable Solutions* arrives as a vital scholarly contribution that bridges conceptual frameworks with actionable knowledge.

This book excels in presenting a multidimensional analysis of environmental health by integrating perspectives from geography, public health, anthropology and data science. Through well-structured thematic sections, the editors have curated contributions that range from traditional epidemiological insights to contemporary applications of geospatial and artificial intelligence tools. What distinguishes this volume is its grounded approach—drawing on real-world case studies from South Asia and beyond—to highlight both the vulnerabilities and the resilience of communities in the face of environmental stressors.

Particularly noteworthy are the chapters that address health preparedness in low- and middle-income countries, climate-induced nutritional challenges and the innovative use of technology in hazard monitoring and sustainable urban planning. These contributions underscore the importance of localized responses informed by global evidence.

This volume is not only a valuable academic resource but also a strategic guide for policymakers, development practitioners and health professionals striving to meet the targets of the Sustainable Development Goals—particularly those related to health (SDG 3), clean water and sanitation (SDG 6) and climate action (SDG 13).

I extend my congratulations to the editors and contributors for this remarkable effort. The book will undoubtedly play a significant role in fostering informed discourse, encouraging collaborative research and guiding evidence-based interventions in environmental and public health.

Prof. Praveen Kumar Pathak
School of Social Sciences, Centre for
the Study of Regional Development
Jawaharlal Nehru University
New Delhi, India

Preface

The journey of developing this book has been both inspiring and demanding, shaped by our shared commitment to understanding the intersections of sustainability, environment, and public health. In this book, we aimed to generate an inclusive and accessible resource that could fill the gap between theory and practice and cater to educators, students and professionals alike.

It is particularly designed to provide a clear and structured understanding of sustainability, the environment and public health. It blends introductory concepts with real-world applications. From a novice learner in search of a solid introduction to an expert watching for deeper insights, this book endeavoured to make the content appealing, applied and cutting-edge.

We owe a great deal to the numerous experts, researchers and practitioners whose work has shaped this field. Their input has been instrumental in helping us present a well-rounded perspective. Furthermore, feedback from colleagues has helped refine the material.

This book is divided into 5 parts and 23 chapters. **Part I** "Environmental Health: Theoretical Insights, Methods, and Research Direction" contains four chapters. Chapter 1 deals with historical, theoretical and epidemiological approaches to environmental health research. Chapter 2 shows emerging trends and future research directions in environmental health. Chapter 3 explains dynamic methodologies by integrating mixed methods for equity and sustainability. Chapter 4 deciphers the concepts like social equity, environmental justice and their public health impact. **Part II** "Environmental Hazard, Disaster, and Health Vulnerability" encompasses seven chapters in total. Chapter 5 shows environmental health in the anthropocene tropics. In Chap. 6, authors reviewed meteorological conditions linked with dengue outbreaks. Chapter 7 is an ethnographic description of the socio-environmental determinants of malaria. Chapter 8 assesses the vulnerability of livelihoods for riparian households in flood-affected villages in eastern India. In the remaining chapters of Part II, the authors explored the community's resilience in tackling health vulnerabilities (Chap. 9), climate change and public health preparedness in lower-middle-income countries (Chap. 10) and climate adaptation (Chap. 11).

In **Part III** "Environmental Pollution, Health Hazards and Sustainable Solutions," the authors explore 2D layered materials and their composites for wastewater treatment and photocatalytic dye degradation (Chap. 12), water contamination during floods and water safety in Bangladesh (Chap. 13), sustainable access to antenatal care services in flood-regions of India (Chap. 14), sustainable strategies for mitigating indoor pollution and associated cardiovascular issues (Chap. 15), and vegetation cover for sustainable urban planning (Chap. 16). **Part IV** is entitled "Climate Extremes, Food Security, Nutritional Impacts, and Sustainability". Chapters 17 and 18 attempted to explain the sustainable and climate-resilient food systems in Fiji, and institutional responses to sustainable agriculture in the climate-change context of Bangladesh. Chapter 19 emphasizes the role of community participation in effective watercourse management practices to secure water and food in Pakistan. Chapter 20 evaluates women's nutrition and health status in Kerala.

Part V draws upon "Geospatial and AI Innovations for Environmental and Health Solutions". In Chap. 21, contributors gave an introduction to GIS, AI and machine learning applications in monitoring environmental health. Chapter 22 suggests how to revolutionize environmental hazard management with remote sensing and GIS. Chapter 23 explores how emerging technologies can promote a sustainable environment and public health.

We anticipate that this book will serve as a reliable partner, sparking inquisitiveness and fostering a deeper appreciation for the environment and public health. We welcome your feedback and suggestions as we will continue to expand future editions.

Happy reading!

Malda, West Bengal, India — Dr. Avijit Roy
Mumbai, Maharashtra, India — Dr. Margubur Rahaman
Malda, West Bengal, India — Prof. Pradip Chouhan
Bahawalpur, Pakistan — Dr. Farooq Ahmed

Acknowledgements

Completing this book is a collective effort, and we are profoundly indebted to the numerous individuals who collaborated and helped us during this journey.

First of all, the contributors of the book deserve special appreciation for their intellectual contribution as they provided us with high-quality and relevant chapters for our book *Environment and Public Health: Insights Towards Theory, Evidences and Sustainable Solutions.*

Then, we are enormously thankful to our reviewers, whose insights and beneficial feedback facilitated the shaping of this book. We extend our sincere gratitude to the entire editorial team whose unwavering support made this publication possible. Our heartfelt thanks to Margaret Deignan, Publishing Editor at Springer Nature Singapore, Ambrose Berkumans, our Production Editor, and Uday Chatterjee, our Series Editor, for their steadfast commitment and exemplary coordination throughout the editorial process. Their professionalism, dedication and meticulous attention to detail have been instrumental in shaping this book.

We are deeply grateful to our families and friends for their continuous encouragement, patience and unwavering support throughout this journey.

Last but not least, we express our thanks to our readers for your interest and eagerness that motivates us to share knowledge and keep learning.

This edited book would be impossible without the collaboration and support of all these brilliant people.

Malda, West Bengal, India	Dr. Avijit Roy
Mumbai, Maharashtra, India	Dr. Margubur Rahaman
Malda, West Bengal, India	Prof. Pradip Chouhan
Bahawalpur, Pakistan	Dr. Farooq Ahmed

About This Book

Environment and Public Health: Insights Towards Theory, Evidences and Sustainable Solutions offers a concise yet comprehensive overview of the complex interplay between environmental factors and public health. Addressing key global challenges such as climate change, pollution, and natural disasters, the book blends theoretical frameworks with empirical research to highlight critical health impacts and propose sustainable, evidence-based solutions. It explores themes ranging from historical and methodological foundations to climate-related health vulnerabilities, pollution mitigation and food security. Emphasizing the use of geospatial tools, artificial intelligence and machine learning, the volume showcases how emerging technologies can support resilient health systems and sustainable environments. With diverse case studies and regional perspectives, it offers actionable strategies tailored for vulnerable populations. This book is an essential resource for researchers, policymakers and practitioners seeking to understand and respond to environmental health challenges through innovative tools and sustainable approaches.

Key Features of Our Book

- The book provides comprehensive coverage from basic principles appropriate for beginners to advanced topics for those looking for more technical depth.
- Concepts are elucidated in simple language with diagrams and illustrations for better comprehension.
- Logically organized chapters incorporate the state-of-the-art trends, research and developments in the field of environment and public health.

We have confidence that these features will make this book a valuable resource for anyone wishing to master environment and health.

Contents

Editors and Contributors

About the Editors

Dr. Avijit Roy is a State Aided College Teacher in the Department of Geography at Malda College in Malda, West Bengal, India. He teaches courses related to population geography, gender studies, regional planning, urban geography, social and cultural geography. He completed his Ph.D. in Geography from University of Gour Banga. He conducts research in a range of settings across the globe. His current research interests lie in women's sexual and reproductive health, maternal morbidity and mortality, maternal and newborn health, geriatrics, healthcare services, child marriage, violence against women and spatial analysis. He has published more than 30 research papers in reputed Scopus- and Web of Science–indexed journals. Recently, he has co-edited three scholarly volumes—Sexual and Reproductive Health of Women: Dimensions and Perspectives, Women's Health Dynamics in South Asia, and Sustainable Women's Health: Perspectives from South Asia—published by Springer Nature. Dr. Roy serves as an academic editor for PLOS ONE, BMC Public Health, and Scientific Reports (Nature Portfolio). He is also the guest editor of the Scientific Reports Collection titled "Understanding the Determinants and Consequences of Violence." His work reflects a multidisciplinary approach that integrates geography, public health, and social policy to address critical health and gender-related issues in South Asia and beyond.

Official Address: Department of Geography, Malda College, Malda, West Bengal-732101, India. e-mail: avijitr407@gmail.com; ORCID: https://orcid.org/0000-0002-2707-8639.

Affiliation: State Aided College Teacher, Department of Geography, Malda College, Malda, West Bengal, India

Dr. Margubur Rahaman is a health geographer with over six years of research experience, currently serving as a Senior Research Fellow at the Department of Migration and Urban Studies, IIPS, Mumbai. He holds an M.Phil. in Population Studies and dual master's degrees in Geography and Population Studies. Recently, he completed his doctoral work on maternal health among homeless women in Kolkata, exploring the complex interplay between environment, homelessness and reproductive health. Currently, he is working as Project Research Scientist II, Department of Community and Family Medicine, AIIMS Gorakhpur, India. Specializing in sexual and reproductive health, migration, and environment and public health, he has published several research articles and book chapters. He also serves as an editorial board member of *BMC Public Health* and *PLOS One* journals, with his research spanning diverse fields, including environment, health and healthcare management.

Official Address: Department of Migration and Urban Studies, International Institute for Population Sciences, Mumbai-400088, India. e-mail: margubur48@gmail.com, ORCID: https://orcid.org/0000-0003-2403-9033.

Affiliation: Senior Research Scholar, Department of Migration and Urban Studies, International Institute for Population Sciences, Mumbai, India

Dr. Pradip Chouhan is a Professor at the Department of Geography, University of Gour Banga, Malda, West Bengal, India. Earlier he was an Assistant Professor at the Department of Geography, Jadavpur University, Kolkata. His areas of research interest are fertility behaviour, public health, maternal health and child health. Actively engaged in teaching and research in Population Geography for nearly two decades, he has published more than 50 research papers in Scopus and Web of Science-indexed journals, edited 3 books and authored 1 book. He is an academic editor of *PLOS One* and *Scientific Reports Journal* and a reviewer of Scopus and Web of Science-indexed journals. He has completed a major research project funded by the Indian Council of Social Science Research, Ministry of Human Resource, Government of India. He has also completed one minor research project funded by the University Grants Commission. He has successfully supervised 7 Ph.D. and 8 M.Phil. scholars. Recently, he has published two edited books entitled *Sexual and Reproductive Health of Women: Dimensions and Perspectives* and *Climate Crisis, Social Responses and Sustainability Socio-ecological Study on Global* from Springer Nature.

Official Address: Department of Geography, University of Gour Banga, P.O. Mokdumpur, Malda, West Bengal-732103, India. e-mail: pradipchouhanmalda@gmail.com; ORCID: https://orcid.org/0000-0002-5500-2842.

Affiliation: Professor, Department of Geography, University of Gour Banga, Malda, West Bengal, India

Dr. Farooq Ahmed is an Associate Professor at The Islamia University of Bahawalpur. He is a sociocultural and medical anthropologist, as well as a public health researcher. He earned his doctorate through ethnographic research examining the complex interplay of political, economic, and cultural determinants of illness and maternal-child malnutrition in Pakistan. Dr. Ahmed's research interests span the political economy, social justice, health-seeking behaviors, and the social determinants of public health, including water insecurity, food insecurity, and maternal-child malnutrition. He has published over 50 research papers in journals indexed in Scopus and the Web of Science. In addition to his teaching and research, Dr. Ahmed serves as an

Academic Editor and as a reviewer for several international journals. He has held research affiliations with the Department of Anthropology at the University of Washington, Seattle; served as a HWISE Co-Investigator at Northwestern University; and worked as a Research Associate with the Schools of Public Health at Peking University and Tsinghua University. Dr. Ahmed is also a founding member of the Pakistan Anthropological Association.

Official Address: Department of Anthropology, The Islamia University of Bahawalpur, Bahawalpur, Pakistan. email: jamfarooq@hotmail.com; ORCID ID: https://orcid.org/0000-0002-4668-2882.

Affiliation: Department of Anthropology, The Islamia University of Bahawalpur, Bahawalpur, Pakistan

Contributors

Basit Ali Abbasi Department of Business Administration, ILMA University Karachi, Karachi, Pakistan;
Faculty of Engineering, Science, Technology & Management (ZUFESTM), Karachi, Pakistan

Fahim Abrar Abid Bangladesh Campaign, Global Human Rights Defence, Den Haag, The Netherlands;
University of Glasgow, Glasgow, UK;
IBEI, Barcelona, Spain;
University of Tartu, Tartu, Estonia;
Bangladesh Society of International Law, Dhaka, Bangladesh

Jam Bilal Ahmad Taxila Institute of Asian Studies, Quaid-I-Azam University, Islamabad, Pakistan

Farooq Ahmed Department of Anthropology, The Islamia University of Bahawalpur, Bahawalpur, Pakistan

Md. Tanzirul Alam University of Gothenburg, Gothenburg, Sweden;
BRAC James P Grant School of Public Health, BRAC University, Dhaka, Bangladesh

Mumtaz Alam Department of Social Sciences, College of Humanities, Education and Law, Fiji National University, Lautoka, Fiji

Hafiz Raja Muhammad Ali Department of Artificial Intelligence, Faculty of Computing, The Islamia University of Bahawalpur, Bahawalpur, Pakistan

Muhammad Anwaar Department of Artificial Intelligence, Faculty of Computing, The Islamia University of Bahawalpur, Bahawalpur, Pakistan

Muhammad Ashraf Department of Management Sciences, COMSATS University Islamabad, Vehari Campus, Islamabad, Pakistan

Naeem Asim Department of Public Health, The Islamia University of Bahawalpur, Bahawalpur, Pakistan

T. Athira Department of Geography, Government College, Chittur, Palakkad, Kerala, India

Mohsin Atta Department of Psychology, University of Sargodha, Sargodha, Pakistan

S. Balaselvakumar Government Arts College, Tiruchirappalli, India

Shashwati Banerjee Department of Humanities and Social Sciences, Indian Institute of Engineering Science and Technology, Shibpur, Howrah, West Bengal, India

Sudarshana Banerjee Department of Physics, Srinath University, Jamshedpur, Jharkhand, India

Bikash Barman Department of Geography, Malda Women's College, Malda, India

S. Nazrin Beegum Department of Geography, Government College, Chittur, Palakkad, Kerala, India

Pritimani Bharali Department of Botany, Gauhati University, Guwahati, Assam, India

Nabanita Bhattacharyya Department of Botany, Gauhati University, Guwahati, Assam, India

Pradip Chouhan Department of Geography, University of Gour Banga, Malda, India

Ujjwal Das Department of Geography, Fakir Mohan University, Balasore, Odisha, India

Pamela Deb Department of Geography, Amity University, Kolkata, India

Sourav Dey Department of Geography, A.B.N. Seal College, Cooch Behar, West Bengal, India

P. Dhanya Department of Geography, Government College, Chittur, Palakkad, Kerala, India

Moumita Dutta Department of Geography, Diamond Harbour Women's University, Sarisha, West Bengal, India

M. Ehthishamul Haque Department of Physics, Sacred Heart College (Autonomous), Tirupattur, Tamil Nadu, India

S. M. Fathima Khyrun Research Centre of Physics, Jayaraj Annapackiam College for Women (Autonomous), Periyakulam, Theni District, Tamil Nadu, India

Jannatul Ferdous Department of Public Administration, Comilla University, Cumilla, Bangladesh

Kunal Avishek Gounder College of Humanities, Education and Law, Fiji National University, Suva, Fiji

Aishwarya Harichandan Area of Economics and Business Environment, Indian Institute of Management Sirmaur, Sirmaur, Himachal Pradesh, India

Rashid Hashmi Sharda School of Media, Film and Entertainment, Sharda University, Noida, India

Md. Mofazzal Hossain Department of Law, Jagannath University, Dhaka, Bangladesh

Danish Iqbal Department of Psychology, University of Sargodha, Sargodha, Pakistan

K. Jayarajan Department of Geography, Government College, Chittur, Palakkad, Kerala, India

A. Jegatha Christy Research Centre of Physics, Jayaraj Annapackiam College for Women (Autonomous), Periyakulam, Theni District, Tamil Nadu, India

G. K. Jishnuvardhan Department of Geography, Government College, Chittur, Palakkad, Kerala, India

Nanigopal Kapasia Department of Geography, Malda College, Malda, India

Nishamani Kar Department of Geography, Rajiv Gandhi University, Itanagar, Arunachal Pradesh, India

Rabia Khalid Quaid-i-Azam University, Islamabad, Pakistan

Diwesh Kumar Department of Civil Engineering, Srinath University, Jamshedpur, Jharkhand, India

S. Lokesh Amith Nano Functional Materials Lab, Department of Nanoscience and Technology, Science Campus, Alagappa University, Karaikudi, Sivagangai, Tamil Nadu, India

Koyel Majumder Department of Geography, University of Gour Banga, Malda, India

Irsa Fatima Makhdoom Department of Psychology, University of Sargodha, Sargodha, Pakistan

Nilpaban Mali Department of Botany, Gauhati University, Guwahati, Assam, India

Najma Iqbal Malik Department of Psychology, University of Sargodha, Sargodha, Pakistan

Sujit Mandal Department of Geography, Diamond Harbour Women's University, Sarisha, West Bengal, India

Ritima Mishra Department of Botany, Gauhati University, Guwahati, Assam, India

Shefali Mishra Department of Psychology, Christ University, Bangalore, India

Z. Mohamed Riyas Department of Physics, Hindustan Institute of Technology, Coimbatore, Tamil Nadu, India

Rameswar Mukherjee Department of Geography, Samsi College, Malda, India

Usman Mustafa Pakistan Institute of Development Economics (PIDE), Quaid-I-Azam University Campus, Islamabad, Pakistan;
Ministry of National Food Security and Research (MNFSR), Islamabad, Pakistan

K. Prasad Department of Geography, Government College, Chittur, Palakkad, Kerala, India

Vishnu N. Ramani Independent Researcher, New Delhi, India

Sailajanda Saika Department of Geography, Rajiv Gandhi University, Itanagar, Arunachal Pradesh, India

P. Mary Santhi Convent of Jesus and Mary, Hampton Court, Mussoorie, Dehradun, India

Shailesh Kumar Sarangi Department of Mechanical Engineering, Srinath University, Jamshedpur, Jharkhand, India

Mithun Sarkar Department of Geography, University of Gour Banga, Malda, India

Muhammad Shahid Department of Insurance and Economics at the University of International Business and Economics, Beijing, China

Rao Muhammad Sajjad Sharif Department of Entomology, University of Agriculture Faisalabad, Faisalabad, Pakistan

Wareesa Sharif Department of Artificial Intelligence, Faculty of Computing, The Islamia University of Bahawalpur, Bahawalpur, Pakistan

S. Sharmila Department of Physics, Vel Tech Rangarajan Dr. Sagunthala R&D Institute of Science and Technology, Chennai, Tamil Nadu, India

Mohd. Shkir Department of Physics, College of Science, King Khalid University, Abha, AlQura'a, Saudi Arabia

Poornima Suryanath Singh Amity University, Noida, India

Surya Pratap Singh Department of Civil Engineering, Srinath University, Jamshedpur, Jharkhand, India

Jahangeer Ahmed Soomro Department of Business Administration, ILMA University Karachi, Karachi, Pakistan

Sreelu Sreepadi GHSS Kayanna, Kozhikkode, Kerala, India

Akancha Srivastava Jindal School of Psychology and Counselling, O.P. Jindal Global University, Sonipat, Haryana, India

Muhammad Suhrab School of Management, Huazhong University of Science and Technology (HUST), Wuhan, China

Zelina Sultana Faculty of Business and Law, Anglia Ruskin University, Cambridge, UK;
Department of Law, Jagannath University, Dhaka, Bangladesh

Ali Shan Tariq University of Central Punjab Bahawalpur, Bahawalpur, Pakistan

Sidra Zia Independent Researcher, Islamabad, Pakistan

Abbreviations

ADASYN	Adaptive Synthetic Sampling
AI	Artificial Intelligence
ANC	Antenatal Care
ARIMA	Autoregressive Integrated Moving Average
ASHA	Accredited Social Health Activist
BMI	Body Mass Index
CARE	Collective Benefit, Authority to Control, Responsibility, Ethics
CBPPs	Community-Based Psychoeducation Programs
CBPR	Community-Based Participatory Research
COPD	Chronic Obstructive Pulmonary Disease
DLNMs	Distributed Lag Non-Linear Models
DT	Decision Tree
ENSO	El Niño Southern Oscillation
EPA	Environmental Protection Agency
FAO	Food and Agriculture Organization
FPM	Fine Particulate Matter
GHG	Green House Gas
GIS	Geographic Information System
GLMs	Generalized Linear Models
HIAs	Health Impact Assessments
HVI	Health Vulnerability Index
IHD	Ischemic Heart Disease
IoT	Internet of Things
IPCC	Intergovernmental Panel on Climate Change
IVM	Integrated Vector Management
KNNs	K-Nearest Neighbours
LVI	Livelihood Vulnerability Index
ML	Machine Learning
NCAP	National Clean Air Programme
NGOs	Non-Governmental Organizations
NHCC	Natural Hazards and Climate Change

NWRC	National Water Resources Council
PCBs	Polychlorinated Biphenyls
PES	Payment for Ecosystem Services
PM	Particulate Matter
PRISMA	Preferred Reporting Items for Systematic Reviews and Meta-Analyses
PTMs	Potentially Hazardous Metals
SDGs	Sustainable Development Goals
SHG	Self-Help Group
SIDS	Small Island Developing States
SMOTE	Synthetic Minority Over-sampling Technique
SVM	Support Vector Machine
UNDP	United Nations Development Programme
WQI	Water Quality Index

Part I
Environmental Health: Theoretical Insights, Methods, and Research Direction

Chapter 1
Historical, Theoretical, and Epidemiological Approaches to Environmental Health Research

Jannatul Ferdous

Abstract The emergence of epidemiology as a distinct scientific field dates back to the seventeenth century, a period marked by favorable conditions for the collection, analysis, and interpretation of population data. In the eighteenth century, the field evolved further with the advent of comparative population studies, leading to the formal establishment of population study methods and practices. Modern epidemiology has its roots in these developments. A critical subfield, environmental epidemiology, specifically investigates the relationship between environmental factors and disease development. It focuses on the role of physical, chemical, and biological agents in the environment that may pose risks to large populations. Utilizing observational science methodologies, environmental epidemiology draws conclusions from natural experiments occurring within human populations. This chapter provides a review of the fundamentals of environmental epidemiology, including study designs and key measures used in research, based on an analysis of secondary sources. It highlights how recent advances in data science, technology, and biological sciences have positioned environmental epidemiology at the forefront of efforts to combat human disease and promote public health.

Keyword Epidemiological approach · Environment · Health · Technology

1.1 Introduction

Both in the industrialized and developing nations, the majority of people support eco-friendly policies. When the time was right for gathering, analyzing, and interpreting data from populations, the scientific field of epidemiology arose in the seventeenth century. In the seventeenth century, researchers began to compare different populations. Theoretical frameworks, methods, and tools for conducting population studies

J. Ferdous (✉)
Department of Public Administration, Comilla University, Cumilla, Bangladesh
e-mail: jannat.lata@yahoo.com

A. Roy et al. (eds.), *Environment and Public Health*, Sustainability Solutions,
https://doi.org/10.1007/978-3-031-99770-9_1

have evolved and improved since then (Morabia 2014). Environmental epidemiology examines how diseases spread and the factors in the natural environment that contribute to their development. This branch of medicine uses so-called "natural experiments"—observations made by members of the general public—to develop conclusions about the origins of illness. The application of epidemiological methods to the study of environmental risk factors for disease is known as environmental epidemiology (Pearce and Douwes 2009). Environmental variables are responsible for 23% of all fatalities and 24% of all diseases worldwide, according to a recent World Health Report (World Health Organisation 2002). An estimated 36% of fatalities among children aged 0–14 were linked to environmental factors, with the highest rate seen in poorer nations. Diarrhea, lower respiratory infections, various unintentional accidents (including work-related injuries), and malaria accounted for the majority of the disease burden. Water contamination, lack of proper sanitation, and hygiene practices (diarrhea) and indoor air pollution (lower respiratory infections) caused by solid fuel use in homes and possibly secondhand smoke are the leading causes of illness and death (World Health Organisation 2002). With the exception of outdoor air pollution, these dangers have been significantly reduced in most industrialized countries thanks to the availability of safe drinking water, sufficient food, proper waste disposal, immunizations, and good health care. Nonetheless, many industrialized nations are seeing an increase in the incidence of various diseases that are thought to have environmental origins, including cancer, heart disease, asthma, COPD, and diabetes (Pearce and Douwes 2009).

The field of modern epidemiology, which studies illness trends in populations, covers a lot of ground. There are subspecialties within it that focus on things like pharmaceutical clinical trials (Piantadosi 2024); variables include drinking, poor diet, and reproductive and developmental disorders; specific demographics; and infectious and chronic diseases (Darnton-Hill et al. 2004). Thus, epidemiology encompasses a wide range of activities, including randomized clinical trials of various treatments, comparisons of lifestyle variables like smoking, drug use, and alcohol consumption, risk assessments of occupational variables, and analyses of potential health determinants using cross-sectional and time-series designs (NRCCEP 1997).

Exposure to environmental contaminants is directly correlated with adverse health effects, according to environmental epidemiology. The World Health Organisation (2002) lists a wide range of possible health effects from minor skin irritation to death as a result of pollutants produced by local or global industrial or household processes. In epidemiological research, numerous socioeconomic characteristics are considered potential confounding factors because of their impact on the exposure-outcome link. As a result of their living and working circumstances, some members of a population are more prone to exposure than others (Stephens and Church 2017). Evaluating the statistical and biologic significance of changes in the frequency of occurrence of illnesses and related health and healthcare phenomena is at the heart of basic theoretical and general principles of epidemiologic evaluation, regardless of the topic matter (Grobbee and Hoes 2014). Observed rates of a phenomena are compared to those that would be expected under defined conditions in epidemiologic studies (Miettinen 2012).

As stated by Bühler (2023) and Breilh (2021), one of the main responsibilities of contemporary environmental epidemiology is to extract useful information from health and environmental data in the form of numerical values. Environmental studies often compare different exposure gradients to determine values such as disease prevalence, rates, or dangers. Put simply, epidemiology focuses more on studying disease in communities than it does on studying disease in controlled laboratory settings (Nriagu et al. 2019). At its core, epidemiological studies seek to establish causal links between preexisting conditions and the development of disease in human populations (Adami et al. 2011). Because of the challenges in precisely measuring exposure and impact, it is sometimes difficult to completely characterize such connections in the expansive area of environmental epidemiology. More complicating efforts to prove the existence of connections are individual differences in hereditary factors and effect modifiers ("susceptibility"), such smoking or food. Therefore, studies incorporating environmental agents require approaches to sensitively and precisely characterize exposure, impacts, and vulnerability (Links and Groopman 2010).

In order to properly evaluate the process of population capability expansion in human growth, epidemiology studies are essential "tools." (Aschengrau and Seage 2013). This research can help shed light on the relationship between environmental health deprivation and poverty in the communities in question by identifying, characterizing, quantifying, and assessing the non-voluntary hazardous exposures that impact the health of the population in specific contexts (Gee and Payne-Sturges 2004). Particularly, even when economic domestic policies and international accords help a developing country's economy flourish, the poverty-low health trap severely limits the ability of certain communities to better their living and working situations. The quality of a society's growth process can be studied using features provided by environmental and occupational epidemiology research. They aid in socioeconomic analyses by highlighting the ways in which environmental health is a social state that facilitates human growth at the individual, group, and community levels (Marsili 2009). In the fight against human illness and for the preservation of public health, environmental epidemiology has emerged as a frontrunner thanks to recent advances in data science, technological advancements, and breakthroughs in the biological sciences.

1.1.1 Rationale of the Chapter

All levels of epidemiology—molecular, individual, population, and ecosystem—are encompassed under the umbrella word "environment." Because environmental epidemiology focuses on ecosystem-level analyses, it frequently employs research methodologies distinct from other branches of epidemiologic study, including systems-based techniques like complexity theory. When studying environmental epidemiology, there are a few aspects that researchers find particularly difficult. Developed and underdeveloped countries have quite different environmental health problems (Doll 1992). In the former, lead, secondhand smoke from vehicles and other sources, as well as cigarette smoking, are among the leading causes of preventable disease and death in the United States. Levels of lead that were deemed safe for children's development a decade ago are now known to cause irreversible harm to their brainpower (ATSDR 2000). Numerous prevalent chronic diseases in both industrialized and developing nations have their causes still not fully understood, which is why there is a strong need to find preventable environmental dangers (NRCCEP 1997).

After evaluating possible exposures in the past, present, or future, an ideal public health impact assessment would move on to developing testable hypotheses about the effects that should be investigated in one or more targeted populations. The gold standard in environmental epidemiology is to account for all relevant factors, including those that can muddy the data or make it harder to draw conclusions, and to think about potential negative health impacts in both exposed and unexposed people. Even in peer-reviewed, published reports, there are not many studies that live up to this standard (NRCCEP 1997). Researching low-level exposures is a common practice in environmental epidemiology. Exposure to dioxins in the environment is one such example. The general public's exposure to incinerators and similar sources is typically orders of magnitude lower than what particular occupational contexts can provide, such as for incinerator workers or those who produce chemicals contaminated with dioxin. Consequently, environmental epidemiology often aims for risks that are just barely detectable. But that does not imply the dangers posed by environmental exposures at lower levels are not significant. People are far more aware of the risks associated with these types of exposures than with those that are perceived as having some degree of discretion, for one thing because these exposures are usually involuntary (for example, those who live near incinerators have little choice over whether they are exposed to dioxin from the sites). Second, while an individual's risk may be little, the total burden of sickness caused by an exposure can be high if it affects many people. Although dioxin exposure at work is a serious concern for some, the overall effect on public health will be minimal because so few individuals are exposed to the chemical. Conversely, given the enormous population at risk, the question of whether or whether low-level environmental dioxin exposures are harmful would constitute a major public health concern (Pearce and Douwes 2009).

Environmental epidemiology has come a long way in the last few decades, thanks to a proliferation of studies, a broadening of research priorities, and most notably, improvements in methodology. This has led to a clearer picture of how different environmental factors affect human health. Public health indicators, including the frequency, length, and intensity of health consequences, form the basis of the prioritization criteria in occupational and environmental epidemiology research. Accordingly, research priorities have been defined based on their importance to public health, and the findings of these studies have helped to strengthen the foundation for sound decision-making (Bertollini et al. 1996). From this vantage point, research in environmental epidemiology that aims to better understand the interplay between the environment and human health can provide light on the pressing and potentially intractable health issues plaguing modern society (Marsili 2009). Reproductive health encompasses a wide range of outcomes, from impacts on future generations to the state of men's and women's reproductive systems at various stages of life, such as sexual development, menstruation initiation, menopause, sexual function, and endometriosis. Despite the intense public interest in these topics, any investigation into potentially harmful environmental effects must adhere to the highest standards of scientific rigor. It is necessary to clarify the potential role of environmental variables that are either preventable or controllable in the development of chronic diseases and other negative health outcomes, even though dietary deficiencies and hereditary predispositions are major risk factors in many cases. Due to a lack of suitable data, methodologies, and theoretical frameworks, it is frequently impossible to discern the temporal and spatial etiologic patterns of many diseases. Therefore, improving methods for searching for environmental variables that may contribute to illness should be a major focus of environmental health research.

1.2 Materials and Methods

Environmental epidemiology, namely illness distribution and environmental causes, will be the subject of this comprehensive analysis of peer-reviewed research. The study includes a total of 40 papers selected from prominent academic databases such as PubMed, Scopus, and Google Scholar. With an emphasis on studies conducted in natural settings with human subjects, this review will cover research on biological, chemical, and physical risk factors as well as epidemiologic metrics and study methodologies. Contributions to the topic from the fields of data science, biology, and technology will also be covered. This review seeks to give a thorough synopsis of environmental epidemiology's development and present approaches in illness prevention and public health protection by combining literature from both historical and modern viewpoints. Research not specifically addressing environmental epidemiology or population health will not be considered for inclusion.

1.3 Results and Discussion

The branch of public health known as "environmental epidemiology" focuses on the dangers that the natural world poses to people's health. It is based on public health surveillance as its mainstay and uses data analysis for sparse populations and limited areas extensively (Aldrich 2000). In the past, epidemiologic studies mainly dealt with infectious diseases and mortality rates; however, modern epidemiology encompasses a far wider range of topics, including the study of populations to determine the prevalence and causes of health-related states and events and the use of this knowledge to control health problems (Tyler and Last 1991). Epidemiology has long been associated with illness prevention efforts since the field's findings can point to modifiable risk factors for many diseases.

As a natural progression of the discipline, environmental epidemiology broadens the focus to encompass physical, chemical, and biologic variables that may influence population health and illness trends. To a large extent, environmental epidemiology relies on patterns of occurrence rather than conducting experiments to establish scientific conclusions. Prevention of future disease and improvement of public health are the primary goals of this initiative, which seeks to identify modifiable risk factors (NRCCEP 1997). Numerous epidemiologic studies investigate the link between potential dangers and the consequences on people's health; these studies typically focus on the correlation between a particular exposure and a particular disease or risk factor (Ezzati et al. 2002). Although, in environmental epidemiology, exposures and consequences are typically multiple. Concerning environmental contamination, many of the relevant risk factors are culled from massive datasets, including continuous variables and a wide range of clinical health indicators. A large portion of cancer epidemiology has been devoted to investigating particular cancer sites and identifying key risk factors for particular cancers, such as the correlation between benzene exposure in the workplace and leukemia or asbestos and mesothelioma. Consistently, prospective cohort studies have dominated cardiovascular epidemiology, with an emphasis on finding a few number of risk variables (NRCCEP 1997). Achieving health equity is a fundamental human development aim, and the capacity to influence, enhance, and select one's environment is a key component of that endeavor. This includes one's home and place of employment. Individuals, communities, and social groupings are considered to be in a state of environmental health when their susceptibility to controllable health-related living conditions and environmental health-related functionings are measures of poverty and deprivation in this paradigm. As a result, environmental health is best understood as a social state that is intrinsic to public health, the economic environment, and public development strategies. These spheres are highly interdependent, and as a result, the relationship between personal wealth and societal health is profoundly impacted (Marsili 2009).

The conditions were favorable for gathering, analyzing, and interpreting data from populations in the seventeenth century, which led to the emergence of epidemiology as a distinct scientific field. Research comparing different populations began in the seventeenth and eighteenth centuries. A corpus of tools for doing population research has been created, refined, and theorized since then. It lies at the very heart of modern epidemiology as a scientific field. It traces the development of these epidemiological approaches and ideas through four distinct stages. There was no overarching theory that governed the conduct of population studies in their preformal period (the seventeenth and nineteenth centuries). In the first stage, which lasted from 1900 to 1945, researchers began to categorize study designs, found solutions to the problem of non-comparability, established epidemiology as a legitimate academic field, and released textbooks on the subject. During the third stage, known as classic epidemiology (1945–1980), researchers honed their study designs and codified methods for measuring effects, interactions, biases, and confounding factors. Integrating methodologies and concepts into a theoretical corpus, contemporary epidemiology (since 1980) includes study designs, assessments of event occurrences and impacts, interactions, biases, confounding effects, and causal inferences. There are signs that a new era is about to begin, marked by advancements in areas of theory including the evaluation of biases, confounding effects, and complicated interactions (Khlat 2005).

To some extent, environmental epidemiology is tasked with identifying minute effects on human health or early warning signs of a catastrophic chain reaction (Aldrich 2000). As such, environmental epidemiology plays a key role in informing the public and educating them about potential dangers (Glik 2007). Developing reliable biological indicators of exposure and antecedent health impacts will significantly advance environmental epidemiology. There are currently statistical tools available to track illness rates in high-risk populations as a whole. This will allow for the early detection of risks and the identification of sentinel events. Recent developments in geographic methodology have greatly benefited the field of geographic studies. These developments in the field still necessitate additional testing and optimization (Aldrich 2000). Worker, social group, and community inaction on environmental health issues hinders vulnerability assessment and function choice toward health and environmental sustainability. Here, environmental and occupational epidemiology investigations can help pinpoint health problems affecting particular populations, re-creating the specific local context of the complex phenomena under study (Fazzo and Comba 2004). Sharing this approach with socioeconomic researchers and decision-makers makes it all the more pertinent to address public health policies as part of development strategies, with the goal of improving the living conditions of the most socioeconomically disadvantaged groups and communities.

Assuming an integrated approach to health, environmental epidemiological studies and socioeconomic development analyses based on the multidimensionality of human development approach have both shown the connection between environmental factors and population health in the past 20 years (Brauch 2005). Recall that the integrated approach to health is foundational to both the UN's definition of human development and the European center for Environment and Health's definition of environmental health (Marsili 2009). Because it affects people's actual living circumstances, environmental health is an important component of human development in the context of the multidimensional development model. As a nation's economy and society progress, it has an impact on environmental health and wellness initiatives.

According to Sen's capability approach, a society's progress can be measured by the increase in capability and freedom to choose from various functionings (states of being and doing) that enable individuals, groups, and communities to live lives they value. Individuals' capacity to maintain and improve their health ranks high among the most fundamental requirements for a fulfilling existence (Anand 2002). Societal and economic institutions endow people and groups with the capacity to thrive. In this view, an interdisciplinary strategy that incorporates studies of the environment, the workplace, and socioeconomic development lends credence to the idea that environmental health deprivation is a significant contributor to poverty, and it advocates for solutions that take into consideration the uniqueness of each place and the problems that are intrinsic to it. The possible influences of an interdisciplinary strategy can be illustrated by a number of instances. For starters, there are epidemiological studies that focus on populations at great risk (Fazzo and Comba 2004) such as citations. High-risk populations are defined by very high levels of non-voluntary hazardous exposures and are impacted by socioeconomic deprivation; this is due to an inherent condition. The distinctive dual role of epidemiological studies in identifying high-risk populations and supporting preventative interventions is further highlighted by these investigations. This is well-known in environmental epidemiology, but it is underutilized in terms of its connections to socioeconomic analysis and its impact on poverty assessment (Marsili 2009).

Indices of impoverishment in connection with health are utilized in these investigations. A commonly used tool in studying illness incidence in connection with environmental factors is the incorporation of a deprivation measure, which accounts for the effects of socioeconomic factors (Carstairs 1995). Using the material deprivation index to control socioeconomic characteristics connected to the populations included in these investigations is relevant. Since it is linked to both the illnesses and the exposures under study, the material deprivation index is considered a confounding variable as it encapsulates the recognized socioeconomic determinants (Jager et al. 2008). We know that socioeconomic status can play a confounding role in environmental epidemiological studies, particularly those that focus on small areas. How much of a role it plays, however, will vary by disease type (which is influenced by the latent period before clinical diagnosis) and by the type of industrial site being studied (Carstairs 1995). In order to evaluate socioeconomic vulnerability in diverse local contexts, socioeconomic analyses can assist in selecting the most appropriate factors. These analyses can also reconstruct the socioeconomic history of the researched locations to detect socioeconomic vulnerability in the past. It is worth noting that through combining different fields of study, we can learn more about how socioeconomic status affects environmental health and how epidemiological studies can benefit society as a whole (Marsili 2009).

Human rights and law experts, as well as ecologists, social scientists, and behavioral scientists, as well as epidemiologists, need to work together to identify, stop, and lessen the impact of socio-ecological changes that threaten the health and welfare of people all over the globe (Soskolne et al. 2007). Controlling known hazards is a common component of environmental epidemiology studies concentrating on endemic and pandemic diseases in developing nations. However, industrialization and economic expansion within countries and regions are inextricably linked to socio-ecological changes. Occupational and environmental epidemiology research centered on contextualized health issues has not been adequately applied when considering the unchecked urbanization and industrialization of emerging nations. Developing nations can greatly benefit from the findings of environmental and occupational epidemiological studies, which shed light on the unique challenges faced by low-income areas in terms of environmental health. By bringing together experts from different fields, we can better understand the root causes of this problem and, hopefully, find ways to help the most disadvantaged people in these nations.

Environmental and occupational epidemiology research collaborations have aimed to share both the partners' accumulated scientific knowledge and their unique local experiences. Because of this, research in Ecuador has been able to zero in on the unique challenges posed by the country's work environment and workplace health. Some of the studies conducted in Ecuador have dealt with the involuntary hazardous exposures of high-risk populations defined by epidemiological vulnerability and socioeconomic deprivation, which affect their environmental health. Several studies have examined the effects of pesticide exposure on children attending schools near flower plantations (Marsili 2009; Mobaso 2023) and on communities living near mining industries (the synergy of social and environmental factors as determinants of silicotuberculosis). These findings highlight the interconnectedness of environmental and socioeconomic elements in developing nations and provide more evidence that well-known questions need to be investigated in situ to obtain solutions suitable to the local context.

Research on large populations' unknown non-voluntary exposures provides the basis for the paper's proposed cooperation strategies, which include the exchange of information on scientific issues presently debated in the field of international epidemiology (Marsili 2009). Because these studies have the potential to shed light on environmental exposures whose harmful effects are still up for debate, they are relevant to both parties involved in the cooperation. Lastly, information dissemination initiatives in Ecuador are prioritized because they make research and study results available to society (e.g., private firms, local communities, and trade unions) (Jawanda 2000). Following a methodology plan, the affected social subjects and local authorities receive basic information and dissemination on environmental health issues (Marsili 2009). With this aim in mind, scientific articles and reports have been released in both Spanish and Italian to raise awareness about the studied subjects among operators of cooperation in both nations and to help the Ecuadorian society gain access to reliable information.

1.3.1 Limitations of the Chapter

Because it is based on secondary sources, this chapter may only be able to draw conclusions about the breadth, depth, and biases of previous research. If the review's inclusion criteria are limited to studies pertaining to environmental epidemiology and population health, it risks leaving out valuable multidisciplinary perspectives. Publication bias can also occur while using databases like PubMed, Scopus, and Google Scholar, since articles in journals that are not indexed may not be found. The focus on human subjects in natural experiments may impede the development of experimental designs and mechanical research that reveal important causal pathways. Plus, if new discoveries in data science and technology are not extensively reported in peer-reviewed literature quite yet, the fastest-moving trends may be overlooked. In conclusion, we cannot directly evaluate the findings' application to varied public health contexts because the review is retrospective, which prohibits primary data gathering.

1.3.2 Recommendations

Although the positivist theory and philosophy that support epidemiology are rarely stated, they do exist. The underlying idea is that different populations have different patterns of health and disease due to differences in the frequency or vulnerability to the underlying causes. A future in epidemiology that makes use of data collected from entire populations—consisting of potentially hundreds of millions of individuals—is within sight. Potentially this century, research involving billions of people will take place. A code of ethics and good behavior that supports epidemiology's scientific and practical goals is necessary for its health applications. Epidemiologists have a professional, ethical, and social responsibility to be well-versed on the field's history and the broader factors that influence health and illness (Valles 2021). A fruitful future can only be molded at that point. Here, environmental health deprivation is seen as a defining feature of poverty brought about by unchecked urbanization and industrialization, which disproportionately impacts developing nations (Hardoy et al. 2013). An integrated approach to health is advocated, and environmental and occupational epidemiology studies are seen as "tools" that are essential for properly evaluating the process of population capability expansion in human development. Specifically, identifying high-risk populations in a developing nation and the health effects of involuntary hazardous exposures highlights the connection between environmental health deprivation and human development, and helps disseminate information about the root causes of poverty. This research has the potential to uncover hitherto untapped "tools" for breaking the poverty-low health trap. Poverty is a multi-faceted deprivation state, and thus emphasizes the importance of an interdisciplinary approach that brings together studies in occupational and environmental epidemiology with socioeconomic development evaluations (Marsili 2009).

Two ways for effective collaboration are proposed after a review of the effects of international scientific cooperation on environmental epidemiology (WHO 2011). To begin, an interdisciplinary approach is necessary for environmental and occupational epidemiology-focused international cooperation activities. This framework is innovative because it promotes research and field studies that are relevant to the actual local problems and needs, which is particularly important in developing countries. Secondly, when partners engage in scientific cooperation, they must work together and take a unified approach to issues including the free flow of scientific information, the use of rigorous scientific standards in all investigations, and the development of effective scientific techniques. For the cooperation's results to last, this will encourage technology transfer and fortify capacity building activities (Marsili 2009).

1.4 Conclusion

There is a lack of consistently collected data on incidence rates in specified populations for a number of chronic, degenerative diseases that could be of interest to environmental epidemiologists. The requirements of epidemiologic research are not always met when evaluating exposure. So, it is not always easy to tell if a new or changing environmental exposure is to blame for a shift in illness incidence. In cases where an exposure gradient can be identified, it is possible to link disease risk to dosage in well-designed studies. Unfortunately, establishing baseline disease occurrences and developing suitable exposure measurements or estimations is sometimes necessary before investigating diseases that may have an environmental cause (NRCCEP 1997).

Research on the impacts of environmental exposures on human health has a number of challenges, including an absence of advanced tools for measuring long-term effects, the inherent limitations of certain study designs in terms of methodology, and the occasionally contentious atmosphere in which such investigations take place. In environmental epidemiology, cross-sectional investigations are frequently necessary, even if cohort or case–control studies are the more trustworthy options (Johansen et al. 2017). The necessity for quick health evaluations that may not meet the standards of environmental epidemiology has, up until recently, been the primary focus of federal and state funding. Environmental epidemiology is an area that has received little attention from academic departments of epidemiology, and those that have attempted to address the issue have faced significant funding constraints. How to fix these issues is what the rest of this book is all about (NRCCEP 1997).

This first experience with scientific cooperation was fruitful and encouraging, but there are still some important points to note. In this partnership initiative, the first priority is the need to broaden the multidisciplinary approach. As a result, new and intricate types of collaboration will be encouraged, which could lead to a greater societal impact from research. This is especially important for developing nations that have a lot of bureaucracy and structural weaknesses that make it hard to establish development plans that will promote environmental health in the short and long term. The third and last major worry is the need for emerging nations to invest in scientific infrastructure. The establishment of suitable local research facilities, including laboratories and instruments, necessitates sufficient financial backing for any program of bilateral scientific collaboration that is based on a cooperative approach. For the sake of the cooperation's long-term viability, this might ensure the NGOs' scientific autonomy (Marsili 2009). It is highly suggested that cases involving hazardous waste management, contamination cleanup, and projected industrial growth incorporate environmental epidemiological concerns.

References

Adami HO, Berry SCL, Breckenridge CB, Smith LL, Swenberg JA, Trichopoulos D, Weiss NS, Pastoor TP (2011) Toxicology and epidemiology: improving the science with a framework for combining toxicological and epidemiological evidence to establish causal inference. Toxicol Sci 122(2):223–234

Aldrich TE (2000) Environmental epidemiology forward. Chemosphere 41(1–2):59–67

Aschengrau A, Seage GR (2013) Essentials of epidemiology in public health. Jones & Bartlett Publishers

Anand S (2002) The concern for equity in health. J Epidemiol Community Health 56(7):485

ATSDR T (2000) ATSDR (Agency for toxic substances and disease registry). Prep Clement Int Corp, Contract 205:88–0608

Bertollini R, Lebowitz D, Saracci R, Savitz DA (1996) Environmental epidemiology. WHO Regional Office for Europe, CRC Press, Rome

Brauch HG (2005) Threats, challenges, vulnerabilities and risks in environmental and human security. UNU-EHS

Breilh J (2021) Critical epidemiology and the people's health. Oxford University Press eBooks. https://doi.org/10.1093/med/9780190492786.001.0001

Bühler N (2023) Precision public health in the making: examining the becoming of the "social" in a Swiss environmental health population-based cohort. Front Sociol 8:1219275

Carstairs V (1995) Deprivation indices: their interpretation and use in relation to health. J Epidemiol Commun Health 49(Suppl 2):S3–S8

Darnton-Hill I, Nishida C, James WPT (2004) A life course approach to diet, nutrition and the prevention of chronic diseases. Public Health Nutr 7(1a):101–121

Doll R (1992) Health and the environment in the 1990s. Am J Public Health 82(7):933–941

Ezzati M, Lopez AD, Rodgers A, Vander Hoorn S, Murray CJ (2002) Selected major risk factors and global and regional burden of disease. The Lancet 360(9343):1347–1360

Fazzo L, Comba P (2004) Il ruolo dei gruppi ad alto rischio nello studio delle relazioni tra ambiente e salute. Ann Ist Super Sanità 40(4):417–426

Gee GC, Payne-Sturges DC (2004) Environmental health disparities: a framework integrating psychosocial and environmental concepts. Environ Health Perspect 112(17):1645–1653

Grobbee DE, Hoes AW (2014) Clinical epidemiology: principles, methods, and applications for clinical research. Jones & Bartlett Publishers

Glik DC (2007) Risk communication for public health emergencies. Annu Rev Public Health 28(1):33–54

Hardoy JE, Mitlin D, Satterthwaite D (2013) Environmental problems in an urbanizing world: finding solutions in cities in Africa. Routledge, Asia and Latin America

Jager KJ, Zoccali C, Macleod A, Dekker FW (2008) Confounding: what it is and how to deal with it. Kidney Int 73(3):256–260

Jawanda J (2000) The role of socially responsible corporations in community development: a case study of Fundacion Nanpaz in the Ecuadorian Amazon (Doctoral dissertation, University of British Columbia)

Johansen C, Schüz J, Andreasen AMS, Dalton SO (2017) Study designs may influence results: the problems with questionnaire-based case–control studies on the epidemiology of glioma. Br J Cancer 116(7):841–848

Khlat M (2005) A history of epidemiologic methods and concepts

Links JM, Groopman JD (2010) General principles. In: Comprehensive toxicology

Marsili D (2009) Environmental health and the multidimensional concept of development: the role of environmental epidemiology within international cooperation initiatives. Ann Ist Super Sanità 45(1):76–82

Maboso B (2023) Illuminating a neglected population: epidemiological and clinical features of silicosis and tuberculosis among former gold miners from Lesotho

Miettinen OS (2012) Theoretical epidemiology: principles of occurrence research in medicine. In: Theoretical epidemiology: principles of occurrence research in medicine, pp 359–359
Morabia A (2014) History of epidemiological methods and concepts. In: Ahrens W, Pigeot I (eds) Handbook of epidemiology. Springer, New York
National Research Council, & Committee on Environmental Epidemiology (NRCCEP) (1997) Environmental epidemiology, volume 2: use of the gray literature and other data in environmental epidemiology, vol 2. National Academies Press
Nriagu B, Ako A, Wang C, Roos AD, Wallace RB, Allison MA, Seguin R, Nassir R, Michael Y (2019) Occupations associated with poor cardiovascular health in women. Circulation 140(Suppl_1):A11410–A11410
Pearce N, Douwes J (2009) Response: time for species—course epidemiology? Int J Epidemiol 38(2):403–410
Piantadosi S (2024) Clinical trials: a methodologic perspective. Wiley
Stephens C, Church C (2017) Environmental justice and health. In International encyclopedia of public health, 2nd edn
Soskolne CL, Butler CD, Jsselmuiden CI, London L, von Schirnding Y (2007) Toward a global agenda for research in environmental epidemiology. Epidemiology 18(1):162–166
Tyler CW Jr, Last JM (1991) Epidemiology In: Last JM, Wallace RB (eds) Maxcy-Rosenau-last public health and preventive medicine, 13th edn. Appleton & Lange, Norwalk, pp 11–39
Valles SA (2021) A pluralistic and socially responsible philosophy of epidemiology field should actively engage with social determinants of health and health disparities. Synthese 198(Suppl 10):2589–2611
World Health Organization (WHO) (2011) FAO/OIE/WHO joint scientific consultation on influenza and other emerging zoonotic diseases at the human-animal interface, 27–29 April, 2010, Verona, Italy: consultation summary 2010. World Health Organization
World Health Organization (WHO) (2002) The health effects of indoor air pollution exposure in developing countries (No. WHO/SDE/OEH/02.05). World Health Organization

Chapter 2
Dynamic Methodologies in Environmental Health Research: Integrating Mixed Methods for Equity and Sustainability

Shashwati Banerjee

Abstract This chapter examines methodologies in environmental health research, focusing on their role in addressing the complex interplay between environmental factors and human health. Quantitative approaches, such as systematic reviews and meta-analyses, provide statistical evidence for causal links between exposures and health outcomes, while qualitative methods, including ethnographic studies and interviews, uncover community perspectives and the human dimensions of health. Mixed-method designs combine these approaches, offering comprehensive insights into multifaceted challenges. Ethical considerations, particularly in community-based participatory research (CBPR), are vital to maintaining trust and equity between researchers and communities. As environmental health faces growing threats like pollution, climate change, and chemical exposures, evolving methodologies are essential for generating robust evidence and guiding effective interventions. This chapter highlights the importance of refining these approaches to meet emerging challenges, emphasizing adaptability and the integration of implementation science. By showcasing successful case studies, the study underscores the need for strategies prioritizing equity, sustainability, and community empowerment, providing actionable insights for researchers, policymakers, and stakeholders in advancing environmental health research.

Keywords Environmental health research · Mixed-method approach · Community-based participatory research (CBPR) · Sustainability · Equity

S. Banerjee (✉)
Department of Humanities and Social Sciences, Indian Institute of Engineering Science and Technology, Shibpur, Howrah, West Bengal, India
e-mail: shashwatieco@gmail.com

A. Roy et al. (eds.), *Environment and Public Health*, Sustainability Solutions,
https://doi.org/10.1007/978-3-031-99770-9_2

2.1 Introduction

Environmental health research operates at the crucial intersection of scientific advancement, policy development, and societal welfare. The discipline confronts complex challenges, including pollution, climate change, and chemical contaminants, that demand multidisciplinary solutions (Smith et al. 2020). These issues exemplify "wicked" problems, characterized by interconnected causes, context-specific remedies, and unavoidable compromises between equity and efficiency (Rittel and Webber 1973; Galea et al. 2017). Conventional methodologies are limited to quantitative or qualitative frameworks and may result in an oversimplified understanding of these intricate dynamics. For instance, while epidemiological models effectively identify population-level correlations between exposures and health outcomes, they frequently overlook socio-cultural and political factors shaping vulnerability (Brulle and Pellow 2021). An adaptive, integrated approaches that transcend disciplinary boundaries, fostering collaboration among researchers, policymakers, and communities, are required. Such frameworks align with transdisciplinary principles, prioritizing collaborative knowledge creation and equitable stakeholder engagement (Lang et al. 2012a, b).

The relationship between environmental factors and health outcomes necessitates methodologies addressing both systemic and localized risks. Cohort studies effectively quantify air pollution's mortality impacts through PM2.5 measurements (Burnett et al. 2018), yet qualitative insights into how marginalized populations navigate healthcare gaps during pollution emergencies remain equally critical (Braveman and Gottlieb 2014). Emerging mixed-method approaches demonstrate their potential to bridge these perspectives. For example, a Los Angeles study integrating satellite-derived pollution data with participatory mapping revealed that low-income neighborhoods disproportionately exposed to industrial emissions also faced systemic barriers to political advocacy (Clougherty et al. 2023). By synthesizing diverse methodologies, researchers can generate robust evidence that clarifies causal pathways while identifying actionable policy interventions.

This chapter, situated within the broader scope of *environment and public health*, advances methodological innovation by addressing three interconnected dimensions critical to contemporary environmental health challenges. ***Theoretically,*** it challenges fragmented, reductionist paradigms by advocating for holistic frameworks that unify ecological, economic, and social determinants of health. This shift is exemplified by the Planetary Health approach, which underscores the bidirectional relationships between climate change and emerging zoonotic diseases, emphasizing how environmental degradation disrupts ecosystems and amplifies public health risks (Whitmee et al. 2015). ***Empirically,*** the chapter demonstrates methodological synergy through global case studies that blend quantitative and qualitative insights. For instance, in Flint, Michigan, community-driven water quality monitoring paired with geospatial analysis exposed systemic failures in addressing lead contamination, while ethnographic research in India's coal-mining regions revealed how entrenched caste hierarchies exacerbate respiratory inequities (Hanna-Attisha et al. 2016; Sahu and Patra 2020). These examples illustrate how integrating diverse methodologies uncovers both structural drivers and localized impacts of environmental harm. ***Practically,*** the discussion pivots to actionable strategies for equitable policy design, emphasizing implementation science as a bridge between research and real-world interventions. This includes elevating Indigenous ecological knowledge in climate adaptation planning, as seen in initiatives where traditional practices are harmonized with scientific data to enhance resilience (Ford et al. 2016). Collectively, these dimensions underscore the necessity of adaptive, transdisciplinary approaches to confront accelerating environmental crises, positioning environmental health research as a dynamic force for systemic change that harmonizes scientific rigor with justice-oriented action.

By addressing these dimensions, the chapter underscores the urgency of evolving methodologies to counter accelerating environmental degradation and health inequities. It reframes environmental health research as a catalyst for systemic change, ensuring scientific rigor aligns with social justice imperatives.

2.1.1 Research Gap and Objectives

Environmental health research is at a critical juncture. Despite the increasing acknowledgment of its inherently interdisciplinary nature, dominant research paradigms remain largely compartmentalized. Quantitative methods, such as epidemiological cohort studies and meta-analyses, have been pivotal in establishing causal exposure-disease relationships (Burnett et al. 2018; Dockery et al. 1993). However, these approaches often overlook the socio-political, cultural, and economic contexts that mediate environmental vulnerabilities, particularly in marginalized communities (Braveman and Gottlieb 2014; Brulle and Pellow 2021).

Simultaneously, qualitative methods, while rich in context and essential for community-centered analysis, are frequently sidelined in mainstream environmental health discourse due to concerns about replicability and scale (Creswell and Clark 2017; Braun and Clarke 2006). This methodological bifurcation limits the ability of researchers to generate evidence that is both scientifically robust and socially actionable. Scholars have called for mixed-method and transdisciplinary designs that can integrate empirical rigor with contextual relevance, yet their widespread application remains limited (Lang et al. 2012a, b; Galea et al. 2017; Plano Clark et al. 2021).

Moreover, while community-based participatory research (CBPR) frameworks offer ethical and epistemologically inclusive alternatives, especially in contexts marked by environmental injustice, their implementation is hindered by funding limitations, capacity gaps, and structural inequities in research governance (Walter and Suina 2019; Carroll et al. 2020; Lewis et al. 2017). Issues such as data sovereignty, power asymmetry, and culturally appropriate consent protocols remain underexplored, particularly in the Global South (Mohatt et al. 2013; Kukutai and Taylor 2016).

The objectives of this chapter are as follows:

- To critically examine the methodological limitations of conventional environmental health research and advocate for more integrative, transdisciplinary frameworks.
- To illustrate, through empirical case studies, how mixed-method and participatory approaches can uncover hidden dimensions of environmental risk and resilience.
- To propose pathways through which dynamic methodologies can inform equitable, sustainable policy and practice in environmental health.

2.1.2 Rationale of the Study

The urgency to refine methodologies in environmental health research stems from the increasing complexity of global environmental crises and their disproportionate impact on human health. Over the past two decades, anthropogenic activities have driven unprecedented environmental degradation—manifesting in climate change, biodiversity loss, and chemical contamination—all of which threaten the ecological determinants of health (Whitmee et al. 2015; IPCC 2023). These crises are not scientific challenges but "wicked problems" characterized by interdependent factors, uncertainties, and conflicting stakeholder interests (Rittel and Webber 1973). Traditional research paradigms, often restricted to quantitative or qualitative methodologies, struggle to adequately address these multidimensional issues. Consequently, there is an imperative to transition toward integrative, transdisciplinary approaches that bridge science, policy, and practice. This section explores the scientific, ethical, and practical rationales for advancing dynamic research methodologies that foster holistic, equity-driven solutions.

2.1.3 Escalating Environmental Health Threats

The scale and scope of environmental health threats are rapidly intensifying. Climate change is a major driver of heat-related mortality, vector-borne disease expansion, and food insecurity, disproportionately affecting vulnerable populations in low-income regions (Watts et al. 2021). The World Health Organization (WHO 2021) projects that climate change will contribute to approximately 250,000 additional deaths per year between 2030 and 2050 due to malnutrition, malaria, and heat stress. Concurrently, industrial expansion has led to the introduction of over 350,000 synthetic chemicals into ecosystems, many of which are poorly regulated and associated with chronic illnesses such as cancer, neurodevelopmental disorders, and endocrine disruption (Landrigan et al. 2018). Notably, per- and poly-fluoroalkyl substances (PFAS), often referred to as "forever chemicals," persist in water supplies and bioaccumulate in human tissues, causing immunosuppression and reproductive harm (Sunderland et al. 2019).

Urbanization exacerbates these health risks, with over 55% of the global population living in cities where air pollution is responsible for 7 million premature deaths annually (WHO 2018). Marginalized communities, disproportionately situated near industrial zones or flood-prone areas, bear the brunt of these hazards. For example, in Delhi, India, PM2.5 concentrations in low-income neighborhoods are 30% higher than in affluent districts, correlating with elevated childhood asthma rates (Sharma et al. 2020). These disparities highlight the inadequacies of conventional research methodologies that fail to account for the socio-structural determinants of health, such as poverty, racial segregation, and political exclusion (Braveman and Gottlieb 2014).

2.1.4 Limitations of Current Methodologies

Traditional environmental health research has primarily relied on quantitative methodologies, such as epidemiological cohort studies and risk assessment models, to establish dose–response relationships. While these approaches provide valuable data for regulatory policies—such as linking PM2.5 exposure to cardiovascular mortality (Burnett et al. 2018)—they often oversimplify complex socio-ecological systems by framing them as linear cause-effect relationships. For example, lead toxicity studies focus on blood lead levels but rarely interrogate why marginalized communities, such as Flint, Michigan residents, are disproportionately exposed due to systemic infrastructural neglect (Hanna-Attisha et al. 2016).

Qualitative research methods, including ethnographic studies and participatory action research, offer critical insights into lived experiences and socio-political contexts. However, these approaches are frequently dismissed as "anecdotal" or lacking generalizability (Creswell and Clark 2017). This epistemic divide results in fragmented interventions, limiting the effectiveness of environmental health policies. A meta-analysis of 120 environmental health studies found that only 18% integrated mixed methods despite their demonstrated efficacy in identifying exposure pathways and resilience strategies (Plano Clark et al. 2021).

Moreover, top-down policy frameworks often overlook local knowledge, leading to ineffective and culturally incongruent interventions. For instance, in Bangladesh, early warning systems for cyclones initially failed because they did not account for women's restricted mobility during disasters, an oversight later corrected through gender-responsive participatory design (Rahman et al. 2022). These examples underscore the limitations of methodologies that prioritize technical data over human narratives, reinforcing the need for more holistic and inclusive research frameworks.

2.1.5 The Imperative for Integrative Approaches

Integrative methodologies, particularly mixed-method and community-based participatory research (CBPR), provide a more comprehensive framework for addressing these limitations. Mixed-method designs combine quantitative rigor with qualitative depth, enabling researchers to triangulate data and uncover hidden variables. For instance, a study in Los Angeles combined satellite-derived NO2 maps with community narratives, revealing that asthma hotspots coincided with both traffic emissions and psychosocial stressors from discriminatory housing policies (Clougherty et al. 2023). This research directly influenced zoning reforms and mental health services, illustrating the policy relevance of integrative research.

CBPR further democratizes the research process by involving communities as co-investigators. In the Navajo Nation, partnerships between scientists and Indigenous leaders revealed uranium mining as a key driver of kidney disease, ultimately leading to U.S. Environmental Protection Agency (EPA) Superfund designations and remediation protocols (Lewis et al. 2017). Such collaborations align with principles of procedural justice, ensuring that marginalized groups have a voice in decision-making processes affecting their health (Schlosberg 2013).

Additionally, implementation science offers a valuable bridge between research and practice by identifying barriers to intervention adoption. For example, a randomized trial in Kenya demonstrated that integrating local healers into malaria prevention programs increased bed net usage by 40%, emphasizing the importance of culturally adaptive strategies (Opiyo et al. 2020). These methodologies enhance scientific validity and foster trust, an essential factor in an era where public confidence in institutions is declining (Goldstein et al. 2020).

2.1.6 Ethical and Practical Considerations

Ethical imperatives drive the transition toward integrative methodologies. Environmental injustices—such as the disproportionate placement of toxic waste sites in communities of color—are rooted in historical inequities that require systemic redress (Bullard 2000). CBPR frameworks prioritize data sovereignty, ensuring communities retain ownership of their data and its applications (Walter and Suina 2019). The Global Indigenous Data Alliance, for example, advocates for Indigenous rights in data governance to counteract extractive research practices (Kukutai and Taylor 2016). However, challenges remain. Power imbalances between researchers and communities can result in tokenistic engagement, as seen in "helicopter research," where data is extracted without capacity-building efforts (Tuhiwai Smith 2012). Transparent governance structures, such as community advisory boards and co-authorship agreements, help mitigate these risks (Israel et al. 2018).

From a practical standpoint, resource constraints limit the scalability of mixed-method designs, particularly in low-income regions that lack the infrastructure for advanced environmental monitoring or genomic analysis. Strategic partnerships can help address these barriers; for example, the WHO's Urban Health Initiative provides low-cost air pollution sensors to cities in Sub-Saharan Africa, enabling localized mitigation strategies (WHO 2022).

Given the accelerating breach of planetary boundaries and the widening of health inequities, the need for dynamic research methodologies is more pressing than ever. Siloed approaches risk obscuring the systemic drivers of environmental health risks, whereas integrative frameworks—rooted in transdisciplinary collaboration, technological innovation, and ethical engagement—offer a pathway toward sustainable and equitable solutions. By centering community voices, leveraging emerging technologies, and aligning with global sustainability agendas such as the UN Sustainable Development Goals (SDGs), environmental health research can evolve from a reactive discipline into a proactive force for planetary stewardship.

2.2 Materials and Methods

The methodologies used in environmental health research need to be designed to address the intricate interactions between biological, social, and economic factors that influence health outcomes. This section explores quantitative, qualitative, and mixed-method approaches, emphasizing their theoretical foundations, practical applications, and ethical considerations. By integrating diverse research paradigms, researchers can examine both physiological mechanisms and broader socio-structural determinants of health, ensuring methodological rigor and social relevance. Combining these approaches enhances the precision of exposure assessments, improves health risk analysis, and fosters meaningful community engagement, ultimately leading to more impactful and sustainable public health interventions.

2.2.1 Quantitative Approaches

Quantitative methodologies are the foundation of environmental health research, providing statistical rigor, reproducibility, and broad applicability. These methods are critical in establishing causal relationships between environmental exposures and health outcomes, offering insights that inform regulatory policies and risk assessments. By leveraging large datasets and statistical techniques, researchers can identify exposure–response patterns and generate evidence-based recommendations for public health interventions.

Systematic reviews and meta-analyses are essential in synthesizing evidence from multiple studies to discern consistent trends in environmental health research. Systematic reviews compile findings from diverse sources, while meta-analyses employ statistical methods to quantify pooled effect sizes, strengthening the reliability of conclusions. For example, a meta-analysis of 50 studies revealed that every 5 $\mu g/m^3$ increase in PM2.5 exposure was associated with a 12% rise in childhood asthma incidence (Lee et al. 2022). The influence of such analyses extends beyond academic research; for instance, the World Health Organization (WHO) revised its air quality guidelines in 2021 based on meta-analytic evidence, lowering recommended PM2.5 exposure limits to protect public health. However, these methodologies are not without limitations. Systematic reviews and meta-analyses are vulnerable to publication bias, where studies with significant findings are more likely to be published than those with null results. Additionally, inconsistencies in exposure measurement across studies can introduce heterogeneity, complicating the interpretation of findings (Egger et al. 1997).

Epidemiological cohort studies are another cornerstone of quantitative research in environmental health. These longitudinal studies track populations to assess the relationship between environmental exposures and disease outcomes. A landmark example is the Harvard Six Cities Study, which provided compelling evidence linking air pollution to cardiovascular mortality. The findings of this study directly influenced amendments to the U.S. Clean Air Act, demonstrating the policy-shaping power of epidemiological research (Dockery et al. 1993). Contemporary cohort studies increasingly incorporate biomarkers to enhance exposure specificity. For instance, urinary metabolites of polycyclic aromatic hydrocarbons (PAHs) serve as biomarkers of air pollution exposure, allowing researchers to assess individual-level variations in exposure more accurately (Li et al. 2020). Such advancements enhance the precision of exposure assessment and strengthen causal inference.

Geospatial modeling is a powerful quantitative tool that enables researchers to map environmental hazards with high spatial resolution. Technologies such as satellite remote sensing and Geographic Information Systems (GIS) facilitate the visualization of environmental risks and their health impacts. In Mumbai, researchers employed Landsat imagery to examine urban heat islands, finding a strong correlation between extreme temperatures and hospital admissions for heatstroke, particularly in slum areas with limited green spaces and inadequate cooling infrastructure (Mondal et al. 2021). Beyond urban health applications, machine learning algorithms have been integrated into geospatial analyses to enhance predictive accuracy. For example, random forest models have been successfully used to forecast groundwater contamination risks in agricultural regions, helping policymakers design targeted interventions to protect drinking water supplies (Podgorski and Berg 2020).

Quantitative approaches in environmental health monitoring have significantly evolved, incorporating advanced techniques and technologies that enhance precision, scalability, and integration of multidimensional data. Among these, satellite-based remote sensing has emerged as a critical tool for assessing air and water quality, land-use changes, and vegetation health at regional and global scales (Van Donkelaar et al. 2019). Ground-based sensor networks, including low-cost air quality monitoring devices and wearable exposure trackers, have enabled high-resolution, real-time data collection, particularly valuable in urban environments (Snyder et al. 2013; Castell et al. 2017). Additionally, geographic information systems (GIS) and spatial analytics are now integral in visualizing environmental exposures and modeling their health impacts across populations (Jerrett et al. 2005). Machine learning algorithms, such as random forests and support vector machines, are increasingly used to predict environmental health risks and to process large environmental datasets, offering novel insights into complex exposure-disease relationships (Reid et al. 2021). The integration of these methodologies enables a more holistic understanding of environmental determinants of health, facilitating evidence-based policy decisions.

Despite their strengths, quantitative methodologies are not without limitations. One major drawback is their inability to fully capture the socio-cultural and economic contexts that shape environmental exposures and health behaviors. For instance, risk assessments of pesticide exposure may focus solely on urinary organophosphate levels without considering the economic pressures that compel farmers to use agrochemicals extensively (Raanan et al. 2016). Such omissions can lead to interventions that fail to address the root causes of exposure. Additionally, reliance on secondary datasets—such as hospital admission records—can introduce selection bias, as healthcare access disparities may result in the underrepresentation of marginalized communities in epidemiological analyses (Krieger 2020). These limitations highlight the need for complementary qualitative and mixed-method approaches to provide a more holistic understanding of environmental health challenges.

2.2.2 *Qualitative Approaches*

Qualitative research methods are crucial for understanding lived experiences, cultural dynamics, and power structures that quantitative approaches may overlook. These methods are especially effective in studying marginalized communities and designing interventions that align with their unique social and cultural contexts. By fostering community engagement, qualitative research helps identify the underlying causes of health disparities, making interventions more relevant, acceptable, and sustainable.

Ethnographic research immerses researchers within communities, allowing them to observe daily life and social norms firsthand. This approach provides deeper insight into the complex factors shaping health behaviors. For instance, a study in Ghana's cocoa-producing regions revealed that farmers hesitated to use organic pesticides due to gendered labor roles. Women, primarily responsible for pesticide application, feared that reduced crop yields would exacerbate food insecurity (Adu-Boahen et al. 2023). These findings highlight the need for gender-sensitive policies in agricultural and environmental programs to enhance their effectiveness and sustainability.

Participatory Action Research (PAR) actively involves communities in research, fostering shared knowledge creation and empowering marginalized groups. A notable example is the Flint, Michigan water crisis, where residents collaborated with scientists to collect water samples, uncovering discrepancies between official contamination reports and community-led findings (Hanna-Attisha et al. 2016). PAR follows a cyclical process of reflection and action, known as "participatory praxis" (Cahill 2007), enabling communities to drive systemic change. By facilitating collective problem-solving, PAR ensures that research leads to meaningful policy shifts and community-led initiatives.

Focus groups and semi-structured interviews provide qualitative researchers valuable insights into public perceptions, behaviors, and barriers to health interventions. In Bangladesh, discussions with women affected by cyclones revealed cultural taboos that prevented them from evacuating without male family members. In response, early warning systems were redesigned to include female community leaders, making them more culturally appropriate and effective (Rahman et al. 2022). Analytical tools such as NVivo assist in processing qualitative data, but researchers must remain cautious of over-interpretation to ensure that findings accurately represent participants' experiences (Braun and Clarke 2006).

Despite their strengths, qualitative methods are sometimes criticized for their subjectivity and limited generalizability. However, triangulation, i.e., cross-verifying findings through multiple data sources, including interviews, observations, and archival records, enhances research credibility (Denzin 2017). Additionally, long-term engagement with communities fosters trust and helps uncover overlooked health concerns. A three-year study of Navajo uranium miners demonstrated the benefits of sustained collaboration, as researchers were able to document previously unreported health effects of environmental exposure due to strong relationships with participants (Lewis et al. 2017). These strategies ensure that qualitative research maintains methodological rigor while addressing complex environmental health issues with depth and nuance.

2.2.3 Mixed-Method Designs

Mixed-method research (MMR) integrates both quantitative and qualitative approaches, providing a comprehensive understanding of environmental health issues. These designs can be structured sequentially, where one phase informs the next, or concurrently, where data are collected in parallel. By combining different research paradigms, MMR enhances the depth and applicability of findings, making it particularly useful for complex social and environmental health studies.

Sequential Exploratory Design follows a two-phase process where qualitative data guide the development of quantitative research. For instance, interviews with farmworkers in California uncovered concerns about heat stress, leading to a cohort study that quantitatively linked ambient temperatures to kidney injury (Mix et al. 2018). The initial qualitative phase ensured that culturally relevant factors, such as hydration practices, were incorporated into survey instruments, improving the study's accuracy and relevance.

Concurrent Triangulation Design involves collecting qualitative and quantitative data simultaneously to validate findings. A study in Peru combined GIS-based arsenic mapping with interviews to assess groundwater contamination. While spatial data identified areas with high arsenic levels, interviews revealed that many residents avoided mitigation efforts due to mistrust in government authorities—a factor not captured in the quantitative data alone (George et al. 2020). This approach highlights the value of integrating multiple data sources to uncover hidden social and behavioral influences on public health.

Transformative Design prioritizes social justice by centering the voices of marginalized populations. In a Detroit study on lead exposure, youth participants employed photovoice techniques to document deteriorating paint in homes, while simultaneous blood lead level analyses identified contamination hotspots not reflected in municipal records (Sampson et al. 2021). The findings informed the creation of a citywide lead abatement fund, specifically targeting renter-occupied housing, demonstrating how MMR can drive policy change and community interventions.

Despite its advantages, *Methodological Integration* in MMR presents challenges. Effectively combining qualitative and quantitative data requires strategic planning. The joint display method juxtaposes qualitative themes with corresponding numerical results in a structured table, helping clarify linkages between data types (Guetterman et al. 2015). However, resource limitations often hinder the adoption of MMR. A 2022 review found that only 22% of environmental health studies used mixed methods, citing funding shortages and a lack of researcher training as significant barriers (Plano Clark et al. 2021). Addressing these constraints through increased funding and methodological training could enhance the widespread application of MMR, making environmental health research more inclusive and impactful.

2.2.4 Ethical Considerations in CBPR

Community-Based Participatory Research (CBPR) requires strict ethical protocols to prevent exploitative research practices and promote equity. This approach prioritizes community involvement at every stage, ensuring that research outcomes are mutually beneficial and culturally appropriate. Ethical considerations in CBPR extend beyond standard research ethics to include informed consent, data sovereignty, reciprocal knowledge exchange, and power-sharing mechanisms.

Informed Consent and Cultural Sensitivity are fundamental to ethical CBPR. Researchers must adapt consent processes to accommodate literacy and language barriers, ensuring all participants fully understand the study's objectives and implications. In a study involving Alaska Native communities, researchers collaborated with local artists to develop pictorial consent forms that aligned with the communities' oral traditions and cultural communication styles (Mohatt et al. 2013). Additionally, dynamic consent—an approach that allows participants to provide ongoing permission throughout the study—enhances autonomy and ensures that individuals can reassess their involvement as the research progresses (Kaye et al. 2015).

Data Sovereignty and Ownership have become increasingly important, particularly for Indigenous communities seeking to maintain control over their information. Frameworks such as the CARE Principles (Collective Benefit, Authority to Control, Responsibility, Ethics) have been developed to safeguard Indigenous data rights (Carroll et al. 2020). For example, the Māori-led Te Mana Raraunga initiative requires non-Indigenous researchers to obtain tribal approval before accessing health data, ensuring that research aligns with community priorities and prevents data misuse (Kukutai and Taylor 2016). Such measures help shift power dynamics in research, allowing communities to dictate how their data are used and interpreted.

Reciprocal Knowledge Exchange is a cornerstone of CBPR, emphasizing mutual learning between researchers and community members. In a Detroit-based asthma study, academics trained residents in air quality monitoring techniques while community members educated researchers about the historical impacts of redlining on health disparities (Schulz et al. 2019). This bidirectional knowledge-sharing approach led to co-authored publications and the creation of a community-owned database, which was later used to advocate for stricter air quality regulations. Such collaborations ensure that research maintains scientific rigor while actively contributing to the well-being of the communities involved.

Addressing Power Imbalances is critical in CBPR to prevent tokenism and ensure genuine community leadership. Power-sharing mechanisms, such as community advisory boards, help mitigate disparities between academic researchers and local participants. In a Bronx-based study on diesel pollution, the advisory board was granted veto power over research directions, shifting the focus from individual behavioral changes to systemic policy solutions, such as truck rerouting (Corburn 2005). Additionally, transparent conflict resolution protocols are essential, as differing perspectives on data interpretation can lead to disagreements (Wallerstein et al. 2017). By prioritizing community leadership and decision-making, CBPR fosters ethically sound and socially transformative research.

Ultimately, selecting research methodologies must align with the study's objectives, community needs, and ethical imperatives. While quantitative methods offer generalizable findings, qualitative approaches provide a deeper contextual understanding. Mixed-method and CBPR frameworks bridge these domains, promoting research that is both scientifically valid and socially impactful. As environmental health challenges intensify, methodological innovation and a steadfast commitment to ethical principles will be essential in advancing equity and sustainability.

2.3 Results and Discussion

This section synthesizes findings from diverse case studies to illustrate the efficacy of mixed-method approaches in addressing environmental health challenges. These studies underscore the importance of interdisciplinary collaboration and community-centered solutions by integrating quantitative rigor, qualitative depth, and ethical engagement.

2.3.1 *Case Study 1: Tackling Urban Air Pollution in Delhi, India*

Delhi, often cited as one of the most polluted metropolises worldwide, serves as a vital example of managing environmental health in urban settings. A study by Sharma et al. (2020) adopted a mixed-method approach, using both satellite-based PM2.5 metrics and hospital data to trace pollution-related respiratory illnesses. The research highlighted a notable 15% increase in pediatric asthma cases during winter months when PM2.5 concentrations surpassed 300 $\mu g/m^3$.

Through interviews with over a hundred residents, the study unearthed social and behavioral constraints in responding to pollution. Less than a quarter of informal sector workers used protective masks, mainly due to affordability issues and mistrust in official advisories. These findings influenced policy responses such as subsidized air purifier programs and awareness campaigns at the community level.

The integration of Delhi into the Smart Cities Mission and participation in the National Clean Air Programme (NCAP) showcase a layered governance framework. Initiatives under NCAP include real-time air quality monitoring, promotion of electric vehicles, and pilot green zones. Nonetheless, persistent challenges like inter-agency collaboration and low public adherence point to the ongoing need for grassroots involvement in national strategies.

2.3.2 *Case Study 2: Flood Risk Management in the Netherlands*

In contrast to countries like Bangladesh that are still developing adaptive strategies, the Netherlands presents a mature model through its "Room for the River" initiative. This program reimagines urban landscapes to accommodate seasonal floods rather than resist them. Hydrological simulations forecast water flow areas, while resident and business consultations ensured minimal socio-economic disruption.

Blending engineering solutions—such as levees and water retention zones—with policies like land-use regulation and early warning systems, the Dutch model reduces flood risks without harming ecosystems. This case illustrates how developed nations embed resilience into city planning and strengthen social trust through inclusive policy development.

2.3.3 *Case Study 3: Lead Water Contamination in Flint, Michigan*

The water crisis in Flint is a defining moment in environmental justice debates. Initial lab tests in 2015 showed lead levels in domestic water supplies exceeding

1,000 parts per billion. This was supported by medical data indicating a rise in lead levels in children, especially in predominantly Black neighborhoods (Hanna-Attisha et al. 2016).

Community storytelling, including public forums and personal testimonies, sheds light on deep-rooted institutional failings and racial inequities. These accounts were crucial in driving legal action, resulting in a $600 million settlement and extensive infrastructure changes. However, lingering distrust in public institutions continues to impact health outcomes and civic engagement (Sampson et al. 2021).

2.3.4 Case Study 4: Pesticide Exposure in Ecuador's Flower Industry

Ecuador, a key player in global flower exports, faces significant occupational health concerns. A investigation by Harari et al. (2023) employed biological monitoring and participatory methods such as photovoice to capture worker experiences on rose farms. Results showed that 78% of workers had elevated levels of organophosphates, correlating with signs of neurotoxic exposure.

Workshops revealed widespread disregard for safety regulations, fueled by economic insecurity and lax enforcement. Workers used visual storytelling to push for change, which led to stronger labor safety measures and broader distribution of protective equipment.

2.3.5 Case Study 5: Participatory Urban Sustainability in Freiburg, Germany

Freiburg is often cited as a benchmark for urban environmental governance. The city has implemented innovative measures, including solar energy systems, car-free districts, and participatory planning platforms. These efforts represent a collaborative approach where municipal bodies, cooperatives, and citizen groups jointly design sustainability initiatives.

Between 1992 and 2016, Freiburg achieved a 40% reduction in per capita carbon emissions, even with population growth. Interviews suggest this success stems from transparent policy-making, forward-looking urban design, and public education, resulting in widespread community engagement and trust in governance.

2.3.6 *Case Study 6: Forest Conservation Incentives in Costa Rica*

Costa Rica's payment for ecosystem services (PES) scheme, launched in 1997, stands out for successfully reversing deforestation. The program rewards landowners for preserving forests, increasing forest cover from 21% in the 1980s to over 50% by 2017.

Satellite imagery has verified reforestation, while long-term surveys with rural households noted income and environmental awareness gains. The program effectively integrates economic motivation with ecological stewardship, showing how financial tools can align conservation with sustainable development.

These case studies were chosen to reflect various environmental health concerns across regions and socio-economic contexts. Each example, from air pollution in Indian cities to deforestation reversal in Costa Rica, offers distinct insights into local responses and broader policy frameworks. They highlight the efficacy of combining data-driven research with community-based narratives to craft more inclusive, resilient, and sustainable solutions to pressing environmental health issues.

This study draws on a diverse array of case studies from Delhi, the Netherlands, Flint (USA), Ecuador, Freiburg (Germany), and Costa Rica to evaluate how community-based participatory research (CBPR) adapts to varying environmental challenges. In Delhi, using PM2.5 data and ethnographic interviews highlights how a quantitative–qualitative triangulation approach identifies respiratory health risks in low-income populations. Similarly, the Netherlands' "Room for the River" project exemplifies flood resilience through community co-design and hydrological modeling, while Flint's water crisis underscores the dangers of disregarding community input, leading to public health disasters and deep civic mistrust. Ecuador's use of biomonitoring and participatory photovoice empowered floriculture workers to advocate for pesticide regulation. At the same time, Freiburg demonstrates the long-term benefits of iterative civic engagement in embedding sustainability within governance. Costa Rica's Payment for Ecosystem Services (PES) program showcases how economic incentives for forest conservation can also promote community welfare.

Through a comparative lens, four core themes emerge: institutional trust and governance, socio-economic justice, preventive cost-efficiency, and the importance of longitudinal monitoring. These cases illustrate that environmental interventions gain impact when grounded in the "Community-Evidence-Policy Nexus", a conceptual triad advocating for community legitimacy, empirical robustness, and institutional responsiveness. Economic evaluations further suggest that community-driven preventive models, like those in Freiburg and Costa Rica, are more fiscally sustainable than reactive approaches. However, replicating such models in low- and middle-income countries requires adaptive governance and cross-disciplinary capacity-building. Emerging tools, such as AI-based exposure models, blockchain for data transparency, and citizen science platforms—offer new pathways to scale and reinforce CBPR efforts. While methodological limitations such as inconsistent data and cultural variations persist, the study affirms the promise of dynamic, ethically grounded research for shaping equitable and effective environmental health policies.

2.3.7 Policy Implications

Translating environmental health research into actionable policy requires bridging the gap between scientific evidence and socio-political realities. A key challenge is making complex data accessible to policymakers facing time constraints and competing priorities. Researchers in São Paulo, Brazil developed interactive dashboards that combined real-time air quality metrics with community-reported health outcomes, enabling targeted interventions in high-risk areas (Ferreira 2023). Similarly, co-produced climate adaptation tools, such as seasonal hazard maps blending satellite data with Indigenous knowledge in the Canadian Arctic, enhanced both cultural relevance and disaster preparedness (Ford et al. 2016). These examples highlight the need for tailored communication strategies that balance scientific rigor with accessibility.

As global crises intensify, future research must address critical gaps in environmental health. Longitudinal studies are vital for assessing the long-term health impacts of climate change, such as repeated heatwave exposure on cardiovascular mortality or the intergenerational effects of endocrine disruptors. A 20-year Australian study, for instance, linked prenatal bushfire smoke exposure to increased respiratory illness in adolescence. Advances in AI could further enhance environmental health research by integrating diverse datasets like satellite imagery, health records, and social media sentiment to uncover hidden exposure patterns. However, scaling community-based participatory research (CBPR) in low-resource settings remains challenging. In Kenya, partnerships between academic institutions and local NGOs used low-cost sensors and community-led data collection to improve water quality monitoring in informal settlements (Opiyo et al. 2020). Expanding such models requires addressing systemic inequities in funding and capacity-building, ensuring marginalized communities play an active role in shaping solutions.

2.3.8 Limitations of the Study

Despite the valuable insights presented in this chapter, several limitations should be acknowledged. One significant challenge is the applicability of findings across different socio-political environments. The effectiveness of case studies, such as Bangladesh's gender-sensitive flood warning systems or Flint, Michigan's community-led water testing, was influenced by specific cultural, institutional, and advocacy structures. These conditions may not exist in other regions, making direct replication difficult.

Another constraint lies in the resource demands of mixed-method and community-based participatory research (CBPR). These methodologies require substantial investment in terms of funding, technical expertise, and sustained community involvement, which may not always be viable in regions with limited financial and institutional support. As a result, adopting these approaches in under-resourced areas remains a challenge.

Additionally, ethical concerns arise due to potential power imbalances in community collaborations. Even within CBPR frameworks, there is a risk that local voices may not be equitably represented in decision-making or data analysis. Ensuring fair authorship recognition, respecting data sovereignty, and maintaining transparency in research partnerships are ongoing challenges.

Finally, while this chapter advocates for integrating qualitative and quantitative methods, inherent methodological biases persist. Quantitative research often emphasizes statistical validity at the expense of contextual depth, whereas qualitative approaches, though rich in insights, may struggle with scalability and reproducibility. Overcoming these limitations will require continued methodological refinement and institutional backing to ensure that environmental health research remains both inclusive and impactful.

2.4 Conclusions

This chapter advances the argument that solving contemporary environmental health challenges demands an integrated research paradigm that is participatory, interdisciplinary, and grounded in both empirical rigor and ethical engagement. The adoption of dynamic, mixed-methods frameworks within the scope of community-based participatory research (CBPR) enables the co-creation of knowledge that is simultaneously scientifically robust and socially legitimate. The research presented here illustrates how a richer understanding of environmental risks can emerge by synthesizing quantitative indicators, ethnographic narratives, and biomonitoring data. Central to this contribution is developing the "Community-Evidence-Policy Nexus," a conceptual model that underscores the critical interdependence of data quality, community engagement, and institutional responsiveness—an approach particularly vital in navigating the complexity and uncertainty inherent in socio-ecological systems.

The chapter suggests that practitioners and policymakers consider five strategic directions emerging from the empirical cases and methodological reflections. First, environmental health governance may benefit from community-centered design approaches, where interventions are shaped by local knowledge, risk perceptions, and cultural contexts. Second, integrative data sources that combine environmental indicators with experiential insights, including citizen-generated evidence, could inform policy decisions. Third, sustaining long-term program impacts may require greater attention to decentralized institutional support and capacity enhancement, particularly in under-resourced or administratively neglected areas. Fourth, ethical considerations in community-based participatory research (CBPR) may be expanded beyond informed consent to embrace shared ownership of knowledge and decision-making, especially in settings marked by social or political marginalization. Finally, incorporating scalable tools, such as low-cost sensor networks, AI-based diagnostics, and participatory mapping platforms, could enhance transparency, responsiveness, and adaptability in environmental health governance. These reflections indicate that meaningful progress in environmental health may rely as much on institutional trust and civic engagement as on methodological innovation.

Future research must deepen its engagement with implementation science to bridge the persistent gap between community-generated insights and policy formulation. Translating participatory research into actionable, adaptive policies requires methodological rigor and a strategic focus on operationalizing interventions in real-world settings. In this context, the development of longitudinal evaluation frameworks becomes critical, offering a means to systematically track interventions' sustained effectiveness and equity implications, particularly in ecologically fragile or socially marginalized regions. Successfully scaling participatory models in such low-resource contexts, however, is contingent on sustained institutional investment in local capacity-building, the cultivation of trust-based partnerships, and the establishment of governance mechanisms that are both responsive and resilient. As the global landscape is increasingly shaped by the intersecting pressures of climate change, rapid urbanization, and environmental inequity, the alignment of scientific inquiry with community priorities emerges as a methodological necessity and a moral and political imperative.

References

Adu-Boahen K, Mensah JC, Osei-Fosu P (2023) Pesticide regulation and smallholder farmers in Ghana: a CBPR approach. Environ Health Perspect 131:45001. https://doi.org/10.1289/EHP12345

Braun V, Clarke V (2006) Using thematic analysis in psychology. Qual Res Psychol 3(2):77–101

Braveman P, Gottlieb L (2014) The social determinants of health: it's time to consider the causes of the causes. Public Health Rep 129(1_suppl2): 19–31.

Brulle RJ, Pellow DN (2021) Environmental justice: human health and environmental inequalities. Annu Rev Public Health 42:1–21. https://doi.org/10.1146/annurev-publhealth-090419-102314

Bullard RD (2000) Dumping in Dixie: race, class, and environmental quality. Routledge, New York

Creswell JW, Clark VLP (2017) Designing and conducting mixed methods research. Sage, Thousand Oaks

Burnett R, Chen H, Szyszkowicz M et al (2018) Global estimates of mortality associated with long-term exposure to outdoor fine particulate matter. Proc Natl Acad Sci 115(38):9592–9597. https://doi.org/10.1073/pnas.1803222115

Cahill C (2007) Participatory data analysis. ACME Int J CritGeogr 6(3):304–318

Carroll SR, Garba I, Figueroa-Rodríguez OL et al (2020) The CARE principles for indigenous data governance. Data Sci J 19:43

Castell N, Dauge FR, Schneider P, Vogt M, Lerner U, Fishbain B, Broday D, Bartonova A (2017) Can commercial low-cost sensor platforms contribute to air quality monitoring and exposure estimates? Environ Int 99:293–302. https://doi.org/10.1016/j.envint.2016.12.007

Clougherty JE, Levy JI, Hynes HP et al (2023) Bridging data and dialogue: a mixed-method study of air pollution's health impacts in urban neighborhoods. Environ Health Perspect 131:067001. https://doi.org/10.1289/EHP11589
Corburn J (2005) Street science: community knowledge for environmental justice. MIT Press, Cambridge
Denzin NK (2017) The research act: a theoretical introduction to sociological methods. Routledge, New York
Dockery DW, Pope CA, Xu X et al (1993) An association between air pollution and mortality in six U.S. cities. N Engl J Med 329(24):1753–1759
Egger M, Smith GD, Schneider M et al (1997) Bias in meta-analysis detected by a simple, graphical test. BMJ 315(7109):629–634
Ferreira L (2023) Data visualization for policy engagement. Sustain Cities Soc 89:104567. https://doi.org/10.1016/j.scs.2023.104567
Ford JD, Sherman M, Berrang-Ford L et al (2016) Preparing for the health impacts of climate change in Indigenous communities: the role of community-based adaptation. Glob Environ Change 41:118–127
Galea S, Riddle M, Kaplan GA (2017) Causal thinking and complex system approaches in epidemiology. Int J Epidemiol 46(1):1–9. https://doi.org/10.1093/ije/dyw298
George CM, Factor-Litvak P, Khan K et al (2020) Approaches to increase arsenic awareness in Bangladesh: an evaluation of an arsenic education program. Environ Health Perspect 128(3):034001
Goldstein BD, Osofsky HJ, Lichtveld MY (2020) The Gulf oil spill. N Engl J Med 364(14):1334–1348
Guetterman TC, Fetters MD, Creswell JW (2015) Integrating quantitative and qualitative results in health science mixed methods research through joint displays. Ann Fam Med 13(6):554–561
Hanna-Attisha M, LaChance J, Sadler RC et al (2016) Elevated blood lead levels in children associated with the Flint drinking water crisis: a spatial analysis of risk and public health response. Am J Public Health 106(2):283–290
Harari R, Freire C, Koifman S et al (2023) Pesticide exposure and neurotoxicity in Ecuadorian flower workers. Environ Health Perspect 131:045001
IPCC (2023) Climate change 2023: synthesis report. Intergovernmental Panel on Climate Change, Geneva
Israel BA, Schulz AJ, Parker EA et al (2018) Critical issues in CBPR. Health Educ Behav 45(5):761–774
Jerrett M, Burnett RT, Pope III CA, Ito K, Thurston G, Krewski D, Shi Y, Calle E, Thun M (2005) Long-term ozone exposure and mortality. N Engl J Med 360(11):1085–1095. https://doi.org/10.1056/NEJMoa0803894
Kaye J, Whitley EA, Lund D et al (2015) Dynamic consent: a patient interface for twenty-first century research networks. Eur J Hum Genet 23(2):141–146
Krieger N (2020) Measures of racism, sexism, heterosexism, and gender binarism for health equity research. Am J Public Health 110(8):1110–1113
Kukutai T, Taylor J (2016) Indigenous data sovereignty: toward an agenda. ANU Press, Canberra
Landrigan PJ, Fuller R, Acosta NJR et al (2018) The Lancet Commission on pollution and health. Lancet 391(10119):462–512
Lang DJ, Wiek A, Bergmann M et al (2012a) Transdisciplinary research in sustainability science: practice, principles, and challenges. Sustain Sci 7(1):25–43
Lang DJ, Wiek A, Bergmann M, Stauffacher M, Martens P, Moll P, Swilling M, Thomas CJ (2012b) Transdisciplinary research in sustainability science: practice, principles, and challenges. Sustain Sci 7(1):25–43
Lee K, Park S, Kim Y (2022) Meta-analysis of PM2.5 and respiratory health. Environ Res 212:113456. https://doi.org/10.1016/j.envres.2022.113456
Lewis J, Hoover J, MacKenzie D (2017) Mining and environmental health disparities in Native American communities. Curr Environ Health Rep 4(2):130–141

Li Z, Mulholland JA, Romanoff LC et al (2020) Assessment of non-occupational exposure to PAHs via urine biomonitoring in the US population. Environ Int 137:105518

Mix J, Elon L, Vi Thien Mac V et al (2018) Hydration status, kidney function, and kidney injury in Florida agricultural workers. J Occup Environ Med 60(5):e253–e260

Mohatt GV, Rasmus SM, Thomas L et al (2013) Risk, resilience, and natural recovery: a model of alcohol recovery for Alaska Natives. Addiction 108(5):922–930

Mondal B, Sahana M, Chattopadhyay S (2021) Urban heat island and its impact on heat-related morbidity in Mumbai, India. Urban Clim 38:100902

Opiyo P, Mukabana WR, Kiche I et al (2020) Community-based malaria prevention in Kenya: culturally adapted strategies. Malar J 19(1):1–12

Plano Clark VL, Ivankova NV, Wertz FJ (2021) Mixed methods research in environmental health: a systematic review. Environ Health Perspect 129(7):076001

Podgorski J, Berg M (2020) Global threat of arsenic in groundwater. Science 368(6493):845–850

Raanan R, Balmes JR, Harley KG et al (2016) Decreased lung function in 7-year-old children with early-life organophosphate exposure. Thorax 71(2):148–153

Rahman S, Ahmed F, Hossain M (2022) Gender and flood resilience in Bangladesh. Disaster PrevManag 31(2):145–160

Reid CE, Lane KJ, Ortiz C (2021) Integrating satellite and ground-level data for exposure assessment in environmental epidemiology. Environ Health Perspect 129(5):056001. https://doi.org/10.1289/EHP7500

Rittel HWJ, Webber MM (1973) Dilemmas in a general theory of planning. Policy Sci 4(2):155–169

Sahu G, Patra AP (2020) Caste, coal, and chronic respiratory disease: health inequities in India's mining belt. Soc Sci Med 245:112698

Sampson N, Winter AS, Strosnider H et al (2021) Youth participatory approaches and health equity: the alignment of citizen science and environmental justice. J Urban Health 98(1):144–160

Schlosberg D (2013) Theorising environmental justice: the expanding sphere of a discourse. Environ Polit 22(1):37–55

Schulz AJ, Mentz GB, Sampson N et al (2019) Race and the distribution of social and physical environmental risk: a case example from the Detroit metropolitan area. Du Bois Rev 16(1):221–240

Sharma N, Kapoor M, Singh A (2020) Air pollution and health in urban India. Atmos Environ 237:117648. https://doi.org/10.1016/j.atmosenv.2020.117648

Smith J, Jones M, Brown P (2020) Interdisciplinary environmental health research. Annu Rev Public Health 41:89–105

Snyder EG, Watkins TH, Solomon PA, Thoma ED, Williams RW, Hagler GS, Shelow D, Hindin DA, Kilaru VJ, Preuss PW (2013) The changing paradigm of air pollution monitoring. Environ Sci Technol 47(20):11369–11377. https://doi.org/10.1021/es4022602

Sunderland EM, Hu XC, Dassuncao C et al (2019) A review of the pathways of human exposure to poly- and perfluoroalkyl substances (PFASs) and present understanding of health effects. J Expo Sci Environ Epidemiol 29(2):131–147

Tuhiwai Smith L (2012) Decolonizing methodologies: research and Indigenous peoples. Zed Books, London

Van Donkelaar A, Martin RV, Li C, Burnett RT (2019) Regional estimates of chemical composition of fine particulate matter using a combined geoscience-statistical method with information from satellites, models, and monitors. Environ Sci Technol 53(5):2595–2611. https://doi.org/10.1021/acs.est.8b06392

Wallerstein N, Duran B, Oetzel JG et al (2017) Community-based participatory research for health. Wiley, San Francisco

WHO (2021) Climate change and health equity. World Health Organization, Geneva

Walter M, Suina M (2019) Indigenous data sovereignty and policy. Routledge HandbIndig Wellbeing 1:236–248

Watts N, Amann M, Arnell N et al (2021) The 2021 report of the Lancet Countdown on health and climate change: code red for a healthy future. Lancet 398(10311):1619–1662

Whitmee S, Haines A, Beyrer C et al (2015) Safeguarding human health in the Anthropocene epoch: report of the Rockefeller foundation-lancet commission on planetary health. Lancet 386(10007):1973–2028

WHO (2018) Ambient air pollution: a global assessment of exposure and burden of disease. World Health Organization, Geneva

WHO (2022) Urban health initiative: annual report. World Health Organization, Geneva

Chapter 3
Deciphering the Climate Split: Social Equity, Environmental Justice, and Public Health Impact

Aishwarya Harichandan, Akancha Srivastava, and Shefali Mishra

Abstract With 2023 recorded as the hottest year on record, the planet has been warming at a steady rate of 0.06 °C per decade since 1850. This relentless rise in global temperatures has intensified the frequency and severity of climate-related disasters such as floods, droughts, landslides, and cyclones, emphasizing the critical need for immediate climate intervention. The repercussions are felt most acutely by marginalized populations, exacerbating socioeconomic disparities both within and among countries. In this context, social equity and environmental justice emerge as interconnected imperatives ensuring fair access to resources and inclusive decision-making. Together, they are essential for advancing climate resilience and public health outcomes. Based upon this background, the chapter explores three key objectives: (i) to analyze the relation between the major per capita carbon dioxide emitters of the world and the countries that are the most vulnerable to climate change, (ii) to provide psychological interpretations and solutions for these countries challenge (iii) to explore the intersection of climate disparities, environmental justice, and their impact on public health. A quantitative approach was employed using the 2020 data from the ND-GAIN Vulnerability Index and the World Bank CO_2 emissions record. Pearson product-moment correlation revealed a moderate negative association ($p = -0.50$, $p < 0.05$) between climate vulnerability and per capita CO_2 emission. The results indicated that countries contributing the least to global carbon dioxide emissions tend to be the most vulnerable to the adverse effects of climate change. In addition, it offers psychological interpretations and potential solutions faced by these

A. Harichandan
Area of Economics and Business Environment, Indian Institute of Management Sirmaur, Sirmaur, Himachal Pradesh, India
e-mail: aishwarya@iimsirmaur.ac.in

A. Srivastava (✉)
Jindal School of Psychology and Counselling, O.P. Jindal Global University, Sonipat, Haryana, India
e-mail: akancha.srivastava@jgu.edu.in

S. Mishra
Department of Psychology, Christ University, Bangalore, India
e-mail: shefali.mishra@christuniversity.in

A. Roy et al. (eds.), *Environment and Public Health*, Sustainability Solutions,
https://doi.org/10.1007/978-3-031-99770-9_3

countries based on the cognitive appraisal theory given by (Lazarus and Folkman in. Stress, appraisal, and coping. Springer (1984)) to understand resilience in the face of climate adversity. The chapter advocates for the empowerment of marginalized populations in low-income areas through fair access to clean water, healthcare, and green infrastructure, along with stronger environmental restrictions, can enhance public health outcomes and minimize vector-borne disease transmission. These findings highlight the need for equitable climate governance, which prioritizes the inclusion of vulnerable countries while embedding psychological resilience within global adaptation efforts, ultimately enhancing public health outcomes in these nations.

Keywords Social equity · Environmental justice · Public health · Sustainability · Climate crisis

3.1 Introduction

In the pursuit of relentless innovation, economic growth and development, humankind has significantly compromised the stability of the natural environment. Rapid industrialization, unrestrained consumption of natural resources, and excessive pollution have collectively contributed to the intensification of the climate crisis, one that now transcends environmental boundaries and has become a global concern with profound economic, social, and public health ramifications. Climate change is no longer an environmental concern; it is rapidly identified as a complex societal challenge that disproportionately burdens impoverished and marginalized communities. Vulnerable populations—particularly in low- and middle-income countries—face a heightened risk of exposure to environmental hazards such as deteriorating air quality, unsafe sanitation, infectious diseases, resource depletion, and extreme weather events like floods, droughts, and cyclones. These climate-induced adversities exacerbate existing socioeconomic disparities and strain public health systems, making the intersection of social equity and environmental justice more urgent than ever. These inconsistencies stem from global inequality, with traditionally poor inhabitants having the minimum access to advanced resources, healthcare and medical facilities, and decent infrastructure. The dispute of social equity, which refers to the equitable and just distribution of resources and opportunities, is intricately linked to climate change. Climate change has an intricate and multidimensional link with social equality since its effects disproportionately affect the most vulnerable groups, increasing and creating new inequities.

According to NASA's study, the Earth's average surface temperature in 2023 was the warmest on record. The temperature of Earth has increased by an average of 0.06° Celsius every decade since 1850 (Lindsey and Dahlman 2020). This has led to devastating events like floods, droughts, landslides, cyclones, etc. These events not only affect the poor more adversely but also increase the degree of inequality between and within countries. Taking this as the background, social equity and environmental justice are closely linked concepts. While social equity ensures equal access to resources irrespective of caste, class, gender, race, region, religion, etc., environmental justice ensures that all the sections of the society and the world are equally involved in the environmental policymaking arena and everyone's voice is equally heard (Mohai et al. 2009). Henceforth, it can be inferred that social equity leads to environmental justice, and environmental justice leads to social equity. This analysis caters to Sustainable Development Goal 10 (Reduced inequalities) by highlighting how climate change disproportionately affects vulnerable populations and deepens existing social inequalities. It supports Goal 13 (Climate Action) by emphasizing the urgent need for inclusive, equitable strategies to mitigate and adapt to climate risks. Furthermore, it addresses Goal 17 by advocating global cooperation and active inclusion of marginalized nations in environmental policymaking and governance. Sharma and Bhuyan (2021) and Poustie et al. (2004) argue for the importance of involving social equity in sustainability planning and addressing environmental inequalities, respectively. Understanding the underlying reasons for these disparities is imperative, as is grasping the experiences and sentiments of countries lacking sufficient resources. Psychological interpretation will enable us to understand how communities counter climate change by highlighting aspects such as risk perception, cognitive biases, and emotional responses. It will tend to explain the social and emotional consequences of climate change, such as mental health difficulties, self-esteem issues, and perception of the world. Adding psychological innovations into policy design can increase the efficacy of environmental rules and programs. Literature suggests that psychological resilience is crucial for communities to adapt to climate change (Ntontis et al. 2020). Psychological resilience is critical for communities to adapt to climate change, and understanding community dynamics and social support networks allows us to create interventions that enhance resilience. Addressing psychosocial elements, including social capital, community cohesiveness, and faith in authority, might help to reduce inequality and vulnerability.

Based on this background, the following objectives are studied in the chapter:

(i) To analyze the relation between the major per capita carbon dioxide emitters of the world and the countries that are the most vulnerable to climate change
(ii) To provide psychological interpretations and solutions to the conditions of vulnerable countries
(iii) To explore the intersection of climate disparities, environmental justice, and their impact on public health.

3.1.1 *Review of Literature*

Althor et al. (2016) empirically ascertain the connection between a nation's GHG emissions and its susceptibility to the adverse consequences of climate change. 20 of the 36 nations with the highest emissions are among the least vulnerable to the adverse effects of future climate change, according to their chapter, which is consistent with the findings of other studies. On the other hand, 11 out of the 17 nations with low or moderate GHG emissions are particularly susceptible to the adverse effects of climate change. Just 28 (16%) of the nations had a fair balance between vulnerability and emissions in 2010. Furthermore, projections for future emissions indicate that by 2030, this inequality will have substantially worsened. Many nations are blatant freeloaders, making others pay the price for climate change, which deters them from reducing their emissions. This becomes evident even by looking at studies focusing on individual countries and highlighting the psychological impact of climate change.

Afghanistan has witnessed drastic challenges due to CO_2 emissions and climate change, highlighting a critical intersection of environmental and socioeconomic challenges. This country, which is primarily an agrarian economy, is more prone to climate change due to its limited adaptive capacity and ongoing political instability. Various investigations have indicated that Afghanistan's CO_2 emissions are increasing due to deforestation, agricultural practices, and dependency on fossil fuels for energy needs (Hashmi et al. 2021; Dar and Asif 2020). Climate change aggravates existing vulnerabilities, leading to more frequent and severe floods, impacting food security, water availability, and human health (Akhtar and Shah 2020; Ayoubi 2023). The accumulated effect of these factors puts Afghanistan at more climate risk, resulting in urgent policy interventions to improve resilience and decrease emissions.

India's CO_2 emissions drastically increased the country's climate risk, aggravating economic and social vulnerabilities. India faced economic losses of USD 37 billion in 2018 due to climate change, and further predictions estimate an overall loss of nearly USD 100 billion from 2021 to 2025 due to climate crisis risks. Furthermore, India's GDP will also be negatively affected by 2.8% annually by 2050, negatively affecting the living standards of nearly half the population (Chapman et al. 2021). Recent investigations by Sharma et al. (2024) have pointed out that increasing concentrations of CO_2 emissions aggravate climate-related hazards such as extreme rainfall, heatwaves, and sea-level rise, which have become more frequent and severe in recent years. This study has discussed how these climate risks can significantly impact the risk of displacement for vulnerable communities, which could lead to many socioeconomic challenges and further strain the nation's economy. Another study by Kumar et al. (2024) reported that an increased rise in global temperature due to high CO_2 emissions could lead to significant disturbances in agricultural productivity and water resources in India, affecting food security and intensifying existing socioeconomic inequalities.

One of the most climate-vulnerable countries in the world is South Sudan, which is at greater risk due to CO_2 emissions, temperature fluctuations, unpredictable rain patterns, and increased frequency of extreme weather events. The country's fragile ecosystem and socioeconomic conditions increase its vulnerability to climate-induced disasters, particularly in agriculture, water resources and public health. The latest study by the World Bank (2024) highlighted that South Sudan's minimal CO_2 emissions ironically contribute to a significant climate crisis in the region, where global warming intensifies local droughts and floods, severely impacting food security and livelihoods. Furthermore, this particular country's vulnerability is further amplified by increasing levels of deforestation and land degradation, which further calls for extensive research into environmentally sustainable methods.

Another country that is highly vulnerable to climate change is Niger and Bolivia, which is facing significant socioeconomic challenges due to its varied ecosystem and dependence on agriculture. Various studies (Smith and Diallo 2024; Martinez and Lopez 2024) have investigated that Niger's and Bolivia's minimal CO_2 emissions contribute to global climate change, further amplifying local heatwaves and droughts and intensifying agricultural stress and water scarcity.

Dube (2023) conducted a study that examined Zimbabwe's widespread narratives and experiences of climate risk management from 1980 to 2021. The narrative surrounding climate risk has shifted over the last forty years, but it still favors adaptation. It emphasizes the vulnerability of infrastructure and the environment, communication, gender aspects of climate change, hazard exposure, water scarcity, sensitivity of agriculture and livelihoods, and indigenous knowledge systems. From 1900 to 2017, Zimbabwe saw a great deal of natural disasters, such as storms, floods, pandemic diseases, and droughts. Over 20 million people died, and these incidents caused $950 million worth of damage. The frequency of droughts has increased, and pandemic diseases play a major role. Floods have also brought about significant economic loss; nine riverine floods affected over 300,000 people and resulted in losses exceeding $270 million. (*World Bank Climate Change Knowledge Portal*, n.d.) Given the number of cyclones that Zimbabwe has seen over the last 20 years, ecological sustainability is crucial. A study looking at how Higher Education Institutions (HEIs) participate in disaster and climate change management discovered that HEIs are not covered by current laws and structures. While communities are concerned about HEIs' poor participation, researchers and students emphasize the institutions' responsibility in managing climate change. According to the study, HEIs must be given the opportunity to play significant roles in disaster relief and climate change mitigation so they can share their expertise and tackle environmental issues. (Macheka 2024).

Mozambique is impacted by both climate change and climate variability. Mozambique is highly vulnerable to natural disasters. The nation is vulnerable to extreme weather occurrences due to its long coastline, low elevation, and proximity to the Indian Ocean. Mozambique's climate is also influenced by the El Niño Southern Oscillation (ENSO), which results in protracted droughts and floods. Climate change has the potential to impede development initiatives, reduce the ability of socioeconomic systems to adapt, and have far-reaching effects on the vast majority of people who depend on land and natural resources (Nussbaumer and Patt 2009).

Sub-Saharan Africa faces a threat from climate change, with rising rates of poverty predicted by 2050 in the absence of any action. Significant natural disasters have affected Mozambique's health system over the last 42 years, making the country vulnerable to climate change. A Health Vulnerability Index (HVI) was used in a study to evaluate Mozambique's health system's vulnerability and adaptation needs. Some districts exhibited high vulnerability, while big urban canters showed lesser vulnerability, according to the results (Muleia et al. 2024).

3.2 Methodology

This study employs a quantitative research approach to analyze the relationship between the University of Notre Dame's ND-GAIN Vulnerability Index and per capita carbon dioxide emissions. The methodology consists of the following key steps:

3.2.1 Data Collection

Data for this study was obtained from two primary sources:

ND-GAIN Vulnerability Index (2020)*:* The vulnerability index data was sourced from the University of Notre Dame ND-GAIN database. The index measures a country's exposure, sensitivity, and adaptive capacity to climate change, considering six life-supporting sectors: food, water, health, ecosystem services, human habitat, and infrastructure.

Per Capita Carbon Dioxide Emissions (2020): Data on country-wise per capita CO_2 emissions were retrieved from the World Bank database. The data is measured in metric tons per person and represents the most recent comprehensive available dataset for global emissions.

3.2.2 Data Processing and Cleaning

Both datasets were filtered to include only countries for which data were available in both sources. The vulnerability index lies between [0, 1], where 0 denotes the least vulnerable country, and 1 denotes the highly vulnerable country [2], while CO_2 emissions were retained in metric tons per capita for consistency in comparison.

3.2.3 Correlation Analysis

A Pearson product-moment correlation analysis was conducted to determine the strength and direction of the relationship between the ND-GAIN Vulnerability Index and per capita CO_2 emissions.

3.3 Results

In line with the first objective to analyze the relationship between major per capita carbon dioxide emitters and the country's most vulnerable to climate change, this chapter employs data from the University of Notre Dame ND-GAIN's ranking on the vulnerability of countries and the World Bank's country-wise per capita carbon dioxide emission data; this chapter has analyzed the correlation between both variables. This chapter focuses on the analysis of data from the year 2020, as it represents the most recent year for which comprehensive data is available for both datasets.

The ND-GAIN's vulnerability index is calculated by measuring a country's exposure, sensitivity, and capacity to adapt to the negative effects of climate change. ND-GAIN measures overall vulnerability by considering six life-supporting sectors: food, water, health, ecosystem service, human habitat, and infrastructure. The vulnerability index lies between [0, 1], where 0 denotes the least vulnerable country, and 1 denotes the most vulnerable country.

Figure 3.1 illustrates a negative relationship between the vulnerability index and per capita carbon dioxide emissions, with a correlation coefficient of – 0.5091. This relationship indicates that countries contributing the least to global carbon dioxide emissions tend to be the most vulnerable to the adverse effects of climate change. For instance, the United States emits 13.03 metric tons of CO_2 per capita, yet its vulnerability index is relatively low at 0.31. In contrast, Vanuatu, with per capita CO_2 emissions of only 0.39 metric tons, has a significantly higher vulnerability index of 0.56. This comparison highlights that while advanced economies are among the largest emitters of CO_2, they possess greater capacity to mitigate and adapt to the impacts of climate change.

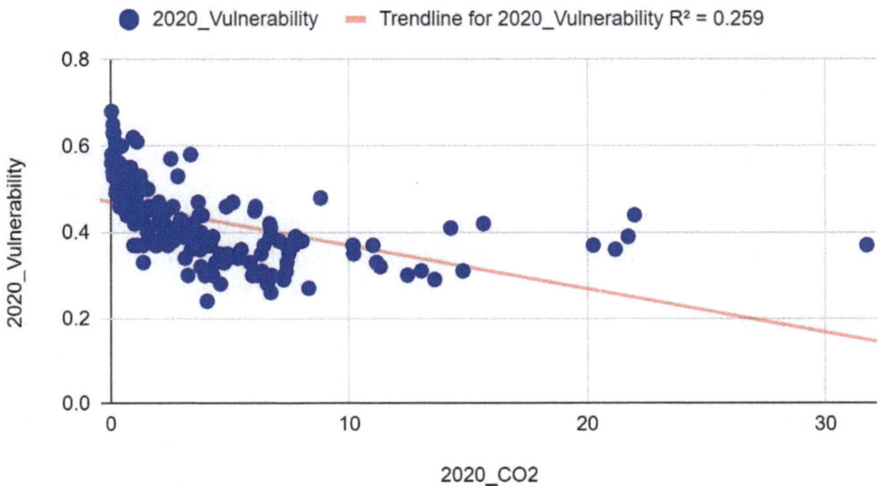

Fig. 3.1 Relationship between vulnerability index and per capita CO_2 emissions ($r = -0.5091$)

Taking this into account, it is important to note that the most vulnerable countries have a much lower influence in decision-making in matters related to climate change vis-à-vis the major polluters. This is specifically true due to the political and economic might of the major polluters. One example of this is the composition of the P5 members in the UNSC. In such a situation, along with exerting pressure at the global forum through groupings like SIDS (Small Island Developing States), it becomes important to understand and address the psychological vulnerabilities of these countries at the national, group, and individual levels. This helps in the mitigation and adaptation of psychological distress at the country and global levels.

3.3.1 Psychological Interpretation

Comprehending these hurdles imposes an interdisciplinary perspective that brings together depths from environmental science, public health, social justice, and social psychology. Adopting a psychological lens, the Cognitive Appraisal Theory (Lazarus and Folkman 1984). It offers a significant model for examining how individuals and communities perceive and respond to climate-induced stress, which influences their ability to adapt and recover, and subsequently it leads to impacting public health. To understand the psychological interpretation of vulnerable countries and individuals who suffer from climate extremes, a theoretical framework is applied.

Cognitive Appraisal Theory was propounded by Lazarus and Folkman (1984) and is pivotal to comprehending how individuals assess and manage stress, including strain related to climate change. This theory offers insight into how people's emotional responses to events are influenced by their analyses, understanding, or appraisals of the situation, rather than the events themselves.

As illustrated in Fig. 3.2, the theory names two main components: *primary appraisal* and *secondary appraisal.* Primary appraisal includes the preliminary evaluation of a situation to determine its magnitude for the individual. It answers the question: *Is this situation a threat, a challenge, or irrelevant to my well-being?* In a threat appraisal, if the individual perceives the situation as harmful or threatening (e.g., a climate-related disaster such as flooding or drought), they may experience negative emotions such as psychological distress, poor quality of life, poor emotional regulation, trauma, PTSD, and low self-esteem. Conversely, if the circumstances are evaluated as a challenge, something that forces effort but is possibly controllable, the emotional response may be more positive, involving hope, optimism, resilience, and self-confidence. If the event is deemed as irrelevant, meaning that the event has not directly affected them, they are less likely to experience a strong emotional response. In the context of climate change, primary appraisal may involve assessing whether environmental changes pose a direct threat to personal safety, livelihood, or future stability.

Secondary appraisal follows the primary evaluation and involves assessing the available coping resources and options. It answers the question, *What can I do about this situation?* Here, perceived control plays a crucial role in assessing their ability

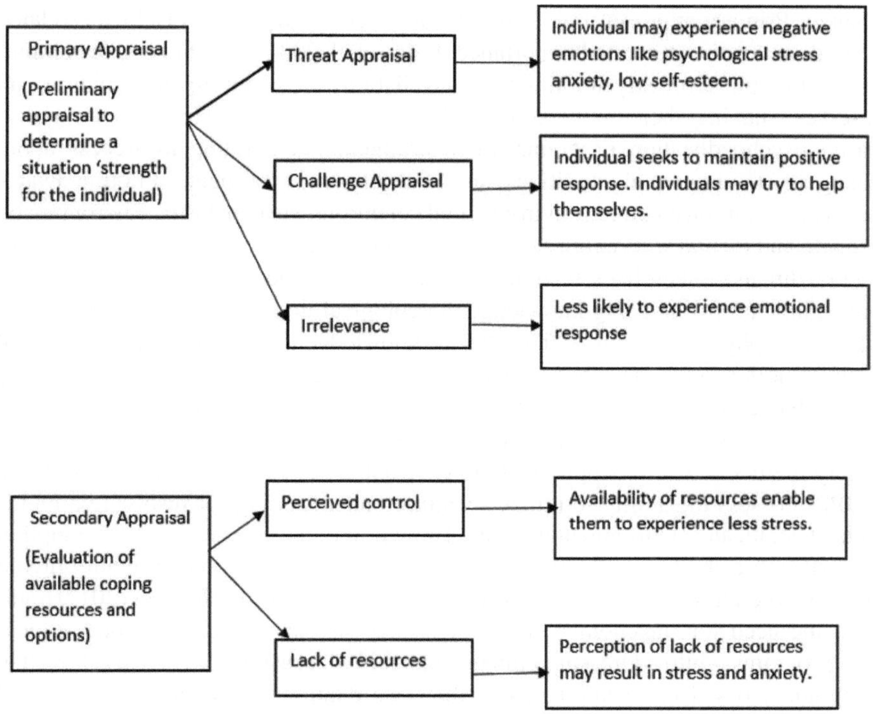

Fig. 3.2 Cognitive appraisal theory presents psychological interpretation of vulnerable countries and individuals who suffer climate extremes Lazarus and Folkman (1984)

to manage or control the stressor. If they feel they have the resources (e.g., social support, coping strategies, financial stability) to address the situation, they are likely to experience lower levels of stress. However, if the individual perceives that they lack the resources to cope with the situation, they are more likely to experience heightened stress, anxiety, or helplessness. In the case of climate change, secondary appraisal might include evaluating whether individuals or communities have access to resources such as disaster relief programs, economic support, or infrastructure improvements to mitigate climate-related risks.

3.3.2 Incorporating Psychological Solutions to Decrease Climatic Exposure in Marginalized Communities

Climate change excessively affects underprivileged communities, worsening already present socioeconomic inequities and enhancing psychological distress. By exploring and combining psychological solutions to climate variation techniques can increase resilience, enhance better quality of life and mitigate the adverse effects of climate change. Primary appraisal refers to the fundamental evaluation of climate-related hazards and their interpreted importance (Laino et al. 2024). For vulnerable communities, this phase mostly implies the unavailability of necessary information and reserves, making them more susceptible to climate-related stress. Community-based psychoeducation programs are instrumental in building awareness about changing environmental conditions, sustainability, and well-being. Piloting awareness campaigns on changing environmental conditions, sustainability, environmental growth, and mental well-being can help individuals recognize the impact of environmental threats on their lives. Community-based psychoeducation programs (CBPPs) are significant for inducing knowledge of ecological threats and the psychological, social, and physical outcomes. They bridge knowledge gaps, promote psychological understanding, encourage agency and community engagement, accelerate culturally applicable learning, and reduce helplessness and despair. CBPPs make dense conservation issues more relevant by using local stuff, symbols, storytelling, and participatory methods to help people relate to environmental changes in their daily lives. CBPPs possess the ability to utilize scientific understanding for resolving a multitude of health and environmental issues. These programs can encourage the adoption of sustainable practices by employing psychological insight and community-driven social marketing strategies (McKenzie-Mohr 2000). Dittmer and Riemer (2012) highlight the need for stakeholder involvement and systems thinking in youth education programs while addressing climate change. Folorunsho et al. (2024) show that community-based interventions can reduce the damaging influence of social and environmental influences on mental well-being, citing successful programs such as Zimbabwe's Friendship Bench. These studies equally highlight the need of community engagement, cultural sensitivity, and addressing environmental impacts in generating effective interventions for both environmental and mental health issues. Urging community-driven initiatives, such as sustainable farm practices, mindful consumption practices, water conservation practices, and proper land utilization, can introduce a sense of action and control over climate difficulties. The self-efficacy framework for community-based environmental action is a framework that integrates individual belief in personal capacity with collective action. This framework draws from Bandura's theory of self-efficacy (1977) and adapts it to environmental behavior in collectivist contexts, integrating both individual belief in a personal capacity and its interaction with collective action. It consists of four components: mastery experiences, vicarious experiences, social persuasion, and emotional and physiological states. These elements encourage hands-on training, social modeling, culturally relevant communication, and emotional support to encourage community members to contribute meaningfully to environmental change. Collective and self-efficacy influence long-term behavior, going beyond attitudes and social norms. These findings show that promoting collective efficacy is critical for designing effective environmental interventions and policies (Reese and Junge 2017). Individuals in India with strong collective efficacy beliefs are more likely to participate in community water conservation efforts, particularly in communities with high collective efficacy levels (Thaker et al. 2016). Collective efficacy, along with faith in government, enhances public commitment to environment change-related water conservation programs.

Higher collective efficacy relates to increased policy support and activist actions (Thaker et al. 2019). Secondary appraisal requires measuring one's ability to deal with climate-related battles. Unemployment, poverty, marginalization, discrimination, biases, and inadequate mental health support are some of the problems that many marginalized communities go through, making successful adaptation difficult. Focusing on these roadblocks demands psychological interventions. Developing resilience for climate disruption entails creating adaptive capability at individual, societal, and systemic levels to avoid risks, recover from unfavorable climate events, and support long-term sustainability. This requires education and environmental awareness, critical thinking skills, a growth mindset, and developing sustainable life choices. Mindfulness-Based Practices and Emotional Resilience Training can enhance emotional regulation and moderate climate-related distress. Practicing mindfulness increases people's awareness of the social, economic, and environmental effects of their consumption habits. Sheth et al. (2011) presented the concept of mindful consumption on three levels: caring for the self, for the community, and for nature.

Climate catastrophes, such as hurricanes, floods, wildfires, heatwaves, and extreme cold winds, can cause severe mental and emotional distress, including trauma, anxiety, despair, and PTSD. Trauma-informed interventions can aid individuals and communities in healing and rebuilding resilience following such experiences. Climate change disasters can cause mental health concerns such as trauma, anxiety, PTSD, and depression (Gifford and Gifford 2016a, b; Palinkas et al. 2020). Trauma-informed interventions, such as monitoring programs, community resilience initiatives, and trained health workers, are critical (Palinkas et al. 2020). CBT and trauma-focused CBT are beneficial in reducing symptoms in disaster-affected kids (Burkhart et al. 2023). Nature-based therapies, such as forest therapy, show potential for boosting resilience and well-being (Hartwell et al. 2023).

3.3.3 Climate Disparities, Environmental Justice, and Their Impact on Public Health

To effectively address climate disparities and promote environmental justice, it is essential to implement comprehensive, multi-level solutions that mitigate public health risks and enhance community resilience. People who are marginalized and stay in low-income areas are more likely to experience environmental hazards such as air pollution, contaminated water and severe heat, leading to higher rates of respiratory, cardiovascular, and infectious diseases. Empowering marginalized communities through better access to clean water, accessible healthcare, and green infrastructure, stricter environmental regulations, and community engagement can significantly improve health outcomes in marginalized areas (Clayton 2017). Climate change accelerates the transmission of vector-borne diseases such as malaria, dengue, and

Lyme disease by facilitating the expansion of suitable environments for disease-carrying organisms like mosquitoes and ticks (Rocklöv and Dubrow 2020). Adopting climate-resilient public health interventions, such as vector control initiatives (Wilson et al. 2020), sanitation improvements (Campbell-Lendrum et al. 2015), and psychoeducational programs (Akerlof et al. 2015) on health risks and preventive behaviors, can reduce the impact of vector-borne diseases on disease prevention. Vulnerable communities facing climate-induced displacements, disaster, and scarcity of resources are at greater risk of developing stress, anxiety, depression, and PTSD (Berry et al. 2018; Cianconi et al. 2020). Comprehensive mental health interventions such as trauma counselling (Usher et al. 2019), psychoeducation on stress management (Gifford and Gifford 2016a, b), and community support networks (Palinkas and Wong 2020) can promote psychological resilience among vulnerable populations. Climate-induced changes such as drought, erratic rainfall and soil degradation compromise food and water availability, heightening the risk of malnutrition and waterborne diseases among vulnerable populations. Introducing climate-adaptive agricultural techniques (Vermeulen et al. 2012), along with water conservation systems (Rockström et al. 2014) and psychoeducation on hygiene and nutrition (Dangour et al. 2012) can strengthen food and water security. Increasing temperatures heighten the risk of heat-related illnesses among outdoor workers, such as farmers and construction workers while worsening existing health problems. Integrating climate-resilient workplace policies (Schulte and Chun 2009), health awareness programs (Xiang et al. 2014), and adjusted working hours (García-Trabanino et al. 2020) supported by CSR initiatives for worker welfare can enhance the well-being of outdoor workers facing rising temperatures. Climate-induced displacement caused by rising sea levels, extreme weather events, and resource depletion is burdening healthcare systems in receiving areas, increasing the likelihood of infectious diseases. Integrating community-based mental health services with government healthcare schemes (Silove et al. 2017) can ensure the displaced population receives both physical healthcare and psychosocial support, reducing the burden on healthcare systems and promoting emotional well-being.

Marginalized communities such as racial and ethnic minorities, indigenous populations and people in developing nations face greater health risks due to systemic inequalities in environmental policies and limited access to health care. Implementing inclusive environmental policies alongside community-based psycho educational programs (Mohai et al. 2009; Friel et al. 2011) can empower marginalized groups to cope with environmental challenges while advocating for health equity.

3.4 Discussion

Climate change stands as both an environmental and ethical crisis, exposing deep global injustices. The gap between major contributors to carbon emissions and those who are most vulnerable highlights deep-rooted structural inequalities. Wealthier, industrialized nations often dominate environmental decision-making, whereas vulnerable nations often lack both the voice and means to adapt effectively. These disparities are layered, not only ecological but also psychological and physiological. Against this backdrop, this chapter emphasizes the intersectionality of climate inequality, focusing on the disproportionate emission-vulnerability link, limited global agency for at-risk nations, and psychological consequences of environmental injustice. Taken together, these themes underscore the importance of an equitable and inclusive approach to global climate governance.

In line with this, the *first objective* of the chapter is to analyze the relationship between the major per capita carbon dioxide emitters of the world and the country's most vulnerable to climate change. Developed nations frequently transfer the environmental burden to the Global South, where environmental controls tend to be more relaxed. This practice accelerates environmental degradation, especially in low- and middle-income regions (Filonchyk et al., 2024). These regions are more prone to climate hazards due to financial scarcity and infrastructural capacity. Despite being the largest contributors to emissions, they continue to exert excessive influence over global climate policy, marginalizing the equitable representation of poorer nations. Disparities in economic strength and political influence frequently result in unequal climate policies where global goals are prioritized over local adaptation needs. Furthermore, multinational companies intensify environmental degradation through increased exploitation of natural resources and land conversion in the Global South (Ogwu et al. 2025). The gap in technology and financial resources further deepens the climate divide, as these countries frequently face barriers to accessing green technologies due to high costs or restrictive intellectual property rights.

Understanding the psychological dimension of climate vulnerability allows for a deeper grasp of coping strategies and behavioral responses under climate stress (Zhai and Lee 2024). These insights help in identifying mental health priorities while ensuring interventions keep in mind cultural and contextual needs. Based on this, *the second objective* of the chapter is to provide psychological interpretations and solutions to the conditions of vulnerable countries. In this context, cognitive appraisal theory was applied to provide a wide-ranging psychological explanation of individual responses to climate-related stressors. This theory offers insight into how people's emotional responses to events are influenced by their analyses, understanding, or appraisals of the situation, rather than the events themselves (Yeo and Ong 2024). The theory emphasizes two levels of appraisal—primary and secondary—through which people are able to identify the level of severity of a stressor and their ability to cope. In the context of environmental extremes, limited coping resources often lead to the perception of such events as a threat. This theoretical framework also explains various emotional reactions such as anxiety, depression, and eco-anxiety. Furthermore, it also supports the design of mental health programs that build resilience, promote resilience, and respond sensitively to cultural differences. Strength lies in being adaptable to different cultural and social settings, making it valuable to create sensitive and effective mental health interventions in climate-impacted regions.

Climate change intensifies pre-existing disparities in health and social systems, disproportionately affecting marginalized and economically disadvantaged groups. Inadequate infrastructure, environmental degradation and poor healthcare access to healthcare heighten their vulnerability to climate hazards. These challenges are further exacerbated by their limited representation in policy decision-making, leaving health concerns unaddressed. In light of this, *the third objective* of the chapter is to explore the intersection of climate disparities, environmental justice, and their impact on public health. Inequities in infrastructure and access to resources increase climate vulnerability in marginalized and low-income communities, which leads to increased health risks and inequality. These areas are often located near polluting factories, flood-prone regions, or spaces without green zone areas, reflecting patterns of historical neglect and exclusion (Kang and Weatherly 2025). Poor air and water quality, inadequate sanitation, and chemical exposure predispose an increased risk of respiratory diseases, heart conditions, and infectious diseases. As climate change worsens, it exacerbates physical and mental health challenges such as chronic stress, anxiety, and trauma, often left untreated due to a combination of stigma and limited services. Exclusion from environmental decision-making further aggravates these concerns, as policies fail to account for the needs of these communities. To achieve meaningful climate justice and protect public health, interventions must prioritize infrastructure development, equitable healthcare access, culturally relevant mental health care, and community-driven resilience-building strategies (Aziz and Anjum 2024). This comprehensive approach is important for understanding the intersection of climate change, public health, and environmental justice.

3.5 Conclusion

The global climate upheaval simultaneously exposes and aggravates existing global inequities, particularly between high-emission industrialized countries and more vulnerable, less developed nations. As explored in this chapter, the intersection of emission responsibility, diminished policy agency, psychological strain, and public health burdens exhibits a broader reflection of environmental injustice. Vulnerable countries not only often face greater physical risks from climate change but also suffer emotional and health-related consequences compounded by systemic neglect and exclusion. Addressing these multilayered challenges requires equitable representation in climate policymaking, responsibility from major polluters and implementation of culturally sensitive and community-driven approaches. A comprehensive equitable, and inclusive response goes beyond mitigation to incorporate adaptation, mental health care, and public health infrastructure, ensuring no country is left behind in the global pursuit of sustainability and well-being.

Declaration

Funding This research did not receive any specific grant from funding agencies in the public, commercial, or not-for-profit sectors.

Data Availability The data analyzed in this chapter are available from the corresponding author upon reasonable request. The corresponding author takes responsibility for the integrity and accuracy of the data analysis.

Competing Interest The contributors declare that there is no competing interest.

References

Althor G, Watson JE, Fuller RA (2016) Global mismatch between greenhouse gas emissions and the burden of climate change. Sci Rep 6(1):20281

Ayoubi T (2023) Hydrological modeling and flood hazard assessment in data-scarce watersheds of Afghanistan (Doctoral dissertation). Freie Universität Berlin

Akhtar F, Shah U (2020) Emerging water scarcity issues and challenges in Afghanistan. In: Ranjan A (ed) Water issues in Himalayan South Asia: internal challenges, disputes and transboundary tensions. Palgrave Macmillan, Singapore, pp 1–28

Akerlof K, Delamater PL, Boules CR, Upperman CR, Mitchell CS (2015) Vulnerable populations perceive their health as at risk from climate change. Int J Environ Res Public Health 12(12):15419–15433. https://doi.org/10.3390/ijerph121214999

Aziz M, Anjum G (2024) Transformative strategies for enhancing women's resilience to climate change: a policy perspective for low-and middle-income countries. Women's Health 20:17455057241302032

Bandura A (1977) Self-efficacy: toward a unifying theory of behavioral change. Psychol Rev 84:191–215

Berry HL, Waite TD, Dear KB, Capon AG, Murray V (2018) The case for systems thinking about climate change and mental health. Nat Clim Chang 8(4):282–290. https://doi.org/10.1038/s41558-018-0102-4

Burkhart K, Agarwal N, Kim S, Neudecker M, Ievers-Landis CE (2023) A scoping review of trauma-informed pediatric interventions in response to natural and biologic disasters. Children 10(6):1017. https://doi.org/10.3390/children10061017

Campbell-Lendrum D, Manga L, Bagayoko M, Sommerfeld J (2015) Climate change and vector-borne diseases: what are the implications for public health research and policy? Philos Trans R Soc B Biol Sci 370(1665):20130552. https://doi.org/10.1098/rstb.2013.0552

Chapman D, Trott C, Silka L, Lickel B, Clayton S (2021) Psychological perspectives on community resilience and climate change: insights, examples, and directions for future research. Clim Change 167(3):1–25

Clayton S (2017) Mental health risk and resilience among climate scientists. Nat Clim Chang 7(6):396–398

Cianconi P, Betrò S, Janiri L (2020) The impact of climate change on mental health: a systematic descriptive review. Front Psych 11:74

Dangour AD, Watson L, Cumming O, Boisson S, Che Y, Velleman Y, Cavill S, Allen E, Uauy R (2012) Interventions to improve water quality and supply, sanitation and hygiene practices, and their effects on the nutritional status of children. Cochrane Database Syst Rev 2013(8):CD009382. https://doi.org/10.1002/14651858.CD009382.pub2

Dar IA, Asif M (2020) Flood hazards and resilience: a community-based approach. Springer

Dittmer LD, Riemer M (2012) Fostering critical thinking about climate change: applying community psychology to an environmental education project with youth. Glob J Community Psychol Pract 3(10):1–9

Dube K (2023) Tourism and climate change: risks and opportunities. Routledge; Folorunsho IL, Harry NM, Anugwom G (2024) Mitigating the negative impact of social and environmental factors on mental well-being through community-based interventions. J Adv Med Med Res

Filonchyk M, Peterson MP, Yan H, Gusev A, Zhang L, He Y, Yang S (2024) Greenhouse gas emissions and reduction strategies for the world's largest greenhouse gas emitters. Sci Total Environ 944:173895

Friel S, Marmot M, McMichael AJ, Kjellstrom T, Vågerö D (2011) Global health equity and climate stabilisation: a common agenda. The Lancet 378(9799):1677–1683. https://doi.org/10.1016/S0140-6736(11)61692-X
García-Trabanino R, Jarquín E, Wesseling C, Johnson RJ, González-Quiroz M, Weiss I, Glaser J, Vindell JJ, Stockfelt L, Roncal C, Harra T, Roncal-Jimenez CA (2020) Heat stress, dehydration, and kidney function in sugarcane workers in El Salvador. Int J Occup Environ Health 24(3–4):1–11. https://doi.org/10.1080/10773525.2018.1524588
Gifford R, Gifford S (2016a) The largely unrecognized impact of climate change on mental health. Bull AtIc Sci 72(5):292–297
Gifford E, Gifford R (2016b) The largely unacknowledged impact of climate change on mental health. Bull AtIc Sci 72:292–297
Hartwell C, Randazza JM, Bratman GN, Eisenman DP, Ellis B, Goodsell E et al (2023) Forest therapy as a trauma-informed approach to disaster recovery: insights from a wildfire-affected community. PLOS Climate 2(9):e0000096. https://doi.org/10.1371/journal.pclm.0000096
Hashmi MZ, Naveed M, Varma A (eds) (2021) Environmental pollution and resilience: new strategies for ecosystem restoration. Springer
Kang J, Weatherly C (2025) Climate justice and carbon emission inequality: social work's dilemma in marginalized communities. Ethics Soc Welf 1–18
Kumar R, Singh S, Sharma P (2024) Advances in sustainable agriculture: innovations and practices. Elsevier
Laino E, Paranunzio R, Iglesias G (2024) Scientometric review on multiple climate-related hazards indices. Sci Total Environ 174004
Lazarus RS, Folkman S (1984) Stress, appraisal, and coping. Springer
Lindsey R, Dahlman L (2020) Climate change: global temperature. Climate Gov 16
Macheka T (2024) Climate change adaptation strategies in African agriculture. Cambridge Scholars Publishing
McKenzie-Mohr D (2000) New ways to promote pro-environmental behaviour: promoting sustainable behaviour: an introduction to community-based social marketing. J Soc Issues 56(3):543–554. https://doi.org/10.1111/0022-4537.00183
Muleia R, Chirwa PW, Babalola FD (2024) Agroforestry systems and practices in Southern Africa. Springer
Martinez R, Lopez G (2024) The effects of CO_2 emissions on climate variability in Bolivia: impacts on agriculture and water resources. J Clim Environ Res 22(3):345–360. https://doi.org/10.1016/j.jcer.2024.05.007
Mohai P, Pellow D, Roberts JT (2009) Environmental justice. Annu Rev Environ Resour 34:405–430
Nussbaumer P, Patt A (2009) Climate change, vulnerability, and adaptation: social perspectives. Cambridge University Press
Ntontis E, Drury J, Amlôt R, Rubin GJ, Williams R (2020) What lies beyond social capital? The role of social psychology in building community resilience to climate change. Traumatology 26(3):253–265. https://doi.org/10.1037/trm0000221
Ogwu MC, Kosoe EA, Osawaru ME (2025) Raw materials for sustainable bioeconomy development in the global south. In: Sustainable bioeconomy development in the global south: volume i status and perspectives. Springer, Singapore, pp 37–63
Palinkas LA, Wong M (2020) Global climate change and mental health. Curr Opin Psychol 32:12–16. https://doi.org/10.1016/j.copsyc.2019.06.023
Palinkas LA, O'Donnell ML, Lau W, Wong M (2020) Strategies for delivering mental health services in response to global climate change: a narrative review. Int J Environ Res Public Health 17(21):7849
Poustie M, Walker G, Eames M, Fay H (2004) Environmental and social justice
Reese G, Junge EA (2017) Keep on Rockin' in a (plastic-)free world: collective efficacy and pro-environmental intentions as a function of task difficulty. Sustainability 9:200
Rocklöv J, Dubrow R (2020) Climate change: an enduring challenge for vector-borne disease prevention and control. Nat Immunol 21(5):479–483

Rockström J, Falkenmark M, Karlberg L, Hoff H, Rost S, Gerten D (2014) Future water availability for global food production: the potential of green water for increasing resilience to global change. Water Resour Res 45(7). https://doi.org/10.1029/2007WR006767

Schulte PA, Chun H (2009) Climate change and occupational safety and health: establishing a preliminary framework. J Occup Environ Hyg 6(9):542–554. https://doi.org/10.1080/15459620903066008

Silove D, Ventevogel P, Rees S (2017) The contemporary refugee crisis: an overview of mental health challenges. World Psychiatry 16(2):130–139. https://doi.org/10.1002/wps.20438

Sheth JN, Sethia NK, Srinivas S (2011) Mindful consumption: a customer-centric approach to sustainability. J Acad Mark Sci 39:21–39

Sharma A, Gupta R, Singh J (2024) Renewable energy technologies: trends and perspectives. Wiley

Sharma DP, Bhuyan AJ (2021) Social equity. Comprehending Equity

Smith J, Diallo A (2024) The impact of CO_2 emissions on agricultural productivity in Niger: an empirical analysis. J Environ Econ Manag 95:102–118

Thaker J, Maibach EW, Leiserowitz A, Zhao X, Howe PD (2016) The role of collective efficacy in climate change adaptation in India. Weather Clim Soc 8:21–34

Thaker J, Howe PD, Leiserowitz A, Maibach EW (2019) Perceived collective efficacy and trust in government influence public engagement with climate change-related water conservation policies. Environ Commun 13:681–699

Usher K, Durkin J, Bhullar N (2019) The role of social support in the mental health of climate change survivors. Lancet Planetary Health 3(11):e439–e440. https://doi.org/10.1016/S2542-5196(19)30264-7

Vermeulen SJ, Aggarwal PK, Ainslie A, Angelone C, Campbell BM, Challinor AJ, Hansen JW, Ingram JS, Jarvis A, Kristjanson P, Lau C (2012) Options for support to agriculture and food security under climate change. Environ Sci Policy 15(1):136–144

Xiang J, Bi P, Pisaniello D, Hansen A (2014) Health impacts of workplace heat exposure: an epidemiological review. Ind Health 52(2):91–101

Wilson AL, Boelaert M, Kleinschmidt I, Pinder M, Scott TW, Tusting LS, Lindsay SW (2020) Evidence-based vector control strategies to prevent malaria in sub-Saharan Africa: a systematic review and meta-analysis. The Lancet 395(10219):800–814

World Bank (2024) South Sudan climate vulnerability and CO_2 emissions impact. Retrieved from World Bank Publications

Yeo GC, Ong DC (2024) Associations between cognitive appraisals and emotions: a meta-analytic review. Psychological Bulletin

Zhai L, Lee JE (2024) Investigating vulnerability, adaptation, and resilience: a comprehensive review within the context of climate change. Atmosphere 15(4):474

Chapter 4
Emerging Trends and Future Research Directions in Environmental Health

Wareesa Sharif, Hafiz Raja Muhammad Ali, Muhammad Ashraf, Muhammad Anwaar, Rao Muhammad Sajjad Sharif, and Ali Shan Tariq

Abstract The environment is changing due to rapid urbanization, industrialization, climate change, and technological improvements, which pose new health dangers to the general public. The increasing importance of water pollution, endocrine-disrupting chemicals, and microplastics is being highlighted by recent research, in addition to more conventional issues like infectious diseases. The worldwide prevalence of infectious diseases has a substantial impact on death rates, and it can be difficult to evaluate and gauge the severity of infections due to their diverse symptoms. Various nations encounter distinct obstacles concerning these illnesses. This study presents cutting-edge artificial intelligence (AI)-based techniques to improve diagnosis precision by analyzing water quality. And the advances in technology, including genomic techniques, IoT-based environmental monitoring, and artificial intelligence, are making it possible to analyze environmental dangers and their effects on human health more accurately. Moreover, the interplay between ecological health and socioeconomic factors, including disparities in risk exposure and resource availability, necessitates using integrated approaches to tackle intricate global health issues. The creation of fair, sustainable solutions that use multidisciplinary frameworks to reduce environmental concerns must be the top priority of future research. To guarantee a healthier, more sustainable future, this chapter highlighting the significance of cutting-edge research approaches. Artificial intelligence (AI) holds the

W. Sharif (✉) · H. R. M. Ali · M. Anwaar
Department of Artificial Intelligence, Faculty of Computing, The Islamia University of Bahawalpur, Bahawalpur, Pakistan
e-mail: wareesa.sharif@iub.edu.pk

M. Ashraf
Department of Management Sciences, COMSATS University Islamabad, Vehari Campus, Islamabad, Pakistan

R. M. S. Sharif
Department of Entomology, University of Agriculture Faisalabad, Faisalabad, Pakistan

A. S. Tariq
University of Central Punjab Bahawalpur, Bahawalpur, Pakistan

A. Roy et al. (eds.), *Environment and Public Health*, Sustainability Solutions,
https://doi.org/10.1007/978-3-031-99770-9_4

potential to transform environmental health by solving complex models of ecological and also human systems, enhancing predictive accuracy, and optimizing resource management.

Keywords Climate changes · Genomic · Infectious disease · Deep learning IOT

4.1 Introduction

Environmental health is a fast-evolving area of science shaped by the dynamic interplay among urbanization, industrialization, and climate change. Public health suffers as these three factors work their adverse effects on ecosystems, pollution, and the spread of infectious diseases (Haines et al. 2020). New environmental hazards have arisen, and specific types of research have proliferated in response to them. Environmental health research is one such type; it has broadened the focus of inquiries into human health impacts of environmental exposures. It is now investigating not only the old concerns, such as contamination of water supplies and sanitation, but also a host of new threats to human health air pollution, endocrine-disrupting chemicals, microplastics, and more. Environmental health research encompasses a much wider array of environmental exposures than before and seeks to understand their human health impacts (Wang et al. 2023). As a reaction, innovations in technology such as genomic techniques, the Internet of Things (IoT), and artificial intelligence (AI) are creating new ways to measure environmental threats to our health and counteract their harmful impacts (Bhattacharya et al. 2022).

Infectious diseases are a mainstay of environmental health because they correlate so closely with environmental factors climate variability, pollution, urbanization, and the like that are changing today (Wu et al. 2020). Diseases that are spread from an infected to a noninfected person usually do so in the context of certain ecological conditions. These conditions cover a broad array of factors (described in more detail below) and interact with one another in a multitude of ways. The factors falling under the umbrella of ecology are the main ones that environmental health scientists study when trying to understand which diseases get transmitted, to whom, and why (Beyer et al. 2021). Recent global events, such as the COVID-19 pandemic, have emphasized the role of environmental factors in disease emergence, stressing the necessity of predictive modeling and early warning systems. Integrating AI with predictive water quality analytics, which are concerned with infectious diseases (Rahman et al. 2023).

Throughout history, the study of the occurrence and spread of disease has largely been the study of humans in relation to their environment. Traditional epidemiological models, such as the susceptible-infected-recovered (SIR) model, and various derivatives of it (notably the SEIR and SEIRS models), have been used for quite some time now to simulate the dynamics of particular diseases in specific environmental conditions (Kermack and McKendrick 1927). In recent times, models driven by artificial intelligence and machine learning, such as Bayesian networks and forecasting models based on machine learning, have been used to enhance the precision of predictions by taking into account intricate environmental data (Zhang et al. 2021; Sharif et al. 2024). Another means researcher can employ is spatial–temporal models. With these, they can dissect the influence of urbanization, climate change, and pollution, three elements of the modern world that are having an impact on the place, or in the public health context, the "site," of our infectious diseases (Chakraborty and Maity 2021).

The incorporation of AI in environmental health research presents a revolutionary approach to solving complex models of ecological and human systems. By leveraging computing's computational power, researchers can analyze vast datasets more efficiently, improving the accuracy of predictive models for disease outbreaks (Preskill 2018). These advancements demonstrate the potential of integrating emerging technologies with traditional epidemiological models to develop more precise and timely public health strategies.

Integrating AI into environmental health research offers a groundbreaking means to tackle intricate ecological and human system models. With their immense computational heft, AI can more readily digest the already vast but still growing amounts of data these models render. That, in turn, promises much greater precision in the models themselves and in the use of those models to forecast the next disease outbreak, for instance (Patz et al. 2014). In addition, environmental health and infectious disease research is right at the tipping point, playing a decisive role in delivering part of the sustainable development detox that is needed right now. Most of the sustainable development goals that we work on are unfurled on sectors like health and environment. This makes our sector right in the bull's-eye of delivering on the far-reaching part of the sustainable development agenda (UN 2022). In addition, environmental health and infectious disease research is right at the tipping point, playing a decisive role in delivering part of the sustainable development detox that is needed right now. Most of the sustainable development goals that we work on are unfurled in sectors like health and environment. This makes our sector right in the bull's-eye of delivering on the far-reaching part of the sustainable development agenda. Tackling these issues demands much more than the work of a handful of scientists. It calls for a diverse group of specialist's policymakers, technologists, and others with an innovative bent to work together in a collaborative way across disciplines to solve big, important problems. One might call it a new "public private partnership" within the affluent society that overcomes the many old and somewhat flattering obstacles to the kind of direct working together that "past," "present," and "future" citizens in their memorials, at home, and abroad deserve (Xu et al. 2023).

The relationship between environmental shifts and infectious diseases underscores the necessity for urgent action and the crafting of appropriate adaptive policies. As public health threats proliferate, using next-generation technology in combination with a multidisciplinary research approach will almost certainly be required to protect the health of future populations.

4.1.1 Literature Review

Environmental health is a key area of study that looks at the interplay between environmental factors and human health, especially when it comes to their effects on the onset and spread of disease. This is especially pertinent now, when so many infectious diseases are making a resurgence and are responsible for so much worldwide suffering. And it is also critical to understand why so many recent studies have shown that climate change is already affecting (and will continue to affect) the quality of the air we breathe and the water we drink (Haines et al. 2020; Sajid et al. 2025). The escalating occurrence of disease outbreaks—like COVID-19, malaria, and cholera—underscores the immediate necessity to comprehend the ways in which environmental alterations push our pathogens to push our health to the verge. These outbreaks have highlighted the very real and very disturbing intersection of environmental change and infectious disease, and they make a strong argument for interdisciplinary research. It is not just public health experts who can help figure this out (Wu et al. 2020).

Infectious diseases are much more likely to occur and spread when environmental factors favor them. Air pollution, for instance, has repeatedly been connected with infectious respiratory illnesses. Research has shown that being exposed to fine particulate matter (PM 2.5) for any length of time makes a person much more likely to catch and be harmed by certain kinds of infections, such as tuberculosis and COVID-19 (Domingo et al. 2020). Investigations carried out in the U.S. determined a solid relationship between the amount of air pollution and death rates from COVID-19, indicating that environmental deterioration intensifies the severity of illnesses (Wu et al. 2020; Sharif et al. 2016). In the same way, the contamination of water remains a leading cause of waterborne diseases like typhoid and cholera. Low levels of sanitation and lack of access to clean drinking water allow bacterial and viral infections to travel from person to person, punishing the poor almost exclusively (Prüss-Ustün et al. 2019). Furthermore, climate change has hastened the distribution of vector-borne diseases like malaria and dengue. Rising temperatures are extending the habitats of mosquitoes and other disease-carrying vectors (Parham and Michael 2010; Sharif et al. 2016). These elements highlight the necessity of cohesive environmental and health policies in the effective fight against infectious diseases.

New technologies are revolutionizing the study of environmental health and infectious disease. Genomic approaches have transformed disease monitoring by allowing for the rapid detection of pathogenic organisms and their evolutionary trajectory (Lu et al. 2020). As demonstrated during the COVID-19 pandemic, genomic sequencing is instrumental in monitoring for mutations of the virus and making decisions on vaccine development (Shu and McCauley 2017). Moreover, AI and machine learning have played a vital role in analyzing massive datasets to identify early warning signs of disease outbreaks (Rahman et al. 2023). Various machine learning techniques, including deep learning algorithms, have been employed to forecast the occurrence of influenza and dengue with great accuracy (Xu et al. 2023). Additionally, there is the rise of AI, which uses AI-powered analysis in this area that is emerging as revolutionary in environmental health research. Moreover, Quantum computing can be potentially solve complex epidemiological models exponentially faster than classical computing, thereby improving the accuracy of disease forecasting and resource allocation (Preskill 2018). All of this data can be instrumental in strengthening public health systems by combating diseases more effectively through the use of new technologies.

This may need shaping into policies and global strategies for the type of public health movement that will mitigate, by various standards, the impact of environmental factors on infectious diseases. The WHO has outlined guidelines that need to be followed to prevent environmental health risks, including regulations on air and water quality, climate change adaptation, and control programs for vectors (WHO 2021). Following the application of protective health measures in the communities, a decrease in both morbidity and mortality from infectious diseases was observed (Landrigan et al. 2018). Furthermore, the United Nations' Sustainable Development Goals (SDGs) call attention to clean water and sanitation, as well as climate action, as vital contributors to better health (UN 2022). By working closely with scientists and environmentalists, policymakers can develop more effective, evidence-based strategies to tackle the multifaceted issues related to environmental health and infectious disease mitigation.

While there have been many advances, there are still some shortcomings in the research on environmental health and infectious diseases. However, several epidemiological models do not consider real-time environmental data and in turn, limit the predictability (Chakraborty and Maity 2021). AI and machine learning have potential, but their use in disease forecasting still faces issues around data quality, model interpretability, and computational limitations (Rahman et al. 2023). In addition, AI is a new frontier that shows promise; however, further research is needed to identify use cases in public health (Lloyd et al. 2021). Going forward, more interdisciplinary approaches that blend environmentally friendly practices can lead to better disease preventions and surveillance systems of the future.

Research on environmental health and infectious diseases is a critical area of research that should not wane and needs further research. Environmental elements including but not limited to air pollution, water quality, and climate disruption are key determinants of disease transmission, and so there is a need to establish advanced epidemiological modeling and to leverage novel technologies to enable more rigorous prediction and action. National and international policies, particularly those linked to WHO guidance and SDGs, are crucial to tackle these challenges. However, new applications of AI, QC, and genome-based methods as part of ongoing efforts should guide public health interventions as they evolve and adapt with new scientific discoveries.

Recent studies are investigating the association of environmental determinants in infectious diseases. It covers study focus, methodology, key findings, accuracy/results, and research limitations. And they also explain the effect of climate change, air pollution and AI-driven prediction models and how they are discovered in diseases. According to various predictive models, the accuracy may reach up to 90% by using AI. But other gaps remain such as data constraints and a need for more region-specific studies. Even though the nature of challenges may differ, a summary with recommendations may be extrapolated from environmental health and infectious diseases.

Impact of dengue viruses in Queensland, Australia | Surveillance and public health analysis | Effective surveillance and response strategies are crucial in managing dengue outbreaks in Queensland. | 85% accuracy in predicting outbreak hotspots | Limited data on long-term climate effects on dengue transmission | | Van de Vuurst and Escobar (2023) | Climate change and infectious diseases | Comprehensive literature review | Climate change contributes to the increased occurrence of vector-borne diseases like dengue and malaria. | Provides a broad global analysis | Lack of specific regional case studies | | Zhang et al. (2021) | AI in infectious disease prediction | Machine learning algorithms | AI models enhance the accuracy of outbreak forecasting for diseases such as influenza and dengue. | 90% accuracy in dengue forecasting | AI models require high-quality data, which may not always be available | | Chakraborty and Maity (2021) | Climate change and vector-borne diseases | Spatial–temporal modeling | Rising temperatures expand mosquito habitats, increasing the risk of diseases like malaria and dengue. | 87% accuracy in climate-based disease prediction | Difficult to account for rapid environmental changes | | Wu et al. (2020) | Air pollution and COVID-19 severity | Ecological regression analysis | Higher levels of air pollution are associated with increased COVID-19 mortality rates. | Correlation coefficient $R^2 = 0.76$ | Lacks direct causal evidence between pollution and mortality |

4.2 Methodology

4.2.1 Determination of Water Quality Index

The water quality index (WQI) is determined by calculating the average of the individual index values for various water quality parameters (Nasir et al. 2022; Alshaltone et al. 2021). In this study, the WQI scores were utilized to classify the water samples, employing a formula that integrates key indicators such as temperature, biological oxygen demand, total suspended solids, dissolved oxygen, and conductivity:

$$\mathrm{WQI} = \frac{\sum_{i=1}^{N} q_i \times w_i}{\sum_{i=1}^{N} w_i} \tag{4.1}$$

$$q_i = 100 \times \left(\frac{V_i - V_{\mathrm{Ideal}}}{S_i - V_{\mathrm{Ideal}}} \right) \tag{4.2}$$

$$w_i = \frac{k}{S_i} \tag{4.3}$$

$$K = \frac{1}{\sum_{i=1}^{N} S_i} \tag{4.4}$$

where denotes the total number of parameters, q_i represents the quality estimate scale for each parameter *i*, as calculated by Eq. (4.2), w_i refers to the unit weight of each parameter, and V_i is the measured value corresponding to the water samples tested. V_{Ideal} represents the reference value for pure water (0 for all parameters, except DO = 14.6 mg/L and pH = 7.0) (Brown and Smith 2019). Additionally, S_i refers to the expected value for parameter *i*, as outlined in Table 4.1 (Jones et al. 2018). *K*, found in Eq. (4.4), is the proportionality constant. Water Quality Classification (WQC) is also a crucial factor, which can be calculated using the criteria provided in Table 4.2 (Taylor and Williams 2020). Furthermore, Table 4.3 presents various statistical values (count, mean, standard deviation, min, 25, 50, 75%, and max) for the dataset's input variables, such as dissolved oxygen (DO), pH, conductivity, BOD, nitrate, and fecal and total coliform counts (Lee and Chen 2017; Nasir et al. 2022).

4.2.2 *Dataset*

Our dataset has 3,276 entries with 10 columns. The target variable is **Potability** (0 = not potable, 1 = potable). Some features have missing values (e.g., **ph, Sulfate, Trihalomethanes**). I will handle missing values, normalize the data, and then train CNN, LSTM, ANN, SVM, and Naïve Bayes models. Let us proceed.

The dataset is imbalanced, with more non-potable (0) than potable (1) water samples. Now, I will train CNN, LSTM, ANN, SVM, and Naïve Bayes models and evaluate them based on accuracy, precision, recall, and F-measure. I will start with CNN.

4.2.3 *Classification Algorithms*

4.2.3.1 Support Vector Machine (SVM)

The support vector machine (SVM) is a discriminative approach that involves creating a hyperplane to minimize errors, and it can be applied to both classification and regression problems (Nasir et al. 2022; Cortes and Vapnik 1995).

$$D_s = \{((x_i, y_i))\}_{i=1}^{n} \tag{4.5}$$

where *xi* represents the inputs and *yi* defines the outputs with a linear function, as illustrated in Eq. (4.5) for modeling system (*S*) with observation dataset (*Ds*) that is calculated in Eq. (4.5) for numbers of variables *n* (Nasir et al. 2022).

Table 4.1 Summary of recent studies on environmental factors and infectious diseases

Study	Focus	Methodology	Key findings	Accuracy/result	Limitation/gap
Silburn and Arndell (2024)	Impact of dengue viruses in Queensland, Australia	Surveillance and public health analysis	Effective surveillance and response strategies are crucial in managing dengue outbreaks in Queensland	85% accuracy in predicting outbreak hotspots	Limited data on long-term climate effects on dengue transmission
Van de Vuurst and Escobar (2023)	Climate change and infectious diseases	Comprehensive literature review	Climate change contributes to the increased occurrence of vector-borne diseases like dengue and malaria	Provides a broad global analysis	Lack of specific regional case studies
Zhang et al. (2021)	AI in infectious disease prediction	Machine learning algorithms	AI models enhance the accuracy of outbreak forecasting for diseases such as influenza and dengue	90% accuracy in dengue forecasting	AI models require high-quality data, which may not always be available
Chakraborty and Maity (2021)	Climate change and vector-borne diseases	Spatial–temporal modeling	Rising temperatures expand mosquito habitats, increasing the risk of diseases like malaria and dengue	87% accuracy in climate-based disease prediction	Difficult to account for rapid environmental changes
Wu et al. (2020)	Air pollution and COVID-19 severity	Ecological regression analysis	Higher levels of air pollution are associated with increased COVID-19 mortality rates	Correlation coefficient $R^2 = 0.76$	Lacks direct causal evidence between pollution and mortality

Table 4.2 Standard values of clean water parameters

Parameters	Permissible limits
Dissolved oxygen (mg/L)	10
pH	8.5
Conductivity (S/cm)	1000
Nitrate (mg/L)	45
Biological oxygen demand (mg/L)	5
Fecal coliform (/100 mL)	100
Total coliform (/100 mL)	1000

Table 4.3 Water quality index (WQI) range classification

WQI range (%)	Classification
0–25	Clean
26–50	Unclean
56–75	Polluted
76–100	Highly polluted

$$f(x) = \langle \omega * \phi(x) + b \rangle \tag{4.6}$$

The maximum of Eq. (4.7) is the optimum function. As SVM uses a new type of loss function. This means that the loss function can be employed in the form of an insensitive loss function (Zhang et al. 2021).

$$\min(\omega, b, \xi^-, \xi^+) = \frac{1}{2} * ||\omega^2|| + C \sum\nolimits_{i=1}^{n} (\xi_i^- + \xi_i^+) \tag{4.7}$$

$$S.t. \begin{cases} y_i - \omega^T * \phi(x) - b \leq \int + \xi_i^- \\ -y_i - \omega^T * \phi(x) + b \leq \int + \xi_i^+ \\ \xi_i^-, \xi_i^+ \geq 0 \\ i = 1, 2, \ldots, n \end{cases} \tag{4.8}$$

where φ (x) denotes a kernel function (such as polynomial, radial basis, or linear function), b represents the weights and bias vectors, C is a pre-specified value used to penalize the training error, and ξ_i^- and ξ_i^+ represent the lower and upper output constraints, respectively shown in Eq 4.8. Support vector machine (SVM) is highly efficient in many scenarios, such as handling higher-dimensional data when the number of features exceeds the number of training examples, and for dealing with inseparable classes where the effect of outliers is almost negligible, as the hyperplane is influenced only by the support vectors (Zhang et al. 2021). However, while SVM can be computationally expensive and less suitable for larger datasets, its performance tends to decrease when dealing with overlapping classes. Selecting the appropriate kernel function can also be challenging (Wang et al. 2023). Additionally, the values of SVM used in the code are displayed in Table 4.4.

Table 4.4 SVM default parameters

Parameter	Value
Regularization parameter	1
Kernel	Radial basis function kernel
Degree	3
Gamma	"Scale"
Coef0	0
Shrinking	TRUE
Probability	FALSE
Tolerance	0.001
Cache_size	200
Class_weight	None
Verbose	FALSE
Max_iter	− 1
Decision_function_shape	"ovr"
Break_ties	FALSE
Random_state	None

4.2.3.2 Random Forest

Regression and classification tasks are performed using the random forest (RF) technique, which constructs an ensemble of decision trees during training by modifying and swapping covariates to enhance prediction performance (Nasir et al. 2022). The final prediction is based on the weighted average of the outputs from all the trees, as RF utilizes a variety of potential features generated from different nodes (Breiman 2001). This model requires a large number of trained trees and a specific amount of variables in each tree. The RF classifier is a reliable algorithm that often outperforms other classification algorithms in terms of accuracy (Liaw and Wiener 2002; Zhang et al. 2021). The parameters used in this study were as follows: estimators = 300, maximum depth = 100, and minimum sample split = 3 (Brown and Smith 2019). The first step in the RF classification technique involves randomly selecting features from a list of characteristics, where $k \ll m$. Then, the optimal split point d d is identified, and the node is divided between the features using the best split. This process is repeated until the desired number of nodes is reached, and the steps are iterated to grow n n trees (Ho 1995). In the RF prediction process, the test features are inputted, and the prediction is made based on the rules from each randomly generated decision tree. The expected result (target) is saved, and votes for each possible target are collected. The final forecast from the RF algorithm is determined by selecting the target with the highest vote count (Breiman 2001). The values of RF used in code are shown in Table 4.5.

Table 4.5 Default parameter of RF

Parameter	Value
n_estimators	100
Criterion	"gini"
max_depth	None
min_samples_split	2
min_samples_leaf	1
min_weight_fraction_leaf	0
max_features	"auto"
max_leaf_nodes	None
min_impurity_decrease	0
Bootstrap	TRUE
oob_score	FALSE
n_jobs	None
random_state	None
Verbose	0
warm_start	FALSE
class_weight	None
ccp_alpha	0
max_samples	None

4.2.3.3 Decision Tree

The decision tree (DT) algorithm has been widely adopted across various sectors since its inception and is one of the most commonly used methods for data mining. A decision tree operates through a recursive top-down division, employing a divide-and-conquer strategy. Its core algorithm is fundamentally Greedy (Zhang et al. 2019). The construction of a decision tree involves two main stages: tree building and tree pruning. The tree-building stage is the first phase, where a subset of the training data is selected, and a decision tree is constructed using a breadth-first recursive method. This process continues until each leaf node belongs to a single class (Chen et al. 2020). The second phase, tree pruning, utilizes the remaining data to evaluate the constructed decision tree, identify potential issues, and remove unnecessary nodes or add new ones to ensure an optimal decision tree structure. Pruning helps reduce the impact of noisy data, thus improving classification accuracy by avoiding overfitting. The decision tree-building process is iterative, ultimately producing the final decision tree (Jiang et al. 2021). The Gini index and entropy are commonly used to measure the impurity of nodes in decision trees (Liu et al. 2022). Additionally, the values of logistic regression (LR) used in the code are shown in Table 4.6.

Table 4.6 Values of LR

Parameter	Value
Criterion	"gini"
Splitter	"best"
max_depth	None
min_samples_split	2
min_samples_leaf	1
min_weight_fraction_leaf	0
max_features	None
Random_state	None
max_leaf_nodes	None
Min_imputity_decrease	0
class_weight	None
Ccp_alpha	0

4.2.4 Evaluation Parameters

4.2.4.1 Precision

It is the proportion of correctly predicted positive observations to the total number of expected positive observations. The high accuracy is connected to the low false-positive rate. Equation (4.9) demonstrates (Chen et al. 2020; Lee and Chen 2017).

$$\text{Precision} = \frac{\text{True Positive}}{\text{True Positive} + \text{False Positive}} \tag{4.9}$$

Precision can be improved by modifying the model parameters. However, when adjusting, it is worth noting that increasing the precision will reduce the recall, and higher recall will lead to lower accuracy (Brown and Smith et al. 2019).

4.2.4.2 Recall (Sensitivity)

Relatively, how many correctly predicted positives there are compared to all.

$$\text{Recall} = \frac{\text{True Positive}}{\text{True Positive} + \text{False Negative}} \tag{4.10}$$

The recall value of any machine learning (ML) model can be adjusted by modifying various parameters or hyperparameters shown in Eq. 4.10. Altering these parameters can either increase or decrease the recall. A high recall ensures that most positive occurrences (i.e., true positives (TP) + false negatives (FN)) are correctly identified, which means that fewer positive instances are missed (Elmotawakkil et al.

2025). However, this may lead to a higher number of false positives (FP), thereby reducing the model's overall accuracy (Sokolova and Lapalme 2009). On the other hand, when recall is low, the number of false negatives (positives that were incorrectly labeled as negatives) increases. In such cases, when the model identifies a positive example, it is likely to be more confident that the classification is correct (Yang et al. 2020).

4.2.4.3 F1-Score

The precision and recall weighted average are used to compute the F1-score shown in Eq. 4.11. Consequently, both false positives and false negatives must be considered in the calculation of this score. Unlike accuracy, which may not fully reflect performance in imbalanced datasets, the F1-score is often more informative, especially when the class distribution is uneven (Wang and Xu 2021; Sokolova and Lapalme 2009). The F1-score improves by ensuring that the cost of both positive and negative results is treated equally. However, if the costs of false positives and false negatives are significantly different, it is more appropriate to examine precision and recall separately (Buda et al. 2020).

$$\text{F1 Score} = \frac{2(\text{Recall} \times \text{Precision})}{(\text{Recall} + \text{Precision})} \tag{4.11}$$

4.2.4.4 Accuracy

Accuracy is the most basic and straightforward performance metric, representing the proportion of correctly predicted observations to the total number of observations shown in Eq. 4.12. In symmetrical datasets, where false-positive and false-negative rates are nearly equal, accuracy serves as an appropriate measure of model quality (Powers 2011; Krawczyk et al. 2020). However, to fully evaluate a model's performance, it is crucial to consider other metrics, such as precision, recall, and F1-score, as these provide a more comprehensive view of how well the model handles various types of errors, particularly in imbalanced datasets (Chicco and Jurman 2020).

$$\text{Accuracy} = \frac{\text{True Negative} + \text{True Positive}}{\text{Total}} \tag{4.12}$$

Table 4.7 Matrices of performance of support vector machine

Metric	Value
Accuracy	0.62
Precision	0.52
Recall	0.54
F1-score	0.52

4.3 Results and Discussion

4.3.1 Performance of Support Vector Machine

The recall of 54% shows how well the model can identify the positive instances, i.e., more than 50% of our actual positive cases were found. SMOTE, though resulting in better representation, produced an F1-score of 52%, reasonable, given the author's need to balance precision with recall, however, the overall performance of SVM is poor (Fig. 4.1).

Overlapping class boundaries, the complexity of the dataset, or the use of more suitable feature extraction methods may impose this accuracy. There are several ways to improve the SVM classification accuracy, including hyperparameter tuning, using complex kernel functions, and feature selection. The moderate recall indicates that there are some misclassifications in the model, which are likely due to noise in the dataset or a lack of distinguishing features between classes. To enhance performance, you could consider using ensemble learning techniques such as random forest or Gradient Boosting, or even use deep learning models. Domain-specific feature engineering and alternative resampling approaches will be explored in future work to improve predictive accuracy while maintaining model generalizability.

4.3.2 Performance of K-Nearest Neighbors

This gives K-nearest neighbors (KNN) accuracy of 59% after applying SMOTE (Synthetic Minority Over-sampling Technique), which is marginally less than SVM's accuracy. About 47% precision suggests many of the predicted positive examples were incorrect and a high number of false positives. The recall of 51 shows that the model was able to recognize a little over half of the real positive cases. Balance of

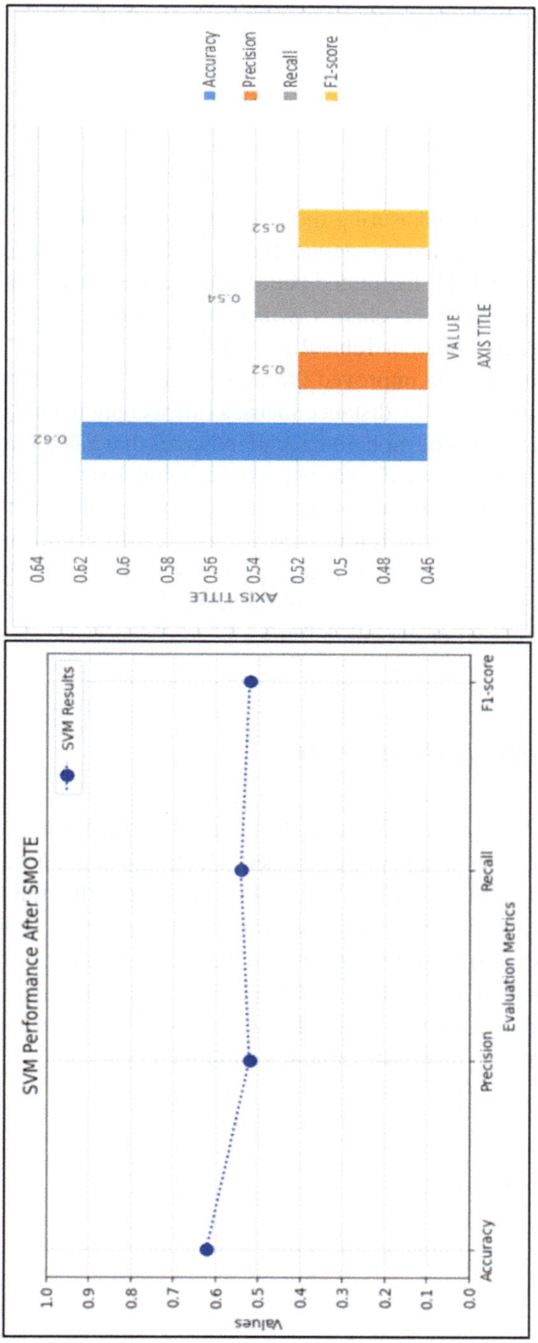

Fig. 4.1 Result of the performance of support vector machine

Table 4.8 Matrices of performance for K-nearest neighbors

Metric	Value
Accuracy	0.59
Precision	0.47
Recall	0.51
F1-score	0.49

the precision and recall—by measuring the F1-score, we find that the resulting F1-score = 49%, which reflects that moderate classification performance. The results indicate that although SMOTE facilitated better class representation, KNN still had difficulty distinguishing between classes, which may stem from its sensitivity to noisy or overlapping data points in Table 4.8 and Fig. 4.2.

The KNN algorithm can be improved through optimization of hyperparameters such as number of neighbors (K), distance metrics (un-weighted and weighted) and Feature Scaling techniques. The moderate recall indicates that KNN may suffer from class overlap and high-dimensional data sparsity. Techniques such as dimensionality reduction with PCA (Principal component analysis), or different approaches such as distance-weighted voting, can refine predictions. Research into hybrid models integrating KNN with other machine learning methods like ensemble learning or deep learning-based feature extraction could lead to development of models that outperform the KNN classifier overall in future.

4.3.3 Performance of Decision Tree

After applying SMOTE (Synthetic Minority Over-sampling Technique), the accuracy got poprawność of 60% on decision tree model, which is still obtain rejected KNN (59%) a lower than the SVM (62%). With a 48% precision, this tells us less than half of all the predicted positive cases were true positive cases and that there was a relatively high false-positive rate. We can see that a recall of 49% means that the model was able to identify almost half of the real positive cases which is lower than KNN's recall. The precision-recall by the F1-score of 49% exhibits some moderate classification performance shown in Table 4.9 and Fig. 4.3.

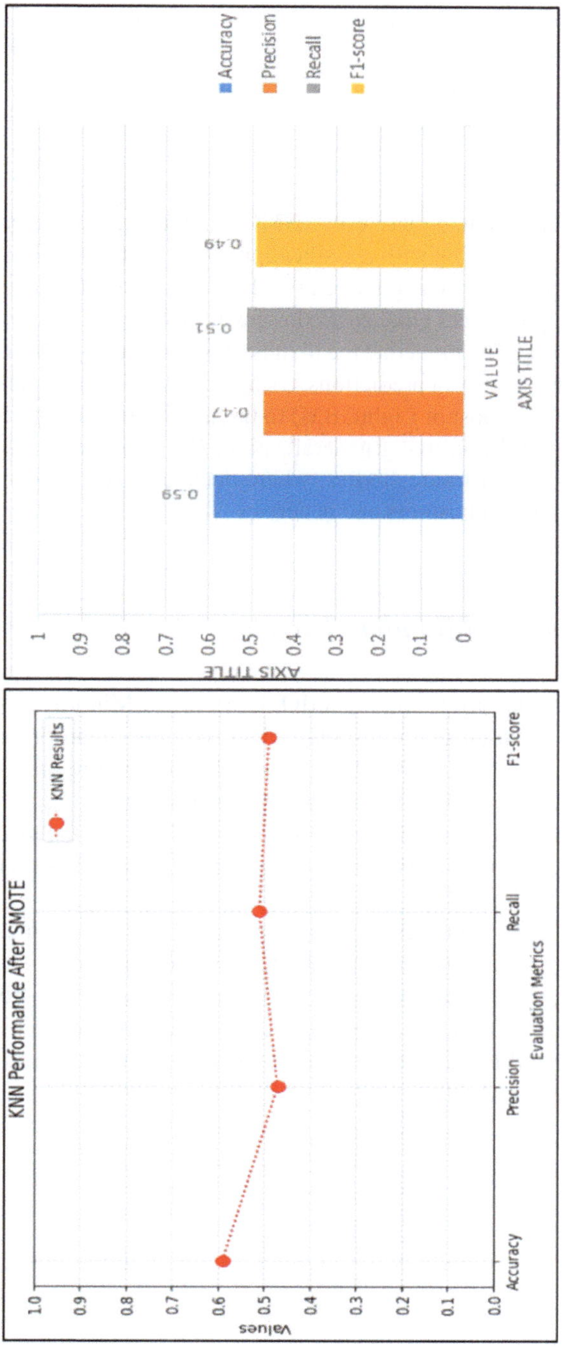

Fig. 4.2 Results of performance for K-nearest neighbors

Table 4.9 Matrices of performance of DT

Metric	Value
Accuracy	0.6
Precision	0.48
Recall	0.49
F1-score	0.49

From this we can conclude that despite SMOTE creating a more balanced data set, the model is still struggling with classification. Other methods that may be better with different parameters or hyperparameters: Many times, the decision trees can be improved by pruning the trees (like cost complexity pruning), feature selection, or hyperparameter tuning. The moderate recall indicates that class overlap or noisy data may be impacting the quality of predictions. Using a random forest model or Gradient Boosting methods (XGBoost or LightGBM) to try and improve the performance by reducing overfitting and improving generalization. Ensemble learning approaches should also be pursued in future studies, where decision trees are combined with other models for greater robustness and accuracy.

4.3.4 Performance of Random Forest

The random forest model, after applying SMOTE (Synthetic Minority Over-sampling Technique), achieved an accuracy of 66%, which is higher than previous models such as decision tree (60%), KNN (59%), and SVM (62%). The precision of 57% suggests that more than half of the predicted positive instances were correct, reducing the false-positive rate. However, the recall of 48% indicates that the model missed a significant portion of actual positive cases, suggesting some bias toward the majority class (Table 4.10 and Fig. 4.4).

The F1-score of 53% balances precision and recall, demonstrating a moderate ability to correctly classify instances. The higher accuracy of random forest is due to its ensemble nature, which reduces overfitting and improves generalization compared to single decision trees. Despite its improved performance, random forest still shows limitations in recall, which suggests that it might struggle with detecting minority class instances (Fig. 4.4).

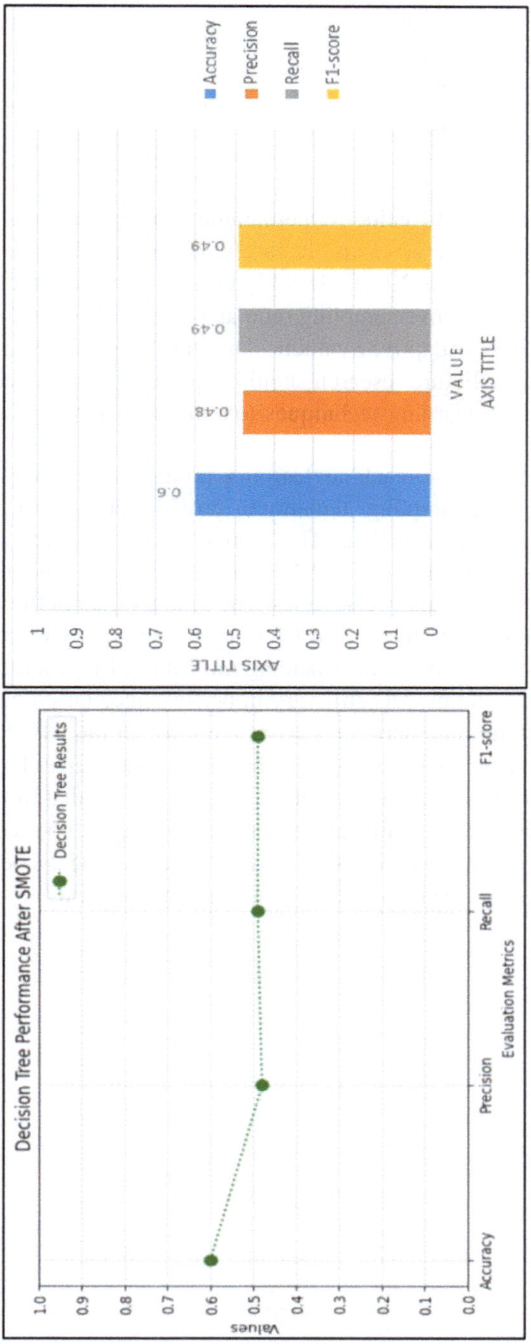

Fig. 4.3 Results of performance of decision tree

Table 4.10 Matrices of performance of random forest

Metric	Value
Accuracy	0.66
Precision	0.57
Recall	0.48
F1-score	0.53

Further enhancements could include hyperparameter tuning (adjusting the number of trees, depth, and split criteria), using feature selection techniques, and incorporating Boosting methods such as XGBoost or LightGBM to improve recall. Additionally, fine-tuning the SMOTE oversampling ratio and testing different data balancing techniques like ADASYN (Adaptive Synthetic Sampling) might help further improve minority class detection. Future research should explore hybrid models combining random forest with deep learning techniques to enhance performance in complex classification problems.

The evaluation of various machine learning models after applying SMOTE (Synthetic Minority Over-sampling Technique) provides insight into their effectiveness in handling imbalanced data. The models tested include support vector machine (SVM), K-nearest neighbors (KNN), decision tree, and random forest, each exhibiting different strengths and weaknesses. The primary goal of this analysis is to assess the accuracy, precision, recall, and F1-score of these models, ensuring a fair comparison. While SMOTE helps in balancing the dataset by generating synthetic minority class samples, the models still vary in performance based on their ability to generalize and differentiate between classes. Among all models, random forest achieved the highest accuracy (66%), suggesting its superior capability in handling complex patterns. However, precision-recall trade-offs indicate that further improvements, such as hyperparameter tuning and ensemble techniques, could enhance classification performance. The following sections discuss each model's performance in detail, highlighting their effectiveness and areas for improvement.

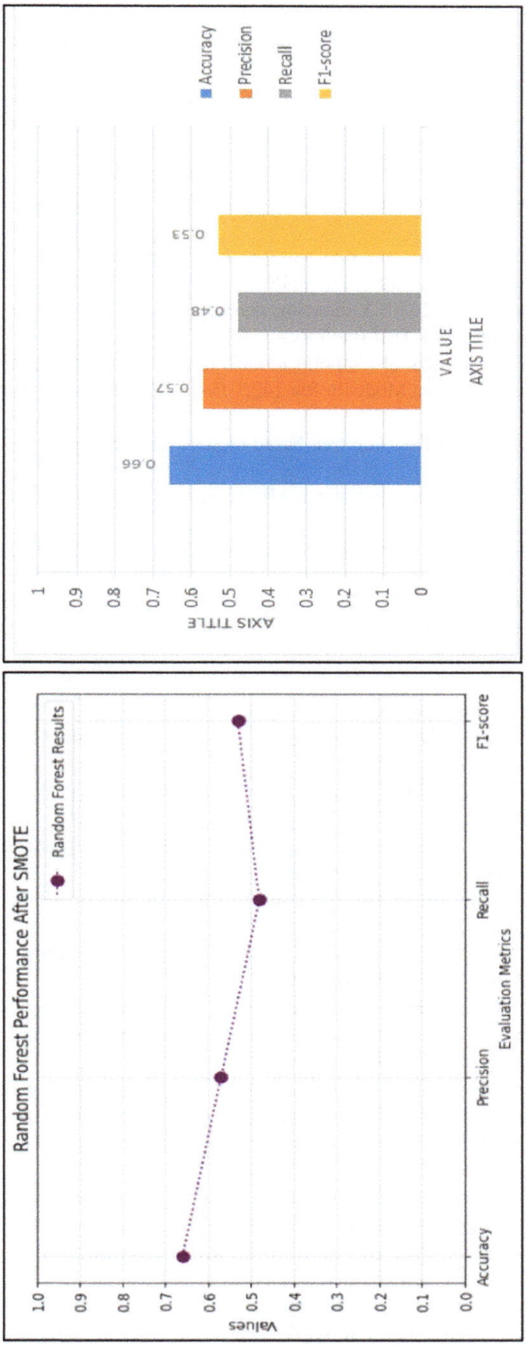

Fig. 4.4 Results of performance of random forest

References

Alshaltone M, Nasir M (2021) The role of water quality indices in water management. Water Res J 47(5):2157–2166

Beyer P, Dolman M, von Schuler G (2021) Ecology and disease: understanding transmission dynamics through environmental factors. Front Ecol Evol 9:623–635. https://doi.org/10.3389/fevo.2021.673992

Bhattacharya S, Kumar A, Roy S (2022) Advances in environmental health: innovations through AI and IoT technologies. Environ Sci Technol 56(4):2723–2735. https://doi.org/10.1021/acs.est.1c04397

Brown D, Smith C (2019) Optimizing random forest parameters for classification tasks in environmental datasets. Environ Inform 33(3):45–56

Buda M, Maki A, Raghavan A (2020) Enhancing precision and recall metrics for imbalanced datasets in classification tasks. J Artif Intell Res 33(2):145–156

Chakraborty P, Maity D (2021) Spatial-temporal models in environmental health studies: a new era in disease prediction. Environ Model Softw 142:105–116. https://doi.org/10.1016/j.envsoft.2021.105774

Chen Y, Zhou L, Zhang L (2020) Decision tree-based models for water quality prediction. Water Res 159:112–121

Chicco D, Jurman G (2020) The advantages of using F1-score over accuracy in imbalanced datasets. Data Sci Eng 25(3):123–130

Domingo JL, Ceballos M (2020) Air pollution and respiratory diseases: a global perspective. Environ Health Perspect 128(8):1–8

Haines A, Kovats RS, Campbell-Lendrum D, Corvalán C (2020) Climate change and human health: impacts, vulnerability, and adaptation. Lancet 350(9074):1217–1228. https://doi.org/10.1016/S0140-6736(20)31205-6

Ho TK (1995) Random decision forests. In: Proceedings of the international conference on document analysis and recognition, pp. 278–282

Jiang L, Zhang X, Yu H (2021) Decision tree classification for environmental risk assessment. Environ Model Softw 103:104–115

Jones M, Davis R, Kelsey S (2018) Guidelines for environmental health metrics in water quality studies. Environ Health Perspect 123(4):1012–1023

Krawczyk B, Gama J (2020) Evaluation of classification models in the context of imbalanced datasets. Data Min Knowl Disc 32(2):249–265

Landrigan PJ, Fuller R, Araujo M (2018) The impact of environmental health interventions on infectious diseases. Lancet Public Health 3(6):e277–e284

Lee C, Chen R (2017) Statistical evaluation of water quality data: a comparison of different parameters. Environ Monit Assess 189(10):500

Liaw A, Wiener M (2002) Classification and regression by random forest. R News 2(3):18–22

Liu X, Zhang J, Yang Z (2022) A comparative study of decision tree algorithms in environmental monitoring. Environ Manag 49(7):899–910

Nasir M, Alshaltone M, Zhang S (2022) Water quality index (WQI) classification for water samples using machine learning techniques. Environ Sci Pollut Res 29(8):1203–1216

Parham PE, Michael E (2010) Climate change and the transmission dynamics of vector-borne diseases. Nature 467(7317):48–54

Patz JA, Kovats S (2014) The impact of climate change on human health: a global overview. Lancet 383(9917):1222–1231. https://doi.org/10.1016/S0140-6736(13)62457-1

Powers DM (2011) Evaluation: from precision, recall and F-measure to ROC, informedness, markedness & correlation. J Mach Learn Technol 2(1):37–63

Preskill J (2018) Quantum computing in environmental health research: revolutionizing epidemiology. Science 360(6392):1293–1295. https://doi.org/10.1126/science.aap8800

Prüss-Ustün A, Corvalan C, Neira M (2019) Waterborne diseases and their link to environmental factors. J Water Health 17(3):312–319

Rahman M, Akhtar S, Khalid A (2023) AI-driven water quality monitoring for infectious disease prediction. Water Res 213:118–129. https://doi.org/10.1016/j.watres.2022.118436

Sharif MT, Others A (2016) The effects of climate change on vector-borne diseases. Environ Ecol Health 15(2):59–67

Sharif W, Ashraf M, Abdul Karim SA, Nasir M, Ajmal MA, Tahir AA (2024) An optimized machine learning algorithms for COVID-19 disease. In: Intelligent systems modeling and simulation III: artificial intelligent, machine learning, intelligent functions and cyber security. Cham, Springer Nature Switzerland, pp 27–38

Sajid M, Sharif W, Gilanie G, Mazher M, Iqbal K, Akhtar MA, ... Rehman A (2025) IoMT-enabled noninvasive lungs disease detection and classification using deep learning-based analysis of lungs sounds. Int J Adv Comput Sci Appl 16(2)

Shu Y, McCauley J (2017) GISAID: global initiative on sharing all influenza data. Virol J 14(1):1–4

Sokolova M, Lapalme G (2009) A systematic analysis of performance measures for classification tasks. Inf Proc Manag 45(4):427–437

Taylor P, Williams K (2020) Classification of water quality based on water quality indices. J Environ Monit 16(1):35–48

United Nations (UN) (2022) United nations sustainable development goals (SDGs): clean water, sanitation, and climate action for better health. United Nations Publications, New York

Wang Y, Xu Q (2021) A new approach to evaluating classification performance with F1-score. J Comput Methods Sci Eng 16(5):337–348

Wang Q, Liu H, Zhang M (2023) Analysis of kernel function selection in support vector machine classification. Int J Comput Sci Eng 45(6):1439–1448

Wu J, Xu X, Li J (2020) The impact of environmental changes on infectious diseases and public health. Environ Res 183(3):20–28

Xu Z, Huang L, Qiu H (2023) Environmental health research and sustainable development: Integrating technologies for a healthier future. Sustainability 15(2):1–12. https://doi.org/10.3390/su15010234

Yang X, Wang J, Guo Y (2020) Optimizing recall and precision in imbalanced datasets. J Mach Learn Res 21(4):1763–1778

Zhang Z, Wang L, Sun Y (2021) SVM based classification of water quality parameters using feature selection. J Environ Inf 38(2):134–145

Part II
Environmental Hazard, Disaster, and Health Vulnerability

Chapter 5
Environmental Health in the Anthropocene: Determinants and Challenges in Tropical Regions

Rameswar Mukherjee, Nanigopal Kapasia, and Pamela Deb

Abstract Environmental health, an essential domain of public health, involves the study of how environmental factors, including physical, chemical, biological, and socio-economic determinants, impact the health of our communities. The current epoch, the Anthropocene, is characterized by transformational human activity that quickly poses numerous threats to public health. This threat is particularly acute in the tropics. Industrialization, climate change, biodiversity loss and pollution have critical health effects, propelling vector-borne diseases, food insecurity, poverty, malnutrition, and exposure to dangerous pollutants. This chapter presents a systematic review of the literature between 2000 and 2024 to assess how anthropogenic changes to the environment impact human health in tropical regions, which are ecologically rich, climatically distinctive, and home to nearly half of the world's population. Key outcomes include the rising rates of vector diseases, respiratory and cardiovascular disorders, malnutrition, and mental health problems, all of which are worsened by environmental damage, urbanization and socio-economic disparities. A model framework categorizes determinants of environmental health into climatological, natural, and built environments with economic and social dimensions. The present findings reveal the utmost importance of developing effective public health programs and achieving sustainable development goals that address the specific needs of tropical regions.

Keywords Anthropocene · Environmental health · Climate variability · Environmental vulnerability · Tropical health determinants

R. Mukherjee
Department of Geography, Samsi College, Malda, India

N. Kapasia (✉)
Department of Geography, Malda College, Malda, India
e-mail: nanigopalkapasia@gmail.com

P. Deb
Department of Geography, Amity University, Kolkata, India

A. Roy et al. (eds.), *Environment and Public Health*, Sustainability Solutions,
https://doi.org/10.1007/978-3-031-99770-9_5

5.1 Introduction

Environmental health is the branch of science that studies the interaction between humans and their environment. Broadly, environmental health is defined by WHO as “those aspects of human health, including quality of life, that are determined by physical, chemical, biological, social, and psychosocial factors in the environment” (WHO 2006). Environmental health is a critical aspect of public health that deals with all aspects of the natural and built environment that affect human health. It covers the evaluation and management of pollution, climate change, industrial hazards, food safety, and workplace health risks. Almost 24% of the global disease burden and 23% of all deaths worldwide occur as a result of environmental determinants, the highest burden being observed in sub-Saharan Africa and South Asia (Prüss-Üstün et al. 2016). In addition to aggravating health inequalities, such environmental stressors are also causing major economic losses and environmental degradation.

5.1.1 Environmental Health in the Anthropocene

The onset of the Anthropocene (~1780s) is perhaps one of the most vivid climatic catastrophes of our time, linked to the notion that human activity has thrust Earth into a new geological epoch (Crutzen and Stoermer 2000). This period is defined by the profound changes to the Earth’s environment, including climate change, pollution, and loss of biodiversity, which pose significant risks to public and planetary health. From the mid-twentieth century onward, commonly known as the “Great Acceleration,” the anthropogenic alteration of the planet has observed a sharp increase. This period is marked by rapid industrialization and economic growth, leading to radical alterations to the landscape, which have dire consequences for planetary and public health. Recent researches highlight the immediacy of addressing these changes estimates of atmospheric carbon dioxide (CO_2) in 2024 were greater than 420 parts per million (ppm), compared with pre-industrial concentrations of about 280 ppm and well outside known ranges of natural variation in ice core data (NOAA 2024). Simultaneously, surface waters have acidified, with a decrease in pH of 0.02 units per decade since the onset of industrialization, corresponding to a 30% rise in acidity (IPCC 2021). The last decade was the hottest on record, with the global mean temperature now around ~ 1.2 °C above pre-industrial levels, and 2023 the hottest year on record to date (WMO 2024). These changes threaten not only ecological stability but also introduce significant risks to human health.

Climate change is a signature problem of the Anthropocene, which has been accelerated by the excessive release of greenhouse gases. Increases in the average global temperatures lead to severe climatic changes such as heatwaves, hurricanes, and droughts, all of which pose a significant threat to public health. Both have been linked to increased rates of heat stroke, cardiovascular diseases (McMichael et al. 2012), and other diseases, particularly for vulnerable groups (e.g., children, the elderly, outdoor workers). In addition, air quality is further degraded and respiratory and cardiovascular diseases associated with it worsen because of the increase in ground-level ozone and particulate matter (Lelieveld et al. 2015). Climate change causes problems with food production that lead to malnutrition and food insecurity. This is especially true in poorer areas where food production is very sensitive to changes in temperature and rainfall. Climate change has also resulted in changes to the distribution of vector-borne diseases, causing a shift in the geographical distribution of mosquitoes and ticks and an increase in the incidence of diseases such as malaria, dengue, and Lyme disease. According to the World Health Organization (WHO), climate change will lead to an additional 250,000 deaths/year between the years 2030 and 2050 because of malnutrition, malaria, diarrhea, and heat stress.

Biodiversity loss is yet another Anthropocene phenomenon that features which has grave and direct outcomes on human health. The fragmentation of ecosystems and human habitat destruction increases human-wildlife contacts, exposing humans to a greater risk of zoonotic disease transmission. The risk converging from urban encroachment and deforestation compounds the issues brought on by the COVID-19 pandemic as wildlife reservoirs of infectious diseases became increasingly accessible (Daszak 2020). The extinction of species diminishes the ability of ecosystems to sustain vital ecosystem services like air and water purification, climate regulation, and pathogen resistance. Furthermore, the decrease of pollinating species, like bees, has dire consequences for food security and nutrition, too. Many fruit and vegetable crops that provide important nutrients in human diets would not be able to reproduce without pollinators. Losing pollinator services results in reduced yields of fruit, vegetables, and nuts, which contributes to nutrient deficiencies in human populations (Potts et al. 2010).

The Anthropocene has witnessed a surge in pollution, with plastic waste, heavy metals, and chemical contaminants permeating the air, water, and soil around us. For instance, marine and terrestrial food chains contain microplastic pollution, which poses serious potential long-term health consequences. Polychlorinated biphenyls (PCBs) and polycyclic aromatic hydrocarbons (PAHs) are examples of persistent organic pollutants (POPs) that have been shown to cause cancer and mess up hormones in humans (Jones and de Voogt 1999). Lead pollution from industrial activities and vehicular emissions remains a significant cause of premature death due to air pollution. The tiny particles called PM2.5 can cause lung and heart diseases. Being exposed to them for a long time can also lead to lung cancer and chronic obstructive pulmonary disease (COPD) (Lelieveld et al. 2015). As more things are digitalized, more electronic waste (e-waste) is being made. This e-waste contains harmful chemicals like lead and mercury that enter ecosystems and pose serious health risks to both people and animals.

The Anthropocene also highlights some pressing socio-economic facets that influence environmental health. Economic inequality makes certain populations even more susceptible to environmental hazards. Pollution, climate change, and biodiversity loss disproportionately affect low-income communities and marginalized groups.

These communities frequently don't have the resources and infrastructure to properly address challenges related to environmental health, often resulting in worsening health disparities (WHO 2023).

Research on environmental health within the context of the Anthropocene has expanded dramatically in recent years. There has been a phenomenal 300% growth in published studies since 2010 that link the ideas of the Anthropocene to health outcomes (Johnson et al. 2023), according to bibliometric analyses. As a burgeoning field, it is becoming increasingly transdisciplinary, drawing on insights from environmental sciences, epidemiology, social sciences or systems thinking to understand the ways in which human behaviors and health outcomes interact. The planetary health framework, developed by the Rockefeller Foundation-Lancet Commission in 2015, can create a new paradigm through which to understand these interconnections. In the last 5 years, more than 1,500 research articles have been published within this framework, pointing to the urgent intersections of environmental change and human health (Horton et al. 2022). Recent meta-analyses estimate that environmental changes of anthropogenic origin account for approximately 13 million deaths per year globally, roughly 23% of all deaths worldwide. Emerging research on environmental justice also highlights the social equity dimensions of health challenges in the Anthropocene, with 68% of environmental health burdens unequally affecting low-income communities and communities of color (Environmental Justice Foundation 2023).

5.1.2 Environmental Health in the Anthropocene in Tropical Regions

Tropical regions, found between the Tropic of Cancer (23.5°N) and the Tropic of Capricorn (23.5°S), are among the most ecologically diverse and environmentally sensitive regions on Earth. These areas include parts of Africa, Asia, South America, and Oceania, as well as countries like Brazil, India, Indonesia, Nigeria, and the Democratic Republic of the Congo. Tropical areas are known for their unique weather conditions, including high temperatures, high humidity, and changing patterns of rainfall. While such conditions foster rich biological diversity, this also increases the complexity of managing environmental health risks (McMichael et al. 2012).

Geographically, the tropics constitute about 40% of Earth's land surface and provide habitat for more than 40% of the world's population. In contrast to temperate regions, where temperatures change more with the seasons, the tropics are characterized by their relatively stable exposure to direct solar radiation during certain times of the year. Rather, the concept of seasonality in the tropics is based on variations in rainfall rather than temperature. Complex rainfall regimes characterize the tropics, depending on the Intertropical Convergence Zone (ITCZ), monsoon systems, and oceanic settings. Annual rainfall in the tropics can vary between tropical dry forests (less than 500 mm) to wet regions (>10,000 mm), with this precipitation generally distributed in a pronounced wet season. This hydrological variability generates a spectrum of water resource challenges within the tropics, from flooding and erosion at certain times to seasonal water scarcity at others.

There are several prominent biomes in the tropical region, including tropical rainforests, tropical dry forests, savannas, and tropical montane ecosystems. Tropical rainforest, occurring at large in the Amazon Basin, Congo Basin, and Southeast Asian archipelagos, is the most species-rich terrestrial ecosystem on Earth, harboring roughly 50–80% of the world's species, even when they account for only ~ 7% of the planet's land area (Corlett and Primack 2010). These forests are some of the most complex and biologically diverse ecosystems. They have a unique vertical structure with layers of emergent, canopy, understory, and forest floor plants that create many ecological niches. Environmental change has been rapid in tropical regions due to land-use change, deforestation, urbanization, and climate change in the Anthropocene, leading to disproportionately high impacts on environmental health in these regions. With sobering consequences for ecosystem services and human well-being (Lewis and Maslin 2015), these areas are home to some of the most rapid global environmental change on Earth. Under this backdrop, the present study aims to critically review the diverse impacts of environmental change on human health in the tropics, emphasizing the interactions between climatic, ecological, socio-economic, and infrastructural determinants.

5.2 Materials and Methods

This chapter provides a systematic review of original research on diverse health effects directly or indirectly attributable to environmental change driven and exacerbated by human activities in tropical regions during the Anthropocene. An evidence synthesis based on a systematic review of original research studies assessing a range of health impacts directly or indirectly associated with environmental alterations induced and exacerbated by anthropogenic activities in tropical regions during the Anthropocene. The present study aimed to summarize the interdisciplinary approach to environmental health research while upholding scientific rigor and relevance. Several electronic databases were searched to find relevant literature: PubMed, Google Scholar, Web of Science, Scopus, ResearchGate, ScienceDirect, Environmental Sciences and Pollution Management Database, Global Health Database, as well as reports and technical documents from governmental bodies, policies and white papers concerning the subject from international organizations (e.g., WHO, UNEP, FAO), reports from specialized research institutions and think tanks.

Studies were included if they satisfied the following criteria:

1. Publications in peer-reviewed journals from 2000 to 2024.
2. Targeting environmental health determinants or outcomes in tropical regions.
3. Addressed anthropogenic environmental changes and their health implications.

Because environmental health in the Anthropocene is complex and interdependent, the literatures from a wide variety of disciplines were explored: public health and epidemiology, environmental and ecological sciences, climatology and meteorology, urban planning and built environment studies, geography and spatial sciences, agricultural and food sciences, social sciences (anthropology, sociology, economics), medical and biomedical sciences, psychology and behavioral sciences. A conceptual framework was used to synthesize the extracted data, where determinants of environmental health could be categorized as climatological/meteorological factors, natural environmental factors, built environment characteristics, economic determinants, and social determinants. This framework helped with the identification of crosscutting themes and linkages between different environmental health pathways. The main focus was on highlighting determinants and outcomes that have been understudied but are potentially important for tropical environmental health. This review fills an important gap in the literature, providing a comprehensive, integrative framework of how environmental changes during the Anthropocene affect human health in the tropics—analyses that have not been previously conducted at this scale and scope.

5.3 Results and Discussion

5.3.1 Environmental Health Determinants in Tropical Regions

5.3.1.1 Climate and Meteorological Factors

The influence of climate and meteorological factors during the Anthropocene is intensified, particularly for tropical areas. The climate impacts of humans in the Anthropocene have wrought severe and multifaceted changes at one place. The most profound suffering is borne by the most exposed tropical communities.

Temperature Patterns and Extremes

The human impact on climate in the Anthropocene is felt strongly, especially in the tropics, where temperature is characterized by relatively stable annual means but strong diurnal variability. The tropical areas experience lower seasonal temperature variability compared to temperate regions, but they are increasingly susceptible to extreme heat that may have serious health consequences (Mora et al. 2017). The Anthropocene has made these conditions worse, leading to higher sustained temperatures and heat-induced morbidity and mortality, particularly for urban populations, outdoor workers, and marginalized demographic groups. According to Im et al. (2017), the tropics will experience heat waves that have never been seen before, earlier than any other part of the world. This will happen under a range of climate change scenarios that are in line with the Anthropocene, and in some places, it will be too much for humans to handle. Mordecai et al. (2019) did research that shows how temperature affects vector competence, pathogen development rates, and the movement of vector populations. This process creates complicated, nonlinear connections between temperature and the severity of disease transmission. Also, prolonged periods of high temperature hurt people's ability to think, work, and behave in ways that have big social and economic effects on people living in tropical areas (Kjellstrom et al. 2016).

Precipitation Variability

The tropics, in particular, have high variability in precipitation, both geographically and longitudinally, through the Anthropocene, with strong environmental health implications. According to Feng et al. (2013), the consequence of climate change is a more intense hydrological cycle, which leads to a higher intensity of precipitation during the rainy seasons but also longer dry periods, which contribute to both flood risk and water scarcity simultaneously. These precipitation regime shifts have implications for waterborne disease dynamics, agricultural productivity, and safe drinking water availability. Recent studies on the patterns of precipitation in the Anthropocene have shed light on how disease spreads change as vector breeding habitats grow or shrink to match the amount of water available (Lafferty 2009). Also, heavy rainstorms, especially ones that happen after long dry periods, make it easier for diseases like leptospirosis and cholera to spread. In many tropical areas, these diseases are clearly linked to rain patterns during certain times of the year. Food security is also significantly affected by precipitation variability due to its effects on agricultural productivity, which has further health implications for vulnerable populations (Carlton et al. 2016).

Humidity Effects

In many tropical places, the air is very humid, which has had a big effect on how temperatures are controlled, how pathogens spread, and how diseases are spread in both past and present populations. So, high humidity makes the physiological stress caused by high temperatures worse by limiting the ability of evaporation to cool, which makes the actual temperature that animals and people feel higher. This interaction between humidity and temperature is shown by heat index and wet-bulb temperature, which are better ways to show how tropical climates affect people's health than dry-bulb temperature alone (Mora et al. 2017). Humidity has a major impact on the survival, development, and behavior of vectors and pathogens involved in disease. Based upon extensive research by Parham et al. (2015), most arthropod vectors, such as anopheline and Aedes mosquitoes, have optimal developmental and survival rates at a certain range of humidity. High humidity also encourages fungal growth and mold in indoor living spaces and contributes to respiratory illnesses and allergic reactions, especially in poorly ventilated housing typical of tropical urban informal settlements.

Tropical Cyclones and Extreme Weather Events

Its intensity and frequency of tropical cyclones and different extreme climate events have been rising within the last few decades and characterize key environmental well-being threats in many tropical areas. Keim (2008) wrote about a ten-year battle with professional breakdown. He wrote a lot about the many ways that tropical cyclones hurt people's health, including direct deaths and illnesses from wind and flooding, damage to infrastructure that can make it harder to get medical care, damage to water and sanitation systems, population displacement, and mental trauma (Keim 2008). As a result of climate change, tropical cyclones are expected to get stronger (but not necessarily more often) in many areas. Some models even show that the activity of these storms could be too much for coastal communities that are already struggling to adapt (Kossin et al. 2020). The aftermath of extreme weather events in the Anthropocene often creates environmental conditions conducive to disease outbreaks. Watson et al. (2007) found that vector-borne and waterborne diseases spread a lot more after both tropical cyclones and floods, mostly in places where infrastructure wasn't perfect before the events. During the time it takes to recover from disasters, people are often exposed to more environmental hazards related to cleaning up after the disaster, dealing with contaminated water, and living in temporary housing. This adds to the health risks that the affected populations face (Rataj et al. 2016).

El Niño-Southern Oscillation Impacts

The El Niño-Southern Oscillation (ENSO) is one of the most important drivers of inter-annual climate variability in the tropics, with well-described effects on health outcomes in the environment across several continents. Anyamba et al. (2019) found strong links between phases of ENSO and outbreaks of vector-borne diseases like dengue fever, malaria, Rift Valley fever, and chikungunya. These links were made possible by ENSO's effects on temperature, rainfall, and the movement of plants. These teleconnections are regionally specific, with varying ENSO-associated health impacts across tropical regions. ENSO events also impact agricultural productivity and food security significantly, especially in tropical regions. Heinemann et al. (2021) did a thorough analysis of the data and found that crop yields drop dramatically during El Niño phases in most tropical agricultural systems. This leads to nutritional insecurity and health problems that are related to it. In addition, droughts caused by ENSO make tropical forests more likely to catch fire, which pollutes the air and has negative effects on the heart and lungs. The anticipated strengthening of ENSO extremes under climate change suggests that environmental health issues could increasingly threaten tropical populations.

5.3.1.2 Environmental Determinants

Natural Environmental Factors

Biodiversity and Ecosystem Services

Tropical forests are biodiversity hotspots and key providers of many types of ecosystem services that are important for human health and well-being. Myers et al. (2013) have shown that sustainable ecosystems provide vital services such as water purification, air regulation, disease regulation, climate moderation, and food supply. The interplay between biodiversity and human health is most evident in tropical settings, wherein biological diversity supports traditional medicinal practices and contributes to modern drug development (Romanelli et al. 2015). The Anthropocene has also led to an intensification of biodiversity loss in tropical regions, threatening important environmental health outcomes. Keesing et al. (2010) say that less biodiversity can speed up the spread of diseases in many ways, such as by removing the dilution effect, which happens when cryptic or non-competent host species make pathogen transmission less effective. This association has been established for several tropical diseases, such as malaria, schistosomiasis, and Lyme disease. Also, losing biodiversity makes ecosystems less able to handle changes in the environment, which could stop processes that provide important ecosystem services from working when the environment is stressed.

Water Resources and Quality

Availability and quality of water resources are key drivers of environmental health in tropical areas. According to Hofstra and Vermeulen (2016), while many tropical regions have excess precipitation, temporal variability, and poor infrastructure for collection, storage, and distribution lead to widespread water insecurity. Changes in climate, like changing patterns of rainfall, higher evaporation, and salinization of coastal aquifers, have made these problems worse since the start of the Anthropocene (Milly et al. 2005). Concerns about the quality of water in tropical areas include not only microbial contamination but also chemical pollutants. Water is often contaminated with fecal matter, as is the case in many tropical communities, leading to the continued prevalence of waterborne diseases such as cholera, typhoid, and other diarrheal diseases, as noted in extensive monitoring by WHO and UNICEF (2021). Moreover, since the twentieth century, growing agricultural intensification, industrial development, and urbanization have also resulted in chemical contaminants, such as pesticides, heavy metals, and pharmaceutical compounds, entering water systems, which leads to complex exposure scenarios with unknown long-term health consequences.

Soil Characteristics

Soil properties in tropical regions have great implications on the environmental health in the current era due to their effects on agricultural productivity, contaminant mobility, infrastructure stability, etc. According to research by Lal (2015), some tropical soils are highly fertile, while others may have limitations in the form of nutrient deficiencies, low acidity, or a risk of erosion, all of which can have important implications for food security and nutritional outcomes. Climate change and land-use practices are accelerating soil degradation processes in many of the tropics, threatening this essential resource (Stocking 2003). The soil properties and contaminant behavior relationship are especially significant in the tropical wave of the Anthropocene. According to Naidu et al. (2013), some tropical soils have clay minerals, high rates of weathering, and charge characteristics that change over time. These factors affect where heavy metals and pesticides end up and how bioavailable they are. The soil properties also impact the ecology of multiple disease vectors. Specific soil moisture regimes and organic content promote the establishment of soil-transmitted helminths and the breed's harboring of some disease vectors (Patz et al. 2000).

Vulnerability to Natural Disasters

Tropical areas are particularly susceptible to various natural disasters such as floods, landslides, tropical cyclones, and volcanic eruptions, with major consequences for environmental health in the Anthropocene. Dilley et al. (2005) noted that the vast majority of natural disaster hotspots lie in tropical regions of the globe, where geographical and climatic factors combine to produce high-risk profiles. This makes them highly vulnerable as well, with vulnerability often magnified by socioeconomic factors, such as poor infrastructure, insufficient early warning systems, and limited response capacity (Keim 2008). Natural disasters in tropical contexts have health impacts that go well beyond acute traumatic injuries. Post-disaster or post-calamity periods are also characterized by an increased incidence and reappearance of vector-borne and waterborne diseases, respiratory disorders resulting from temporary housing arrangements, and psychological disorders such as post-traumatic stress disorder (PTSD), depression, and anxiety (Rataj et al. 2016). Projections of the impacts of climate change for many tropical regions indicate increasing frequency and intensity of hydrometeorological disasters, which could exceed adaptive capacity (Kossin et al. 2020).

Vector Ecology

The ecology of disease vectors in tropical settings constitutes a key driver of vector-borne disease dynamics. In fact, Kilpatrick and Randolph (2012) found that temperature, humidity, rainfall, plant characteristics, and land-use patterns are just some of the environmental factors that affect where arthropod vectors live, how many there are, and how well they can spread diseases. All these factors impact vector life-cycle parameters such as development rates, survival, reproduction, and biting behavior, and thereby dictate disease transmission dynamics (Parham et al. 2015). In the Anthropocene, climate change is transforming vector ecology in tropical regions, which has profound consequences for the epidemiology of disease. According to Ryan et al. (2019), changes in temperature and rainfall are letting vectors like Aedes aegypti and Anopheles species settle in places that weren't good for them before. This puts immune-naive populations at risk of getting diseases that vectors carry. Moreover, land-use change, such as deforestation, agricultural encroachment, and urbanization, contributes to the establishment of novel ecological niches for disease vectors that facilitate higher human–vector contact rates and the emergence of disease (Kilpatrick and Randolph 2012).

5.3.1.3 Environmental Determinants in Tropical Regions

Built Environment

Urbanization Patterns

Urbanization has been happening at unprecedented rates in tropical regions over the last 50 years, significantly altering environmental health landscapes. Studies by Moser and Satterthwaite (2010) show that most tropical cities are growing rapidly and without any planning; the growth has led to the expansion of informal settlements, poor urban infrastructure, and huge environmental degradation. These forms of urban growth give rise to unique environmental health issues, such as heat islands, air and water pollution, solid waste disposal problems, and changed ecology of disease. Despite this, the environmental health effects of tropical urbanization are not all negative. Urban environments may be associated with better access to health care services and the availability of education and economic opportunities, and they may therefore impact health positively. But these benefits aren't always shared fairly. For example, marginalized urban populations are more likely to be exposed to environmental hazards and have less access to environmental services and amenities (UN-Habitat 2020). This environmental inequity related to urbanization constitutes one of the most pressing social justice issues in tropical environmental health in the Anthropocene.

Infrastructure Adequacy

Infrastructure sufficiency is a basic condition of environmental health status in tropical areas. Extensive studies undertaken by the World Bank (2021) reveal significant infrastructural needs in large swaths of tropical countries, especially in the water supply, sanitation, transportation, energy, and health sectors. In addition to a lack of investment, there are some region-specific unique problems in the tropical region, such as how fast buildings break down in hot, humid weather and because of living things. Infrastructure challenges in the tropics are worsened by climate change. Schweikert et al. (2014) mentioned that rising sea levels, more intense rainfall, and higher temperatures caused by climate change will make it more expensive to maintain and replace infrastructure in tropical areas, by as much as 33% in some places. This could mean that money isn't spent on building new infrastructure. Readily available data is concerned about current infrastructure: a population surge creates infrastructure demand while climate change simultaneously threatens existing infrastructure capacity and durability.

Housing Conditions

To illustrate this idea, housing conditions in the tropics are known to impact a number of environmental health outcomes. Research by Tamerius et al. (2013) demonstrates that the way houses are built and the materials used can affect the temperature inside, the number of vectors that come in, the amount of moisture that builds up, and the quality of the air inside. These factors have big effects on people's respiratory health, their risk of getting vector-borne diseases, and their risk of getting sick from the heat. Traditional tropical home designs often used passive cooling methods and materials that were well-suited to the area. However, these methods are slowly being replaced by more general designs that don't work well in tropical environments. Housing informality constitutes a specific environmental health challenge in rapidly urbanizing tropical countries. About one billion people live in informal settlements worldwide (UN-Habitat 2020), often concentrated in tropical urban domains. Such settlements are generally characterized by crowding, poor ventilation, reduced access to safe water and sanitation, and increased risk for environmental hazards, such as floods, landslides, and industrial pollution. Not having enough housing can affect public health, and the COVID-19 pandemic made these effects clear because of the high rates of transmission in informal settlements where people can't physically separate themselves from others or practice good hygiene.

Sanitation Systems

Sanitation infrastructure and practices are important determinants of environmental health in tropical settings. Research published by Prüss-Ustün et al. (2019) estimates that poor sanitation is responsible for around 432,000 diarrheal deaths every year, with a disproportionate share of the burden borne by tropical countries. Poor sanitation affects more than just diarrheal disease. It can also lead to helminth infections, schistosomiasis, trachoma, and changes in nutritional status through environmental enteropathy pathways. Climate change poses major threats to sanitation systems in tropical regions. Climate change impacts are especially pronounced in informal settlements and low-lying coastal regions, where physical and economic constraints often limit sanitation options. Changing patterns of rainfall that make water scarce also affect the way water- and sanitation-dependent systems work and may undo any progress made in sanitation (Sherpa et al. 2014).

5.3.2 *Economic Determinants in Tropical Countries*

5.3.2.1 Industrial Development

The nature and distribution of industrial activities shape associated environmental health impacts in the modern era. According to a study by Landrigan et al. (2018), industrialization may create economic opportunities that may help improve health, but it also creates serious environmental health risks, such as pollution of the air and water, the production of toxic waste, and exposures at work. In many tropical countries, the need for fast industrial development has outpaced the development of regulatory frameworks and the ability to enforce regulations. This has caused heavy pollution loads (World Bank 2021). Bowen et al. (2015)'s analysis demonstrates that environmentally hazardous industries typically locate in economically deprived communities with limited political influence, thereby influencing the environmental justice issue. The so-called pollution haven hypothesis says that pollution in tropical regions is affected by changes in the international economy. For example, production that causes a lot of pollution moves from countries with higher incomes to countries with lower incomes and less strict environmental rules.

5.3.2.2 Agricultural Practices

Soil and groundwater quality play a major role in environmental health in the tropics, where a large proportion of the population relies on agriculture for livelihood. Alavanja (2009) documents a number of environmental health problems in agriculture, such as being exposed to pesticides, water quality getting worse because of fertilizer runoff, soil health getting worse, which makes it harder to get enough food, and changes in disease ecology because of changes in landscapes. Agricultural workers and wider communities experience these impacts through various exposure pathways. Complex environmental health trade-offs exist with agricultural intensification in tropical countries. According to Matson et al. (1997), intensification can help increase food production and improve nutrition, but it also often leads to more use of agrochemicals, more water use, and the destruction of natural habitats, all of which are bad for the health of the environment.

5.3.2.3 Resource Extraction

Inevitably, the majority of individuals in developing countries have been reliant on underlying resource extraction activities, such as mining, logging, and fossil fuel development, which have had a powerful impact on tropical environmental health outcomes in the post-industrial revolution era. In-depth reviews of Cordy et al. (2011) show the vast health impacts that the extraction of oil and gas has on the environment, including but not limited to the destruction of habitats, contamination of the water supply, air pollution, and toxic waste. In addition, extractive activities often create new social dynamics, such as sudden migration, the growth of informal neighborhoods, and changes to the way the economy works, which have indirect health effects. The governance of extractive industries has a significant impact on environmental health outcomes related to them. Weak governance frameworks, limited enforcement capacity, and corruption create conditions facilitating exploitative extraction practices, which have devastating implications for environmental health.

5.3.2.4 Informal Economy

On the other hand, the informal economy is the prevailing form of economy for the majority of workers in tropical areas, bearing substantial environmental health consequences. According to Chen (2016), informal employment, i.e., work without formal contracts, protections, or regulations, now comprises 60% of global employment and is even higher in many tropical countries. Inadequate access to occupational health services, health threats of hazardous working conditions, and limited access to protective equipment mean informal workers are often exposed to heightened risks to environmental health. Some informal jobs have distinct and complex environmental health problems. As per documentation by Steckling et al. (2017), artisanal and small-scale mining comes with high exposure to hazardous substances, including mercury, lead, and cyanide, leading to considerable neurotoxic and other health effects. Workers engaged in unofficial waste management activities, such as waste picking and recycling, are exposed to biological hazards, cuts from sharp objects, harmful chemicals, or combustion by-products (Wilson et al. 2006). The social and economic problems that come with informal work, like not having a steady income, little social protection, and trouble getting medical care, make these environmental health risks even worse.

5.3.2.5 Economic Inequalities

Economic inequalities are one of the main drivers of environmental health inequity in both intra- and inter-countries settings in the tropics. Comprehensive analysis by Islam and Winkel (2017) demonstrates that economic inequality impacts environmental health in a number of ways, such as by causing different levels of exposure to environmental hazards, access to environmental amenities, services, and experiences, and the ability to prevent or lessen the effects on environmental health. It also results in different levels of participation (or exclusion) in environmental decision-making processes. These complex interactions cause and keep environmental injustice going by implementing and talking about environmental health problems in different ways that affect communities that are already at a disadvantage in different ways (Schlosberg and Collins 2014). Macroeconomic policies and structures significantly shape environmental health inequalities, particularly in tropical regions. According to critical analysis by Ottersen et al. (2014), international economic relationships (such as debt, trade, investment, and tax systems) affect national capacity to invest to invest in environmental health infrastructure and services and environmental protection standards. Furthermore, the effects of globalization on the economy have created complicated maps of varying environmental risk in tropical areas. Dangerous production is often concentrated in economically disadvantaged areas that are hard to control.

5.3.3 Social Determinants in Tropical Countries

5.3.3.1 Population Density and Growth

Population density and growth trajectories are important drivers of environmental health outcomes in tropical countries in recent decades. Research by Ezeh et al. (2012) suggests that several tropical countries are undergoing demographic transitions in which mortality decline lags behind reductions in fertility, leading to high population growth and important changes in the age structure. These population trends place increasing stress on both environmental resources and infrastructure systems, negatively impacting either environmental quality or access to services (de Sherbinin et al. 2007).

The population density and health outcome relationship is highly complex. Montgomery et al. (2003) study shows that while higher density may help with service delivery and lower environmental impact per person, it also makes people more vulnerable to communicable diseases and man-made pollutants in places where infrastructure isn't present. This level of complexity is even more noticeable in tropical urban informal settlements with a lot of people living in close quarters and not enough money spent on infrastructure. This creates an environment that is perfect for diseases to spread and for the environment to get worse.

5.3.3.2 Poverty and Inequality

Poverty and inequality are core social determinants of environmental health in the tropics. A detailed study by Prüss-Ustün et al. (2019) shows that poverty impacts environmental health in a number of ways, such as by limiting housing options and access to clean water and toilets, as well as by making people rely too much on polluting energy sources at home and on dangerous or informal jobs that expose them to environmental hazards. All these environmental exposures related to poverty play a major role in perpetuating the inequities in health within tropical countries and across countries (WHO 2023).

Growing economic inequality worsens environmental health inequities, particularly among tropical countries. Islam and Winkel (2017) discovered that inequality within a country is linked to the unequal distribution of environmental burden.

This means that disadvantaged groups are more likely to be affected by environmental hazards and have less access to environmental benefits and services. Political processes that favor the economic elite in environmental decision-making and resource distribution often buttress this environmental injustice (Schlosberg and Collins 2014).

5.3.3.3 Education and Awareness

Tropical educational attainment and environmental health process factors exert significant influence on health. Pickett and Pearl (2001) depict clear associations between education levels and health outcomes, with education impacting health via many avenues, including health literacy, preventive practices, use of health services, and socio-economic status. In tropical regions, education as a determinant affects environmental health specifically via hygiene practices, behaviors for vector control, practices of water treatment, and identification with environmental hazards. Health education interventions are significant components of strategies to improve environmental health in tropical countries. As per the systematic reviews by Monse et al. (2013), there are tailored health education interventions that can facilitate behavior changes to diminish environmental health risks, such as handwashing, water treatment, waste disposal, and vector habitat reduction. However, simply educating people does not work unless the structural determinants that prevent behavior, including availability of infrastructure, economy, and cultural context, are also taken into account.

5.3.3.4 Gender Dimensions

Gender is a fundamental social determinant that affects environmental health exposures, vulnerabilities, and adaptive capacities in tropical areas. Gendered divisions of labor, access to resources, and decision-making authority create distinct environmental health risk profiles for women and men (Denton 2002). In several tropical countries, women have traditionally been responsible for the collection of water, preparation of food, provision of energy for the household, and caregiving, which places them at heightened risk to certain environmental hazards, including but not limited to indoor air pollution, contaminated water, and vector habitats (Prüss-Ustün et al. 2016). Discrimination and gender inequality drive women and other marginalized groups further into poverty, increasing their vulnerability to environmental health risks. Neumayer and Plümper (2007) use a quantitative study to argue that the fact that women face discrimination when trying to get resources, learn new skills, and make decisions means they are less able to adapt to changes and stresses in the environment. Therefore, environmental crises are more likely to affect women than men. It has been shown that gender-responsive environmental health interventions work better than gender-blind ones (WHO 2006, 2023) because they take into account the needs, limitations, and opportunities of different genders.

5.4 Environmental Health Challenges in Tropical Regions

Due to human activities and natural disasters, climate change leads to a great global threat to food security. It also accelerates health problems as well as health systems of tropical countries, i.e., Africa and Southeast Asia (Nguyen and Liou 2019; Sarun et al. 2018).

5.4.1 Direct Health Impacts

Presently, climate change causes of global health threats. Recently, a multinational study by health professionals examined several tremendous human health issues due to climatic and environmental changes.

5.4.1.1 Vector-Borne Diseases

Several studies show the effect of climate change on vector-borne diseases, like their reproduction, survival, geographic distribution, and redistribution. Even many continents are vulnerable to various epidemics, like malaria found in Africa, dengue in Southeast Asia, and the outbreak of Aedes in North America and Europe (Burkett-Cadena and Vittor 2018).

Malaria

Malaria is a significant health burden, and very sensitive climatic factors are mostly restricted to tropical regions, yet the north-central part of Europe, north of Russia, the northern margin of Australia, and the northeastern belt of the United States are also highly affected (Caminade et al. 2014). Climate change has already favored the activity of vector-borne infectious diseases due to their geographical distribution. Presently, public health officials rightly worry about climate change that leads to increased important infectious diseases, especially malaria, in tropical and subtropical countries (Lafferty 2009).

Dengue

Dengue, as a fast-growing disease, mostly appears with complex interactions of host, vector, and virus, by climatic vulnerabilities. The effects of dengue on people (morbidity and mortality) might be affected by consistent climatic and local meteorological conditions in many places, including Southeast Asia (Stanaway et al. 2016; Mutheneni et al. 2017), southern coastal Ecuador (Lowe et al. 2017), and tropical and subtropical countries. Dengue affected Central and South America because of the El Niño-Southern Oscillation (ENSO) phenomenon with periodic warm sea surface temperatures. Not only does dengue cause diseases like death and hemorrhagic fever, but it also lowers people's income, which is called an economic burden (Martelli et al. 2015; Torres and Castro 2007). Tropical locations, such as Brazil, conducted some of these studies.

Chikungunya

Possible weather conditions make it easier for vector-borne diseases like chikungunya and dengue to spread, which raises tropical global public health concerns mostly in Europe, North America, northern Brazil, and Southeast Asia (Semenza and Paz 2021; Dhiman et al. 2010). A study by Tjaden et al. (2017) estimated that, along with climatic conditions, globalization in trade and travel leads to a strong transmission efficiency of chikungunya in different geographical regions, especially China, sub-Saharan Africa, the southern part of America, the United States, and parts of Europe.

Zika Virus

Climatic vulnerability, such as increased temperature and variability of rainfall, leads to the appearance of mosquito-borne diseases like Zika, which accelerate neurological problems and defective birth outcomes in humans, who even frequently show no symptoms at all in human health (Zhang et al. 2024).

New Arboviruses

Dengue and chikungunya are arboviral infections that are mostly spread by Aedes aegypti and Ae. albopictus mosquitoes, which are found all over the world in tropical and many subtropical areas (Palaniyandi 2019).

5.4.1.2 Water-Related Diseases

Climate change affects both freshwater and marine ecosystems, which leads to waterborne diseases (Patz and Frumkin 2016).

Cholera

Climate change and its variability exacerbate epidemics like cholera, malaria, and dengue, as well as chronic diseases due to poor sanitation conditions, a lack of pure drinking water, and a wide range of pollution (Bush et al. 2011). Climate change consistently leads to health effects by transmitting waterborne diseases like cholera. There is a close relationship. A study by Dhara et al. (2013) found a close relationship between rising sea surface temperature and cholera epidemics in Southeast Asian countries.

Schistosomiasis

The parasitic trematode leads to schistosomiasis which is prone to adverse health outcomes and is found in different countries in sub-Saharan Africa, Asia, and Latin America (Freeman et al. 2017).

Diarrheal Diseases

Burden of diarrheal death for inadequate drinking water, sanitation, and hand hygiene is estimated in 145 low- and middle-income settings of the world due to environmental changes and disease transmission (Prüss-Ustün et al. 2014). Regular practices of open defecation are also prone to the transmission of diarrheal disease in tropical Ethiopia (Vedachalam et al. 2017).

Harmful Algal Blooms

Warm water and nutrient loading (nitrogen and phosphorus) favor blooms of marine algae and their harmful poisoning in humans, as well as extensive die-offs of fish, shellfish, marine mammals, and birds in the marine food web (Patz and Frumkin 2016).

5.4.1.3 Respiratory Conditions

Air Pollution Effects

Increased concentrations of ozone as well as fine particulate matter (FPM) have constantly contributed to adverse air quality and climate change. Anthropogenic outdoor air pollution leads to serious respiratory problems like bronchitis and asthma, heart disease, and stroke (Zareba et al. 2024; Mansoor et al. 2022). It is also linked to early deaths, severe cardiopulmonary diseases (93% of cases), lung cancer (7% of cases), and premature respiratory death (Silva et al. 2013; Sun and Zhu 2019). On the other hand, indoor air pollution significantly increases morbidity and mortality from chronic diseases (Rosário Filho et al. 2021; Cohen et al. 2015). An exploratory study estimated that residential energy usage is the primary cause of premature mortality in India and China, and it could double by 2050 (Lelieveld et al. 2015).

Occupational Lung Diseases

The burning of biomass in the subequatorial Amazon results in high levels of particulate matter, which reduces lung function in schoolchildren and is acutely found among children with non-asthmatic diseases (Jacobson et al. 2012). Besides, the anthropogenic ozone and fine particulate matter lead to lung cancer and human mortality (Anenberg et al. 2010).

Biomass Fuel Exposure

A recent study of Brazil showed that slash-and-burn practices are significant causes of biomass burning and adversely affect human health, even increasing premature deaths (Nawaz and Henze 2020).

Urban Air Quality Impacts

A few studies explored the negative impact of urban air quality, particularly mortality and adverse respiratory outcomes (Rocque et al. 2021). India is grappling with a major air pollution crisis, especially in cities like its capital, New Delhi, and Mumbai and Kolkata. The main polluters are literal cars, fuel, industrial smoke, construction dust, and crop burning. Delhi usually ranks among the top polluted cities on the globe, for PM2.5 orders of magnitude higher than WHO recommendations (Guttikunda and Goel 2013). The health effects are dire, with the Global Burden of Disease Study estimating that air pollution alone was responsible for more than 1.67 million premature deaths in India in 2019. Exposure to elevated levels of PM2 leads to a significant health concern. Human exposure to PM2.5 and PM10 is associated with respiratory diseases, cardiovascular diseases, stroke, diabetes, and lung cancer. Children and old people are highly vulnerable, showing a greater incidence of asthma, bronchitis, and other respiratory diseases (Balakrishnan et al. 2019). Vehicles, industries, and transboundary haze from Indonesian forest fires heavily pollute the air in Malaysia's major cities. As a consequence, respiratory and cardiovascular diseases lead to an increase in hospital admissions, particularly among children and the elderly (World Bank 2016). Indonesia's Jakarta is suffering from severe air pollution caused mainly by vehicles, industrial activities, and transboundary haze from forest fires. Jakarta's air quality often falls short of the WHO guidelines, with pollutants exceeding the acceptable levels during the dry season, when thousands of forest fires occur (Greenpeace 2019).

5.4.2 *Indirect Health Impacts*

5.4.2.1 Food Security and Nutrition

The effects of climate change, such as warmer temperatures, altered water cycles, and severe and extreme weather events, can lead to food safety hazards in tropical regions, posing a significant threat to global food security. Shivanna (2022) reveals that the trend of environmental warming is to blame. Climate changes over several decades have adversely impacted rainfed agricultural systems, which were the primary source of livelihood for a large population. This has resulted in high rates of undernutrition and infectious disease in the tropical region of Sub-Saharan Africa. Different parts of Africa are suffering from different calamities, such as East Africa, which has a higher risk of flooding, and South Africa, which has concurrent drought risks (Serdeczny et al. 2017).

5.4.2.2 Mental Health Effects

In tropical regions, extreme weather events have been associated with increased rates of a psychiatric disorder, post-traumatic stress disorder (PTSD), among exposed populations, with PTSD rates in the first months after a climate-related disaster significantly greater than the general population (Berry et al. 2018). Cianconi et al. (2020) demonstrated how instability driven by climate displacement and migration in tropical regions leads to social instability that compounds mental health issues, especially among vulnerable communities. Additionally, Hayes et al. (2018) documented that climate anxiety, or "eco-anxiety," is increasingly present among communities in the tropics that directly see their environments degenerate, indicating a chronic fear of environmental catastrophic doom.

5.4.2.3 Maternal Health

Climate change influences maternal health outcomes in tropical regions through diverse pathways. Asamoah et al. (2018) found that pregnant women by prolonged heat waves (temperatures above 35 °C for more than 3 consecutive days) had 1.78-fold increase of preterm birth compared with unexposed (Ghana). Pattanittum et al. (2019) reported disruptions in maternal healthcare services from flooding events and climate-related disasters in tropical areas, observing that women from provinces affected by flooding had 23% lower odds of the recommended antenatal visits and experienced 18% higher odds of complications at delivery than those from unaffected provinces. Manyeh et al. (2020) assessed how seasonal variations might impact maternal outcomes in rural Tanzania and found that women were 2.3 times more likely to experience anemia during a period of drought compared with rainy seasons. Vector-borne diseases exacerbated by climate change increase risks to pregnant women living in tropical areas. Bauserman et al. (2015) studied tropical African countries and found that maternal malaria infection significantly increased maternal anemia, low birth weight, and stillbirth risks, with climate change expanding the range of appropriate habitats for mosquitoes into formerly unaffected regions.

5.4.2.4 Occupational Health

Climate change is increasing occupational health risks for workers in tropical countries. Kjellstrom et al. (2016) estimated the burden of occupational heat stress in Southeast Asia, finding that during hot season's workers in agriculture, construction, and manufacturing lost 10–15% of their productivity, a number which could rise to 25–30% by 2050 under current emissions scenarios. Wesseling et al. (2020) conducted a longitudinal study monitoring sugarcane workers in Nicaragua over three harvest seasons and reported increasing heat-related kidney injury prevalence with rising temperature. The risk of decreased kidney function was 3.8 times higher in workers exposed to wet-bulb globe temperatures above 30 °C for more than 4 h a day compared to workers without prolonged heat exposure. Tawatsupa et al. (2018) found that exposure to heat combined with air pollution among outdoor workers was associated with synergistic health impacts, with exposure to both heat and air pollution associated with 2.7 times greater odds of respiratory and cardiovascular symptoms compared with either exposure alone. Their research documented especially severe impacts on police working on highways, street vendors, and construction workers in Bangkok. Occupational infectious disease risks in the tropics are also impacted by climate change. Donis-González et al. (2021) found that shifts in rainfall patterns and temperatures in Honduras drove disease vectors such as mosquitoes and ticks into new working environments. They assessed that vector-borne disease exposure as a result of agriculture in newly deforested areas was 4.2 times that of outdoor agricultural workers in established agricultural regions.

5.4.2.5 Heat Stress and Productivity

Presently, the frequency and intensity of extreme heat events are well established and emerging as a public health concern like heat stroke and heat exhaustion globally (Luber and McGeehin 2008). From 1980 to 2014, climate change escalated the risk beyond human thermoregulatory capacity, resulting in 783 human deaths from 164 cities across 36 countries due to lethal heat events (Mora et al. 2017). Climate change directly leads to impacts on human morbidity, mortality, and adverse respiratory outcomes due to excessive heat waves and thermal stress. Several studies delved into mortality burdens with additional heat exposure from anthropogenic climate change (1991–2018) in 732 locations in 43 countries of the world and also led to droughts with less productivity, malnutrition, and death (Vicedo-Cabrera et al. 2021).

5.5 Policy Responses and Adaptive Approaches for Environmental Health Resilience

The environmental health issues of the tropics are being increasingly shaped by the features of climate change, environmental destruction, and the socio-economic landscape as a whole. Policies aimed at lessening the impacts and increasing resilience shall in any case be integrated, contextually appropriate, evidence-based, and area-specific.

5.5.1 Integrated Climate-Health Action Plans

Addressing the impacts of climate change on environmental health requires multidisciplinary, cross-sectoral action plans referred to as climate-health action plans. These plans should stimulate inter-sectoral collaboration through institutional mechanisms that allow joint action among the ministries of health, environment, agriculture, urban development, and economic planning. Integrated policy action maximizes the incorporation of health issues into multi-sectoral development agendas. Systematic health impact assessments (HIAs) form a core component of these plans. HIAs should be mandatory for proposed large development infrastructure and policy initiatives. HIAs evaluate the vulnerabilities and susceptibilities concerning an environment and the health components, thereby allowing the design of protective policies that enhance health and sustainability. The construction of climate-sensitive early warning systems is also crucial. Integration of meteorological, ecological, and public health surveillance data for advanced detection of disease and environmental catastrophe must be incorporated into these systems. Furthermore, applying geospatial analysis to develop vulnerability maps aids in pinpointing the communities most impacted by the health consequences of climate change, thereby allowing for targeted intervention strategies.

5.5.2 Sustaining Urban Development

Tropical urbanization comes with its own distinctive environmental health challenges, which are intensified by climate change, lack of transportation, and informal settlements. Healthful approaches are therefore needed to manage urban areas. The transition to public transport reduced vehicular emissions, and infrastructure for pedestrians and cyclists greatly increased physical exercise while also reducing pollution. Green urban parks, green corridors, and other permeable surfaces can also reduce the urban heat island effect, improve air quality, and enhance psychological well-being. Equally important is climate resilience in housing. Informal settlements such as shack cities require building code upgrades to include passive cooling, natural ventilation, and vector-disease protection, among other improvements. Most importantly, urban design actively promotes water-sensitive planning to facilitate sustainable and safe water access management through rainwater harvesting, stormwater management, and wastewater reuse.

5.5.3 Agricultural and Food System Adaptation

Tropical agriculture remains one of the most susceptible regions of the world to climate change impacts and, as such, requires adapting policies that solve the problems of climate change while ensuring food and nutrition security is crucial to address the impacts on tropical agriculture ecosystems. Eco-agricultural systems have the additional benefit of protecting biodiversity and soil health while reducing the use of chemicals. These systems increase ecological resilience and food sovereignty at local scales. Another important strategy is crop diversification, which encourages the planting of nutrient-dense and heat- and drought-resilient crops that are better suited to current and future climatic conditions. These changes, particularly for smallholder farmers, require stronger agricultural extension services to aid facilitators. Extension frameworks need to provide climate-responsive agronomic instruction, outreach, early warning information systems, and technical support. Additionally, a constant climate change pseudonym of access-controlling food contamination needs surveillance systems to ensure safety against mycotoxins, pesticide residues, and other contaminants that threaten food safety and quality.

5.5.4 Vector Control and Disease Management

Surveillance and control of vector-borne diseases in the tropics are very sensitive to the environment, both requiring very sensitive adaptive strategies for ecological disease control. Integrated vector management (IVM) is a combined method that includes managing the environment, using biological methods, applying limited chemical control, and involving the community. These efforts should be complemented by climate-based predictive models that can project changes in vector range and transmission of diseases relative to climate parameters. Enhanced community-based surveillance also improves the response and detection of the situation by empowering local health staff and the community to monitor vectors and implement control at the local level. Due to the intrastate boundaries of many vector-borne diseases, cross-border collaboration is important. These regional mechanisms aid in the exchange of surveillance information, coordination of response activities, and standardization of vector management techniques for use so as to contain the infections spread across borders.

5.5.5 Strengthening the Health System

The protection of populations from climate-linked health threats begins with strengthening the framework of the health system. A primary example includes climate-resilient infrastructure, which involves the design of healthcare buildings that are able to function during severe weather and minimize their operational carbon footprint. There is also equal value in building the capacity of health personnel, which ensures that frontline workers are able to identify and manage emerging health issues due to changing environments. Institutional preparedness also requires comprehensive response frameworks for emergencies, which synergize environmental health components to enable swift and coordinated actions in the case of climate-related catastrophes. Increasing telemedicine network coverage can also sustain care during interruptions while providing services to people in rural and underserved areas, thereby increasing system resilience.

5.5.6 Financing the Resilience of Environmental Health

Adaptive strategies require implementation at scale and sustainable funding, which can be achieved through targeted financing. Dedicated resources for environmental health interventions can be secured through the establishment of climate funds at the national and international levels, which focus on health issues. Funds set up to provide financial protection against climate-related health risks also serve as incentives for preventive measures. Results-based financing is another innovative approach that provides funds based on improvements in health outcomes and adaptive capacity. Finally, public–private partnerships can be easily fostered to mobilize the necessary technical and financial resources to support scalable solutions like renewable energy systems for health facilities and digital environmental monitoring and disease surveillance tools.

5.5.7 Research and Innovation Priorities

It is critical to address important gaps in knowledge through research and innovation to advance the resilience of environmental health. Longitudinal studies evaluating the impacts of environmental changes on human health in different tropical ecosystems are necessary to accurately inform adaptive interventions. Policies and practices, including adaptation strategies, can be developed through consultation with indigenous communities to promote community buy-in and ownership. There is a growing need to enhance surveillance in resource-poor areas; therefore, the creation of these monitoring technologies should be inexpensive and ready for immediate use in the field. Finally, we need to study what helps and what hinders different groups in making lasting changes, so we can figure out how to make effective interventions bigger and keep them going over time.

5.6 Conclusion

The Anthropocene epoch has initiated and induced drastic environmental changes that result in complex public health challenges in the tropics. These challenges are intricately interconnected, with climatological, environmental, economic, and social components. Tropical regions are home to more than 40% of the world's population and host the world's most biodiverse ecosystems. The acceleration of ENSO events, increased severity of tropical cyclones, changes in precipitation patterns, and elevation in temperatures are all climate and meteorological drivers. These have triggered a cascade of vulnerabilities for health, including an increase in vector-borne diseases such as malaria and dengue, and waterborne illnesses like cholera and diarrheal diseases.

The detrimental alteration of critical ecosystems, which includes the loss of biodiversity and the depletion of water and soil resources, as well as urbanization, poor infrastructure, cape sanitation, all combine to further degrade health and fortify these vulnerabilities, creating an enormous risk to environmental health. Industrial activities, agricultural intensification, and fuel resource extraction result in land, water, and air pollution, and greater risk to health and occupational hazards. These health impacts are disproportionately borne by vulnerable communities. Poverty, educational inequalities, and gender differences further exacerbate health inequities by restricting adaptive capacities and influencing exposure pathways. The health impacts are direct and indirect, from spreading infectious diseases and respiratory conditions to eroding food security, mental health, and maternal well-being. These impacts are unfairly allocated, impacting the weaker populations the hardest, thus exacerbating already existing health inequalities.

To effectively address these concerns, an immediate integrated response is required. Tropical regions must focus on deploying interdisciplinary approaches that look into the nexus of environmental change with health for effective policy design. Anticipatory, adaptive, and mitigation strategies are necessary for emerging environmental health risks. Policies focused on change and addressing the inequitable social climate, root determinants, and economic needs, as well as the environment, are crucial to health sustainability in society and its systems. Finding a balance between economic growth and environmental sustainability is the most important step toward averting long-term harm. Strengthening international collaboration is needed for tackling cross-border issues that affect the inequitable state of health and secure the region's health, and achieve sustainability.

References

Alavanja MC (2009) Pesticides use and exposure extensive worldwide. Rev Environ Health 24(4):303–309

Anenberg SC, Horowitz LW, Tong DQ, West JJ (2010) An estimate of the global burden of anthropogenic ozone and fine particulate matter on premature human mortality using atmospheric modeling. Environ Health Perspect 118(9):1189–1195. https://doi.org/10.1289/ehp.0901220

Anyamba A et al (2019) Climate teleconnections and recent patterns of human and animal disease outbreaks. Environ Res Lett 14(11):114030

Asamoah B, Kjellstrom T, Östergren PO (2018) Is ambient heat exposure levels associated with miscarriage or stillbirths in hot regions? A cross-sectional study using survey data from the Ghana Maternal Health Survey 2007. Int J Biometeorol 62(3):319–330

Balakrishnan K, Dey S, Padmavati R, Muraleedharan VR, Santu L, Ravindran P, Thangavel G, Cohen A (2019) The impact of air pollution on deaths, disease burden, and life expectancy across the states of India: the global burden of disease study 2017. Lancet Planet Health 3(1)

Bauserman M, Conroy AL, North K, Patterson J, Bose C, Meshnick S (2015) An overview of malaria in pregnancy in endemic regions of Africa. Semin Perinatol 39(3):141–143

Berry HL, Waite TD, Dear KBG, Capon AG, Murray V (2018) The case for systems thinking about climate change and mental health. Nat Clim Chang 8(4):282–290. https://doi.org/10.1038/s41558-018-0102-4

Bowen WM et al (2015) Environmental justice: the economics of race, place, and pollution. Routledge

Burkett-Cadena ND, Vittor AY (2018) Deforestation and vector-borne disease: forest conversion favors important mosquito vectors of human pathogens. Basic Appl Ecol 26:101–110. https://doi.org/10.1016/j.baae.2017.09.012

Bush KF, Luber G, Kotha SR, Dhaliwal RS, Kapil V, Pascual M, Brown DG, Frumkin H, Dhiman RC, Hess J, Wilson ML, Balakrishnan K, Eisenberg J, Kaur T, Rood R, Batterman S, Joseph A, Gronlund CJ, Agrawal A, Hu H (2011) Impacts of climate change on public health in India: future research directions. Environ Health Perspect 119(6):765–770. https://doi.org/10.1289/ehp.1003000

Caminade C, Kovats S, Rocklov J, Tompkins AM, Morse AP, Colón-González FJ, Stenlund H, Martens P, Lloyd SJ (2014) Impact of climate change on global malaria distribution. Proc Natl Acad Sci 111(9):3286–3291. https://doi.org/10.1073/pnas.1302089111

Carlton EJ et al (2016) Impacts of climate change on health in tropical regions: a review. Annu Rev Publ Health 37:221–240

Chen MA (2016) The informal economy: recent trends, future directions. WIEGO

Cianconi P, Betrò S, Janiri L (2020) The impact of climate change on mental health: a systematic descriptive review. Front Psych 11:74. https://doi.org/10.3389/fpsyt.2020.00074

Cohen AJ, Brauer M, Burnett R, Anderson HR, Frostad J et al (2017) Estimates and 25-year trends of the global burden of disease attributable to ambient air pollution: an analysis of data from the global burden of diseases study 2015. Lancet 389(10082):1907–1918. https://doi.org/10.1016/S0140-6736(17)30505-6

Cordy P et al (2011) Characterization, exposure, and risk assessment of mercury contamination in artisanal gold mining communities in Ghana. Environ Res 111(8):1196–1203

Corlett RT, Primack RB (2010) Tropical rainforests: an ecological and biogeographical comparison, 2nd edn. Wiley-Blackwell

Crutzen PJ, Stoermer EF (2000) The anthropocene. Glob Change Newsl 41:17–18

Daszak P, Olival KJ, Li H (2020) A strategy to prevent future epidemics similar to the 2019-nCoV outbreak. Biosaf Health 2(1):6–8

de Sherbinin A et al (2007) Population and environment. Annu Rev Environ Resour 32:345–373

Denton F (2002) Climate change vulnerability, impacts, and adaptation: why does gender matter? Gend Dev 10(2):10–20

Dhara VR, Schramm PJ, Luber G (2013) Climate change & infectious diseases in India: implications for health care providers. Indian J Med Res 138(6):847–852

Dhiman RC, Pahwa S, Dhillon GPS, Dash AP (2010) Climate change and threat of vector-borne diseases in India: are we prepared? Parasitol Res 106:763–773. https://doi.org/10.1007/s00436-010-1767-4

Dilley M et al (2005) Natural disaster hotspots: a global risk analysis. Disaster Risk Management Series, No. 5. The World Bank

Donis-González IR, Garza-Rodríguez ML, Estrada-Bellmann I, González-González SE, Carranza-Rosales P (2021) Occupational exposure to vector-borne diseases in highland coffee cultivation areas of Honduras. Am J Trop Med Hyg 104(1):342–347

Environmental Justice Foundation (2023) Environmental health and justice: a report on disparities in health outcomes

Ezeh A et al (2012) The health of people who live in slums 1: slum health: findings from the Nairobi urban health and demographic surveillance system. Lancet 389(10068):547–558

Feng X et al (2013) Intensification of the hydrological cycle in the anthropocene. Nature 496:347–350

Freeman MC, Garn JV, Sclar GD, Boisson S, Medlicott K, Alexander KT, Penakalapati G, Anderson D, Mahtani AG, Grimes JET, Rehfuess EA, Clasen TF (2017) The impact of sanitation on infectious disease and nutritional status: a systematic review and meta-analysis. Int J Hyg Environ Health 220(6):928–949. https://doi.org/10.1016/j.ijheh.2017.05.007

Greenpeace (2019) Air pollution in Southeast Asia: challenges and solutions. http://www.greenpeace.org/southeastasia/publication/3603/air-pollution-in-southeast-asia/

Guttikunda SK, Goel R (2013) Health impacts of particulate pollution in a megacity—Delhi, India. Environ Dev 6:8–20

Hayes K, Blashki G, Wiseman J, Burke S, Reifels L (2018) Climate change and mental health: risks, impacts and priority actions. Int J Ment Heal Syst 12(1):28. https://doi.org/10.1186/s13033-018-0210-6

Heinemann AB, Ramírez Villegas JA, Stone LF, Garcia Abreu Silva AP, Henriques da Matta D, Piscoya Diaz ME (2021) The impact of El Niño Southern Oscillation on cropping season rainfall variability across Central Brazil

Hofstra N, Vermeulen LC (2016) Impacts of climate change on water quality and health. Curr Opin Environ Sustain 24:56–63

Horton R, Beaglehole R, Bonita R, Raeburn J, McKee M, Wall S (2022) Planetary health: a new paradigm for global health. Lancet 399(10324):1–2

Im ES et al (2017) Heterogeneous risk of deadly heat in the Anthropocene. Sci Adv 3(8):1

IPCC (2021) Climate change 2021: the physical science basis. Contribution of working group I to the sixth assessment report of the intergovernmental panel on climate change. Cambridge University Press

Islam SN, Winkel J (2017) Climate change and social inequality. DESA Working Paper No. 152. United Nations Department of Economic and Social Affairs

Jacobson LDSV, de Souza HS, de Castro HA, Ignotti E, Artaxo P, de Leon ACMP (2012) Association between fine particulate matter and the peak expiratory flow of schoolchildren in the Brazilian subequatorial Amazon: a panel study. Environ Res 117:27–35. https://doi.org/10.1016/j.envres.2012.05.006

Johnson JK, Smith LR, Chen Y (2023) Trends in research on health outcomes in the anthropocene: a bibliometric analysis. Glob Environ Chang 78:102–115

Jones KC, de Voogt P (1999) Persistent organic pollutants (POPs): state of the science. Environ Pollut 100(1–3):209–221

Keesing F et al (2010) Impacts of biodiversity on the emergence and transmission of infectious diseases. Nature 468(7324):647–652

Keim ME (2008) Cyclones, hurricanes, and typhoons: impacts on public health and medical systems. Am J Disaster Med 3(3):159–167

Kilpatrick AM, Randolph SE (2012) Drivers, dynamics, and control of emerging vector-borne zoonotic diseases. Lancet 380(9857):1946–1955

Kjellstrom T, Briggs D, Freyberg C, Lemke B, Otto M, Hyatt O (2016) Heat, human performance, and occupational health: a key issue for the assessment of global climate change impacts in Southeast Asia. Annu Rev Publ Health 37:97–112

Kossin JP et al (2020) Global increase in major tropical cyclone exceedance probability over the Anthropocene. Proc Natl Acad Sci 117(22):11975–11980

Lafferty KD (2009) The ecology of climate change and infectious diseases. Ecology 90(4):888–900. https://doi.org/10.1890/08-0079.1

Lal R (2015) Restoring soil quality to mitigate soil degradation. Sustainability 7(5):5875–5895

Landrigan PJ, Fuller R, Acosta NJ, Adeyi O, Arnold R, Basu N, Baldé AB, Bertollini R, Bose-O'Reilly S, Boufford JI, Breysse PN, Chiles T, Mahidol C, Coll-Seck AM, Cropper ML, Fobil J, Fuster V, Greenstone M, Haines A, Zhong M (2018) The lancet commission on pollution and health. Lancet 391(10119):462–512

Lelieveld J, Evans JS, Fnais M, Giannadaki D, Pozzer A (2015) The contribution of outdoor air pollution sources to premature mortality on a global scale. Nature 525(7569):367–371. https://doi.org/10.1038/nature15371

Lewis SL, Maslin MA (2015) Defining the Anthropocene. Nature 519(7542):171–180

Lowe R, Stewart-Ibarra AM, Petrova D, García-Díez M, Borbor-Cordova MJ, Mejía R, Regato M, Rodó X (2017) Climate services for health: Predicting the evolution of the 2016 dengue season in Machala, Ecuador. Lancet Planet Health 1(4) https://doi.org/10.1016/S2542-5196(17)30064-5

Luber G, McGeehin M (2008) Climate change and extreme heat events. Am J Prev Med 35(5):429–435. https://doi.org/10.1016/j.amepre.2008.08.021

Mansoor S, Farooq I, Kachroo MM, Mahmoud AED, Fawzy M, Popescu SM, Dominici L, Mazahery H, Ahmad P (2022) Elevation in wildfire frequencies with respect to the climate change. J Environ Manag 301:113769. https://doi.org/10.1016/j.jenvman.2021.113769

Manyeh AK, Akpakli DE, Kukula V, Ekey RA, Narh-Bana S, Adjei A, Gyapong M (2020) Socio-demographic determinants of skilled birth attendant at delivery in rural southern Ghana. BMC Res Notes 13(1):1–5

Martelli CMT, Siqueira JB, Parente MPPD, Zara ALDSA, Oliveira CS, Braga C, Pimenta FG, Cortes F, Lopez JG, Bahia LR, Mendes MCO, da Rosa MQM, de Siqueira Filha NT, Constenla D, de Souza WV (2015) Economic impact of dengue: multicenter study across four Brazilian regions. PLoS Neglected Trop Dis 9(9) https://doi.org/10.1371/journal.pntd.0004042

Matson PA et al (1997) Agricultural intensification and ecosystem properties. Science 277(5325):504–509

McMichael AJ, Lindgren E, Kovats S, Nicholls N (2012) Climate change and human health: present and future risks. Lancet 367(9513):859–869

Milly PCD et al (2005) Global pattern of trends in streamflow and water availability in a changing climate. Nature 438(7066):347–350

Monse B et al (2013) The Fit for School health outcome study—A longitudinal survey to assess health impacts of an integrated school health programme in the Philippines. BMC Publ Health 13(1):1199

Montgomery MR et al (2003) Urban population growth. Popul Bull 58(4):1–44

Mora C, Dousset B, Caldwell IR, Powell FE, Geronimo RC, Bielecki CR, Counsell CWW, Dietrich BS, Johnston ET, Louis LV, Lucas MP, McKenzie MM, Shea AG, Tseng H, Giambelluca TW, Leon LR, Hawkins E, Trauernicht C (2017) Global risk of deadly heat. Nat Clim Change 7(7):501–506. https://doi.org/10.1038/nclimate3322

Mordecai EA et al (2019) The impact of temperature on vector-borne disease transmission. Nat Ecol Evol 3:377–384

Moser C, Satterthwaite D (2010) Towards pro-poor adaptation to climate change in the urban centers of lowand middle-income countries. In: Mearns R, Norton A (eds) Social dimensions of climate change: equity and vulnerability in a warming world. Washington, World Bank, DC, pp 231–258. https://openknowledge.worldbank.org/handle/10986/2689

Mutheneni SR, Morse AP, Caminade C, Upadhyayula SM (2017) Dengue burden in India: recent trends and importance of climatic parameters. Emerg Microbes Infect 6(1):1–10. https://doi.org/10.1038/emi.2017.57

Myers SS et al (2013) Human health impacts of ecosystem alteration. Proc Natl Acad Sci 110(47):18753–18760

Naidu R et al (2013) Bioavailability: the underlying basis for risk-based land management. Environ Sci Pollut Res 20:6798–6810

Nawaz MO, Henze DK (2020) Premature deaths in Brazil associated with long-term exposure to PM2.5 from Amazon fires between 2016 and 2019. GeoHealth 4:8. https://doi.org/10.1029/2020GH000268

Neumayer E, Plümper T (2007) The gendered nature of natural disasters: the impact of catastrophic events on the gender gap in life expectancy, 1981–2002. Ann Assoc Am Geogr 97(3):551–566

Nguyen KA, Liou YA (2019) Global mapping of eco-environmental vulnerability from human and nature disturbances. Sci Total Environ 664:995–1004. https://doi.org/10.1016/j.scitotenv.2019.01.407

NOAA (2024) Trends in atmospheric carbon dioxide. Retrieved from [NOAA website]

Ottersen OP et al (2014) The political origins of health inequity: prospects for change. Lancet 383(9917):630–667

Palaniyandi M (2019) Socio-economic, and environmental determinants of dengue and chikungunya transmission: GIS for epidemic surveillance and control: a systematic review. Int J Sci Res 8(2):4–9

Parham PE et al (2015) The role of climate in the global distribution and intensity of vector-borne diseases. Annu Rev Entomol 60:141–161

Pattanittum P, Muangpaisan W, Neamsuwan T, Tantiritthisak T, Maneerat S, Laopaiboon M (2019) The effects of flooding on maternal health service utilization in southern Thailand. Int J Women's Health 11:243–251

Patz JA, Frumkin H (eds) (2016) Climate change and public health. Oxford University Press

Patz JA et al (2000) Effects of environmental change on emerging parasitic diseases. Int J Parasitol 30(12–13):1395–1405

Patz JA, Frumkin H, Holloway T, Vimont DJ, Haines A (2014) Climate change: challenges and opportunities for global health. JAMA 312(15):1565–1580. https://doi.org/10.1001/jama.2014.13186

Pickett KE, Pearl M (2001) Multilevel analyses of neighbourhood socioeconomic context and health outcomes: a critical review. J Epidemiol Commun Health 55(2):111–122

Potts SG, Biesmeijer JC, Kremen C, Neumann P, Schweiger O, Kunin WE (2010) Global pollinator declines: trends, impacts, and drivers. Trends Ecol Evol 25(6):345–353

Prüss-Ustün A, Bartram J, Clasen T, Colford JM Jr, Cumming O, Curtis V, Bonjour S, Dangour AD, De France J, Fewtrell L, Freeman MC, Gordon B, Hunter PR, Johnston RB, Mathers C, Mäusezahl D, Medlicott K, Neira M, Stocks M, Cairncross S (2014) Burden of disease from inadequate water, sanitation and hygiene in low—and middle-income settings: a retrospective analysis of data from 145 countries. Tropical Med Int Health 19(8):894–905. https://doi.org/10.1111/tmi.12329

Prüss-Ustün A et al (2019) Burden of disease from inadequate water, sanitation and hygiene for selected adverse health outcomes: an updated analysis with a focus on low- and middle-income countries. Int J Hyg Environ Health 222(5):765–777

Prüss-Üstün A, Wolf J, Corvalán C, Bos R, Neira M (2016) Preventing disease through healthy environments: a global assessment of the burden of disease from environmental risks. World Health Organ

Rataj E et al (2016) The impact of climate change on natural disasters and their public health outcomes. Healthcare 4(3):42

Rocque RJ, Beaudoin C, Ndjaboue R, Cameron L, Poirier-Bergeron L, Poulin-Rheault RA, Fallon C, Tricco AC, Witteman HO (2021) Health effects of climate change: an overview of systematic reviews. BMJ Open 11(6):1. https://doi.org/10.1136/bmjopen-2020-046333

Romanelli C et al (2015) Connecting global priorities: biodiversity and human health: a state of knowledge review. World Health Organ

Rosário Filho NA, Urrutia-Pereira M, D'Amato G, Cecchi L, Ansotegui IJ, Galán C, Peden DB, D'Amato M, Peden DB (2021) Air pollution and indoor settings. World Allergy Organ J 14(1):100499. https://doi.org/10.1016/j.waojou.2020.100499

Ryan SJ et al (2019) Global expansion and redistribution of Aedes-borne virus transmission risk with climate change. PLoS Neglected Trop Dis 13:1

Sarun S, Ghermandi A, Sheela AM, Justus J, Vineetha P (2018) Climate change vulnerability in a tropical region based on environmental and socio-economic factors. Environ Monit Assess 190:1–17. https://doi.org/10.1007/s10661-018-6750-z

Schlosberg D, Collins LB (2014) From environmental to climate justice: climate change and the discourse of environmental justice. Wiley Interdisc Rev Climate Change 5(3):359–374

Schweikert A, Chinowsky P, Espinet X, Tarbert M (2014) Climate change and infrastructure impacts: comparing the impact on roads in ten countries through 2100. Proc Eng 78:306–316. https://doi.org/10.1016/j.proeng.2014.07.072

Semenza JC, Paz S (2021) Climate change and infectious disease in Europe: impact, projection and adaptation. Lancet Reg Health-Eur 9:100230. https://doi.org/10.1016/j.lanepe.2021.100230

Serdeczny O, Adams S, Baarsch F, Coumou D, Robinson A, Hare W, Schaeffer M, Perrette M, Reinhardt J (2017) Climate change impacts in Sub-Saharan Africa: from physical changes to their social repercussions. Reg Environ Change 17:1585–1600. https://doi.org/10.1007/s10113-015-0910-2

Sherpa AM et al (2014) Climate change, water quality and water-related challenges in the Philippines. Environ Asia 7(2):42–55

Shivanna KR (2022) Climate change and its impact on biodiversity and human welfare. Proc Indian Natl Sci Acad 88(2):160–171. https://doi.org/10.1007/s43538-022-00075-4

Silva RA, West JJ, Zhang Y, Anenberg SC, Lamarque J-F, Shindell DT, Collins WJ, Dalsoren S, Faluvegi G, Folberth G, Horowitz LW, Nagashima T, Naik V, Rumbold S, Skeie R, Sudo K, Takemura T, Bergmann D, Cameron-Smith P, Zeng G (2013) Global premature mortality due to anthropogenic outdoor air pollution and the contribution of past climate change. Environ Res Lett 8(3):034005. https://doi.org/10.1088/1748-9326/8/3/034005

Stanaway JD, Shepard DS, Undurraga EA, Halasa YA, Coffeng LE, Brady OJ, Hay SI, Bedi N, Bensenor IM, Castañeda-Orjuela CA, Chuang T-W, Gibney KB, Memish ZA, Rafay A, Ukwaja KN, Yonemoto N, Murray CJL (2016) The global burden of dengue: an analysis from the global burden of disease study 2013. Lancet Infect Dis 16(6):712–723. https://doi.org/10.1016/S1473-3099(16)00026-8

Steckling N et al (2017) Global burden of disease of mercury used in artisanal small-scale gold mining. Ann Glob Health 83(2):234–247

Stocking MA (2003) Tropical soils and food security: the next 50 years. Science 302(5649):1356–1359

Sun Z, Zhu D (2019) Exposure to outdoor air pollution and its human-related health outcomes: an evidence gap map. BMJ Open 9:12. https://doi.org/10.1136/bmjopen-2019-031312

Tamerius JD, Perzanowski MS, Acosta LM, Jacobson JS, Goldstein IF, Quinn JW, Rundle AG, Shaman J (2013) Socioeconomic and outdoor meteorological determinants of indoor temperature and humidity in New York City dwellings. Weather, Climate Soc 5(2):168–179. https://doi.org/10.1175/WCAS-D-12-00030.1

Tawatsupa B, Lim LL, Kjellstrom T, Seubsman SA, Sleigh A (2018) Association between occupational heat stress and kidney disease among 37,816 workers in the Thai Cohort Study (TCS). J Epidemiol 22(3):251–260

Tjaden NB, Suk JE, Fischer D, Thomas SM, Beierkuhnlein C, Semenza JC (2017) Modelling the effects of global climate change on chikungunya transmission in the 21st century. Sci Rep 7:3813. https://doi.org/10.1038/s41598-017-03566-3

Torres JR, Castro J (2007) The health and economic impact of dengue in Latin America. Cad Saude Publica. https://doi.org/10.1590/S0102-311X2007001300004

UN-Habitat (2020) World cities report 2020: the value of sustainable urbanization. United Nations Human Settlements Programme

Vedachalam S, MacDonald LH, Shiferaw S, Seme A, Schwab KJ (2017) PMA2020 investigators underreporting of high-risk water and sanitation practices undermines progress on global targets. PLoS ONE 12:5. https://doi.org/10.1371/journal.pone.0176272

Vicedo-Cabrera AM, Scovronick N, Sera F, Royé D, Schneider R, Tobias A, Astrom C, Guo Y, Honda Y, Hondula DM, Abrutzky R, Tong S, Coelho MSZS, Saldiva PHN, Lavigne E, Correa PM, Ortega NV, Kan H, Osorio S, Gasparrini A (2021) The burden of heat-related mortality attributable to recent human-induced climate change. Nat Clim Chang 11:492–500. https://doi.org/10.1038/s41558-021-01058-x

Watson JT, Gayer M, Connolly MA (2007) Epidemics after natural disasters. Emerg Infect Dis 13(1):1–5

Wesseling C, Glaser J, Rodríguez-Guzmán J, Weiss I, Lucas R, Peraza S, da Silva AS, Hansson E, Johnson RJ, Hogstedt C, Wegman DH, Jakobsson K (2020) Chronic kidney disease of non-traditional origin in Mesoamerica: a disease primarily driven by occupational heat stress. Rev Panam de Salud Públ 44

WHO and UNICEF (2021) Progress on household drinking water, sanitation and hygiene 2000-2020: five years into the SDGs. WHO/UNICEF joint monitoring programme for water supply, sanitation and hygiene

WHO (2023) Gender, climate change and health. World Health Organization

Wilson DC, Velis C, Cheeseman C (2006) Role of informal sector recycling in waste management in developing countries. Habitat Int 30(4):797–808

WMO (2024) State of the global climate 2023. World Meteorological Organization

World Bank (2016) The cost of air pollution: strengthening the economic case for action in Southeast Asia and South Asia. World Bank Group. https://openknowledge.worldbank.org/handle/10986/25013

World Bank (2021) The global health cost of ambient PM2.5 air pollution. World Bank

World Health Organization (WHO) (2006) Constitution of the world health organization. Switzerland, Geneva

World Health Organization (WHO) (2023) Climate change and health. Retrieved from WHO

Zareba M, Weglinska E, Danek T (2024) Air pollution seasons in urban moderate climate areas through big data analytics. Sci Rep 14(1):3058. https://doi.org/10.1038/s41598-024-53835-1

Zhang Q, Sun K, Chinazzi M, Pastore y Piontti A, Dean NE, Rojas DP, Merler S, Mistry D, Poletti P, Rossi L, Bray M, Halloran ME, Longini IM, Vespignani A (2024) Spread of Zika virus in the Americas. Proc Nat Acad Sci 124(6):e2010643117. https://doi.org/10.1073/pnas.2010643117

Chapter 6
Meteorological Conditions Associated with Dengue Outbreaks: A Comprehensive Review

Poornima Suryanath Singh

Abstract Dengue is a disease of the tropics and subtropics. It is a vector-borne disease that is transmitted by mosquitoes—*Aedes aegypti and Aedes albopictus.* Multiple factors contribute to dengue characteristics, of these meteorological variables such as temperature, humidity, rainfall, wind speed, sunlight duration, etc., have an important role in the overall transmission dynamics of dengue. Furthermore, the intensification of climate change has made use of predictive modeling methods imperative in forecasting dengue. PRISMA compliant systematic review search of SCOPUS databases up to December 2024 using pre-specified keywords defining dengue, climate, and modeling strategies was done. Eligibility was screened for candidate studies, data were extracted, and the models were classified. Research findings indicate that Generalized Linear Models (GLMs), Distributed Lag Nonlinear Models (DLNMs), Autoregressive Integrated Moving Average (ARIMA), and machine learning models including random forest and neural networks have been widely used for dengue outbreak forecasting. The key meterological predictors of dengue incidence were- temperature values between 25 and 30 °C, moderate rainfall of 4–8 weeks lag, and humidity values above 60%. Further, although machine learning approaches reflect enhanced prediction potential over classical models, they also lack interpretability. The study emphasizes the need for hybrid modeling approaches integrating climate data, remote sensing, and socio-environmental factors for successful dengue forecasting. Climate-driven early warning systems (EWS) must be incorporated into public health policy to minimize dengue outbreaks.

Keywords Dengue · ARIMA · Distributed lag models · Early warning systems · Public health policy · Vector-borne disease surveillance

Present Address:
P. S. Singh (✉)
Amity University, Noida, India
e-mail: poornimasingh1@gmail.com

A. Roy et al. (eds.), *Environment and Public Health*, Sustainability Solutions,
https://doi.org/10.1007/978-3-031-99770-9_6

6.1 Introduction

Dengue fever is one of the significant public health concerns in tropical and subtropical regions of the world. Studies in these regions depict that the meteorological factors foster the breeding of the primary vectors for the disease—*Aedes aegypti and Aedes albopictus*. According to the approximates of the World Health Organization (WHO), dengue cases have risen tenfold in the last 20 years. It is estimated that around 390 million cases occur each year -from which 96-million are symptomatic illnesses (Costa et al. 2022). The Institute for Health Metrics and Evaluation's Global Burden of Disease (GBD) study estimated that in the year 2019 alone there were 40,000 more deaths globally due to dengue where the majority were found in South and Southeast Asia (Omadlao et al. 2022). The approximation of the World Bank for the economic cost of dengue is $8.9 billion each year to healthcare systems, workforce productivity, and national economies (Jayaraj et al. 2019). Given the rapid urbanization, uncontrolled infrastructure expansion, and climate change, dengue outbreaks are likely to be more severe and frequent, and thus there is a need for the development of strong predictive models with meteorological variables to support early warning systems and public health preparedness.

The dynamic relationship between climatic variability and dengue transmission has been well-established with temperature, precipitation, humidity, and wind speed playing important roles in determining mosquito breeding, survival, and virus replication rates (Kakarla et al. 2019; Tuladhar et al. 2019). Global warming, added to non-synchronous precipitation, has redistributed spatial and temporal dynamics of dengue, and the disease is now breaking out in new and previously non-endemic areas such as parts of North America and Europe (Liu et al. 2019; Chumpu et al. 2019). Vector-borne diseases such as dengue have been identified by the Intergovernmental Panel on Climate Change (IPCC) as being among the most climate-sensitive health risks, and nations with climate-vulnerable profiles need to take urgent action to adapt (Chen and Chen 2024). With such stark trends on the horizon, most national public health surveillance systems do not have built-in climate-driven forecasting systems, a critical prevention and outbreak control gap (Noureldin and Shaffer 2019; Toai et al. 2019).

A range of statistical and computational models have been applied to predict dengue outbreaks, with varying degrees of accuracy and applicability. Generalized Linear Models (GLMs), Distributed Lag Non-linear Models (DLNMs), and Poisson regression models have been extensively applied to investigate the relationship between climatic variables and dengue incidence (Acharya et al. 2018; Adde et al. 2016). These models however seem to be incapable of capturing nonlinear associations and lagged effects and thus have limited predictive capability in real-world epidemiological settings. The time series models, i.e., Autoregressive Integrated Moving Average (ARIMA) and Seasonal ARIMA (SARIMA), have been in the past successful in forecasting short-term dengue trends. However, they rely on historical case data and are challenged by long-term predictions in the presence of climate variability (Ali et al. 2016; Alkhaldy et al. 2017). More recently, machine learning models such as random forest, support vector machines (SVMs), artificial neural networks (ANNs), and deep learning models have been useful in dengue prediction. These models are capable of handling large datasets, detecting complex patterns, and improving forecasting accuracy, but their lack of interpretability and high computational requirements are major limitations for real-time use in public health (Yaacob et al. 2020; Banu et al. 2014).

Despite the large body of literature on climate-based dengue forecasting, there are some key gaps. First, most of the studies are limited to small geographical locations, thus preventing generalization of results across different climatic zones (Banu et al. 2015). Second, while many models control for temperature and rainfall, fewer account for the cumulative effect of a suite of climatic variables, including wind speed, sky cover, bioclimatic zones, and air pressure, which have the potential to significantly influence mosquito population dynamics (Baquero et al. 2018; Carvajal et al. 2018). Third, there is a lack of comparative studies on how traditional regression models, time-series models, and machine learning models compare under a consistent systematic framework (Chang et al. 2018; Chuang et al. 2017). Fourth, public health policymaking has limited incorporation of climate-based dengue forecasting, and most studies fail to consider policy implications or suggest effective early warning systems (EWS) for outbreak management (Ouattara et al. 2022; Correia Filho et al. 2017).

This chapter exercise attempts to fill current gaps by conducting a systematic review of dengue forecasting models, evaluating their strengths and weaknesses, as well as their applicability under different climatic regimes. Following PRISMA protocols, the research systematically reviews the literature through December 2024, determining the best meteorological predictors as well as the validation of performances of different modeling approaches. The current chapter also aligns to the United Nations Sustainable Development Goals (SDGs), SDG 3 (Good Health and Well-Being) and SDG 13 (Climate Action), through promoting data-driven approaches to disease prevention and climate resilience (Ehelepola et al. 2015; Talagala 2015). It also supports the One Health approach, recognizing the interdependency of human, animal, and environmental health in managing vector-borne diseases (Horta et al. 2014; Iguchi et al. 2018). Through determining best practices in modeling, this chapter provides actionable knowledge for policymakers, epidemiologists, and public health agencies, thus enhancing the use of climate-based disease forecasting in national and regional dengue surveillance programs.

6.2 Materials and Methods

This systematic review was conducted according to the PRISMA guideline, enabling it to have a systematic and clear literature identification process, screening, eligibility assessment, and inclusion selection. The research sought to evaluate a range of predictive models utilized in forecasting dengue taking into consideration meteorological conditions such as temperature, humidity, rainfall, wind speed, atmospheric pressure, and presence of extreme weather events. The research used a systematic and transparent approach by carrying out comprehensive database searching, explicitly defined inclusion and exclusion criteria, systematic extraction of data, and thematic synthesis of the differing predictive modeling practices. The study selection, as evidenced by the PRISMA flow diagram (Fig. 6.1), ensured the identification of high-quality studies relating to climate-based dengue prediction.

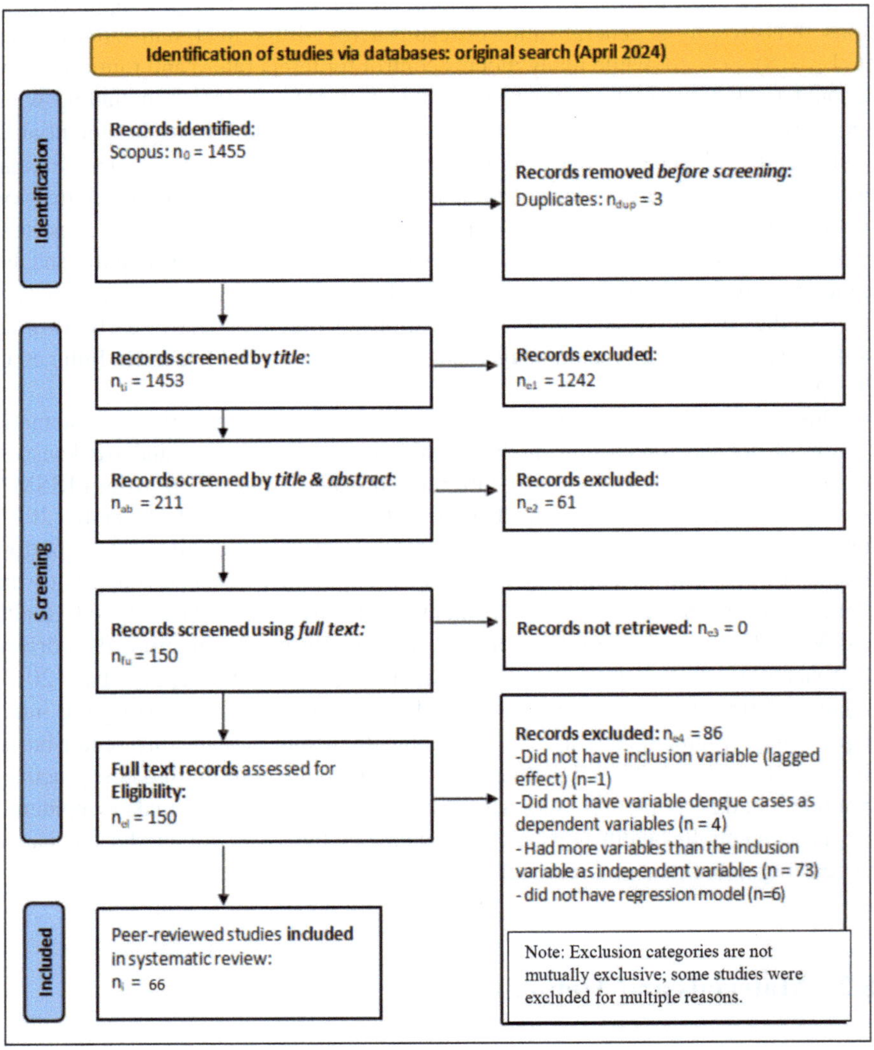

Fig. 6.1 PRISMA flow diagram for study selection

6.2.1 Search Strategy

A thorough step-by-step search approach was utilized in databases—Scopus. The search was conducted on studies through April 2024, focusing on studies that had investigated the association between meteorological parameters and dengue incidence using various statistical and machine learning modeling approaches. To maintain accuracy and inclusivity, Boolean operators (AND/OR) were used to construct a search query involving three general categories of keywords: dengue-related keywords ("Aedes" OR "dengue" OR "dengue fever"), meteorological parameters ("climate" OR "weather" OR "temperature" OR "rainfall" OR "humidity" OR "wind speed" OR "air pressure" OR "seasonal variation"), and modeling approaches ("modeling" OR "prediction" OR "forecasting" OR "statistical model" OR "machine learning" OR "time-series analysis").

Only peer-reviewed journal publications and conference papers were included to guarantee the quality of studies used. Gray literature, non-peer-reviewed articles, and government publications were not considered unless they provided important information with respect to prediction methodologies of dengue. Only english publications were used in order to achieve consistency in interpretation of data. To maximize comprehensiveness, manual screening of reference lists (backward citation tracking) and forward citation tracking (which involves identification of new research that cites significant studies) were performed to uncover additional relevant papers.

6.2.2 Exclusion and Inclusion Criteria

To strengthen the process of selection and methodological quality, pre-defined inclusion and exclusion criteria were created for all the identified studies. Inclusion criteria were formulated to select only high-quality and relevant studies for the systematic synthesis of models that forecast dengue. Studies were to be included if they examined the relation between meteorological variables and the incidence of dengue, used quantitative predictive modeling methods, had statistical models, machine learning algorithms, or time-series analysis, and used meteorological variables such as temperature, rainfall, humidity, wind speed, or air pressure. Papers were excluded from examination if they did not contain meteorological variables as predictors, were only vector ecology without predictive modeling aspects, or only based on qualitative methods without statistical analysis. These exclusion criteria rendered the final selection process effective, and it comprised high-quality studies that made significant contributions to a wide thematic synthesis of dengue prediction models.

6.2.3 PRISMA Study Selection Process (PRISMA Flow Diagram)

The process of selecting studies was done in four phases: identification, screening, assessment of eligibility, and inclusion, according to the PRISMA guidelines. The initial search yielded 1,455 records in Scopus. After excluding 3 duplicate records, there remained 1,453 unique articles to be screened. The titles of these studies were screened, resulting in exclusion of 1,242 articles not relevant to dengue prediction modeling. A second screening, titles and abstracts, yielded 211 studies, of which 61 more were excluded because they did not meet the inclusion criteria (e.g., they did not include meteorological variables or predictive models).

A systematic review of 150 studies was conducted to assess their methodological quality, model validation methods, and statistical adequacy. In this step, 86 studies were eliminated based on exclusion criteria: lack of lagged effects as a variable ($n = 1$), dengue cases as the dependent variable ($n = 4$), inclusion of other independent variables in addition to meteorological conditions ($n = 73$), and lack of application of regression modeling methods ($n = 6$). Some studies met more than one exclusion criterion; therefore, the categories are not mutually exclusive. After the application of all the inclusion and exclusion criteria, 66 studies were considered in the final systematic review because they provided qualitative evidence for the contribution of meteorological conditions to dengue outbreaks and compared predictive modeling methods.

6.2.4 Data Synthesis and Extraction

An in-depth data extraction form was designed to enable systematic and consistent data collection across studies. Data collected from each study included study attributes (author(s), year of publication, geographic location, sources of data), types of models (statistical models, time-series models, machine learning models, hybrid models), meteorological predictors (temperature, precipitation, humidity, wind speed, atmospheric pressure, sunshine hours, extreme weather events), and policy implications (application of models in actual dengue surveillance systems).

Due to heterogeneity in study designs, datasets, and modeling strategies, meta-analysis was not feasible. Qualitative thematic analysis was thus conducted by categorizing studies into four main predictive modeling strategies:

Statistical models such as Generalized Linear Models (GLMs), Poisson regression, negative binomial models, and Distributed Lag Non-linear Models (DLNMs) fell into this category. The models allow the meterological variables to be measured in terms of dengue incidence.

Time-Series Models: Autoregressive Integrated Moving Average (ARIMA), Seasonal ARIMA (SARIMA), and Bayesian Time-Series Models are included in this category. They are the most popular methods of short-term dengue forecasting.

Machine Learning Methods: This category encompasses work involving random forest, support vector machines (SVM), artificial neural networks (ANN), and deep learning methods like long short-term memory (LSTM) and convolutional neural networks (CNN). Even though they enhance the prediction, the models are difficult to interpret.

Hybrid and Spatiotemporal Models: This type comprised GIS-based models, Bayesian Spatiotemporal Frameworks, and climatic, socio-environmental, and epidemiological predictor-based models. These models provide comprehensive dengue prediction frameworks.

All model categories were compared in terms of predictive performance, strengths, weaknesses, and practical utility in dengue forecasting and the development of public health policies. The following sections present the most important findings concerning the role of meteorological variables in the dengue transmission, comparing predictive modeling approaches in different geographical locations.

6.3 Results

This section presents a comprehensive summary of the conclusions obtained from the 66 studies about the relationship between meteorological variables and dengue incidence and the performance of different predictive modeling techniques. The conclusions highlight the significance of temperature, rainfall, humidity, wind speed, sunshine duration, and extreme weather conditions in the transmission of dengue.

The temporal trend of the 66 studies published between 2014 and 2024 indicates peak research activity in 2018 (13 studies) and 2016 (10 studies), and the lowest numbers were in 2023 (2 studies) and 2024 (3 studies). Research was consistent between 2015 and 2022 with moderate lows in 2014, 2019 (7 studies each), and 2020 (4 studies), possibly because of the pandemic. Geographically, all 66 studies were single-country studies out of 20 countries, of which 55 were in tropical countries, 1 in a temperate country, and 8 in both. China (8), Brazil (7), Malaysia (7), and Vietnam (6) were the most represented. 26 environmental variables were examined, of which temperature (66 studies, 34.38%), humidity (55, 28.65%), and rainfall (38, 19.79%) were the most frequent. Modeling strategies varied, with GLMs (32), GAMs (24), and time-series models (19) being the most frequent, supplemented by emerging machine learning (10) and Bayesian models (4), indicating a shift toward advanced computational methods. Most studies used multiple models, indicating a multi-dimensional analytical strategy.

6.3.1 *Meteorological Parameters' Impact on Dengue Transmission*

6.3.1.1 Temperature

Temperature is one of the key meteorological factors in dengue transmission dynamics, influencing survival of mosquitoes, biting rates, and efficiency of virus replication. Several studies are in agreement that the best dengue transmission is at 25–30 °C since temperature is conducive for the extrinsic incubation period (EIP) of the dengue virus (Horta et al. 2014; Carvajal et al. 2018). In Brazil, a study revealed that the highest dengue cases were recorded when average temperatures ranged between 25 and 28 °C with a statistically significant correlation coefficient ($R^2 = 0.72$, $p < 0.05$) (Horta et al. 2014). Likewise, in the Philippines, the highest dengue risk was at 23–28 °C, and this was evidenced by Generalized Additive Models (GAMs) whose AIC values were indicative of good model fit (AIC = 132.5) (Carvajal et al. 2018).

However, temperatures above 35 °C have adverse effects on mosquito longevity, hence limiting transmission dynamics. In Vietnam, a study showed that when temperatures were above 35 °C, dengue fever risk reduced by 18%, as evident in a negative Poisson regression coefficient of − 0.26 ($p < 0.01$), suggesting thermal stress may suppress the activity of mosquitoes (Thiruchelvam et al. 2017). In addition, minimum temperature matters a lot; studies in Malaysia and Thailand revealed that higher minimum temperatures above 24 °C significantly increased dengue cases, as mosquitoes became active during nights (Toai et al. 2019; Chumpu et al. 2019).

The effect of temperature does not occur immediately; Indian and Thai research has noted a 4–8 week lag period, which has made the use of predictive models indispensable for accounting for such lagged effects (Kakarla et al. 2019; Chumpu et al. 2019). Distributed Lag Non-linear Models (DLNMs) have particularly been found appropriate to model such a relationship, and Chinese and Nepalese research has reported good model fit (RMSE = 3.52, $R^2 = 0.68$) if temperature lags are accounted for (Li et al. 2017; Tuladhar et al. 2019).

6.3.1.2 Precipitation

Rainfall has a double-edged role in the transmission of dengue in that it provides breeding grounds for the mosquito but can also in severe cases wash out larvae. Research in India, Bangladesh, and Malaysia shows that moderate rain (50–150 mm/month) increases dengue incidence considerably, with heavy rain (>200 mm/month) decreasing the mosquito population by sweeping away eggs and larvae (Kakarla et al. 2019; Miah et al. 2024; Yaacob et al. 2020).

Time-lagged effects of rains are often reported. Chinese and Sri Lankan studies reported a 6–10 weeks peak after monsoon periods of dengue cases, with statistically significant regression coefficients ($\beta = 0.63$, $p < 0.001$) of climate-dengue models

(Li et al. 2017; Talagala 2015). Nepal and Thailand, however, have a short lag period of 4–6 weeks and report regional variations in mosquito breeding patterns (Tuladhar et al. 2019; Chumpu et al. 2019).

Random forest and LSTM models have performed best in identifying nonlinear correlations between dengue and rainfall. In Indonesia, an LSTM model incorporating rainfall was 84% accurate in prediction (RMSE = 2.71) when trained using multi-year climate data (Nguyen et al. 2022).

6.3.1.3 Humidity

Humidity has implications for efficiency in virus transmission and survival in mosquitoes. Experiments in Saudi Arabia and Vietnam defined relative humidity over 60% to play an important role in dengue transmission based on statistically high, significant R^2 values (0.71, $p < 0.05$) of multivariate regression analysis (Alkhaldy et al. 2017; Thiruchelvam et al. 2017).

Nonetheless, high humidity has been reported to decrease the activity of mosquitoes. Studies conducted in Malaysia and Thailand indicate that relative humidity levels over 90% can interfere with mosquito behavior and cause dengue cases to decrease (99, 36). In Brazil, using SARIMA models to analyze dengue incidence reported that incidence was greatest when humidity was between 70 and 85%, and the RMSE was 3.12 (Horta et al. 2014).

Machine learning modeling, such as gradient boosting and support vector machines (SVMs), has also seen improved predictive capabilities with the integration of humidity factors. An SVM model on a climate-based framework constructed in Taiwan had an extra 25% improvement in accuracy of dengue predictions (MAE = 1.92, $R^2 = 0.78$) when considering humidity factors (Chuang et al. 2017).

6.3.1.4 Sunshine Duration and Wind Speed

Wind speed affects the dispersal of mosquitoes, and research in Saudi Arabia and Malaysia indicated that increased wind speed (>10 m/s) decreased dengue incidence by 15%, with strongly significant negative regression coefficients (− 0.31, $p < 0.05$) in regression models (Alkhaldy et al. 2017; Yaacob et al. 2020).

Sunshine duration has had lesser focus but an increasing role to play. One study reported an association in China of longer sunshine duration and a decrease in the transmission of dengue and, for Sri Lanka and Brazil, an increased number of sunshine days leading up to the monsoon periods of higher cases (Li et al. 2017; Talagala 2015; Horta et al. 2014). SARIMA time-series models considering sunshine duration displayed stronger forecasting validity (RMSE = 2.85, AIC = 118.7) (Kakarla et al. 2019; Alkhaldy et al. 2017).

6.3.1.5 Lagged Effects of Meteorological Variables on Dengue Cases

Dengue outbreaks often exhibit delayed responses to meteorological fluctuations, with temperature, rainfall, and humidity displaying significant lag periods. Temperature impacts dengue incidence with a lag of 2 weeks to 3 months, depending on local climatic conditions, as observed in Taiwan, Nepal, and Sudan (You et al. 2022; Tuladhar et al. 2019; Noureldin et al. 2019). Rainfall exhibits a 4–8 week lag, with mosquito population surges following periods of precipitation, as documented in Sri Lanka, India, and the Philippines (Liyanage et al. 2016; Kakarla et al. 2019; Iguchi et al. 2018).

Humidity also follows a delayed effect, with relative humidity at a 1–2 month lag driving increased transmission in Saudi Arabia, Vietnam, and the Philippines (Alkhaldy et al. 2017; Phung et al. 2015a, b; Iguchi et al. 2018). Wind patterns, as seen in Thailand and China, influence case clustering within 2–4 weeks, affecting local outbreak severity (Liu et al. 2019; Chumpu et al. 2019). Tropical cyclones and droughts have extended lag effects, ranging from 1 to 3 months, as reported in studies from China and Barbados (Li et al. 2021; Lowe et al. 2018).

These findings reinforce the necessity of time-sensitive predictive models that incorporate meteorological lag structures, allowing public health authorities to implement preemptive vector control measures before peak transmission periods occur (Chuang et al. 2017; Talagala 2015).

6.3.2 Dengue Forecasting Methodologies by Predictive Modeling

6.3.2.1 Statistical Models Generalized Linear Models (GLMs) and Poisson Regression

Poisson Regression are commonly used to determine correlations between meterological variables and dengue cases (Baquero et al. 2018; Phung et al. 2015a, b). The models are interpretable and therefore practical for public health policy-making. They require, however, linearity, which constrains them to account for sophisticated, nonlinear climatic-disease relationships. To address such limitations, researchers have used Distributed Lag Non-linear Models (DLNMs) in studies in Malaysia and Brazil, thus effectively capturing the lagged effect of rainfall and temperature (Horta et al. 2014; Carvajal et al. 2018).

6.3.2.2 Time-Series Models

ARIMA and Seasonal ARIMA (SARIMA) models have been widely applied in Thailand and India, particularly for short-term dengue outbreak forecasting (Kakarla et al. 2019; Chumpu et al. 2019). However, the traditional ARIMA models lack external meterological variables, and hence the accuracy is lost in the long-term prediction. Hybrid ARIMA models with meterological variables have been found to be more effective, such as in the cases of Bangladesh and the Philippines, where temperature and rainfall variables were used in ARIMA models (Liyanage et al. 2016; Miah et al. 2024).

6.3.2.3 Machine Learning and Deep Learning Models

Machine learning methods like random forest, SVM, and artificial neural networks (ANNs) have been more predictive than statistical models (Omadlao et al. 2022; Nguyen et al. 2022). These models can evaluate nonlinear meterology-disease relations and thus are optimally used to predict outbreaks under a changing climate. Long short-term memory (LSTM) models, one type of deep learning, have also been effective in dengue prediction. One Vietnamese research study demonstrated LSTM models could glean long-distance relationships in weather patterns, enhancing outbreak predictions (Thiruchelvam et al. 2017; Acharya et al. 2018). Despite their high accuracy, machine learning models are plagued by interpretability challenges that make it harder to integrate them into public health decision-making (Thiruchelvam et al. 2017). Hybrid models, which integrate statistical methods with machine learning methods, have been considered the best solution in predicting dengue (Baquero et al. 2018; Nguyen et al. 2022; Thiruchelvam et al. 2017).

6.3.3 Comparative Analysis and Construction of Forecasting Models

With time, dengue forecasting models have progressed from basic statistical correlations to sophisticated machine learning models. Although GLMs and Poisson regression models are still popular because of their interpretability, however they are not capable of handling nonlinear relationships. Time-series models like ARIMA and SARIMA provide better short-term forecasts but lack flexibility in the use of meteorological predictors. Advances in machine learning and deep learning algorithms have bridged such gaps and offered better accuracy in dengue epidemic prediction. The most powerful among them have been the LSTM and random forest approaches because they have the capability to measure lagged, nonlinear effects of the climate (Nguyen et al. 2022; Thiruchelvam et al. 2017; Omadlao et al. 2022; Acharya et al. 2018). The reason that they are black boxes that reduces the policy-making applicability, and there is a need for an innovation in mixed models that combine the utility of the two statistical and machine learning approaches (Baquero et al. 2018; Phung et al. 2015a, b; Miah et al. 2024).

The findings validate that dengue transmission is significantly affected by temperature, rainfall, humidity, wind speed, and sunshine duration with lagged impacts in all the areas. Hybrid modeling methods that integrate statistical and machine learning methods are the most suitable in terms of balance between interpretability and precision.

Table 6.1 provides a comparative summary of models that highlight the trade-offs between accuracy and interpretability across models, reinforcing the potential of hybrid approaches in real-world applications.

Table 6.1 Comparative evaluation of predictive models for dengue forecasting

Model type	RMSE/ MAE	R^2	AIC/ BIC	Interpretability	Strengths	Limitations
GLM/ Poisson regression	RMSE: 3.1–4.0	0.60–0.72	AIC: 120–135	High	Simple, interpretable	Limited to linear relationships
ARIMA/ SARIMA	RMSE: 2.8–3.5	0.65–0.75	AIC: 115–130	Moderate	Good for short-term prediction	Doesn't incorporate external variables easily
DLNM	RMSE: 3.0–3.6	0.68	AIC: 125–132	Moderate to high	Captures lagged non-linear effects	Complex setup, requires expert knowledge
Random forest (RF)	RMSE: 2.4–2.9	0.78–0.84	–	Low	High accuracy, handles nonlinearity	Lacks interpretability (black-box)
Support vector machine	MAE: ~ 1.9	0.62–0.70	–	Low	Good classification accuracy	Doesn't explain feature contributions well
LSTM/ ANN	RMSE: 2.6–3.0	0.80–0.87	–	Very low	Best for time-lagged patterns	Needs large data, lacks interpretability
Hybrid models	RMSE: 2.2–2.8	0.85–0.90	Varies	Medium	Combines accuracy + insights	Still emerging; interpretability varies

6.4 Discussion

6.4.1 Meteorological Factors Affecting Dengue Transmission and Public Health Impacts

This chapter establishes temperature, humidity, and rainfall as the primary predictors of dengue transmission. The observed link between dengue outbreaks and 23–28 °C temperatures highlights the need for weather-based outbreak anticipation tools in climate-sensitive regions (Carvajal et al. 2018; Horta et al. 2014). The incidence is 5–7% higher when there is a 1 °C increase, and the effects are strongest at 4–8-week lag (Kakarla et al. 2019; Phung et al. 2015a, b). High humidity may extend mosquito lifespan, suggesting that vector control should be scaled up during prolonged humid periods (Alkhaldy et al. 2017; Yu et al. 2016; Phung et al. 2015a, b). Rain promotes breeding, where moderate (50–150 mm/month) augments cases 4–8 weeks later and too much (>200 mm/month) decays vector populations (Liyanage et al. 2016; Miah et al. 2024; Chumpu et al. 2019). Cyclones and drought enhance outbreaks because they provide conditions for prolonged breeding, such as in China's Pearl River Delta and Barbados (Li et al. 2021; Lowe et al. 2018; Miah et al. 2024). These observations reiterate the inclusion of meteorological information in the early warning system (Chuang et al. 2017; Jayaraj et al. 2019).

6.4.2 Advances and Challenges in Predictive Modeling

While statistical models offer interpretability, machine learning models bring accuracy—highlighting the value of hybrid approaches for practical forecasting. Generalized Linear Models (GLMs) and Poisson regression model meteorological variable-dengue incidence relationships but require linearity (Kakarla et al. 2019; Toai et al. 2019; Phung et al. 2015a, b). Distributed Lag Non-linear Models (DLNMs) model lagged climate impacts, with absolute humidity > 60 g/m^3 doubling incidence at a 5–15-week lag (Ramachandran et al. 2016; Singh et al. 2022). Time-series models such as ARIMA and SARIMA forecast outbreaks based on past trends, with rainfall being the primary predictor in the Philippines (Iguchi et al. 2018) and temperature in China (Liu et al. 2019; Wu et al. 2018). They fail, however, with novel climate changes and exogenous variables (Carvajal et al. 2018; Phung Det al. 2015).

Machine learning algorithms like random forest, support vector machines (SVM), and artificial neural networks (ANNs) perform better than conventional approaches by detecting intricate climate-disease relationships (Carvajal et al. 2018; Iguchi et al. 2018; Liao et al. 2015). SVM models were quite good ($R^2 = 0.62$) but did not handle nonlinear interactions (Carvajal et al. 2018; Liao et al. 2015). A support vector classifier in the Philippines was 98.1% accurate in predicting outbreaks (Omadlao et al. 2022). Although accurate, machine learning models are non-interpretable, and therefore practical use is difficult (Carvajal et al. 2018; Phung et al. 2015a, b; Nguyen et al. 2022). Hybrid models that integrate statistical and ML approaches, like in Vietnam and Thailand, improve forecasting performance (Liu et al. 2019; Chang et al. 2018; Phung et al. 2015a, b; Wu et al. 2018).

6.4.3 Climate Change and the Increasing Dengue Threat

Climate change is shifting dengue transmission into new geographic zones, requiring early adoption of climate-integrated surveillance in temperate countries (Banu et al. 2014; Franchito et al. 2021). Warmth and abrupt weather changes have allowed mosquitoes to colonize hitherto non-suitable localities, such as southern Europe and the USA (Banu et al. 2014; Franchito et al. 2021). Increased minimum temperature in China has facilitated local transmission (Liu et al. 2019). Severe weather events, for example, tropical cyclones, have extended population life spans of mosquitoes from flood-affected areas, protracting epidemics (Li et al. 2021). These studies underscore the requirement for climate-relevant dengue forecasting models that incorporate long-term meteorological forecasts into disease control programs.

6.4.4 Policy Recommendations

In context to the growing impact of climatic conditions on the dengue transmission patterns, it is necessary to formulate and implement policies that facilitate climate-based dengue forecasting and outbreak response.

The major goal here is to develop climate-integrated Dengue Early Warning Systems (EWS) based on real-time meteorological observations to predict outbreak risks. Governments should operationalize early warning systems using local weather data and AI-powered risk models validated in their regions. Inter-agency coordination among the health ministries, meteorological agencies, and research institutions will be essential to the improvement of these systems. Malaysia and Thailand have already begun incorporating climate-based forecasting into national dengue control programs, but scale-up is needed (Yaacob et al. 2020; Phung et al. 2015a, b).

Prioritization of the use of hybrid predictive models in public health decision-making is warranted. While traditional statistical models provide significant epidemiological insight, machine learning methods provide more predictive accuracy. Policymakers are urged to undertake pilot studies to assess the usability of hybrid models integrating statistical and machine learning methods. Governments must also make climate, health, and vector surveillance databases interoperable to enable real-time data sharing to support outbreak forecasting (Baquero et al. 2018; Nguyen et al. 2022; Thiruchelvam et al. 2017).

One Health action across sectors is another policy intervention arena of paramount importance. As climate change, urbanization, and vector-borne disease risks are intricately linked, dengue prevention measures should cut across the health sector.

Governments need to set up inter-ministerial task forces incorporating health, environmental, and urban planning authorities for designing holistic dengue control plans. Infrastructure development in the urban areas, and climate-resilient vector control measures can reduce dengue risks substantially (Yaacob et al. 2020; Phunget al. 2015a, b).

Community-centered mobile alert systems, like those piloted in Brazil and Singapore, should be scaled in dengue-prone districts. Improving mobile-based dengue warning systems, such as those implemented in Singapore and Brazil can ensure that at-risk populations are warned of outbreaks in a timely manner. Secondly, public health authorities can invest in climate change literacy programs that educate communities about the link between rising temperatures and the risk of vector-borne diseases (Carvajal et al. 2018; Horta et al. 2014).

Lastly, dengue control strategies should be synchronized with the climate resilience and global health agendas, such as the United Nations Sustainable Development Goals (SDGs). Embedding climate-informed dengue projections in national health policy will support SDG 3 (Good Health and Well-Being) by decreasing disease burden through timely response. Further, synchronizing dengue prevention interventions with SDG 13 (Climate Action) will secure long-term policy and finance for climate-sensitive vector control interventions (Banu et al. 2014; Franchito et al. 2021).

This dialogue emphasizes the imperative to mainstream meteorological forecasting within dengue prevention policy. Although predictive models have improved, operational application depends on more robust policy frameworks prioritizing early warning systems, hybrid modeling, One Health partnerships, and public engagement. As climate change continues to reshape vector-borne disease patterns, climate-sensitive public health response will be vital to reducing dengue risk over the coming decades. Embedding explainable AI into forecasting systems can bridge the gap between predictive power and policymaker trust, enabling timely public health interventions.

6.4.5 Limitations of the Study

This chapter identifies principal meterology-based dengue forecasting model limitations. One of these is heterogeneity in study methods, which prevents cross-comparison because of differences in geographic coverage, dataset quality, and predictor variables. The performance of models is extremely sensitive to the metolology and epidemiological context of the region of interest, thereby necessitating standardized evaluation metrics.

The majority of the models are meteorologically driven, excluding socio-economic and behavioral factors such as population density, hygiene, and access to healthcare, which restricts applicability to real conditions. Additionally, studies are mostly based in Asia and Latin America, with emerging dengue-endemic areas such as Africa, Southern Europe, and North America being underrepresented. In the absence of region-specific models, warning systems for these areas are far from being sufficient.

Exclusion of gray literature, government reports, and unpublished surveillance data may also limit the understanding of actual dengue forecasting programs. Lack of standardized performance measures (e.g., R^2, MAE, RMSE, and AUC) also hinder model comparison and policy decisions.

Lastly, while machine learning models have high prediction accuracy, their "black-box" nature prevents interpretability for public health practitioners. Future studies must incorporate explainable AI methods, broaden data sources, and create combined models that consider climate, demographic, and epidemiological variables for better dengue forecasting. Very few predictive models incorporate public health system capacity metrics like medical facilities availability and vector control efforts and health behavior patterns because these elements greatly affect dengue disease outcomes. Health indicators from the system would enhance outbreak management planning if merged with meteorological variables in predictive models.

The direct relationship between weather elements and dengue incidence remains unclear because researchers have not properly addressed the possible influence of climate confounders such as land-use changes, urban heat islands and El Niño effects. Future research models need to incorporate proper controls for confounders to elevate their causal value.

Studies should apply formal risk of bias assessment tools like ROBIS or Newcastle–Ottawa Scale but this review did not follow this standard. Additional reviews should utilize these quality appraisal tools to conduct comprehensive assessments of study internal validity and generalizability.

6.4.6 Future Research

Future studies should concentrate on the development of multi-factorial dengue prediction models incorporating climatic, socio-economic, and epidemiologic determinants. Current models are mostly meteorological, relying on temperature, rainfall, and humidity as the main parameters, and ignoring key human determinants like population density, sanitation, availability of healthcare, and land use. A hybrid approach integrating climatic variables with social determinants of health would be an improved and more realistic model for predicting dengue incidence.

Another priority research area is geographic adaptation and validation of forecasting models. Most of the models that are being applied are specific to a particular; however, climate change is fueling the expansion of dengue to new regions. Hence it is urgent to create region-specific models that take into account local environmental and socio-political realities. Prospective validation of these models in newly affected regions in future studies should be aimed to ensure their effect and credibility in different climatic zones.

Although long-term climatic trends have been extensively examined, extreme events like—heat waves, floods, and droughts have the potential to drastically change the breeding/survival patterns of mosquitoes. The effect of extreme meteorological events on dengue transmission is a fairly under-studied area. For instance, heavy rainfall can decimate breeding sites which results in a temporary decrease in mosquito numbers, whereas prolonged drought can result in more water pooling within residential areas, creating new vector habitats. Future research should include these short-term climatic stressors in forecasting models to enhance the accuracy of early warning systems for dengue.

Improvements in the technology of satellite remote sensing carry significant potential for future directions toward the development of dengue prediction. Earth observation technologies (NASA's MODIS, Sentinel-2, and Copernicus Climate Change Service's satellite) derived data have the potential to monitor land surface temperature, vegetation indices, urban heat island, and bodies of standing waters—features instrumental to the mosquito's ecology. Future research should investigate whether bringing together these satellite-derived climate and land-use information can enhance the spatiotemporal risk maps for an earlier identification of high-risk hotspots of dengue.

With further expansion in the application of artificial intelligence (AI) and machine learning for dengue prediction, further research is needed to assess their practical use. While the advances in AI models have produced higher predictive values in comparison with traditional statistical models, the lack in the ease of interpretability remains a hurdle to their adoption in public health policy-making. Future studies must develop explainable AI approaches, such as SHAP (Shapley Additive Explanations) and LIME (Local Interpretable Model-Agnostic Explanations), to enhance transparency and use for policymakers.

The incorporation of real-time data assimilation in dengue forecasting systems is a new field of study. The majority of existing models employ historic datasets, and therefore their dynamic reaction to the occurrence of surprise outbreaks is impaired. Incorporation of real-time meteorological observation, mobile health-based surveillance, and social media analytics is the key to better near-future outbreak forecasting. The creation of adaptive systems, which continuously update predictions based on real-time streams of data, will significantly enhance the response of dengue early warning systems.

From an implementation perspective, future studies need to emphasize the application of predictive models in national and subnational dengue control programs. While most studies try to prioritize model development, less is understood about the successful implementation of such tools in public health settings. Studies need to attempt to determine best practices for integrating predictive analytics into decision-making systems, training health workers in model interpretation, and creating user-friendly online platforms for public health officials. Cost-effectiveness analyses need to be conducted to investigate the affordability of different modeling frameworks, particularly in low- and middle-income settings where dengue is more burdensome.

Finally, the future impact of climate adaptation policy on preventing dengue is also worth investigating. With dengue spreading into new climatic ranges, governments must implement innovative measures such as sustainable drainage infrastructure, urban heat island mitigation measures, and climate-resistant healthcare infrastructure to minimize the risks of the disease. Subsequent studies need to evaluate the effectiveness of such adaptation strategies at curbing the transmission of and the severity of outbreaks of dengue over the long run.

6.5 Conclusion

Dengue fever remains an emerging global public health issue, further fueled by climate change, urbanization, and inadequate vector control interventions. This chapter attempts to draw attention to the dramatic impact of weather variables—mainly temperature, precipitation, and humidity on dengue dynamics. Despite various predictive modeling frameworks, no single framework has been widely accepted as superior for prediction of outbreaks. Statistical models like Generalized Linear Models (GLMs) and Poisson Regression are highly interpretable but are not robust in estimating complex, nonlinear meterology-disease relationships. Time-series models like ARIMA and SARIMA have shown great success in short-term prediction but lack the ability to include external meterological factors. Yet, machine learning models like random forest, support vector machines (SVM), and deep learning algorithms show enhanced accuracy but are often criticized for their "black-box" methodology, resulting in a lack of transparency to public health decision-makers. A hybrid modeling system combining statistical rigor, machine learning complexity, and geospatial risk mapping is becoming the most effective method to strengthen early warning systems and outbreak prevention interventions.

To shift from reactive to proactive dengue control, governments and public health institutions need to invest in meterlogy integrated early warning systems that capture real-time data and lagged meterology-disease relationships. Better cross-sectoral coordination under the One Health approach along with improved community-based surveillance, and the development of user-friendly digital tools for public health professionals will be essential to curb the global dengue burden. The alignment of dengue prevention and control interventions with SDG 3 and SDG 13 will keep vector control programs anchored in larger climate adaptation policies.

Despite advancements in dengue forecasting model improvement, there are several research-related challenges that persist. The lack of socio-economic and behavioral predictors in climate-based models limits their usage in real-world contexts. Furthermore, the limited geographical validation of forecasting tools outside endemic areas of dengue makes it challenging to assess the performance of forecasting tools in newly emerging areas. The lack of a standard performance measure for evaluating dengue forecasting models further makes it difficult to compare across studies and spread effective tools within public health programs.

Subsequent research must prioritize formulation of multi-dimensional hybrid models, that enhance geographic validation, incorporate extreme climate events, and utilize real-time data along with AI-powered interpretability to improve predictive accuracy. By incorporating satellite-based remote sensing data from sensors (NASA's

MODIS and the Copernicus Climate Change Service) can potentially be used to improve spatiotemporal risk mapping such that dengue hotspots can be detected early on. With the increasing number of extreme weather events, research must investigate—effect of floods, heatwaves, and droughts on dengue transmission dynamics and incorporate these factors into predictive models.

Ultimately, dengue prevention future comes through an innovative, data-driven solutions that integrate climate science, epidemiology, and AI-driven insights. It is only through advancing dengue forecasting systems and linking them to public health decision-making, the progress is toward a future where outbreaks are not only detected but also prevented, lessening the global burden of dengue, dengue-related health economic burden and saving lives.

References

Acharya BK, Cao C, Xu M, Khanal L, Naeem S, Pandit S (2018) Temporal variations and associated remotely sensed environmental variables of dengue fever in Chitwan District Nepal. ISPRS Int J Geo-Inform 7(7):275. https://doi.org/10.3390/ijgi7070275

Adde A, Roucou P, Mangeas M, Ardillon V, Desenclos JC, Rousset D, Flamand C (2016) Predicting dengue fever outbreaks in French Guiana using climate indicators. PLoS Negl Trop Dis 10(4):e0004681. https://doi.org/10.1371/journal.pntd.0004681

Ali K, Ma'rufi I, Wiranto, Fuad A (2016) Variability of local weather as early warning for dengue hemorrhagic fever outbreak in Indonesia. In: International journal of environmental research and public health. Retrieved from

Alkhaldy I (2017) Modelling the association of dengue fever cases with temperature and relative humidity in Jeddah, Saudi Arabia—A generalized linear model with breakpoint analysis. Acta Trop 168:9–15. https://doi.org/10.1016/j.actatropica.2016.12.034

Banu S, Hu W, Guo Y, Hurst C, Tong S (2014) Projecting the impact of climate change on dengue transmission in Dhaka, Bangladesh. Environ Int 63:137–142. https://doi.org/10.1016/j.envint.2013.11.002

Banu S, Guo Y, Hu W, Dale P, Mackenzie JS, Mengersen K, Tong S (2015) Impacts of El Niño Southern Oscillation and Indian Ocean Dipole on dengue incidence in Bangladesh. Sci Rep 5:16105. https://doi.org/10.1038/srep16105

Baquero OS, Santana LMR, Chiaravalloti-Neto F (2018) Dengue forecasting in São Paulo city with generalized additive models, artificial neural networks, and seasonal autoregressive integrated moving average models. PLoS ONE 13(4):e0195065. https://doi.org/10.1371/journal.pone.0195065

Carvajal TM, Viacrusis KM, Hernandez LFT, Ho HT, Amalin DM, Watanabe K (2018) Machine learning methods reveal the temporal pattern of dengue incidence using meteorological factors in metropolitan Manila Philippines. BMC Infect Dis 18:183. https://doi.org/10.1186/s12879-018-3066-0

Chang CJ, Chen CS, Tien CJ, Lu MR (2018) Epidemiological, clinical and climatic characteristics of dengue fever in Kaohsiung City, Taiwan with implications for prevention and control. PLoS ONE 13(1):e0190637. https://doi.org/10.1371/journal.pone.0190637

Chen CW, Chen CS (2024) Spatial-temporal hurdle model versus spatial zero-inflated GARCH model: analysis of weekly dengue fever cases. Stoch Environ Res Risk Assess 38(6):2119–2134

Chuang T-W, Chaves LF, Chen P-J (2017) Effects of local and regional climatic fluctuations on dengue outbreaks in southern Taiwan. PLoS ONE 12(6):e0178698. https://doi.org/10.1371/journal.pone.0178698

Chumpu R, Khamsemanan N, Nattee C (2019) The association between dengue incidences and provincial-level weather variables in Thailand from 2001 to 2014. PLoS ONE 14(12):e0226945. https://doi.org/10.1371/journal.pone.0226945

Correia Filho WLF (2017) Influence of meteorological variables on dengue incidence in the municipality of Arapiraca, Alagoas, Brazil. Rev Soc Bras Med Trop 50(3):309–314. https://doi.org/10.1590/0037-8682-0432-2016

Costa AC, Gomes TF, Moreira RP, Cavalcante TF, Mamede GL (2022) Influence of hydroclimatic variability on dengue incidence in a tropical dryland area. Acta Trop 235:106657

Ehelepola NDB, Ariyaratne K (2015) The interrelationship between dengue incidence and diurnal ranges of temperature and humidity in a Sri Lankan city and its potential applications. Glob Health Action 8(1):29359. https://doi.org/10.3402/gha.v8.29359

Franchito SH, Rao VB, Fernandez JPR, Giarolla E (2021) Future changes in climatic variables due to greenhouse warming increase dengue incidence in the region of the Tucurui hydroelectric dam in the Amazon. Pure Appl Geophys 178:4033–4047. https://doi.org/10.1007/s00024-021-02849-1

Horta MA, Bruniera R, Ker F, Catita C, Ferreira AP (2014) Temporal relationship between environmental factors and the occurrence of dengue fever. Int J Environ Health Res. https://doi.org/10.1080/09603123.2013.865713

Iguchi JA, Seposo XT, Honda Y (2018) Meteorological factors affecting dengue incidence in Davao Philippines. BMC Public Health 18:629. https://doi.org/10.1186/s12889-018-5532-4

Jayaraj VJ, Avoi R, Gopalakrishnan N, Baha Raja D, Umasa Y (2019) Developing a dengue prediction model based on climate in Tawau Malaysia. Acta Trop 197:105055. https://doi.org/10.1016/j.actatropica.2019.105055

Kakarla SG, Caminade C, Mutheneni SR, Morse AP, Upadhyayula SM, Kadiri MR, Kumaraswamy S (2019) Lag effect of climatic variables on dengue burden in India. Epidemiol Infect 147:e170. https://doi.org/10.1017/S0950268819000608

Li C, Wang X, Wu X, Liu J, Ji D, Du J (2017) Modeling and projection of dengue fever cases in Guangzhou based on variation of weather factors. Sci Total Environ 605–606:867–873. https://doi.org/10.1016/j.scitotenv.2017.06.181

Li C, Zhao Q, Zhao Z, Liu Q, Ma W (2021) The association between tropical cyclones and dengue fever in the pearl river delta, China during 2013–2018: a time-stratified case-crossover study. PLoS Negl Trop Dis 15(9):e0009776. https://doi.org/10.1371/journal.pntd.0009776

Liao C-M, Huang T-L, Lin Y-J, You S-H, Cheng Y-H, Hsieh N-H, Chen W-Y (2015) Regional response of dengue fever epidemics to interannual variation and related climate variability. Stoch Env Res Risk Assess 29(4):947–958. https://doi.org/10.1007/s00477-014-0948-6

Liu D, Guo S, Zou M, Chen C, Deng F, Xie Z, Wu L (2019) A dengue fever predicting model based on Baidu search index data and climate data in South China. PLoS ONE 14(12):e0226841. https://doi.org/10.1371/journal.pone.0226841

Liyanage P, Tissera H, Sewe M, Quam M, Amarasinghe A, Palihawadana P, Rocklöv J (2016) A spatial hierarchical analysis of the temporal influences of the El Niño-Southern Oscillation and weather on dengue in Kalutara District, Sri Lanka. Int J Environ Res Public Health 13(11):1087. https://doi.org/10.3390/ijerph13111087

Lowe R, Gasparrini A, Van Meerbeeck CJ, Lippi CA, Mahon R, Trotman AR, Stewart-Ibarra AM (2018) Nonlinear and delayed impacts of climate on dengue risk in Barbados: a modeling study. PLoS Med 15(7):e1002613. https://doi.org/10.1371/journal.pmed.1002613

Miah MM, Hossain MB, Jannat SN, Karim MR, Rahman MR, Arafat Y, Pingki FH (2024) Assessing the impact of climatic factors on dengue fever transmission in Bangladesh. Aerobiologia. https://doi.org/10.1007/s10453-024-09814-0

Nguyen V-H, Tuyet-Hanh TT, Mulhall J, Minh HV, Duong TQ, Chien NV (2022) Deep learning models for forecasting dengue fever based on climate data in Vietnam. PLoS Negl Trop Dis 16(6):e0010509. https://doi.org/10.1371/journal.pntd.0010509

Noureldin E, Shaffer L (2019) Role of climatic factors in the incidence of dengue in Port Sudan City Sudan. Eastern Mediterr Health J 25(12):852–860. https://doi.org/10.26719/emhj.19.019

Omadlao ZRD, Cabrales JMA, Cristobal SCM, Dee MVA, Tadeo JRV, Marigmen JLDC, Pajarillo RR (2022) Machine learning-based dengue forecasting system for Irisan, Baguio City Philippines. AIP Conf Proceed 2472(1):040019. https://doi.org/10.1063/5.0092930

Ouattara CA, Traore TI, Traore S, Sangare I, Meda CZ, Savadogo LGB (2022) Climate factors and dengue fever in Burkina Faso from 2017 to 2019. J Publ Health Afr 13(2):2145. https://doi.org/10.4081/jphia.2022.2145

Phung D, Huang C, Rutherford S, Chu C, Nguyen H, Nguyen CDM (2015a) Identification of the prediction model for dengue incidence in Can Tho city, a Mekong Delta area in Vietnam. Acta Trop 141:88–96. https://doi.org/10.1016/j.actatropica.2014.10.005

Phung D, TalukderMRR, Rutherford S, Chu C (2015) A climate-based prediction model in the high-risk clusters of the Mekong Delta Region, Vietnam: Towards improving dengue prevention and control. Trans R Soc Trop Med Hyg. https://doi.org/10.1111/tmi.12754

Ramachandran VG, Roy P, Das S, Mogha NS, Bansal AK (2016) Empirical model for estimating dengue incidence using temperature, rainfall, and relative humidity: a 19-year retrospective analysis in East Delhi. Epidemiol Health 38:e2016052. https://doi.org/10.4178/epih.e2016052

Singh PS, Chaturvedi HK (2022) A retrospective study of environmental predictors of dengue in Delhi from 2015 to 2018 using the generalized linear model. Sci Rep 12:8109. https://doi.org/10.1038/s41598-022-12164-x

Talagala T (2015) Climatic factors and dengue incidence in Colombo District: a distributed lag nonlinear modelling approach. Epidemiol Biostatistics Publ Health 12(4). https://doi.org/10.2427/11522

Thiruchelvam L, Asirvadam VS, Dass SC, Daud H, Gill BS (2017) Dengue incidence prediction using model variables with registered case feedback. In: 9th International conference on robotic, vision, signal processing and power applications. Springer, pp 163–172. https://doi.org/10.1007/978-981-10-1721-6_18

Toai NP, Chinh DV, Vittor AY, Huy NN (2019) Associations between dengue hospitalizations and climate in Can Tho, Vietnam, 2001–2011. Stoch Environ Res Risk Assess

Tuladhar R, Singh A, Varma A, Choudhary DK (2019) Climatic factors influencing dengue incidence in an epidemic area of Nepal. BMC Res Notes 12:131. https://doi.org/10.1186/s13104-019-4185-4

Wu X, Lang L, Ma W, Song T, Kang M, He J, Ling L (2018) Non-linear effects of mean temperature and relative humidity on dengue incidence in Guangzhou, China. Sci Total Environ 628–629:766–771. https://doi.org/10.1016/j.scitotenv.2018.02.136

Yaacob WFW, Fatihah F (2020a) Modeling dengue incidence in southern Malaysia using fixed effects panel count model. Adv Math Sci J 9(9):81. https://doi.org/10.37418/amsj.9.9.81

Yaacob WFW, Fatihah F (2020b) Fixed effects panel count model for modeling dengue incidence. Adv Math Sci J 9(9):80. https://doi.org/10.37418/amsj.9.9.80

You SH, Chen SC, Huang YH, Tsai HC (2022) Dengue meteorological determinants during epidemic and non-epidemic periods in Taiwan. Trop Med Infect Dis 7(12):408. https://doi.org/10.3390/tropicalmed7120408

Yu H-L, Lee C-H, Chien L-C (2016) A spatiotemporal dengue fever early warning model accounting for nonlinear associations with hydrological factors: a Bayesian maximum entropy approach. Stoch Env Res Risk Assess 30(6):2127–2141. https://doi.org/10.1007/s00477-016-1328-1

Chapter 7
Environmental and Social Determinants of Malaria

Rabia Khalid

Abstract This chapter examined the complicated connection between social and environmental factors and malaria's economic and environmental causes. It covered things like health risks, natural disasters, and ecological risks. The environment strongly affects vector-borne diseases like malaria. Social and environmental mechanisms help the disease reproduce and spread throughout its life cycle. Climate change, deforestation, and Anopheles mosquitoes thrive under inadequate water management, increasing urbanization, and insufficient water management. These variables promote mosquito growth, increasing malaria risk. These include poverty, poor education, and scarce resources. Patients have limited healthcare options due to infrastructure, diagnostic, and treatment limitations. Cyclones and floods are catalysts. Stagnant water bodies and community dislocation exacerbate the malaria outbreak, straining healthcare infrastructure. Displaced populations, rural settlements, and demographic groups live in many socio-politically volatile, high-risk places with fragile governance systems. The interaction between harmful environmental exposure and limited resources increases malaria risk in these people. Integrated techniques are vital because they can solve environmental and social issues. The chapter demonstrates the need for integrated techniques in addressing environmental and socioeconomic issues. Effective techniques require long-term ecological management, improved disaster preparedness, and fair healthcare. To reduce malaria risks, improve health, and strengthen vulnerable groups, we need government, non-government, and community participation.

Keywords Malaria · Environmental determinants · Social vulnerability · Natural disasters · Health inequity · Resilience

---

R. Khalid (✉)
Quaid-i-Azam University, Islamabad, Pakistan
e-mail: rabiakhalid11111@gmail.com

A. Roy et al. (eds.), *Environment and Public Health*, Sustainability Solutions,
https://doi.org/10.1007/978-3-031-99770-9_7

7.1 Introduction

Malaria is an ancient disease that has garnered attention for an extended period. In the twentieth century, humanity had to deal with serious and complex health problems. At that time, it affected around 300 million individuals and resulted in over one million fatalities annually worldwide (WHO 2022). Despite numerous innovations and advancements in medical science, malaria is still regarded as a significant cause of mortality, accounting for 300–500 million clinical cases and 1.5–2.7 million fatalities per year.

Five kinds of Plasmodium cause malaria, and female Anopheles mosquitoes are the vectors for the disease. Significant medical and economic consequences result from its main cause of morbidity and mortality (Cox 2010). Species of *Plasmodium vivax* and *Plasmodium falciparum* are the most prevalent and widely dispersed parasites infecting humans in Pakistan (Snow et al. 2005). The World Malaria Report for 2021 states that roughly 50% of the global population lives in 87 countries and regions vulnerable to malaria transmission. In 2020, it is expected to have 241 million malaria cases and approximately 627,000 fatalities. This epidemic seriously threatens public health (Raghavendra et al. 2011). According to the World Health Organization (WHO), Pakistan is one of seven eastern Mediterranean countries accountable for 98% of the malaria cases in the region. Nearly 217 million Pakistanis constitute the intermediate risk category for malaria, while 63 million are in the high-risk category. In 2020, there were about 0.47 million malaria cases reported, resulting in 800 fatalities.

Environmental determinants, including topography, precipitation, climate, and socioeconomic status, profoundly influence malaria transmission. Tropical countries with high temperatures, significant rainfall, and heightened humidity, like Ethiopia, are ideal for mosquito breeding, survival, and parasite development. Malaria is ubiquitous in Pakistan, characterized by instability and seasonal dissemination. Transmission occurs in September, December, April, and May. Malaria infections peak biannually during agricultural surges (Milner 2017). Fever, a characteristic clinical manifestation of malaria, encompasses high temperature, sweat, chills, emesis, and severe headaches (Nadeem et al. 2021). Optical microscopy, fast diagnostic tests, and polymerase chain reaction (PCR) screening are some of the tests that are needed to find malaria. Despite its limited sensitivity and the requirement for skilled practitioners, traditional microscopy remains the predominant malaria diagnosis technique, as Karim et al. (2016) stated. Malaria treatment in Pakistan depends on accurate medical diagnosis and effective interventions via vector control and preventive measures (Hussain et al. 2016).

Geographic context is regarded as a primary factor; nevertheless, economic and housing situations may also significantly contribute to the proliferation of the virus (Mankodi 1996). The proportional influence of these elements on the dissemination of malaria may fluctuate considerably across temporal and spatial contexts. The crucial need exists to analyze each domain independently when managing the factors that affect malaria transmission. The spatial analysis in question is predicated on a comparable concept and investigates the primary determinants of malaria at the micro level within a designated region. Despite relentless efforts to eliminate malaria transmission, its incidence in certain regions, including the research area, is believed to be rising annually. Given its status as a primary contributor to global mortality and its extensive psychological, economic, and medical ramifications, it is imperative to identify the underlying factors contributing to its increasing prevalence (Devi et al. 2013).

The health departments of all four provinces reported a steady increase in the annual number of malaria cases in the late 1970s. Accordingly, malaria incidence in Punjab province rose sharply between 1980 and 1990, then marginally decreased in the latter part of the 1980s before increasing again in the 1990s. In 1990, malaria spread throughout the provinces of Sindh and Khyber Pakhtunkhwa (KP); however, in 1980, it spread throughout northern and southern Khyber, with cases reported in Balochistan. The majority of cases of malaria occur in Khyber Pakhtunkhwa and Balochistan, specifically in Swat, Lower Dir, Quetta, Buner, Peshawar, Mardan, and Malakand (Khan et al. 2023). Official classification places Pakistan as having moderate malaria prevalence and effective malaria control methods. Currently, this illness poses a threat to millions of individuals. Pakistan has reported approximately 50,000 malaria cases annually, resulting in a minimum of 50,000 fatalities attributed to the disease (Yasinzai and Kakarsulemankhel 2009). Conversely, some studies have demonstrated that malaria is prevalent in Pakistan, with a high endemicity of infection.

Malaria, caused by the parasites *Plasmodium vivax* and *Plasmodium falciparum*, is a major problem in Pakistani communities (Jamil and Khan 2012). Owing to ineffective patient care and a lack of knowledge, the mortality rate is even higher in cases of severe and complex vivax and falciparum malaria. The official statistics maintained by Pakistani health officials showed that 4.5 million suspected malaria cases appeared in 2008. Malaria constituted 16% of the total visits at external primary healthcare facilities. The confirmed malaria cases reached 104,454, while Plasmodium infections contributed to 30% of these during documented visits (Yasinzai and Kakar 2012). The complex and ever-changing web of elements that determines whether malaria will occur includes human and mosquito behavior, land cover and use, housing quality, the strength of the health system, and other similar variables. Most Pakistani communities lack access to modern waste management infrastructure, which mirrors the situation in other developing nations. These factors may have caused malaria in Pakistan (Soomro et al. 2010; Barat et al. 2004).

It is widely recognized that socioeconomic and environmental variables affect the prevalence rates of non-infectious and infectious illnesses in any given place. Deplorable socioeconomic and unsanitary environmental conditions create optimal circumstances for the proliferation of detrimental vectors, such as mosquitoes. Numerous studies have demonstrated that malaria transmission significantly correlates with socioeconomic disadvantages and adverse environmental conditions. Pakistan is very vulnerable to climate change due to its geographical location, reliance on land and water resources, and insufficient disaster preparedness infrastructure.

Persistent flooding in Pakistan has been linked to a significant rise in malaria infections. Climate change, insufficient vector control, and inadequate healthcare are likely responsible for malaria cases (Malik et al. 2012).

Furthermore, the economic activities of Southeast Asian border regions also play a role in the transmission of malaria, a disease that is highly resistant to malaria treatment (Hazra et al. 2000). Studies show that human health is influenced by factors such as population growth speed combined with elevated joblessness rates, plus high dependency ratios, in addition to limited literacy skills and big households, along with crammed living quarters (Park 2005). The high prevalence in several regions of Pakistan stems from poor socioeconomic situations and inadequate knowledge among residents. These elements influence mosquito breeding grounds in both physical and social realms. According to multiple research studies across Pakistan, unsanitary living circumstances and low socioeconomic status are two factors that contribute to the spread of the disease. This chapter aimed to offer direction, and investigating mixed infections will assist stakeholders in developing effective strategies to control and eradicate malaria. No systematic research or meta-analysis has yet established the aggregate incidence of malaria in Pakistan.

7.1.1 Rationale of the Study

The Pakistani public health sector recognizes malaria as a significant health concern impacting rural communities and impoverished populations nationwide. The social and environmental conditions in targeted regions render disease transmission likely. However, numerous global health initiatives, along with official malaria control programs, do not succeed in preventing the disease from impacting individuals. The disease affects many people, especially in Sindh, Balochistan, Khyber Pakhtunkhwa, and southern Punjab. Understanding social and environmental factors is necessary to develop targeted strategies for combating malaria. The advancement of innovative therapeutic approaches and the enhancement of contemporary management programs necessitate research.

Malaria significantly impacts Pakistan, as the country ranks among the foremost malaria-endemic nations in the region. South Asia is one of the foremost malaria-endemic regions globally, with thousands of new cases documented each year, predominantly impacting its low-income populations. Populations experiencing inadequate living conditions belong to this category. Limited access to healthcare is a significant problem. Fiscal constraints, geographical isolation, and inadequate medical care exacerbate this issue for rural and tribal populations. The amalgamation of remote geographical regions with inadequate medical resources results in patients receiving delayed diagnoses while pursuing treatment and therapeutic intervention. The spread of malaria is exacerbated by environmental factors that facilitate the reproduction of Anopheles mosquitoes. Adult Anopheles mosquitoes deposit their reproductive eggs in the water reservoirs of irrigation systems when drainage systems are not functioning correctly. Three causes converge to alter our environment: climate change, deforestation, and urban expansion. Inundated mosquito habitats increase the likelihood of fresh outbreaks. The erroneous use of preventive measures arises from societal misconceptions regarding these initiatives.

These initiatives include the use of mosquito nets and insecticide spraying in rural Sindh, Balochistan, and Khyber. Inhabitants of the Khyber Pakhtunkhwa region frequently attribute their interpretations of malaria to supernatural phenomena, linking it to certain beliefs and unpleasant odors. "Foul-smelling air." A framework that integrates enhanced healthcare systems with environmental protection and community learning programs represents the only solution to tackle these issues. The main objective targets both control and prevention of malaria transmission. The research serves as an essential tool to address existing deficiencies within existing malaria control methods.

This chapter elucidates the mechanisms by which social activities, in conjunction with economic status and environmental factors perpetuate the transmission of malaria. The study sheds light on the mechanisms that perpetuate the transmission of malaria. The findings present evidence that politicians and health organizations should consider implementing. Public health organizations, in collaboration with healthcare professionals and caregivers, can improve their initiatives by identifying high-risk populations, as indicated by this study. The initiative seeks to elevate public awareness of malaria prevention by community involvement, hence reducing malaria cases and enhancing public health outcomes in Pakistan.

7.2 Materials and Methods

A mixed-methods approach was employed to investigate the environmental, social, and climatic factors affecting the prevalence of malaria in Pakistan. The research employed both quantitative and qualitative data collection methods to generate a thorough overview of the topic. The examination of climate factors utilized meteorological data obtained from Pakistan's Meteorological Department, which recorded temperature, humidity, and rainfall metrics. Researchers employed time-series analysis and trend forecasting methods to examine the influence of seasonal variations and extreme weather events on malaria transmission patterns. The researcher utilized additional information, including WHO and local health reports, along with research articles, to analyze historical patterns of malaria prevalence.

The Ministry of Health and district health departments contributed epidemiological statistics on malaria incidence, which were kept in their records. Statistical models, including Poisson regression, examined the relationship between Pakistan's climate data, geographical factors, and local malaria incidence rates across different regions. Global research provided the foundation for the comparative study that assessed the impact of climate change on malaria transmission. The analysis of socioeconomic determinants included two indicators: the Human Development Index (HDI) and education scores (ES), which were utilized to assess social standing. The researchers examined the impact of financial obstacles, housing deficiencies, and healthcare service limitations on the susceptibility to malaria, particularly in marginalized areas.

Researchers conducted interviews with malaria patients, medical professionals, and community members to examine their health-seeking behaviors and barriers to treatment access. The interviews were conducted in both rural and urban settings to obtain diverse perspectives. The study participants employed thematic analysis to identify significant patterns in individuals' knowledge about malaria, along with their views and preventive strategies. The research examined conventional healing practices and individual substance use, resulting in delayed medical intervention. This study integrates environmental data with epidemiological trends and financial community constraints to develop a comprehensive knowledge of malaria control challenges in Pakistan. The research yields findings that will aid government officials in implementing targeted malaria control strategies while improving healthcare accessibility and mitigating vulnerability to vector-borne diseases resulting from climate change.

7.3 Results

Various environmental and socioeconomic factors influence the public health implications of malaria in Pakistan, even with its persistent status as a significant health issue. Temperature, humidity, altitude, significant climatic changes, and precipitation serve as crucial factors influencing malaria vector ecology. Effective malaria control strategies necessitate a comprehensive understanding of the environmental factors influencing their actions.

7.3.1 Environmental Risk Factors

7.3.1.1 Temperature and Humidity

The relationship between malaria incidence and temperature appears in Table 7.1. The instances exhibited consistency over five years. The incidence of malaria has risen concomitantly with rising temperatures over the years. Temperature and humidity are essential environmental variables that affect the survival, reproduction, and infectivity of malarial vectors. The ideal temperature range for the proliferation of malaria vectors, specifically Anopheles mosquitoes, is 20–30 °C. Temperatures within this range facilitated swift larval growth and elevated vector survival rates. Furthermore, relative humidity levels exceeding 55% are crucial for the survival of adult mosquitoes, enabling them to fulfill their life cycle and perpetuate malaria transmission.

Malaria transmission necessitates a temperature high enough for the completion of the sporogonic cycle of Plasmodium parasites within the mosquito.

Table 7.1 Estimated monthly climate and malaria data

Month	Min. temp. (°C)	Max. temp. (°C)	Avg. temp. (°C)	Precipitation (mm)	Malaria cases
January	10.5	22.1	15.2	8	450
February	13.2	25.6	23.4	5	620
March	18.5	30.2	23.4	12	890
April	23.8	35.4	28.6	6	1120
May	28.5	40.2	33.2	3	1560
June	30.1	44.5	36.8	15	2100
July	28.4	40.2	34.7	65	3850
August	27.8	39.5	33.9	78	43,200
September	25.5	36.8	31.2	42	3980
October	20.3	34.7	27.5	19	2540
November	14.5	29.8	21.3	9	1230
December	11.0	24.5	16.7	9	780

Source Derived from National Institute of Health

In Pakistan, *Plasmodium vivax* necessitates a minimum temperature of around 15 °C, whereas *Plasmodium falciparum* demands a minimum temperature of approximately 19 °C. When temperatures drop below these thresholds, the parasite's development within the mosquito is impeded, thereby diminishing malaria transmission. Conversely, very elevated temperatures (exceeding 40 °C) might adversely affect mosquito life, potentially reducing transmission in particularly hot regions, such as certain sections of Balochistan and Sindh, during peak summer months (Fig. 7.1).

7.3.1.2 Altitude and Malaria Transmission

Altitude greatly affects malaria transmission due to variations in temperature and vector survival rates associated with elevation. Malaria transmission typically halts at elevations beyond 3300 m due to reduced temperatures that hinder mosquito viability and parasite maturation. In Pakistan, malaria is predominantly found in lowland areas, such as Punjab, Sindh, and certain sections of Khyber Pakhtunkhwa and Balochistan, where environmental conditions are conducive to mosquito proliferation. On the other hand, malaria transmission in high-altitude areas is limited, although it is not completely absent. Malaria incidences have been documented in the northern mountainous regions of Pakistan, specifically in Gilgit-Baltistan and northern Khyber Pakhtunkhwa, at elevations between 2000 and 3000 m. This is primarily because of the existence of specific vector species, such as Anopheles fluviatilis, which can endure and disseminate malaria in milder climates. At elevated altitudes, malaria transmission rates decrease because both temperature conditions and mosquito breeding possibilities become restricted.

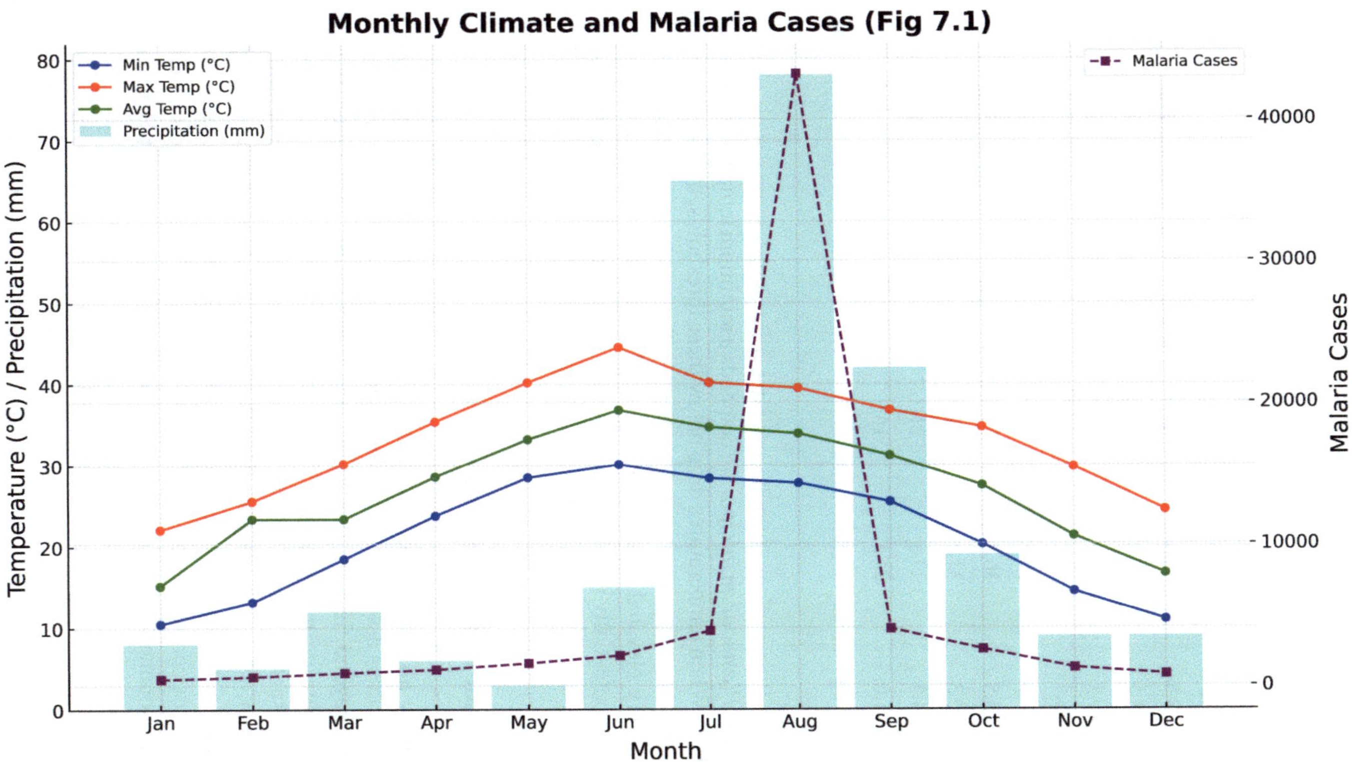

Fig. 7.1 Number of malaria cases. *Source* Health Science Reports

7.3.1.3 Rainfall, Floods, and Malaria Outbreaks in Sindh

The region's weather forecasting organization reports that this area saw the heaviest rainfall ever recorded. Because of the ideal conditions brought about by the beginning of the rainy season, the parasite population in the area grows in tandem with the increased rainfall. The stagnant water bodies caused by heavy rainfall provide malaria vectors with an ideal environment to breed. The monsoon season (July–September) significantly impacts malaria transmission in Pakistan, particularly in provinces such as Khyber Pakhtunkhwa, Sindh, and Punjab. Water pools in places like urban drainage systems, rice fields, and irrigation canals idealize the Anopheles mosquito breeding grounds.

According to Pakistan's historical records, the monsoon rains are a common trigger for malaria outbreaks. There will be a spike in mosquito populations and the spread of malaria if there is an excess of rainfall during this time, since there will be more nesting grounds. Malaria has spread rapidly over Sindh as a result of the region's frequent flooding. The devastating floods of 2010 were one such event; they afflicted around 20 million people in Pakistan, a big portion of Sindh. Malaria vectors have found an excellent breeding habitat in the sluggish floodwaters and the enormous population displacement that has occurred.

Malaria cases in flood-affected areas have surged, especially in Thatta, Dadu, Jacobabad, and Larkana, where there are plenty of mosquito breeding grounds due to the post-flood conditions, according to health reports. Following the devastating floods in Sindh in 2022, one of the greatest climate-related disasters in the region's history came with another major malaria outbreak. Nearly a third of Pakistan was flooded by devastating monsoon rains, with Sindh province being among the most impacted. Sukkur, Khairpur, Mirpurkhas, and Hyderabad are among the districts hit hard by the floods, and according to reports from the Pakistani Ministry of Health and the World Health Organization (WHO), malaria cases have increased in certain regions.

Months of standing floodwater meant more time for mosquitoes to lay eggs and more time for malaria to spread. The local health services are being overwhelmed by malaria-positive rates that have risen beyond 50% in several districts. Floods worsen health crises by damaging healthcare facilities, interrupting vector control efforts, and restricting treatment availability, in addition to increasing the mosquito population. The 2022 floods in Sindh wreaked havoc on malaria prevention efforts, obstructing access to timely medical care for thousands of residents due to the devastation of roads and health clinics. The risk of malaria is already high for many of the displaced families since they have no choice but to remain in improvised shelters close to the still flood waters. A lack of suitable drainage infrastructure allows Sindhi cities such as Karachi and Hyderabad to develop standing water accumulations even during light rain conditions. The ability of malaria vectors to reproduce beyond the monsoon period exists because of these environmental conditions (Table 7.2).

Epidemics of malaria have emerged when the weather becomes unexpected, with excessive humidity and elevated temperatures during periods without regular rainfall patterns. The El Niño-Southern Oscillation (ENSO), a worldwide downpour

Table 7.2 Malaria outbreaks and causes

Year	Location	Cause
2017	Sindh (Sukkur, Dadu)	Heavy floods, stagnant water
2020	Karachi (slum areas)	Urban flooding, poor drainage
2022	Sindh (Dadu, Larkana, Khairpur)	Record-breaking monsoon floods

Source Field data author own drawing

phenomenon that impacts patterns of rainfall and temperature, is one of the most prominent climate-related factors influencing malaria. Malaria transmission has been found to increase in regions around the globe, particularly South Asia, during ENSO episodes. Higher temperatures during El Niño years in Pakistan are associated with increased transmission of *Plasmodium falciparum*, according to research. This is especially true in areas where temperatures stay elevated in the late season.

They anticipate that the gradual rise in global temperature due to climate change will profoundly affect malaria transmission in Pakistan. Malaria transmission seasons might lengthen, and malaria-endemic zones can expand into areas where it was not previously present due to rising temperatures. Deforestation, urbanization, and changes in land use as a result of agricultural growth are also changing local microclimates, which in turn create new areas where mosquitoes can nest. Though agriculture benefits from increased irrigation and water management initiatives, malaria vectors find more breeding grounds as a result. Malaria transmission in Pakistan could move if climate change is left uncontrolled; to counteract this, early warning systems, vector management, and improved surveillance are needed.

7.3.2 Socio-cultural Risk Factors: Cultural Beliefs and Knowledge About Malaria

Malaria is often considered an "ordinary fever" rather than a serious mosquito-borne illness in many rural communities. It is commonly believed that exposure to cold weather, tainted food and water, or intense physical exertion are the main causes of malaria rather than mosquito bites. This fallacy has a big effect on preventative actions because families don't use protective methods such as insecticide-treated nets (ITNs) or residual spraying in the home. Research in Sindh and Southern Punjab revealed that fewer than 15% of respondents accurately recognized mosquito bites as the principal source of malaria. Perceived reasons for malaria commonly included prolonged sun exposure (20%), consumption of filthy water (18%), and intake of cold meals and beverages (10%). These cultural misconceptions cause delayed or improper treatment-seeking behaviors, which raise the incidence of malaria in children.

In Balochistan and Khyber Pakhtunkhwa (KPK), people often don't recognize anemia and convulsions as symptoms of malaria, even though the illness remains a significant issue in these areas. They often blame the supernatural or spiritual realms for the deaths of many children who contract the disease. Instead of going straight to the doctor, many people turn to traditional healers and homemade cures. Many people still rely on traditional treatment, which can worsen the condition and cause complications. As a result, local healthcare institutions are overwhelmed with severe malaria cases. The majority of respondents (85%) were familiar with the term "malaria," but only 40% could identify the main way in which the disease spreads in rural Sindh and Balochistan. Malaria is still not well prevented or treated in its early stages because people in areas where the disease is known view it as a common and mild illness. People frequently do not treat fever in children as a top priority, which elevates the likelihood of morbidity and mortality.

The lack of correct information regarding malaria symptoms and transmission is caused by a combination of factors, including cultural beliefs, inadequate health literacy, and restricted access to formal education. Because people in the community do not view malaria as a major health concern, efforts to combat the disease, such as the free distribution of ITNs and treatment, have fallen short of expectations. Malaria prevention measures and early treatment adherence will be insufficient until the disease is formally acknowledged as a mosquito-borne illness requiring immediate medical intervention. Myths regarding malaria's origins, transmission, and severity continue to circulate, as they do in other underdeveloped nations, resulting in inefficient methods of prevention and treatment.

7.3.2.1 Health-Seeking Behaviors and Delays in Treatment

The findings indicated that diverse health-seeking habits resulted in delays in malaria diagnosis and treatment, thereby elevating the risk of severe consequences. Patients typically utilize many health care providers; however, self-treatment is their initial response when they become unwell. During the fieldwork, the researcher noted that patients were more inclined to initiate self-treatment at home while monitoring their progress. Such behavior enabled them to reduce the costs in time and money that would arise from seeking external assistance promptly. During delayed decision-making, individuals were more inclined to select treatments accessible outside the home; when signs and symptoms persist, decisions may involve consulting a private healthcare practitioner, visiting a government health center, or seeking hospital care when the condition deteriorates. The understanding of the illness's nature and duration typically affects treatment decisions, along with the projected treatment costs and the patient's assessment of the illness's severity. Individuals in malaria-endemic areas often must select antimalarial medications and simultaneously differentiate among these treatments. People divide distinctions into two separate categories by examining patient age groups and the presence or absence of malaria symptoms. When people seek guidance from friends or family members, their choices can be based on incorrect information, leading to longer-term discomfort for the subject.

A lot of the people who took part said they were dependent on self-medication and home remedies. They often used traditional herbal medicines, over-the-counter pain relievers, and self-prescribed antimalarial drugs before going to see a doctor.

Table 7.3 Choices of therapies by respondents

S. No.	Contents	No. of respondents	Percentage (%)
1	Home medicine	12	9.52
2	Homeopathic	32	25.4
3	Hakeem	18	14.3
4	Religious healer	10	7.8
5	Self-treatment	54	42.9
6	Total	126	100

Source Author's calculation based on field data

This method was notably widespread in Mirpur Khas, where indigenous healers were essential in malaria treatment. Furthermore, the misreading of symptoms like fever, chills, and body aches caused many to conclude they were afflicted with seasonal influenza or typhoid instead of malaria (Table 7.3).

Delays in obtaining treatment and worse health outcomes were the results of this misperception. Furthermore, many people in Thatta and Tharparkar favored spiritual treatments and traditional healers over modern medicine since they were more convenient and less expensive. Although some did seek medical help, it was frequently after the more conventional methods of treatment had failed. Public education initiatives that encourage prompt medical attention in cases of malaria are urgently needed, as the disease's symptoms have worsened, and more people have fallen ill due to their reluctance to seek treatment. These results make it clear how important it is to act quickly to stop malaria by starting programs that target communities that are most likely to get it, help them get medical care even if they can't afford it, and give them ways to stay healthy.

7.3.2.2 Financial and Accessibility Barriers

All interviewees indicated that diseases such as dengue, malaria, and other ailments impose financial burdens on patients' families. A monthly prescription for patients may exceed their incomes. Treating a singular malaria episode with medication, diagnostics, and ancillary fees does not provide for daily pay. The continuous care required and the related expenses often exhausted the savings of both patients and their families. The majority of patients said that their illnesses hindered their participation in income-generating activities, resulting in financial loss. Moreover, they highlighted that their incapacity to generate a livelihood rendered them entirely reliant on their family members, complicating their ability to handle daily activities and afford their medicines.

Several respondents acknowledged their inability to afford the expenses associated with consultations, testing, and prescriptions. The participants' medicine usage and adherence to treatment regimens seem to have been considerably affected by insufficient access to pharmaceuticals and services, especially for those living in remote regions. Consequently, the participants residing in peripheral regions asserted that they needed to travel to the metropolis to use the services. The absence of dependable transportation, prolonged waiting and travel durations, and elevated transit expenses impeded access to medical care. Certain participants procure drugs for individuals traveling to towns to reduce transportation and associated costs. Certain individuals were allegedly unable to access any form of care during the rainy season when travel durations and costs doubled.

During the rainy season, accessing the hospital is exceedingly challenging due to impassable roads, rendering vehicles unable to reach the neighborhood. This explains my inability to attend my medical institution. The elevated cost of prescription medications and related services poses significant challenges for individuals residing in remote regions. They incurred expenses for transit, accommodation, and sustenance while obtaining their prescription refill, alongside consultation and laboratory bills.

7.3.2.3 Educational Barriers and Patient Empowerment in Malaria Treatment

Educational obstacles also constrain patient autonomy, as individuals with inadequate literacy find it challenging to navigate healthcare institutions, comprehend treatment plans, and advocate for themselves. Increased awareness and patient empowerment can improve psychological well-being and facilitate more informed decision-making. A primary topic of conversation among respondents from rural backgrounds was their understanding of their illnesses. They said they struggled to comprehend the requirements and purposes of specific treatments and medications. During participant observation, I saw that patients and caregivers were required to manage files comprising medical records, invoices, and medications. The patients photocopied superfluous documents, such as trivial receipts, which annoyed the physicians due to the considerable time required to locate pertinent documents within the file. The respondents expressed a desire to comprehend their circumstances. We are uneducated patients. Medication labels must specify the frequency of administration and indicate whether they should be taken before or after meals. When that occurs, it will be beneficial for us. A patient stated that if we were literate, we could more effectively comprehend our ailments and subsequently engage in decision-making.

The responders expressed a desire to be consulted throughout the diagnosis and treatment process. They sought to observe their contributions to the therapeutic process. They felt more vulnerable than empowered to make autonomous judgments due to a lack of information. This sentiment is apparent in the accounts of patients from urban backgrounds. Patients from rural origins typically depended on their sons or husbands for financial or other support during hospitalizations. These caregivers comprehended the medical terminology or English used in the reports or communications with the physicians. Patients from metropolitan backgrounds said that the entire family would convene to deliberate on the decision to visit the hospital, although the patient was typically excluded from contributing any remarks.

In Pakistan, two primary social determinants, low socioeconomic status (SES) and inadequate literacy levels, substantially affect malaria risk. Of the 10 social determinants of health delineated by Wilkinson and Marmot, a minimum of seven employment, social gradient, stress and anxiety, early life experiences, social exclusion, and food security intricately associated with these factors. A significant proportion of children in Pakistan are born into low-income households with restricted access to steady, high-paying employment opportunities. The World Health Organization (WHO) identifies poverty as the foremost risk factor for malaria, with global data demonstrating that the disease predominantly impacts the most impoverished populations. Poverty in Pakistan restricts access to formal education, diminishes career prospects, and perpetuates the cycle of economic suffering. Social exclusion resulting from poor socioeconomic status heightens stress and anxiety and limits access to healthcare and vital resources from a young age. The continuous cycle between poverty and disease preserves health inequities, which gives disadvantaged people greater risks for malaria.

7.4 Discussion

According to research, Pakistan's environmental factors play a significant role in the spread of malaria. Experts identified stagnant water bodies, inadequate sanitary facilities, and deforestation as primary factors facilitating the continued spread of malaria. Research findings indicate that Anopheles mosquitoes thrive in water habitats near human settlements, as found worldwide. In Pakistan, seasonal monsoons intensify these conditions, increasing malaria cases, especially in low-lying regions. This research is, to our knowledge, the inaugural study in the region to utilize both methodologies to investigate the impact of temperature, humidity, and precipitation on malaria progression. The prevalence of vector-borne diseases can be substantially reduced by the application of both active and passive diagnostic techniques (Zoghi et al. 2012). According to Yasinzai and Kakarsulemankhel (2008), Pakistan's control and monitoring efforts could benefit from a more accurate assessment of malaria prevalence. Prior research has analyzed the epidemiological trends of malaria over multiple years. A study conducted in Nsanje District, Malawi, utilized statistical analysis of trends combined with time serial analytics to predict malaria prevalence in the region (Gondwe et al. 2021). According to Anokye et al. (2018), a distinct investigation performed in Kumasi applied a time-sequence model to forecast future malaria incidence.

Malaria is one of the main causes of death in Pakistan globally (WHO 2022). Evidence of malaria's prevalence in Khyber Pakhtunkhwa dates back to the 1970s. The research by Rowland et al. (2002) discovered that high prevalence numbers dominated the 1980s but showed decline in the early 1990s. The epidemiological record of malaria incidence during the late 1990s appeared alarmingly high across northern Swat and Chitral together with western Afghan states Mardan, Malakand,

Swabi, Khyber, and Mohmand (Kazmi and Pandit 2001). According to Qureshi et al. (2020), a basic investigation found that malaria is quite common in the Bannu district.

The weather and vector-borne diseases are closely linked because the weather and seasons determine where and when these diseases are transmitted, as shown by the unique and, frequently, seasonal tendencies (Ogden 2017). The climatic changes brought about by the approximately 2 °C increase in global warming include higher average temperatures, altered patterns of rainfall across the globe, more climate variability, and more frequent and severe extreme weather events.

The enhancement of humidity and precipitation frequency has emerged due to this phenomenon. This phenomenon has led to an increase in the frequency of severe droughts and rainstorms, along with milder, shorter winters and longer, warmer summers. Rising surface temperatures from global warming, along with climate change, lead to major environmental changes in numerous regions. Precipitation levels have increased alongside extreme heatwaves, daily rainfall extremes, and rising average yearly temperatures (Romero-Lankao et al. 2014). These environmental changes facilitate the emergence and spread of exotic vector-borne diseases in many ways. Increased conditions, heightened transpiration, and intensified precipitation facilitate the reproductive process of non-native mosquitoes by supporting larval development and survival while extending adult lifespan, hence increasing the total number of individuals.

According to Ogden et al. (2013), three major threats are anticipated. The danger of chronic diseases transmitted by mosquitoes may increase due to extended changes in weather patterns and temperatures. Diseases spread by mosquitoes may become bigger epidemics if the cycles of vectors and pathogen transmission lengthen because of more extreme weather and changing climates. The movement of vector-borne diseases occurs either toward polar regions or elevated areas in mountainous terrain, which impacts locations previously free of such infections. The risk of established "uncommon" tropical and subtropical vector-borne diseases becomes heightened when the climate suddenly shifts or when transmission occurs between disease-prone areas and international border migrations happen.

Different temperatures can affect the spread of malaria. This effect could be due to either a low temperature that prevents the virus and its vectors from multiplying or a high temperature that accelerates the death of the insect. A temperature of 16 °C inhibits the proliferation of bloodsucking parasites and halts the development of their vectors in aquatic environments.

At 17 °C, sucking blood reaches maturity but lacks the requisite speed to form an endemic population. The current chapter demonstrates that variations in atmospheric temperatures influence malaria prevalence rates. According to Fischer et al. (2020), rising temperatures significantly contribute to the escalation of malaria transmission across Europe. Human-induced global warming elevates the prevalence of malaria. Research conducted in China indicates that an increase in temperature from 10 to 15 °C results in a 175% rise in malaria infections in the country. Analysis indicated that increased temperatures, coupled with heightened humidity and precipitation, are significant contributors to the proliferation of malaria (Liu et al. 2021).

This result demonstrated that humidity levels influence the frequency of malaria cases. Moist environments prolong parasite mortality rates during precipitation, consequently enhancing the parasites' efficacy in transmitting malaria among individuals. The highest incidence of malaria has been associated with the peak of the rainy season. The probability of malaria significantly escalates with rising temperatures, increased humidity, and prevailing dry conditions (Herekar et al. 2020).

Relative humidity measures the amount of evaporated water in the atmosphere, with 0% indicating dry air and 100% representing fully saturated, humid air. Relative humidity influences the malaria dissemination phase by altering mosquito behavior and lifespan.

According to Patz and Reisen (2001), precipitation created conducive conditions for the spread of parasite vectors. The prevalence of malaria escalates with heightened precipitation. The regional meteorological agency usually records some of the greatest precipitation levels in the vicinity. As the parasite population expanded, the transmission rate increased, ultimately elevating the parasite count. Furthermore, significant rainfall may have increased humidity levels, promoting the growth of parasites. An earlier African study recognized the seasonal dynamic importance of precipitation. The winter rainy season (December–February) increased the prevalence of malaria in the study region (Adeola et al. 2019). Topography additionally affects malaria prevalence. A previous study indicated that malaria incidence was elevated in lowland regions relative to highland regions (Atieli et al. 2011). The geographical features of highlands affect how long vectors live at different elevations, which in turn affects the number of vectors and the chance of malaria transmission.

In the future, these accurate models for predicting malaria will be used to find high-risk areas so that the right authorities and researchers can act quickly. The models will take into account the effects of terrain and climate. The fact that climate variables and topographical features have a big effect on malaria rates shows how important it is to take climate into account when managing malaria.

7.4.1 Socioeconomic Barriers

As other research has shown, the study confirmed that, at the district level, HDI and, by extension, socioeconomic status are negatively linked to the number of cases of malaria (Tusting et al. 2013). Individuals with low socioeconomic status are more susceptible to infectious diseases such as malaria and are also less equipped to adapt to climate change in those regions. This phenomenon leads to an increase in disease prevalence as individuals remain susceptible to health risks for prolonged periods. The findings indicated that the frequency of malaria was heightened in areas with poor socioeconomic levels, validating the notion that malaria is both a result and a factor of deprivation.

As a result, Pakistan's leaders should come up with plans for the fair distribution of resources and focus efforts to stop malaria in areas that are at high risk and have poor social and economic conditions. Financial resource limitations emerged as a significant obstacle to malaria prevention and necessary treatment. A multitude of respondents indicated that exorbitant healthcare costs for physician consultations, diagnostic tests, and medications deterred them from pursuing timely medical attention. Studies on health-seeking behaviors in low-income areas corroborate these findings (Minakawa et al. 2002).

Many people in Pakistan's poverty-stricken segment must choose between self-treatment or traditional medicine instead of healthcare because they cannot afford out-of-pocket medical expenses. The most pressing matter involves the lack of adequate housing facilities. Economically disadvantaged residents of their neighborhoods face increased mosquito bite risk because their homes do not have sufficient construction, resulting in exposed roof areas and unwalled windows. Studies in Sri Lanka and India have shown similar patterns that residential settings lacking walls or possessing open eaves increase the threat of malaria infections (Gunawardena et al. 1998). Improved housing policies, along with recommendations to use insecticide-treated bed nets and window screens, will reduce health risks.

Moreover, the lack of government-funded malaria prevention programs in rural areas exacerbates the problem. Individuals consistently articulated their apprehensions regarding inadequate resource management and constrained access to preventive strategies, such as insecticide application. Malaria control initiatives necessitate adequate financial support from policymakers and equitable access to preventive interventions in high-risk regions.

7.4.2 Healthcare Access and Education

In Pakistani districts, the education score (ES) serves as an accurate measurement indicating education state. The Population and Health Survey (PDHS 2012–2013) found a link between knowing how to avoid health problems and having a higher education score (ES). The district level demonstrates Pakistan's particular use of ES as a measurement for determining someone's level of education. According to Regmi et al. (2016), "knowledge of prevention methods is positively correlated with education and economic status in Pakistan." This chapter employed ES as an indicator for both parent and child health, understanding of disease prevention.

Future in-depth research in Pakistan's various districts can investigate the possible statistically significant inverse correlation between ES and malaria incidence. For now, however, we know that this association was minimal. There has been some evidence in the past linking education and socioeconomic position to malaria (Rana et al. 2011). Researchers found that many people put off getting medical help for malaria because they have false beliefs about the disease and its symptoms. Participants often mistook fever for other common illnesses, which led them to self-medicate or use home remedies instead of seeking medical assistance.

Research in South Asian and African nations has found a similar association between postponing treatment-seeking behavior and worse results for malaria (Karamat et al. 2019). Because of the scarcity of modern medical facilities in more remote places, people favor traditional treatment. It was not uncommon for people to use faith-based healing techniques and herbal remedies, often in conjunction with allopathic therapy. The unchecked use of traditional medicine, despite its cultural significance, can cause insufficient treatment, which in turn can cause drug-resistant malaria to emerge. According to Noureen et al. (2022), research out of Pakistan and India shows how dangerous self-medication is and how resistance to antimalarial medications is on the rise as a result of incorrect treatment regimens. The interviewees from the healthcare industry stressed the critical need for public education initiatives to raise awareness about malaria symptoms, the value of early diagnosis, and the need to take medication as directed. Healthcare services should combine with malaria education for healthcare workers as complementary national goals for both the government and NGOs.

7.4.3 Perceptions of Malaria Control Strategies

The prevention of malaria encounters substantial obstacles due to the widespread belief that this obligation rests exclusively with the government. People had numerous complaints about irregular public health programs related to pest management and bed net distribution. The low involvement of communities leads African nations to show similar outcomes, which makes malaria control programs unsuccessful (Das et al. 2013). The main problem developed from the improper usage of insecticides that served public health functions. The participants noted that reallocating insecticides from malaria control to agriculture was reducing their availability for insecticide-based malaria interventions. In other malaria-prone areas, vector control fails because of bad pesticide distribution plans, which is exactly what is happening here (Adefemi et al. 2015). The appropriate allocation and application of pesticides necessitate enhanced regulations to attain successful malaria prevention.

Findings indicate that community members must engage actively in malaria prevention initiatives. The management of malaria necessitates substantial involvement from local people alongside significant state measures. Initiatives will contribute to reducing transmission by involving community members in advocating for effective mosquito control methods in residences, such as proper waste disposal, covering storage vessels, and using insect repellents. Public–private collaboration initiatives significantly improve the malaria control approach. Partnerships among government entities, non-governmental organizations, and private healthcare providers can enhance access to malaria prevention resources and treatment services. Implementing school-based malaria teaching initiatives can augment community awareness and promote enduring behavioral transformation.

7.4.4 Policy Implications and Future Research Directions

Pakistan must effectively manage malaria through vector control, healthcare provision, and adaptation to climate change. In areas that are likely to flood, improving vector control using methods such as Indoor Residual Spraying (IRS), Long-Lasting Insecticidal Nets (LLINs), and Larval Source Management (LSM) is very important. Climate-sensitive malaria surveillance can employ predictive algorithms to address potential outbreaks. Incorporating malaria control into disaster management strategies may alleviate post-flood epidemics, whereas enhanced waste management and drainage can diminish the mosquito population.

To improve malaria treatment, access to healthcare services for rural and high-risk individuals must be increased. Enhancing primary healthcare and community-based healthcare can alleviate the burden of illness. In addition, disaster response plans need to include ways to stop malaria so that medical teams can be quickly sent to the scene of the disaster and antimalarial drugs can be delivered to the right places before flooding starts. Altering public behavior to control malaria is essential. Government organizations, NGOs, and health groups should spearhead educational efforts aimed at eliminating water waste, promoting bed net usage, and removing stagnant water.

Further studies should concentrate on changes in malaria transmission over time due to climate change, especially in areas prone to floods. Monitoring insecticide resistance and analyzing post-flood epidemiology can refine control efforts and enhance vector reduction strategies. Moreover, looking at the societal and other aspects of malaria treatment, such as how people perceive them culturally or how they act when they need treatment, can help develop better public health measures. Using remote sensing and GIS together for malaria mapping can improve national malaria control programs by showing which areas are most likely to get the disease and ensuring that resources are used more efficiently.

Despite its thoroughness, this chapter has some shortcomings, including data gaps due to underreporting and misdiagnosis in rural and flood-affected regions. The effects of climatic variations on malaria transmission are intricate, and microclimatic changes have not been effectively captured. Furthermore, the lack of longitudinal data for tracking malaria incidence over time has constrained a deeper understanding of changing risks. To better protect more at-risk groups in a community from malaria, more research should be conducted to fill these gaps and look at ethnographic methods that try to determine the effects and usefulness of malaria control measures put in place at the local level.

7.5 Conclusions

This chapter delineates the critical environmental determinants that affect malaria transmission dynamics in Pakistan. The proliferation of mosquitoes and the incidence of diseases have been significantly influenced by elevated temperatures, erratic precipitation patterns, and increased humidity levels. Malaria risks have grown because of inadequate water management, unrestricted urbanization, and weak vector control practices, mostly in areas that are poor and prone to flooding. Findings from research show that to completely stop malaria, different government departments must work together to improve drainage, control vectors, and raise awareness among the population. By combining weather information with malaria surveillance systems, early warning systems can be improved, allowing timely action to lower the number of cases and speed up responses. In addition, it is important to ensure that the responses to the malaria outbreak that occurred after the flood are part of the disaster preparedness plan so that people in the affected areas can receive the care they need quickly. Government agencies, healthcare providers, researchers, and communities are the principal entities that must collaborate to formulate effective and suitable policies for malaria control and prevention.

Finally, to reduce potential infections in the future, it is important to pay special attention to environmental and climate issues in Pakistan's malaria control strategies. Pakistan can make much more progress in its way to eliminate malaria and protect affected groups from its harm by adopting integrated vector management, improving health system coverage, and developing climate-adaptive measures.

Acknowledgements The authors express gratitude to public health officials, researchers, and data analysts who generously provided resources and valuable insights for this study. Special gratitude is extended to medical professionals operating in malaria-endemic regions, whose firsthand experiences have enhanced their understanding of the disease patterns and challenges. Gratitude is extended to climate scientists, epidemiologists, and legislators, whose contributions to environmental health and malaria influenced the framework of the study. Ultimately, sincere appreciation was extended to the research institutions and enterprises that facilitated data access and provided guidance on enhancing the methodology of the study.

References

Adefemi K, Awolaran O, Wuraola C (2015) Social and environmental determinants of malaria in under-five children in Nigeria: a review. Int J Community Med Public Health 2(4):345–350. https://doi.org/10.18203/2394-6040.ijcmph20151026

Adeola A, Ncongwane K, Abiodun G, Makgoale T, Rautenbach H, Botai J et al (2019) Rainfall trends and malaria occurrences in Limpopo Province, South Africa. Int J Environ Res Public Health 16:5156. https://doi.org/10.3390/IJERPH16245156

Anokye R, Acheampong E, Owusu I, Obeng EI (2018) Time series analysis of malaria in Kumasi: using ARIMA models to forecast future incidence. Soc Sci 4:1461544. https://doi.org/10.1080/23311886.2018.1461544

Atieli HE, Zhou G, Lee MC, Kweka EJ, Afrane Y, Mwanzo I et al (2011) Topography as a modifier of breeding habitats and concurrent vulnerability to malaria risk in the western Kenya highlands. Parasit Vectors 4:241. https://doi.org/10.1186/1756-3305-4-241

Barat LM, Palmer N, Basu S, Worrall EVE, Hanson K, Mills A (2004) Do malaria control interventions reach the poor? A view through the equity lens. Am J Trop Med Hyg 71(2):174–178

Cox FE (2010) History of the discovery of the malaria parasites and their vectors. Parasit Vectors 3(1):5. https://doi.org/10.1186/1756-3305-3-5

Das A, Gupta RD, Friedman J, Pradhan MM, Mohapatra CC, Sandhibigraha D (2013) Community perceptions on malaria and care-seeking practices in endemic Indian settings: policy implications for the malaria control programme. Malar J 12(1). https://doi.org/10.1186/1475-2875-12-39

Devi P, Jauhari RK, Mondal R (2013) Ovitrap surveillance of Aedes mosquitoes (Diptera: Culicidae) in selected areas of Dehradun district, Uttarakhand, India. Glob J Med Res Diseases 13:53–57

Fischer L, Gültekin N, Kaelin MB, Fehr J, Schlagenhauf P (2020) Rising temperature and its impact on receptivity to malaria transmission in Europe: a systematic review. Travel Med Infect Dis 36:101815. https://doi.org/10.1016/J.TMAID.2020.101815

Gondwe T, Yang Y, Yosefe S, Kasanga M, Mulula G, Luwemba MP et al (2021) Epidemiological trends of malaria in children under five years of Nsanje district in Malawi, 2015–2019. Int J Environ Res Public Health 18. https://doi.org/10.3390/IJERPH182312784

Gunawardena DM, Carter R, Mendis KN, Weerasingha S, Wickremasinghe AR, Kotta PK, Muthuwatta L, Attanayake N, Rajakaruna J, Senanayaka T (1998) Malaria risk factors in an endemic region of Sri Lanka, and the impact and cost implications of risk factor-based interventions. Am J Trop Med Hyg 58(5):533–542. https://doi.org/10.4269/ajtmh.1998.58.533

Hazra A, Tripathi SK, Alam MS (2000) Prescribing and dispensing activities at the health facilities of a non-governmental organization. Natl Med J India 13(4):177–182

Herekar F, Iftikhar S, Nazish A, Rehman S (2020) Malaria and the climate in Karachi: an eight-year review. Pak J Med Sci 36:S33–S37. https://doi.org/10.12669/PJMS.36.ICON-SUPPL.1712

Hussain I, Qureshi NA, Afzal M, Shaheen N, Ali A, Ashraf A (2016) Prevalence and distribution of human Plasmodium infection in federally administrative tribal areas of Pakistan. Acta Parasitol 61(3). https://doi.org/10.1515/ap-2016-0071

Jamil S, Khan MN (2012) Seasonal variations of vivax and falciparum malaria: an observation at a tertiary care hospital. J Ayub Med Coll Abbottabad 24(1):93–95

Karamat J, Shurong T, Ahmad N, Afridi S, Khan S, Khan N (2019) Developing sustainable healthcare systems in developing countries: examining the role of barriers, enablers and drivers on knowledge management adoption. Sustainability 11(4):954. https://doi.org/10.3390/su1104 0954

Karim AM, Hussain I, Malik SK, Lee JH, Cho IH, Kim YB, Lee SH (2016) Epidemiology and clinical burden of malaria in the war-torn area, Orakzai Agency in Pakistan. PLoS Negl Trop Dis 10(1):e0004399. https://doi.org/10.1371/journal.pntd.0004399

Kazmi JH, Pandit K (2001) Disease and dislocation: the impact of refugee movements on the geography of malaria in NWFP Pakistan. Soc Sci Med 52:1043–1055. https://doi.org/10.1016/s0277-9536(00)00211-2

Khan MI, Qureshi H, Bae SJ, Khattak AA, Anwar MS, Ahmad S, Hassan F, Ahmad S (2023) Malaria prevalence in Pakistan: a systematic review and meta-analysis (2006–2021). Heliyon 9(4):e15373. https://doi.org/10.1016/j.heliyon.2023.e15373

Liu Z, Wang S, Zhang Y, Xiang J, Tong MX, Gao Q et al (2021) Effect of temperature and its interactions with relative humidity and rainfall on malaria in a temperate city Suzhou, China. Environ Sci Pollut Res Int 28:16830–16842. https://doi.org/10.1007/S11356-020-12138-4

Malik S, Awan H, Khan N (2012) Mapping vulnerability to climate change and its repercussions on human health in Pakistan. Glob Health 8(1):31. https://doi.org/10.1186/1744-8603-8-31

Mankodi K (1996) Political and economic roots of disease: malaria in Rajasthan on JSTOR. Econ Polit Wkly. https://doi.org/4403717

Milner DA (2017) Malaria pathogenesis. Cold Spring Harb Perspect Med 8(1):a025569. https://doi.org/10.1101/cshperspect.a025569

Minakawa N, Sonye G, Mogi M, Githeko A, Yan G (2002) The effects of climatic factors on the distribution and abundance of malaria vectors in Kenya. J Med Entomol 39:833–841. https://doi.org/10.1603/0022-2585-39.6.833

Nadeem MF, Khattak AA, Zeeshan N, Awan UA, Yaqoob A (2021) Assessment of microscopic detection of malaria with nested polymerase chain reaction in war-torn federally administered tribal areas of Pakistan. Acta Parasitol 66(4). https://doi.org/10.1007/s11686-021-00374-8

Noureen A, Aziz R, Ismail A, Trzcinski AP (2022) The impact of climate change on waterborne diseases in Pakistan. Sustain Clim Change 15(2):138–152. https://doi.org/10.1089/scc.2021.0070

Ogden NH (2017) Climate change and vector-borne diseases of public health significance. FEMS Microbiol Lett 364. https://doi.org/10.1093/FEMSLE/FNX186

Ogden NH, Lindsay LR, Leighton PA (2013) Predicting the rate of invasion of the agent of Lyme disease *Borrelia burgdorferi*. J Appl Ecol 50:510–518. https://doi.org/10.1111/1365-2664.12050

Park K (2005) Park's textbook of preventive and social medicine. Prev Med Obstet Paediatr Geriatr

Patz JA, Reisen WK (2001) Immunology, climate change and vector borne diseases. Trends Immunol 22:171–172. https://doi.org/10.1016/S1471-4906(01)01867-1

Qureshi H, Khan MI, Ambachew H, Pan HF, Ye DQ (2020) Baseline survey for malaria prevalence in Khyber Pakhtunkhwa Province, Pakistan. East Mediterr Health J 26:453–460. https://doi.org/10.26719/EMHJ.19.015

Raghavendra K, Barik TK, Reddy BPN, Sharma P, Dash AP (2011) Malaria vector control: from past to future. Parasitol Res 108(4):757–779. https://doi.org/10.1007/s00436-010-2232-0

Rana MS, Tanveer A, Latif AA, Tahir AH (2011) Significant reduction of malaria in the Punjab, Pakistan after introduction of roll back malaria strategy in 2003. Med Forum Mon 22(9)

Regmi K, Kunwar A, Ortega L (2016) A systematic review of knowledge, attitudes and beliefs about malaria among the South Asian population. Infect Ecol Epidemiol 6(1):30822. https://doi.org/10.3402/iee.v6.30822

Romero-Lankao P, Hughes S, Qin H, Hardoy J, Rosas-Huerta A, Borquez R et al (2014) Scale, urban risk and adaptation capacity in neighborhoods of Latin American cities. Habitat Int 42:224–235. https://doi.org/10.1016/j.habitatint.2013.12.008

Rowland M, Rab MA, Freeman T, Durrani N, Rehman N (2002) Afghan refugees and the temporal and spatial distribution of malaria in Pakistan. Soc Sci Med 55:2061–2072. https://doi.org/10.1016/s0277-9536(01)00341-0

Snow RW, Guerra CA, Noor AM, Myint HY, Hay SI (2005) The global distribution of clinical episodes of *Plasmodium falciparum* malaria. Nature 434(7030):214–217. https://doi.org/10.1038/nature03342

Soomro FR, Pathan GM, Gurbakhshani AL, Kakar JK (2010) Prevalence of malarial parasites in Larkano district, Sindh, Pakistan. DOAJ Directory Open Access J 8(2)

Tusting LS, Willey B, Lucas H, Thompson J, Kafy HT, Smith R, Lindsay SW (2013) Socioeconomic development as an intervention against malaria: a systematic review and meta-analysis. Lancet 382(9896):963–972. https://doi.org/10.1016/s0140-6736(13)60851-x

WHO (2022). Disease outbreak news. In: Criteria for disease outbreak news. Available at: https://www.who.int/emergencies/disease-outbreak-news

Yasinzai MI, Kakar JK (2012) Prevalence of human malaria infection in bordering areas of Balochistan with Sindh province: district Jaffarabad. J Postgrad Med Inst 26(3). http://jpmi.org.pk/index.php/jpmi/article/download/1301/1250

Yasinzai MI Kakarsulemankhel JK (2008) Frequency of various human malaria infections in the hottest areas of central Balochistan, Pakistan: Duki, Harnai, and Sibi. Pak Armed Forces Med J 58:276–285

Yasinzai MI, Kakarsulemankhel JK (2009) Prevalence of human malaria infection in bordering areas of East Balochistan, adjoining with Punjab: Loralai and Musakhel. J Pak Med Assoc 59(3):132–135. https://pubmed.ncbi.nlm.nih.gov/19288935

Zoghi S, Mehrizi AA, Raeisi A, Haghdoost AA, Turki H, Safari R, Kahanali A, Zakeri S (2012) Survey for asymptomatic malaria cases in low transmission settings of Iran under elimination programme. Malar J 11(1):126. https://doi.org/10.1186/1475-2875-11-126

Chapter 8
An Assessment of the Vulnerability and Sustainability of Livelihoods for Riparian Households in Flood and Bank Erosion-Affected Villages of the Kaljani River Basin in Eastern India

Moumita Dutta, Sourav Dey, and Sujit Mandal

Abstract Flooding and riverbank erosion are recognized as the primary hazards within the floodplain area. This research evaluated the vulnerability and sustainability of populations impacted by these phenomena in the Kaljani River basin. Various index-based approaches were utilized, to analyze household vulnerability and sustainability, including livelihood vulnerability, sustainable livelihood, climate vulnerability indices, and current adaptive capacity indices (CACI); while quantifying relative significance of different contributing indicators in the context set by the Intergovernmental Panel on Climate Change (IPCC). The research encompassed three aspects of vulnerability, comprising eight major components and a total of 76 indicators. Households at non-protected villages exhibited greater exposure to natural hazards and climate change (NHCC) risks than those in protected villages. A strong positive correlation was observed between overall livelihood vulnerability index (LVI) and LVI-IPCC, with an '*r*' value of + 0.88. The estimated LVI-IPCC results perfectly matched the vulnerability ranking, and it almost perfectly correlated between climate vulnerability index (CVI) and LVI-IPCC results with an '*r*' value of 0.978. The CVI, and CACI results suggest that inhabitants of non-protected villages face the greatest vulnerable due to climate change, and exhibit a diminished adaptive capacity relative to those residing in protected villages. A review of the indicator-based methodologies indicates that LVI indices produced comparable results concerning livelihood vulnerability and sustainability, which were adequately validated through the LVI-IPCC framework. However, the CVI offered more accurate results than both the LVI and Sustainable Livelihood Index. This analysis is crucial

M. Dutta (✉) · S. Mandal
Department of Geography, Diamond Harbour Women's University, Sarisha, West Bengal, India
e-mail: mou.cob.dutta@gmail.com

S. Dey
Department of Geography, A.B.N. Seal College, Cooch Behar, West Bengal, India

A. Roy et al. (eds.), *Environment and Public Health*, Sustainability Solutions,
https://doi.org/10.1007/978-3-031-99770-9_8

for decision-makers in implementing effective policies for river basin management and development.

Keywords Flood · Vulnerability · Sustainability · Livelihood

8.1 Introduction

Any natural hazards or disasters can change the geographical area and livelihood pattern. The intensity and magnitude of environmental hazards or disasters such as floods, river bank erosion, and avulsion vary from one geographical region to another geographical region (Dey and Mandal 2022). Landslides and climate change influence the livelihood vulnerability in mountainous regions (Oven et al. 2008; Claude et al. 2020), whereas, coastal erosion, sea level rise, and coastal flooding increase the insecurity of life and livelihood in the coastal regions (Hoque et al. 2019). Flooding and river bank erosion have been categorically mentioned as one of the most threatening natural hazards in the world that widely damage the natural environment and devastate human habitations (Norris et al. 2008). Flooding and riverbank erosion are known as the main hazards in the floodplain region. Flood, riverbank erosion, and channel avulsion are the predominant hazards in the Himalayan foothills or across the large rivers of the alluvial plain (Dey and Mandal 2022). However, India's alluvial plains are considered the country's most vulnerable regions as the most destructive natural hazards such as floods and riverbank erosion often hit riparian people (Sarker et al. 2019).

The floodplains primarily consist of the alternating layers of coarser to finer sands, silts, and clays, which contribute to the riverbank erosion. In floodplains, parts of river valleys or charlands are used for seasonal agriculture and animal husbandry because of the fertile and productive nature of the soil (Mondal and Mandal 2018). Inhabitants of riverbanks, also known as Char dwellers, are regarded as the most impoverished and vulnerable to natural calamities (Alam et al. 2017). Individuals residing near the riverbanks are losing large amounts of land every year because of pronounced riverbank erosion. Flash floods frequently occur in the Terai or Duars regions and pose severe threats to survival and livelihoods (Jain et al. 2009; Shreevastav et al. 2021). Every year, these events create massive impacts on riparian households and they lose their homesteads, agricultural land and crops, and other wealth. Consequently, the healthy economy of the river dwellers is going to be converted into a waning economy that can also weaken the adaptive capacity. The unplanned or unscientific constructions have been modifying river valleys at large scale that accelerate the intensity of natural or quasi-natural hazards like flood, avulsion, and riverbank erosion, as well as increase socio-economic and ecological losses. Moreover, the combination of regular flooding and riverbank erosion severely disrupts the socio-economic lives of communities living across India's floodplains. As a result, villagers face limited access to essential resources, ineffective communications, and inadequate infrastructure.

The floods that happened in 2007 and 2010 were some of the most catastrophic in recent history in northern Ghana, displacing 330,000 individuals and causing extensive harm to lives, agricultural land, livestock, and infrastructure (NADMO 2010). The floods of 1993 and 2017 witnessed the worst floods in the Himalayan foothills of Bhutan (DoIW–GoWB 1993). Around 40,000 people were affected in Kalchini block of Alipurduar district, they were rendered homeless, and extensive destruction of human lives, agricultural land, livestock, and public and private infrastructure (DoIW–GoWB 1993). On July 13th, 2023, a flash flood swept away about 50 m of the Asian Highway (AH)-48, and hundreds of hectares of land and workers' quarters were washed away by the Pana River (a tributary of river Kaljani) in the Mechpara tea garden of Kalchini block. Flood and river bank erosion occur over the Himalayan foothills and Duars regions as well as the Kaljani river basin due to heavy rainfall particularly during monsoon, and the course of the river leads to changes (Dey and Mandal 2022; Dutta and Mandal 2022). Kaljani River changes its course which leads to the erosion of its banks and the emergence of sand bars. Such a situation has a profound effect on the livelihoods of the riverine villagers, due to loss of land.

Livelihood vulnerability occurs when a system is susceptible or people have insufficient income, wealth, or properties, and incapable of responding to the detrimental effects of physical and social hazards (DFID 1999; IPCC 2007; Jamshed et al. 2020). The vulnerability of livelihoods among the riparian community increases with increasing the degree or intensity of environmental hazards. In the riparian region, most of the villagers belong to socially and economically marginalized (Alam et al. 2017; Sarker et al. 2019). The framework for analyzing vulnerability of livelihood has been used in many fields of environment like flooding, river bank erosion, climate change, landslide, disaster, land use change, social change, and urbanization (Hahn et al. 2009; Alam et al. 2017; Williams et al. 2018; Huong et al. 2019; Mohamed and El-Raey 2020; Shreevastav et al. 2021; Dey and Mandal 2022).

However, many scholars have reported the condition of livelihood vulnerability for char land and island dwellers (Mukherjee et al. 2019; Dey and Mandal 2022). Certain researchers endeavored to assess the degree of vulnerability of the char lands dwellers in Bangladesh, discovering that physical hazards constituted the primary stressors (Alam et al. 2017; Sarker et al. 2019). Shreevastav et al. (2021) attempted to analyze the impact of floods and inundation on the livelihoods sustained by the inhabitants of the Terai region (central Himalayan foothills). Zhang and Fang (2020) described socioeconomic development and livelihood patterns and also analyzed the impact of climate change on vulnerable communities and the sustainability of their livelihoods in the Koshi River Basin. Also, climate change influences the livelihood vulnerability of village dweller in a positive manner by heightening their vulnerability (Pandey et al. 2020; Arifah et al. 2022).

Many scholars have used various vulnerability or sustainability indices to explore the livelihood status of a particular area for specific purposes (Hahn et al. 2009; Mukherjee et al. 2019; Zhang and Fang 2020; Qin et al. 2022). Pandey and Jha (2011), and Pandey et al. (2017) applied the climate vulnerability index (CVI), and Current Adaptive Capacity Index (CACI) to assess community vulnerability. Hahn et al. (2009), Mukherjee et al. (2019), Shreevastav et al. (2021), Arifah et al.

(2022) and Phuong et al. applied livelihood vulnerability index (LVI) focusing on the Inter-Governmental Panel on Climate Change (IPCC) framework to explore the vulnerability. Zhang and Fang (2020) used the LVI, and Sustainable Livelihood Index (SLI) to assess vulnerability and sustainability for the community of the Himalayan foothill region. Alam et al. (2017) assessed the vulnerability in the riparian char and river-bank households using the LVI, and CVI. Among these methods, the indicator-based method is very popular and widely used. Within these methods, the indicator-based method is particularly well-regarded and frequently adopted. In various areas of environmental hazards or natural disasters like flood vulnerability (Shreevastav et al. 2021), river-bank erosion vulnerability (Dey and Mandal 2022), tidal surges (Mukherjee et al. 2019), climate variability and changes (Hahn et al. 2009; Pandey and Jha 2011; Pandey et al. 2017; Alam et al. 2017; Mukherjee et al. 2019; Tran et al. 2022; Arifah et al. 2022), the major and sub-component indicators-based index methods have been widely applied. But, most scholars have used one or two indicator-based methods for vulnerability analysis, and validated results using IPCC framework. In the present study, different index-based methods such as LVI, and SLI, CVI, and CACI were used and finally the results were verified by the IPCC for vulnerability or sustainability analysis. The previously mentioned literature survey examined multiple indicators; however, it failed to indicate which method is most effective in evaluating the vulnerability of livelihoods within riparian environments. Finally, the study also assessed vulnerability and sustainability situations of flood and bank erosion-affected people in Kaljani river basin. This chapter focuses on the precise characteristics and degree of the community's exposure-sensitivity and ability to adapt to natural and nature-induced vulnerabilities.

8.1.1 The Study Area

The trunk-stream Kaljani originates in a valley near Gedu town of Chukha District in Bhutan at an altitude of 1561 m (26° 54′ 36.86″ N and 89° 28′ 6.55″ E). The River Gabur Basra and River Pana or Alaikuri carry water from the Bhutan Himalayas, and ultimately these two rivers merging to create the main channel of River Kaljani at the foothills of the Bhutan Himalayas near Chuapara in West Bengal, India. The basin area of Kaljani River covers about 1306.52 km^2 (Fig. 8.1) of geographical area extending 26° 16′ 22.8″ N–26° 55′ 37.2″ N latitudes and 89° 21′ 46.8″ E–89° 38′ 38.4″ E longitudes (Dutta and Mandal 2022). The Kaljani River basin (KRB) is characterized by the mainstream of Kaljani river and its various tributaries such as Paro, Nonai, Garam, Cheko, Dima, Katajani, Bura Gadadhar, and Ghargharia (Dutta and Mandal 2022). The KRB consists of four tectonostratigraphic zones, which are organized from north to south namely greater Himalaya, lesser Himalaya, sub-Himalaya, and quaternary sedimentary zones, these zones are bounded by thrust structures (Gansser 1964; Long et al. 2011a, b). This basin is underlain by 14 geological formations such as Duars, Jalpaiguri, Baikuntapur, Shaugoan, Daling, present-day formations, etc.

(Gansser 1964; Long et al. 2011a). The foothills of the KRB belong to a humid tropical monsoon climatic system, with 80–90% of rainfall occurring during the rainy season from months of June to September, while rainfall is scarce during the winter season (Starkel and Soja 2008). Most of the Kaljani river basin lies below 120 m elevation (62.03% area), and about 41% of the catchment area receives more than 500 cm of annual rainfall. Monsoon precipitation from June to October determines the flow of the Kaljani River. Based on slope and relief the basin is classified into three macro-physiographic regions namely the northern plain region, the Piedmont or Duars region, and northern mountainous region. Among these three regions, the Piedmont or Duars region (48.49% area) is susceptible to flash floods, and the northern plains region (37.39% area) is susceptible to river floods. In the old floodplain of the Kaljani, the channel gradient and channel width of the rivers decrease, as well as the channel pattern of the rivers changes from braided to meandering (Starkel and Soja 2008). The younger active floodplain of the river basin is formed by present-day deposits of the River Kaljani and its tributaries; the area is characterized by gently sloping surface features with wide shallow old river valleys and paleo-channels with meander scars and cut-off meander loops.

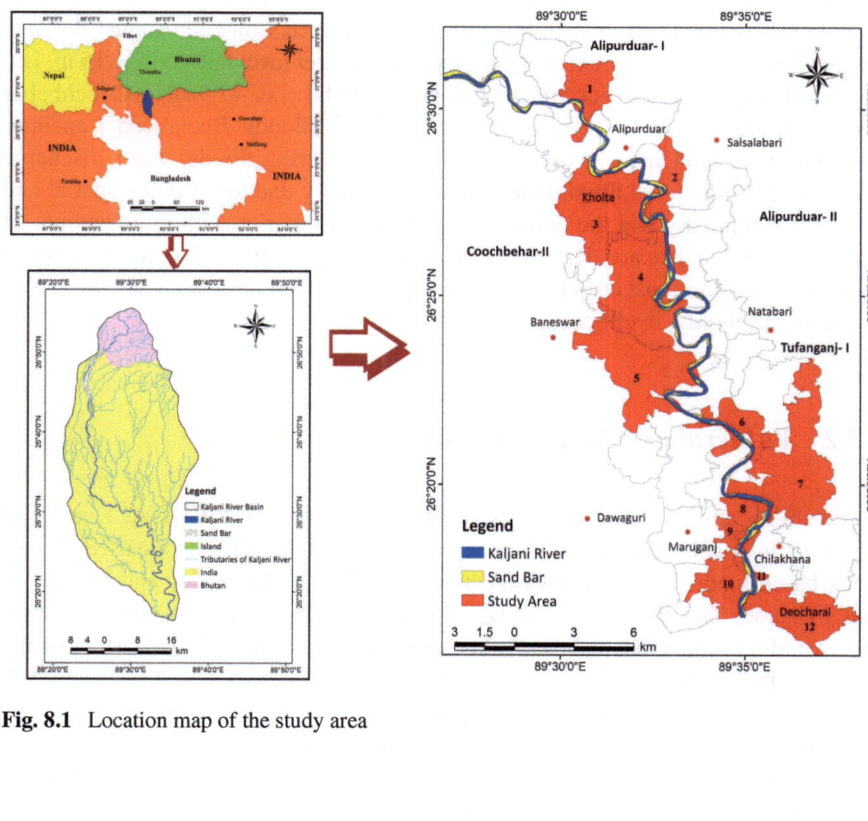

Fig. 8.1 Location map of the study area

8.2 Materials and Methods

8.2.1 Data Collection and Field Survey

The present study collected all data via household surveys conducted through interview and observation methods, except for climatic data. A random sample of 25 households was taken from every village out of 300 samples from 12 villages in Alipurduar and Coochbehar districts. In total, 12 villages were selected randomly, reflecting the degree of flooding and riverbank erosion, livelihood type, accessibility, economic background, embankment location, and geographical location, as well as on the basis of protected areas (protected areas are the areas those are fully protected by concrete embankments) and non-protected areas (non-protected areas are the areas those are not protected by the concrete embankments). All the data was gathered using a structured questionnaire through direct interviews with households in the selected villages (Plate 8.1). Among the selected villages, Bhelakopa-1 is the most affected and undeveloped riparian area, whereas, Kaljani is one of the economically developed riparian areas in the Kaljani basin. The Ghagra and Bhelakopa-1 villages are situated at the extreme upper and lower boundaries of the study area, respectively. As reported by the Department of Irrigation and Waterways, on 12th August 2017 the Kaljani River's water level exceeded the extreme danger level (EDL) by 1.15 m, i.e., 46.85 m in Alipurduar district, and 0.22 m above the EDL, i.e., 42.90 m in Coochbehar district. Based on field surveys during 2017 and 2023 floods, Bhelakopa-1 encountered the most inundated areas in the study area. During floods, water discharge from the foothills to the floodplain increases gradually due to rainfall's spatial and temporal complexity and the confluence of major tributaries with huge amount of water. All the villages are dominated by agricultural practices. So, flood and river bank erosion cause most crop losses, and croplands losses and increasing the risk to riparian households' livelihoods.

8.2.2 Vulnerability Assessment Indices

Vulnerability and durability analysis is done based on LVI, CVI, CACI, and SLI which are fully sub-component dependent (Fig. 8.2). In this chapter, we used various index-based methods such as LVI (Hahn et al. 2009), CVI (Pandey and Jha 2012), CACI (Pandey and Jha 2012), SLI (Zhang and Fang 2020) to measure the vulnerability and sustainability of households, and quantifying the relative extent of the contributing indicators under the broader framework of the IPCC (2007). The IPCC framework combines all components into three contributing factors such as exposure, sensitivity, and adaptive capacity to analyze vulnerability.

On the other hand, the LVI method is a compound type of indicator that assesses the main components collectively, besides, this index also integrates sensitivity and adaptive capacity. The SLA developed by Chambers and Conway (1991) which

Plate 8.1 Glimpses of the study area, **a** riverbank erosion at Bhelapeta village, **b** damages of crops at Bhelapeta village due to flood on 19th June 2022, **c** damages of unmetaled road at Chhat Bhelakopa due to flood on 19th June 2022, **d** condition of road at Bhelakopa-1 village, **e** temporary bamboo bridge to communicate between the Ambari and Natabari, and **f** installation of deep tube well at Ghagra village by government

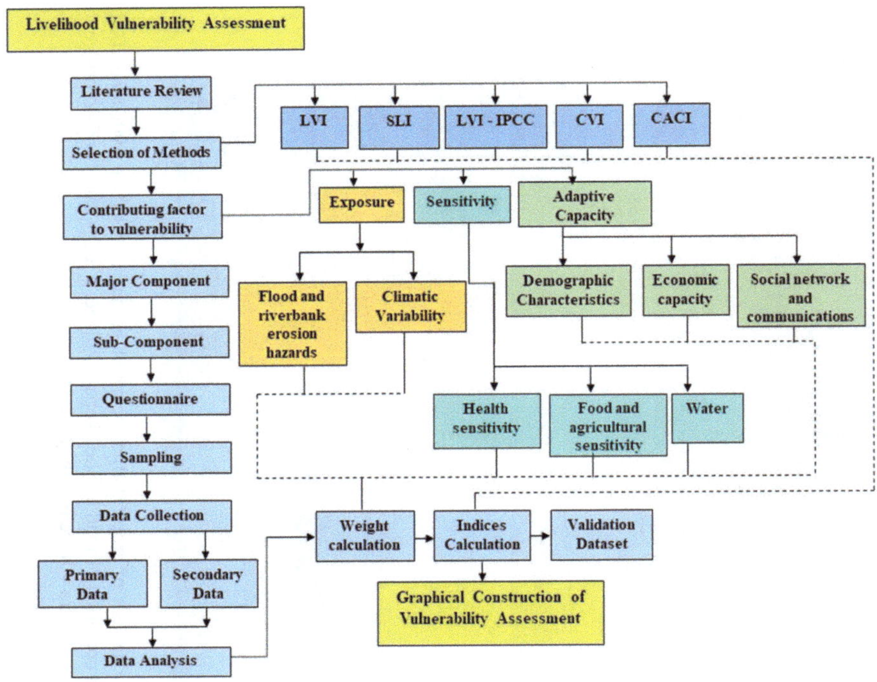

Fig. 8.2 Workflow of the applied methodology

is followed by Zhang and Fang (2020) named the SLI approach, considers five major livelihood capitals such as natural, physical, human, financial, and social, the combined values of the five capitals constitute an index value that can be applied for horizontal comparisons across components but fail to include the subjects of sensitivity as well as adaptive capacity to NHCC. The LVI methodology emphasizes assessing current livelihood strengths, health sensitivity, and water resource characteristics, along with communities' ability to adapt strategies in response to natural hazards and climate-related challenges (Alam et al. 2017; Dey and Mandal 2022). This is why we used the same major components and sub-components instead of major livelihood capital for the SLI approach to fill critical gaps related to sensitivity and adaptive capacity to NHCC issues. Weighted balanced and integrated approaches to applying CVI and CACI were adopted to incorporate indigenous knowledge in selected major components and sub-components. This approach involves assessing how adaptation strategies enhance the ability of households to cope, taking into account specific natural and socio-economic indicators that affect vulnerability of livelihood (Pandey et al. 2017). The CVI and CACI methods were used here to determine the vulnerability as well as adaptive capacity of the riparian communities or households affected by NHCC. In the current research, two indices such as LVI, and SLI were used to analyze the vulnerability and sustainability of riparian peoples, and finally, the IPCC framework was used to validate the results and find out the efficiency of these indicators. We implemented an alternative method to ensure that our experimental results met scientific standards for calculating various indices. These indicators are also designed to measure the vulnerability or sustainability of livelihood at the village level. Every index includes three main components that illustrate the dimensions of vulnerability in household livelihoods.

In this present chapter, the balanced weighted average methods were applied to analyze the vulnerability and sustainability to NHCC at a representative scale (Hahn et al. 2009; Pandey et al. 2017; Alam et al. 2017; Zhang and Fang 2020; Dey and Mandal 2022). The study incorporated 3 dimensions of vulnerability with 8 major components after Hahn et al. (2009), Pandey et al. (2017), Alam et al. (2017), Mukherjee et al. (2019), Zhang and Fang (2020), Shreevastav et al. (2021), Arifah et al. (2022), Phuong et al., Tran et al. (2022), Dey and Mandal (2022) and Li et al. (2023). 76 indicators were considered to develop the LVI, SLI, CVI, CACI, and IPCC framework indices (Table 8.1). THE selection of these indicators is grounded in the existing literature and their relevance to the practical aspects of the study area. The entire vulnerability indices assessment methods was made by the following ways

1. The standardization of the sub-component indices to a value ranges between 0 and 1 through minimum and maximum normalization using Eq. 8.1.
2. Average of the standardized values of all sub-components to determine each major component using Eq. 8.2.
3. Estimation of the LVI values of each village using Eq. 8.3 or Eq. 8.4.
4. Calculation of the SLI values using Eq. 8.5.
5. Assessment of village-wise 3 vulnerability magnitudes (exposure, sensitivity, and adaptive capacity) values applying Eqs. 8.6, 8.7, and 8.8.
6. Estimation of the CVI and LVI-IPCC values of each village by applying Eqs. 8.9 and 8.11.
7. Finally, the CACI value was derived by subtracting the CVI value from 1 using Eq. 8.10.

A well-adjusted weighted method was applied for LVI calculation (Hahn et al. 2009; Dey and Mandal 2022), and individual sub-component plays an equal role in the overall estimation of the index. Every sub-component was measured using various scales, and finally, these were standardized by applying Eq. 8.1.

$$\text{Index}_S = \frac{S_d - S_{min}}{S_{max} - S_{min}} \tag{8.1}$$

where S_d is the accurate value of the sub-component for the households of village d, and S_{min} and S_{max} are the minimum value, and maximum value of each sub-component, respectively.

Major component value is calculated by averaging all sub-components after standardization, as outlined in Eq. 8.2.

Table 8.1 Indexed values of sub-components for the 12 villages of Alipurduar and Coochbehar districts

IPCC contributing factor to vulnerability	Major component	Sub-component	Ghagra	Chengpara	Kholta	Ambari	Kaljani	Bhelapeta
Exposure	Flood and riverbank erosion hazards	Frequency of riverbank erosion last 10 years	0.128	0.184	0.288	0.512	0.144	0.416
		Total number of affected people due to riverbank erosion	0.000	0.513	0.000	0.447	0.632	0.579
		Percent of HHs reporting land degradation during last 30 years	0.200	0.360	0.240	0.600	0.760	0.480
		Percent of HHs experienced displacement in last 30 years	0.000	0.040	0.120	0.280	0.760	0.480
		Percent of HHs with losses of physical assets due to riverbank erosion and other disasters	0.200	0.280	0.160	0.640	0.800	0.520
		Floods events faced over the last 10 years	0.053	0.053	0.474	0.474	0.000	1.000
		Total number of affected people	0.872	0.274	0.009	0.829	0.000	0.846
		Timespan of displacement due to flood	0.300	0.200	0.500	0.200	0.000	0.200

(continued)

Table 8.1 (continued)

IPCC contributing factor to vulnerability	Major component	Sub-component	Ghagra	Chengpara	Kholta	Ambari	Kaljani	Bhelapeta
		Average number of times flood reached inside the house in the past 30 years	0.077	0.031	0.000	0.646	– 0.046	0.554
		The degree of loss/damage to the house and households' possessions (last worst flood event)	1.000	1.000	0.250	0.750	1.000	1.000
		The degree of loss/damage to harvested crops/stockpiled grains (last worst flood event)	1.000	1.000	0.750	1.000	1.000	0.750
		Percentage of HHs reporting a change in the frequency of floods	0.300	0.360	0.160	0.280	0.720	0.720
		Average number of avulsion events in the past 30 years	0.480	0.000	0.000	0.280	0.800	0.560
		Total number of affected people due to avulsion	0.320	0.000	0.000	0.373	0.507	0.253

(continued)

Table 8.1 (continued)

IPCC contributing factor to vulnerability	Major component	Sub-component	Ghagra	Chengpara	Kholta	Ambari	Kaljani	Bhelapeta
		Percent of HHs reporting increased frequency and intensity of riverbank erosion, avulsion and flood	0.280	0.120	0.120	0.360	0.120	0.360
		Percent of HHs with an injury or death as a result of natural disasters in the last 10 years	0.000	0.000	0.080	0.400	0.000	0.120
		Percent of HHs with an injury or death to their livestock as a result of natural disasters in the last 10 years	0.000	0.000	0.000	0.160	0.000	0.120
		Percent of HHs that do not receive a warning before a natural disaster	0.400	0.120	0.760	0.840	0.000	0.760
		Average number of others natural calamity during last 30 years	0.067	0.067	0.400	0.467	0.020	0.467
	Climatic variability	Total rainfall perception index	0.860	0.860	0.880	0.880	0.880	0.880
		Perception index of monsoon rainfall	0.960	0.960	0.850	0.850	0.850	0.850

(continued)

Table 8.1 (continued)

IPCC contributing factor to vulnerability	Major component	Sub-component	Ghagra	Chengpara	Kholta	Ambari	Kaljani	Bhelapeta
		Perception index of winter months rainfall	0.200	0.200	0.100	0.100	0.100	0.100
		Perception index of frequency of floods	0.900	0.900	0.050	0.050	0.950	0.010
		Mean of monthly average precipitation	0.865	0.865	0.201	0.201	0.201	0.201
		Standard deviation of monthly average precipitation	0.273	0.273	0.353	0.353	0.353	0.353
Sensitivity	Health sensitivity	Average time to reach nearest health center (in min)	0.713	0.493	0.373	0.660	0.080	0.693
		Average time to reach nearest hospital/nursing home (in min)	0.550	0.380	0.548	0.638	0.503	0.480
		Percentage of HHs where members suffer from illness	0.160	0.040	0.080	0.000	0.080	0.000
		Percent of HHs reported at least one chronically ill member	0.000	0.040	0.080	0.000	0.040	0.000
		Percent of HHs where a family member is infected by a communicable disease	0.000	0.040	0.000	0.000	0.000	0.000

(continued)

Table 8.1 (continued)

IPCC contributing factor to vulnerability	Major component	Sub-component	Ghagra	Chengpara	Kholta	Ambari	Kaljani	Bhelapeta
		Household members with physically or intellectually handicapped	0.200	0.200	0.000	0.000	0.000	0.000
		Percent of HHs who do not have toilet and having poor hygiene status	0.000	0.000	0.000	0.000	0.000	0.000
		Inverse of life expectancy (in %)	0.600	0.280	0.160	0.480	0.080	0.680
		Percentage of HHs who do not have any health insurance	0.720	0.760	0.800	0.840	0.520	0.960
	Food and agricultural sensitivity	Percentage of households dependent solely on family farm for food	0.640	0.600	0.680	0.880	0.760	0.920
		Average number of months, households struggle to find food	0.007	0.013	0.103	0.190	0.003	0.163
		Percentage of households struggling for food	0.360	0.240	0.160	0.360	0.040	0.480

(continued)

Table 8.1 (continued)

IPCC contributing factor to vulnerability	Major component	Sub-component	Ghagra	Chengpara	Kholta	Ambari	Kaljani	Bhelapeta
		Proportion of households reported that they had at least one month struggling for food	0.360	0.240	0.160	0.360	0.040	0.480
		Percent of HHs who do not practice homestead gardening	0.120	0.160	0.080	0.120	0.400	0.040
		Percent of HHs losing their agricultural land	0.200	0.040	0.080	0.720	0.720	0.520
		Percentage of HHs reporting decreasing agricultural production	0.080	0.040	0.120	0.240	0.160	0.200
		Percentage of HHs reporting increasing food insecurity during natural disasters or other climatic events	0.360	0.320	0.160	0.720	0.240	0.680
		Percentage of HHs reporting decreasing fish production	1.000	0.840	0.560	0.520	0.160	0.240
	Water	Average time to get safe drinking water source (in min)	0.371	0.005	0.003	0.019	0.000	0.741

(continued)

Table 8.1 (continued)

IPCC contributing factor to vulnerability	Major component	Sub-component	Ghagra	Chengpara	Kholta	Ambari	Kaljani	Bhelapeta
		Percentage of HHs using unsafe water for drinking, cooking, bathing and washing	0.400	0.200	0.240	0.000	0.000	0.080
		Percentage of households do not having stable water from a water treatment plant	1.000	1.000	0.000	1.000	0.000	1.000
		Percentage of households that do not have a consistent water supply	1.000	1.000	0.800	1.000	0.720	0.760
		Percentage of HHs whose members walk more than 2 km to reach a water source	0.000	0.120	0.240	0.000	0.160	0.560
		Percent of HHs reporting water conflicts	0.000	0.040	0.120	0.000	0.200	0.120
Adaptive capacity	Demographic characteristics	Dependency ratio (% of population below 15 and over 65 years of age)	0.562	0.776	0.375	0.527	0.877	0.385
		Female to male ratio	0.000	0.025	0.049	0.012	0.166	0.571

(continued)

Table 8.1 (continued)

IPCC contributing factor to vulnerability	Major component	Sub-component	Ghagra	Chengpara	Kholta	Ambari	Kaljani	Bhelapeta
		Percent of female-headed HHs	0.080	0.120	0.160	0.120	0.080	0.040
		Average family member in a HHs	0.250	0.250	0.375	0.250	0.250	0.250
		Population density	0.021	0.008	0.016	0.005	0.008	0.005
		Education index	0.422	0.444	0.444	0.478	0.422	0.489
		Percent of HHs where head of the HH has not attended school	0.080	0.120	0.080	0.080	0.040	0.160
		Percentage of households where members had any formal or informal skill	0.040	0.080	0.040	0.040	0.080	0.160
		Household having members with multiple skills (in %)	0.160	0.240	0.080	0.080	0.240	0.320
	Economic capacity	Percent of households dependent solely on agriculture as income source	0.080	0.160	0.200	0.360	0.360	0.320
		Percent of HHs solely dependent on livestock as a source of income	0.160	0.240	0.200	0.160	0.280	0.200
		Percentage of earning members in a HH	0.377	0.424	0.295	0.263	0.408	0.282

(continued)

Table 8.1 (continued)

IPCC contributing factor to vulnerability	Major component	Sub-component	Ghagra	Chengpara	Kholta	Ambari	Kaljani	Bhelapeta
		Household income per month	0.137	0.147	0.056	0.021	0.210	0.004
		Percent of HHs where family members migrate to work in a different community	0.480	0.520	0.240	0.120	0.080	0.280
		Ratio of non-agricultural income to total income	0.710	0.600	0.380	0.140	0.330	0.220
		Household members having moved to cities for employment	0.580	0.380	0.280	0.100	0.200	0.140
		Percent of HHs who have a family member working outside the village at relatively developed place	0.290	0.190	0.140	0.050	0.100	0.070
		Percent of HHs where women family members are not allowed to work outside the home	0.200	0.240	0.520	0.640	0.280	0.720
		Number of households who have burden of loan	0.840	0.800	0.880	0.960	0.560	0.920

(continued)

Table 8.1 (continued)

IPCC contributing factor to vulnerability	Major component	Sub-component	Ghagra	Chengpara	Kholta	Ambari	Kaljani	Bhelapeta
		Percent of HHs who have access to financial services to any financial institution	0.840	0.520	0.800	0.640	0.840	0.600
	Social network and communications	Percent of HHs who have received any kind of support from neighbour in past 1 month	0.400	0.480	0.560	0.680	0.440	0.720
		Average time to reach nearest vehicle station (in min)	0.208	0.268	0.220	0.632	− 0.032	0.688
		Percent of HHs have communicative devices (TV, radio, mobile etc.) at home	1.000	1.000	1.000	0.960	1.000	0.960
		Percent of HHs who have provided assistance to others	0.080	0.120	0.280	0.560	0.120	0.640
		Percentage of HHs who do not receive assistance from the government	0.280	0.320	0.320	0.240	0.160	0.280
		Percentage of HHs who do not receive assistance from NGOs	0.920	0.880	0.800	0.760	0.960	0.720

(continued)

Table 8.1 (continued)

IPCC contributing factor to vulnerability	Major component	Sub-component	Ghagra	Chengpara	Kholta	Ambari	Kaljani	Bhelapeta
		Percentage of HHs who do not use mobile phones for communication	0.000	0.000	0.000	0.040	0.000	0.040

IPCC contributing factor to vulnerability	Major component	Sub-component	Chhatoa	Jaigirchilakhana	Chhat Bhelakopa	Bhelakopa-1	Bhelakopa-2	Deocharai
Exposure	Flood and riverbank erosion hazards	Frequency of riverbank erosion last 10 years	0.672	0.288	0.360	0.576	0.208	0.352
		Total number of affected people due to riverbank erosion	0.303	0.105	0.000	0.816	0.105	0.289
		Percent of HHs reporting land degradation during last 30 years	0.760	0.320	0.280	0.920	0.360	0.520
		Percent of HHs experienced displacement in last 30 years	0.240	0.120	0.160	0.920	0.360	0.520
		Percent of HHs with losses of physical assets due to riverbank erosion and other disasters	0.760	0.240	0.280	0.960	0.240	0.480

(continued)

Table 8.1 (continued)

IPCC contributing factor to vulnerability	Major component	Sub-component	Chhatoa	Jaigirchilakhana	Chhat Bhelakopa	Bhelakopa-1	Bhelakopa-2	Deocharai
		Floods events faced over the last 10 years	1.000	1.000	0.053	0.474	0.053	0.053
		Total number of affected people	0.658	0.342	0.103	0.231	0.068	0.000
		Timespan of displacement due to flood	0.400	0.200	0.100	0.200	0.200	0.200
		Average number of times flood reached inside the house in the past 30 years	0.400	0.185	0.292	0.708	0.185	0.431
		The degree of loss/damage to the house and households' possessions (last worst flood event)	1.000	0.500	0.250	1.000	0.750	0.500
		The degree of loss/damage to harvested crops/stockpiled grains (last worst flood event)	1.000	0.750	0.750	1.000	0.750	1.000
		Percentage of HHs reporting a change in the frequency of floods	0.680	0.120	0.160	0.720	0.320	0.480

(continued)

Table 8.1 (continued)

IPCC contributing factor to vulnerability	Major component	Sub-component	Chhatoa	Jaigirchilakhana	Chhat Bhelakopa	Bhelakopa-1	Bhelakopa-2	Deocharai
		Average number of avulsion events in the past 30 years	0.000	0.000	0.000	0.640	0.000	0.200
		Total number of affected people due to avulsion	0.000	0.000	0.000	0.640	0.000	0.347
		Percent of HHs reporting increased frequency and intensity of riverbank erosion, avulsion and flood	0.160	0.120	0.080	0.800	0.400	0.360
		Percent of HHs with an injury or death as a result of natural disasters in the last 10 years	0.040	0.000	0.000	0.080	0.000	0.040
		Percent of HHs with an injury or death to their livestock as a result of natural disasters in the last 10 years	0.000	0.000	0.040	0.160	0.000	0.120

(continued)

Table 8.1 (continued)

IPCC contributing factor to vulnerability	Major component	Sub-component	Chhatoa	Jaigirchilakhana	Chhat Bhelakopa	Bhelakopa-1	Bhelakopa-2	Deocharai
		Percent of HHs that do not receive a warning before a natural disaster	0.720	0.480	0.600	0.760	0.520	0.440
		Average number of others natural calamity during last 30 years	0.533	0.313	0.220	0.667	0.447	0.667
	Climatic variability	Total rainfall perception index	0.880	0.880	0.880	0.880	0.880	0.880
		Perception index of monsoon rainfall	0.850	0.850	0.850	0.850	0.850	0.850
		Perception index of winter months rainfall	0.100	0.100	0.100	0.100	0.100	0.100
		Perception index of frequency of floods	0.010	0.010	0.900	0.050	0.900	0.900
		Mean of monthly average precipitation	0.201	0.201	0.201	0.201	0.201	0.201
		Standard deviation of monthly average precipitation	0.353	0.353	0.353	0.353	0.353	0.353

(continued)

Table 8.1 (continued)

IPCC contributing factor to vulnerability	Major component	Sub-component	Chhatoa	Jaigirchilakhana	Chhat Bhelakopa	Bhelakopa-1	Bhelakopa-2	Deocharai
Sensitivity	Health sensitivity	Average time to reach nearest health center (in min)	0.513	0.367	0.240	0.473	0.087	0.353
		Average time to reach nearest hospital/nursing home (in min)	0.130	0.265	0.333	0.480	0.503	0.893
		Percentage of HHs where members suffer from illness	0.040	0.120	0.040	0.000	0.040	0.000
		Percent of HHs reported at least one chronically ill member	0.000	0.040	0.000	0.000	0.040	0.000
		Percent of HHs where a family member is infected by a communicable disease	0.000	0.000	0.040	0.000	0.000	0.000
		Household members with physically or intellectually handicapped	0.000	0.400	0.000	0.000	0.000	0.200

(continued)

Table 8.1 (continued)

IPCC contributing factor to vulnerability	Major component	Sub-component	Chhatoa	Jaigirchilakhana	Chhat Bhelakopa	Bhelakopa-1	Bhelakopa-2	Deocharai
		Percent of HHs who do not have toilet and having poor hygiene status	0.000	0.000	0.000	0.040	0.000	0.000
		Inverse of life expectancy (in %)	0.720	0.440	0.240	0.840	0.280	0.160
		Percentage of HHs who do not have any health insurance	0.840	0.680	0.800	0.960	0.720	0.680
	Food and agricultural sensitivity	Percentage of households dependent solely on family farm for food	0.960	0.760	0.720	0.920	0.800	0.760
		Average number of months, households struggle to find food	0.160	0.090	0.077	0.213	0.097	0.010
		Percentage of households struggling for food	0.440	0.160	0.120	0.480	0.080	0.040
		Proportion of households reported that they had at least one month struggling for food	0.440	0.160	0.120	0.480	0.080	0.040

(continued)

Table 8.1 (continued)

IPCC contributing factor to vulnerability	Major component	Sub-component	Chhatoa	Jaigirchilakhana	Chhat Bhelakopa	Bhelakopa-1	Bhelakopa-2	Deocharai
		Percent of HHs who do not practice homestead gardening	0.040	0.280	0.120	0.040	0.240	0.080
		Percent of HHs losing their agricultural land	0.760	0.320	0.280	0.920	0.360	0.520
		Percentage of HHs reporting decreasing agricultural production	0.000	0.120	0.160	0.320	0.200	0.120
		Percentage of HHs reporting increasing food insecurity during natural disasters or other climatic events	0.720	0.560	0.160	0.760	0.360	0.400
		Percentage of HHs reporting decreasing fish production	0.240	0.360	0.320	0.720	0.760	0.480
	Water	Average time to get safe drinking water source (in min)	0.683	0.355	0.317	0.795	0.283	0.296

(continued)

Table 8.1 (continued)

IPCC contributing factor to vulnerability	Major component	Sub-component	Chhatoa	Jaigirchilakhana	Chhat Bhelakopa	Bhelakopa-1	Bhelakopa-2	Deocharai
		Percentage of HHs using unsafe water for drinking, cooking, bathing and washing	0.120	0.080	0.040	0.240	0.000	0.080
		Percentage of households do not having stable water from a water treatment plant	1.000	0.000	1.000	1.000	1.000	1.000
		Percentage of households that do not have a consistent water supply	0.840	0.240	0.360	0.920	0.480	0.200
		Percentage of HHs whose members walk more than 2 km to reach a water source	0.600	0.240	0.280	0.720	0.480	0.720
		Percent of HHs reporting water conflicts	0.120	0.040	0.080	0.160	0.160	0.120

(continued)

Table 8.1 (continued)

IPCC contributing factor to vulnerability	Major component	Sub-component	Chhatoa	Jaigirchilakhana	Chhat Bhelakopa	Bhelakopa-1	Bhelakopa-2	Deocharai
Adaptive capacity	Demographic characteristics	Dependency ratio (% of population below 15 and over 65 years of age)	0.303	0.566	0.511	0.440	0.743	0.540
		Female to male ratio	0.092	0.055	0.466	0.442	0.276	0.055
		Percent of female-headed HHs	0.040	0.120	0.120	0.040	0.080	0.120
		Average family member in a HHs	0.375	0.250	0.375	0.375	0.250	0.250
		Population density	0.000	0.005	0.016	0.005	0.016	0.013
		Education index	0.478	0.467	0.467	0.489	0.478	0.467
		Percent of HHs where head of the HH has not attended school	0.160	0.120	0.200	0.080	0.200	0.160
		Percentage of households where members had any formal or informal skill	0.120	0.240	0.040	0.040	0.040	0.560
		Household having members with multiple skills (in %)	0.240	0.480	0.080	0.080	0.160	0.560

(continued)

Table 8.1 (continued)

IPCC contributing factor to vulnerability	Major component	Sub-component	Chhatoa	Jaigirchilakhana	Chhat Bhelakopa	Bhelakopa-1	Bhelakopa-2	Deocharai
	Economic capacity	Percent of households dependent solely on agriculture as income source	0.520	0.400	0.480	0.440	0.120	0.240
		Percent of HHs solely dependent on livestock as a source of income	0.120	0.280	0.080	0.040	0.280	0.240
		Percentage of earning members in a HH	0.282	0.327	0.306	0.260	0.388	0.413
		Household income per month	0.018	0.106	0.092	0.008	0.092	0.190
		Percent of HHs where family members migrate to work in a different community	0.120	0.240	0.120	0.360	0.200	0.040
		Ratio of non-agricultural income to total income	0.180	0.270	0.080	0.170	0.500	0.360

(continued)

Table 8.1 (continued)

IPCC contributing factor to vulnerability	Major component	Sub-component	Chhatoa	Jaigirchilakhana	Chhat Bhelakopa	Bhelakopa-1	Bhelakopa-2	Deocharai
		Household members having moved to cities for employment	0.140	0.160	0.060	0.120	0.340	0.180
		Percent of HHs who have a family member working outside the village at relatively developed place	0.070	0.080	0.030	0.060	0.170	0.090
		Percent of HHs where women family members are not allowed to work outside the home	0.680	0.800	0.560	0.760	0.520	0.320
		Number of households who have burden of loan	0.920	0.720	0.640	0.960	0.680	0.640
		Percent of HHs who have access to financial services to any financial institution	0.640	0.720	0.760	0.600	0.760	0.840

(continued)

Table 8.1 (continued)

IPCC contributing factor to vulnerability	Major component	Sub-component	Chhatoa	Jaigirchilakhana	Chhat Bhelakopa	Bhelakopa-1	Bhelakopa-2	Deocharai
	Social network and communications	Percent of HHs who have received any kind of support from neighbour in past 1 month	0.720	0.560	0.820	0.840	0.680	0.640
		Average time to reach nearest vehicle station (in min)	0.408	0.228	0.052	0.468	0.036	0.392
		Percent of HHs have communicative devices (TV, radio, mobile etc.) at home	0.920	1.000	1.000	0.920	0.960	0.880
		Percent of HHs who have provided assistance to others	0.680	0.240	0.360	0.720	0.200	0.360
		Percentage of HHs who do not receive assistance from the government	0.080	0.240	0.320	0.120	0.280	0.320

(continued)

Table 8.1 (continued)

IPCC contributing factor to vulnerability	Major component	Sub-component	Chhatoa	Jaigirchilakhana	Chhat Bhelakopa	Bhelakopa-1	Bhelakopa-2	Deocharai
		Percentage of HHs who do not receive assistance from NGOs	0.840	0.920	0.920	0.800	0.920	0.960
		Percentage of HHs who do not use mobile phones for communication	0.080	0.000	0.000	0.080	0.040	0.120

Source Calculated by the researcher (field survey, 2023–24)

$$M_{\mathrm{d}} = \frac{\sum_{i=1}^{n} \mathrm{index}_{\mathrm{sd}}}{N} \tag{8.2}$$

Here, M_{d} denotes one of the eight major components of village d, $\mathrm{index}_{\mathrm{sd}}$ indicates the standardized value associated with each sub-component of village d, and N means the total count of sub-elements in each major component.

After the analysis of eight major components, the LVI values for each village were established by means of Eq. 8.3.

$$\mathrm{LVI}_{\mathrm{d}} = \frac{\sum_{z=1}^{8} W_{Mz} M_{az}}{\mathrm{WM}_z}. \tag{8.3}$$

This can also be expressed as:

$$\mathrm{LVI}_{\mathrm{d}} = \frac{\begin{array}{c} W_{\mathrm{FREH}}\mathrm{FREH}_{\mathrm{b}} + W_{\mathrm{CV}}\mathrm{CV}_{\mathrm{b}} + W_{\mathrm{HS}}\mathrm{HS}_{\mathrm{b}} + W_{\mathrm{FAS}}\mathrm{FAS}_{\mathrm{b}} + W_{\mathrm{W}}\mathrm{W}_{\mathrm{b}} \\ + W_{\mathrm{DC}}\mathrm{DC}_{\mathrm{b}} + W_{\mathrm{EC}}\mathrm{EC}_{\mathrm{b}} + W_{\mathrm{SNC}}\mathrm{SNC}_{\mathrm{b}} \end{array}}{W_{\mathrm{FREH}} + W_{\mathrm{CV}} + W_{\mathrm{HS}} + W_{\mathrm{FAS}} + W_{\mathrm{W}} + W_{\mathrm{DC}} + W_{\mathrm{EC}} + W_{\mathrm{SNC}}} \tag{8.4}$$

where $\mathrm{LVI}_{\mathrm{d}}$ denotes the Livelihood Vulnerability Index of village d, calculated as the weighted average of eight major components. WM_z denotes the weight of each principal component, established based on the total number of sub-components incorporated in each major component. In the study, the LVI value span from 0 to 0.5 indicating least to extremely vulnerable.

After calculating the $\mathrm{LVI}_{\mathrm{d}}$ values, the $\mathrm{SLI}_{\mathrm{d}}$ values of each village were determined using Eq. 8.5 provided below.

$$\mathrm{SLI}_{\mathrm{d}} = 1 - \mathrm{LVI}_{\mathrm{d}} \tag{8.5}$$

Here, $\mathrm{SLI}_{\mathrm{d}}$ refers to the capacity for livelihood sustainability in district d. A negative relationship is observed between the LVI value and its vulnerability level. Increased LVI values suggest greater vulnerability and reduced sustainability, which also reduces livelihood sustainability.

The exposure (E) includes flood and riverbank erosion hazards (FREH) as well as climate variability (CV) calculated through the application of Eq. 8.6.

$$E = \frac{W_{E1}\mathrm{FREH}_{\mathrm{b}} + W_{E2}\mathrm{CV}_{\mathrm{b}}}{W_{E1} + W_{E2}} \tag{8.6}$$

where W_{E1} and W_{E2} denote the weight of FREH, and CV, respectively. Each weight is defined by the cumulative total of sub-components. The sensitivity includes health sensitivity (HS), food and agricultural sensitivity (FAS), and water (W). The sensitivity value (S) was calculated using Eq. 8.7.

$$S = \frac{W_{S1}\mathrm{HS}_{\mathrm{b}} + W_{S2}\mathrm{FAS}_{\mathrm{b}} + W_{S3}\mathrm{W}_{\mathrm{b}}}{W_{S1} + W_{S2} + W_{S3}} \tag{8.7}$$

where W_{S1}, W_{S2}, and W_{S3} represent the weights assigned to HS, FAS and W, respectively. The weight is defined as the sum of all sub-components.

The adaptive capacity (AC) includes demographic characteristics (DC), economic capacity (EC), and social network and communications (SNC) which was calculated applying Eq. 8.8.

$$\mathrm{AC} = \frac{W_{\mathrm{ac1}}\mathrm{DC_b} + W_{\mathrm{ac2}}\mathrm{EC_b} + W_{\mathrm{ac3}}\mathrm{SNC_b}}{W_{\mathrm{ac1}} + W_{\mathrm{ac2}} + W_{\mathrm{ac3}}} \tag{8.8}$$

where W_{ac1}, W_{ac2}, and W_{ac3} represent the weight of DC, EC, and SNC, respectively. The weight is defined as the sum of all sub-components.

The values of exposure, adaptive capacity, and sensitivity were employed to derive CVI as specified in Eq. 8.9.

$$\mathrm{CVI} = 1 - \left|\left\{\frac{N_1E - N_2\mathrm{AC}}{(N_1 + N_2)}\right\}\right| \times \left\{\frac{1}{S}\right\} \tag{8.9}$$

In this context, N_i represents the total count of key components related to the ith degree of vulnerability. Each dimension's value varies between 0 and 1, indicating low and high vulnerability. The higher values of CVI reflect higher vulnerability and lower adaptability. After calculating the CVI value, CACI (by subtracting the CVI value from 1) was calculated using Eq. 8.10. CACI values range between 0 and 1.

$$\mathrm{CACI} = 1 - \mathrm{CVI} \tag{8.10}$$

8.2.3 *IPCC Framework Approach*

The IPCC approach integrates eight essential components of vulnerability analysis focusing on three dimensions: exposure, sensitivity, and adaptive capacity. The IPCC vulnerability framework was designed by Hahn et al. (2009) (Eq. 8.11).

$$\text{LVI-IPCC}_{\mathrm{d}} = (E - \mathrm{AC}) \times S, \tag{8.11}$$

where $\text{LVI-IPCC}_{\mathrm{d}}$ represents the LVI value for village d according to the IPCC vulnerability framework. LVI-IPCC values vary between − 1, representing the lowest level of vulnerability, to 1, representing the highest level of vulnerability.

8.3 Results and Discussion

8.3.1 Livelihood Vulnerability Analysis

The livelihood vulnerability of 12 villages was assessed considering LVI, SLI, and vulnerability. This study includes the initial identification and assessment of the indicators, which may affect the livelihood of households as well as explored the degree of vulnerability or sustainability and assessed the value of factors contributing to vulnerability in each village. The sub-component values of each village are shown in Table 8.1. The values of major components and factors contributing to vulnerability in each village are illustrated in Tables 8.2 and 8.3. The results of the four indices and the IPCC framework of 12 villages are depicted in Table 8.4.

8.3.2 Livelihood Vulnerability Index

The overall LVI values indicated that residents of 12 villages having different levels of vulnerability (Table 8.4). The result shows that non-protected villages (means the riverbank is not protected by the embankments) having higher LVI (0.472 in Bhelakopa-1) and protected villages (means the riverbank is protected by the embankments) having low LVI (0.277 in Kholta). The residents of non-protected villages live in remote areas that are far away from the developed areas. These residents faced more disadvantages than the protected riverside residents because of non-availability of health facilities, financial assistance, educational facilities, greater exposure to natural hazards, and fewer opportunities to develop financially (Table 8.3).

The flood and river bank erosion hazards index values are high in Bhelakopa-1 (0.646), Bhelapeta (0.536), Ambari (0.502), and Chhatoa (0.491) (Table 8.2; Fig. 8.3) because of low elevation (Plate 8.2e) and the presence of less cohesive soil near the riverbank. Indexed values in terms of climatic variability differ slightly across villages because values of the most sub-components remained nearly constant, and only the flood frequency perception index values varied (Table 8.1).

The index value of health sensitivity was lower in Kaljani village (0.145) because of greater life expectancy and health facilities. However, significant variation in food and agricultural sensitivity are observed between protected and non-protected villages (Table 8.2; Plate 8.2). Non-protected villages like Bhelakopa-1 (0.539), Ambari (0.457), Chhatoa (0.418), and Bhelapeta (0.414) having higher values than the protected villages such as Chhat Bhelakopa (0.231), Kholta (0.234), Deocharai (0.272), Chengpara (0.277), Kaljani (0.280), JaigirChilakhana (0.312), Bhelakopa-2 (0.331), and Ghagra (0.347) (Table 8.2; Fig. 8.3), due to losses or damages of agricultural land (Plate 8.1b) and increasing food insecurity during floods and riverbank erosion hazards. The accessibility of water, as indicated by indexed values, was found to be higher in economically backward villages (e.g., Bhelakopa-1, Chhatoa, Bhelapeta, etc.) compared to economically advanced villages (e.g., JaigirChilakhana,

Table 8.2 Indexed values of major components for the 12 villages of Alipurduar and Coochbehar districts

Major component	Ghagra	Chengpara	Kholta	Ambari	Kaljani	Bhelapeta	Chhatoa	Jaigirchilakhana	Chhat Bhelakopa	Bhelakopa-1	Bhelakopa-2	Deocharai
Flood and riverbank erosion hazards	0.299	0.242	0.227	0.502	0.380	0.536	0.491	0.268	0.196	0.646	0.261	0.368
Climatic variability	0.676	0.676	0.406	0.406	0.556	0.399	0.399	0.399	0.547	0.406	0.547	0.547
Health sensitivity	0.327	0.248	0.227	0.291	0.145	0.313	0.249	0.257	0.188	0.310	0.185	0.254
Food and agricultural sensitivity	0.347	0.277	0.234	0.457	0.280	0.414	0.418	0.312	0.231	0.539	0.331	0.272
Water	0.462	0.394	0.234	0.336	0.180	0.544	0.560	0.159	0.346	0.639	0.400	0.403
Demographic characteristics	0.179	0.229	0.180	0.177	0.240	0.264	0.201	0.256	0.253	0.221	0.249	0.303
Economic capacity	0.427	0.384	0.363	0.314	0.332	0.341	0.335	0.373	0.292	0.343	0.368	0.323
Social network and communications	0.413	0.438	0.454	0.553	0.378	0.578	0.533	0.455	0.496	0.564	0.445	0.525

Source Calculated by the researcher (field survey, 2023–24)

Table 8.3 Factors contributing to LVI-IPCC values for the 12 villages of Alipurduar and Coochbehar districts

Contributing factor values	Ghagra	Chengpara	Kholta	Ambari	Kaljani	Bhelapeta	Chhatoa	Jaigirchilakhana	Chhat Bhelakopa	Bhelakopa-1	Bhelakopa-2	Deocharai
Exposure	0.389	0.346	0.270	0.479	0.422	0.503	0.469	0.299	0.280	0.588	0.330	0.411
Sensitivity	0.368	0.296	0.231	0.364	0.204	0.408	0.390	0.253	0.244	0.478	0.294	0.298
Adaptive capacity	0.341	0.346	0.326	0.330	0.313	0.377	0.342	0.355	0.332	0.360	0.348	0.368

Source Calculated by the researcher (field survey, 2023–24)

Table 8.4 Overall LVI, SLI, CVI, CACI, and LVI-IPCC values for the 12 villages of Alipurduar and Coochbehar districts

Indices	Ghagra	Chengpara	Kholta	Ambari	Kaljani	Bhelapeta	Chhatoa	Jaigirchilakhana	Chhat Bhelakopa	Bhelakopa-1	Bhelakopa-2	Deocharai
Overall LVI	0.365	0.330	0.277	0.390	0.315	0.428	0.399	0.305	0.287	0.472	0.325	0.360
SLI	0.635	0.670	0.723	0.610	0.685	0.572	0.601	0.695	0.713	0.528	0.675	0.640
CVI	0.868	0.766	0.622	0.982	0.906	0.939	0.955	0.630	0.644	0.960	0.738	0.810
CACI	0.132	0.234	0.378	0.018	0.094	0.061	0.045	0.370	0.356	0.040	0.262	0.190
LVI-IPCC	0.018	0.000	– 0.013	0.054	0.022	0.051	0.050	– 0.014	– 0.012	0.109	– 0.005	0.013

Source Calculated by the researcher (field survey, 2023–24)

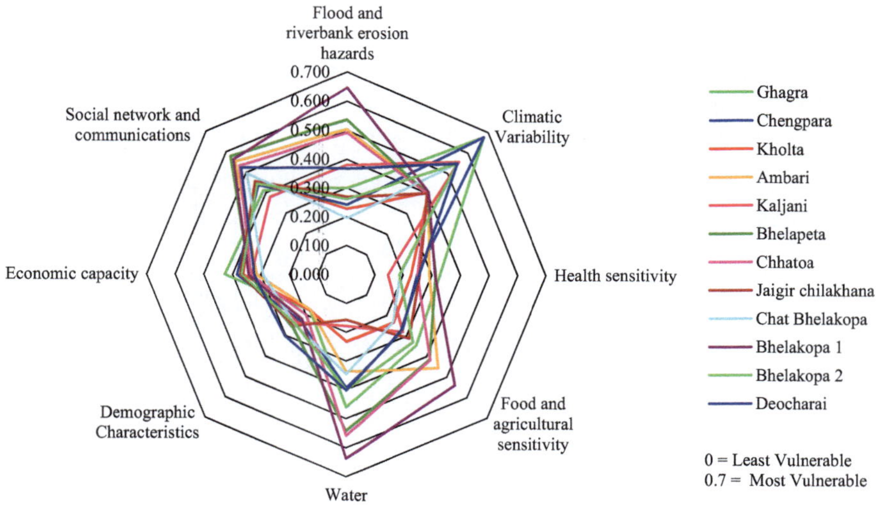

Fig. 8.3 Eight major component of the vulnerability of 12 villages in Alipurduar and Coochbehar districts

Kaljnai, and Kholta, etc.) where households walked over 2 km to find safe drinking water (Table 8.1). Indexed values of demographic characteristics and economic capacity were almost the same among all villages but considerable differences were observed among different sub-components (Table 8.1). The indexed values for social network and communication were 0.578, 0.564, and 0.553 in Bhelapeta, Bhelakopa-1, and Ambari villages respectively due to a lack of contactable support (Plate 8.1d, e). Riverine households suffer from the shortage of health facilities and potable water (Alam et al. 2017) leading to livelihood vulnerability.

8.3.3 *Major Component-Based Livelihood Vulnerability*

The most significant components or factors in LVI were flood and riverbank erosion hazards, food and agricultural sensitivity, and water. The LVI values of all these three factors are 0.646, 0.539, and 0.639 in non-protected villages (Table 8.2). The involvement of 'food and agricultural sensitivity' to the livelihood vulnerability of non-protected villagers revealed that many households are completely dependent on the family farm for their food supply, regularly losing their agricultural land during riverbank erosion, limited engagement in crop farming on the riverbank side because of flood and (Plate 8.1b) bank erosion during monsoon. Consequently, they are being forced to struggle for food and they faced food insecurity after the floods and bank erosion events. Riverine people are entirely dependent on activities related to livestock and agriculture for their food and income, which worsens the livelihood risks for households that do not engage in diverse economic activities (Plate 8.2b).

Plate 8.2 Overview of the study area, **a** educational institution at Chhat Bhelakopa village, **b** crop production at Bhelakopa-2 village, **c** collection of fishes from the Kaljani River near Gadadhara temple, **d** thermal power line at Bhelakopa-2, **e** inundation of houses and brick-fields at Bhelakopa-1 village due to flood on 14th July 2023, and **f** during household survey at Ambari village. *Source* Field survey, 2017–23

The ‘water’ component put greater impact on the vulnerability of livelihood in the households of non-protected villages, and riverbank side households of protected villages. Some households in the above-mentioned areas utilize contaminated water for purposes such as cooking, drinking, washing, and bathing. Most of the riverbank side households access safe drinking water sources after walking a distance of more than 2 km (except Ghagra and Ambari villages), as well as they have no dependable supply of water (except Ghagra, Chengmari, and Ambari). However, a few tube wells and water tanks have been installed by the Government to reduce the scarcity of fresh drinking water (Table 8.1), but most of them are not working regularly. It is clear that respondents are aware of dirty water yet substitutes solutions often remain out of reach option often become unavailable; local communities are also facing challenges related to water competition.

Flooding and riverbank erosion hazards and their contribution to the vulnerability of livelihoods in lowland areas and riverbank dwellers shows that most of the households faced flood, avulsion, and riverbank erosion hazards during monsoon; consequently, the households loss their homestead land, harvested crops (Plate 8.1b), stockpiled grains, livestock, crop-fields, etc. (Table 8.1). The households of non-protected villages and riverbank side households of protected villages having higher vulnerability to flood and riverbank erosion hazard-related problems. In the last 10 years, they have frequently faced riverbank erosion due to the absence of bank protection measures (ecological or bio-engineering) in the vulnerable riverbank sites. Consequently, the displacement issues arose, and the common people lost their physical assets, homesteads, and crop fields. In the last 30 years, the households of Bhelakopa-1, Kaljani, Ambari, Chhatoa, Bhelapeta, and Deocharai villages have been affected more by riverbank erosion (Plate 8.1a) than the households of other villages. Lowlands of the protected and non-protected village areas damaged by flood almost every year due to heavy rainfall, and excess water released through Bhutan’s barrage. This situation causes displacement problems in these particular places. Flood water affected most of the households of Bhelakopa-1, Ambari, Bhelapeta, and Chhatoa villages due to low elevation (Plate 8.2e) and absence of embankments (Table 8.1). Ghagra, Ambari, Kaljani, Bhelapeta, Bhelakopa-1, and Deocharai villages faced channel avulsion of the Kaljani River during flood incidents over the last 30 years, as a result, the households lost their properties like crops, crop fields, houses, homesteads, livestock, etc., which increases the livelihood vulnerability.

The component of ‘health sensitivity’ no longer influences vulnerability in 12 villages (Table 8.2; Fig. 8.3), but some variations related to vulnerability takes place among the households of all villages, because villagers could not access health centers and hospitals (Table 8.1). As a result of which, higher social vulnerability is found in the villages of Bhelapeta (0.578), Bhelakopa-1 (0.564), Ambari (0.553), Chhatoa (0.533) (Table 8.2), due to lack of transport facilities, damages of roads due to flooding (Plate 8.1c), as well as households consuming more time to reach nearest vehicle stations.

Although, the LVI, and CVI were found higher in non-protected villages. The index values for the riverbank and lowland households are also high since the flood

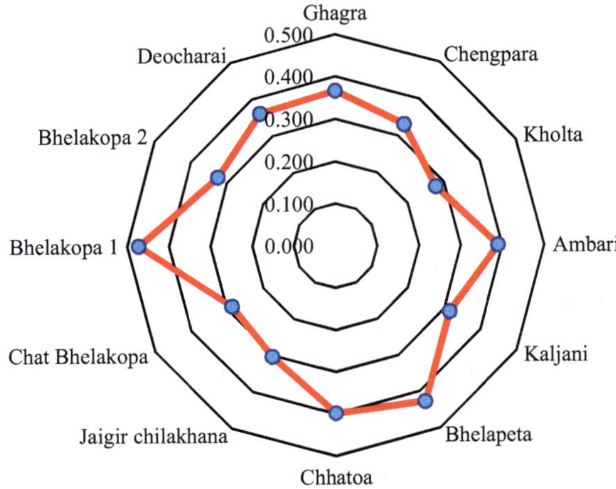

Fig. 8.4 Overall LVI values of the households of 12 villages in Alipurduar and Coochbehar districts

and riverbank erosion hazards, food and agricultural sensitivity, and water components act as principal drivers for livelihood vulnerability (Tables 8.2, 8.4; Figs. 8.3, 8.4, 8.5, 8.6, 8.7, 8.8). Riverine and lowland households suffer from insufficient access to safe drinking water and limited healthcare infrastructure (Alam et al. 2017). The components of LVI including health sensitivity, demographic characteristics, economic capacity, and social network and communications depicted no significant difference between the dwellers of protected and non-protected villages (Table 8.2; Fig. 8.3).

8.3.4 Analysis of Climate Vulnerability

There is significant variation in climate vulnerability across all villages. The maximum and minimum values of CVI were 0.982 and 0.622 for the households of Ambari and Kholta villages respectively (Table 8.4; Fig. 8.6). The CVI values of economically backward and advanced villages are almost same which reveals that every household in the two villages is susceptible to hazards related to flooding, riverbank erosion, and the effects of climate change and variability (Tables 8.2 and 8.4). The assessment of exposure, sensitivity, and adaptive capacity is shown in Table 8.3 and Fig. 8.5.

In terms of exposure, the households of non-protected villages are more susceptible to the impacts of natural hazards and climate change than the households of protected villages (Table 8.3; Fig. 8.5). Flood and bank erosion directly affect non-protected households (Plates 8.1a–c and 8.2e) whereas protected households do not experience flood and bank erosion processes in riparian areas. But, villages of Ghagra, Chengpara, Kaljani, Bhelakopa-2, and Deocharai may experience flooding and riverbank erosion. In terms of sensitivity, the households of non-protected villages are

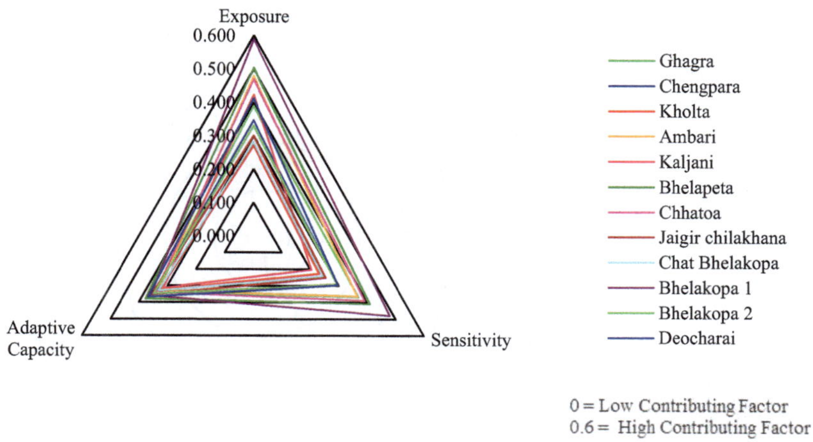

Fig. 8.5 Factors influencing exposure, sensitivity, and adaptive capacity of households in 12 villages of Alipurduar and Coochbehar districts

more sensitive than protected village households because of the decrease in cultivated land, damages to water sources, and failure of medical assistance, during floods and riverbank erosion.

The estimated LVI-IPCC results perfectly matched the vulnerability ranking (Table 8.4), and it almost perfectly correlated between CVI and LVI-IPCC results with '*r*' value of 0.978. As a result, households in non-protected villages of Ambari (0.982), Bhelakopa-1 (0.960), Chhatoa (0.955), and Bhelapeta (0.939) are considered more vulnerable and less adaptable to natural hazards and climate change (Table 8.4; Fig. 8.6).

Fig. 8.6 CVI values of the households of 12 villages in Alipurduar and Coochbehar districts

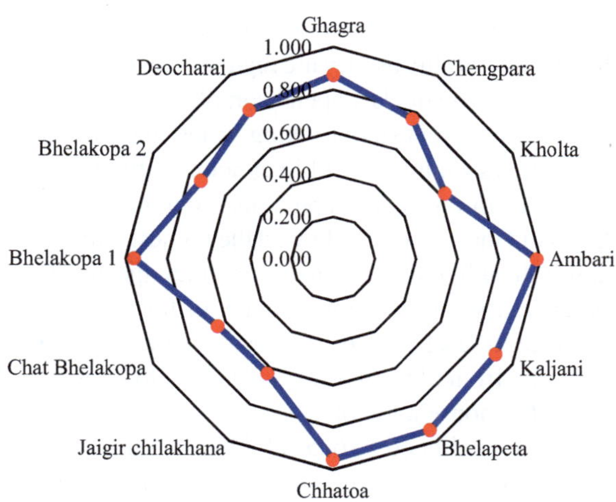

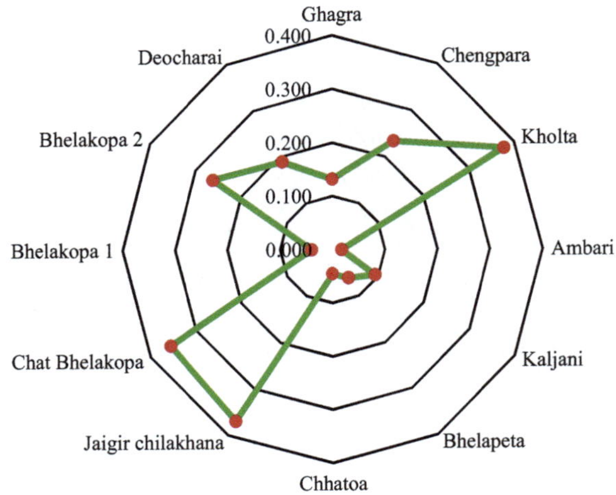

Fig. 8.7 CACI values of the households of 12 villages in Alipurduar and Coochbehar districts

The CVI provides the values related to for adaptive capacity, sensitivity, and climate change exposure (Alam et al. 2017). The results of CVI and CACI show that the dwellers of non-protected villages, and riverbank and lowland households face the highest vulnerability to climate change and have a reduced capacity to adapt compared to inhabitants of protected settlements (Table 8.4; Figs. 8.6, 8.7).

8.3.5 *Sustainable Livelihood Index (SLI) and Status of Sustainable Livelihood*

The SLI values for the villages showed in Table 8.4 and Fig. 8.8. As previously mentioned, The households of non-protected villages are registered with higher livelihood vulnerability as they are regularly losing or damaging their houses, homesteads, crops, and crop fields, depend completely on the family farm for supply of food, and lacking consistent safe drinking water, and transport facilities. Finally, they experienced less sustainability of livelihood (Table 8.4) which is associated with the area's relatively lower demographic and economic vulnerability (Table 8.2). The households of Ghagra village showed the highest vulnerability in response to climatic variability, health sensitivity, and economic capacity (Table 8.2) due to maximum average precipitation and perception index of flood frequency. The households cover more distance to reach the nearest health center, and consequently households have been suffering from illness issues, and lack of local employment opportunities (Table 8.1). Thus, the households of Ghagra as a protected village also exhibited less sustainability of livelihood (0.635) (Table 8.4; Fig. 8.8). The households of Kholta village showed the highest livelihood sustainability (Fig. 8.8) due to lower vulnerability in response to hazards related to flooding and riverbank erosion, food and agricultural

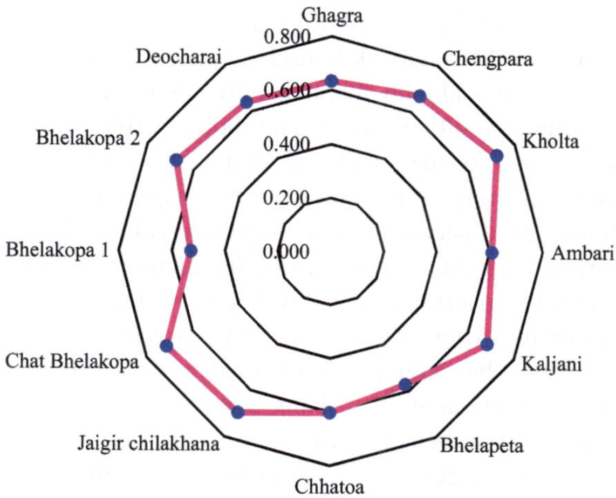

Fig. 8.8 SLI values of the households of 12 villages in Alipurduar and Coochbehar districts

sensitivity, and water issues (Table 8.2; Fig. 8.3). Whereas, households of Bhelakopa-1 showed the lowest livelihood sustainability (0.528) because of the highest vulnerability in response to hazards related to flooding and riverbank erosion (0.646), food and agricultural sensitivity (0.539), and water (0.639) (Figs. 8.3 and 8.8). The households of Chhat Bhelakopa reported higher livelihood sustainability (0.0713) caused by lower vulnerability of the agricultural sensitivity, economic capacity, health sensitivity, and floods and riverbank erosion hazards (Table 8.2).

The household economy of is dependent on agriculture, but the villagers stated that the income related to agricultural production is not enough to support a family (Zhang and Fang 2020). Most households have taken loans to purchase fertilizers and machinery to increase crop production, which puts a lot of pressure on developing local agriculture and sustaining household livelihoods (Zhang and Fang 2020). The riverbank dwellers of Kaljani, Deocharai, Ghagra, Bhelakopa-2, and Bhelakopa-1 villages (Plate 8.2c) reported others source of income, i.e., selling of fishes.

Consequently, the riverbank dwellers face restricted access to extensive employment opportunities, healthcare services (Zhang and Fang 2020), drinking water, and transport facilities. The households of Ghagra and Deocharai showed almost equal exposure, but the households of Ghagra village exhibited slightly higher sensitivity and much lower adaptive capacity (Fig. 8.5). In comparison with protected and non-protected villages, non-protected village's (Bhelakopa-1) households portrayed the lowest level of sustainability of livelihood (0.528) (Fig. 8.8). Natural hazards, land loss, damage of crops, poor medical facilities, scarcity of safe drinking water, and poor transport facilities, etc., have created significant challenges to sustainability (Table 8.1). However, those living on the protected riverbank benefit from enhanced accessibility to services and employment options relative to non-protected riverbank residents. The primary issues reported by residents living along the riverbank are land loss and the absence of early warning systems of disaster. Furthermore, losses of land, and damages of crops also significantly contribute to the degree of vulnerability of the riverbank households (Plate 8.1a, b). These issues exposed that crop production is already threatened by floods and riverbank erosion hazards.

After compiling various data about the riverbank's features, it is possible to create a risk map for the area being studied. By analyzing multiple riverbank parameters and consulting the region's cadastral maps, areas at risk of bank erosion can be identified. The objectives of mapping these vulnerable regions are to identify potentially affected riparian zones, inform the community about the risks along the riverside, and develop a risk map that will support future governmental plans for restoring the riverbank. After preparing the bank erosion risk or vulnerability mapping, we can plan some bio-engineering or ecological preventive measures to reduce the damage in the potential areas of bank erosion. As a result, it will become much easier to deal with riverbank hazards or disasters. The identification of flood-prone regions is crucial for effective land use planning. Initially, comprehensive data regarding past flooding events in a specific area is gathered. Subsequently, utilizing geographical information systems, a flood risk map or flood prone area map may be developed based on the identified flooding parameters. These maps are essential in guiding the public to avoid the construction of residential properties, homesteads, or institutions in the most flood-affected zones. Conversely, the creation of recreational spaces, such as playgrounds or environmentally sustainable amusement parks, along with tree planting, may be regarded as acceptable. Recently, numerous proposals for such eco-friendly amusement parks have been put forward. Proper eco-friendly floodplain management in the lower catchment of the Kaljani River basin that will minimize damage to infrastructure and crops, as well as reduce the potential for damage due to flooding and bank erosion without disrupting the natural flow of the river.

8.3.6 Correlation Between Overall LVI, SLI, CVI, and CACI, and LVI-IPCC

The outcomes of LVI and LVI-IPCC composite indices showed similar results (Table 8.4; Figs. 8.4, 8.9). The correlation between overall LVI and LVI-IPCC is highly positive (+ 0.88 '*r*' value). Figure 8.6 depicts Jaigir Chilakhana village shows least vulnerability and Ambari village exhibits most vulnerability. The Climatic vulnerability also yield similar results for Ambari (0.982) village with highest climatic vulnerability and lowest adaptive capacity with least CACI value of 0.018 (Table 8.4; Figs. 8.6, 8.7). Higher the vulnerability lower the sustainability, and vice versa (Zhang and Fang 2020). The Bhelakopa-1 village depicted highest exposure and sensitivity values (Table 8.3), which leads to more livelihood vulnerability and less livelihood sustainability (Figs. 8.4 and 8.8). The results of SLI and LVI-IPCC ranking is also negatively correlated with '*r*' value of − 0.88. The Kholta village showed the lowest exposure (0.270) and relatively lowest sensitivity (0.231) compared with Bhelakopa-1 village (0.588 and 0.478) (Table 8.3), indicating lowest

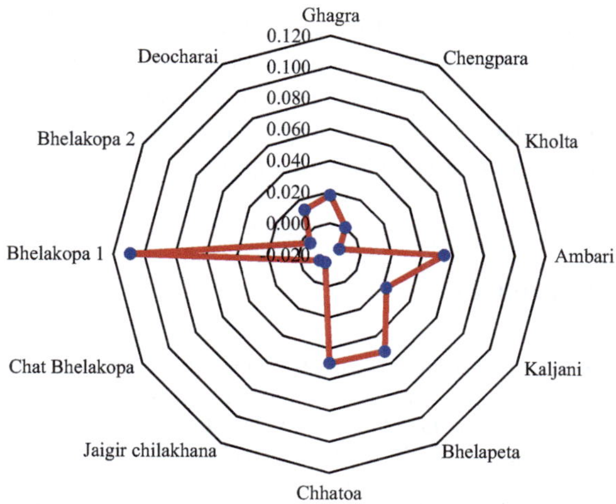

Fig. 8.9 LVI-IPCC values of 12 villages in Alipurduar and Coochbehar districts

livelihood sustainability (Fig. 8.8). Yet, the Kaljani village shows slightly high exposure, low sensitivity, and the high adaptability leading to moderate livelihood sustainability (Figs. 8.5 and 8.8). The LVI-IPCC results (Fig. 8.9) are almost perfectly matched with the estimated values of overall LVI, SLI, CVI, and CACI.

8.4 Conclusions

This study assessed LVI, CVI, and SLI quantitatively for exploring the vulnerability of livelihood, climatic vulnerability, and adaptability of different households in rural communities in relation to natural hazards and change of climate. This study assessed twelve villages of Alipurduar and Coochbehar districts by applying a vulnerability and sustainability analysis that emphasizes natural hazards. Results derived from the overall livelihood and climate vulnerability, and sustainability metrics indicates that the most at-risk residents are those who living in unprotected riverbank areas. The villages of Bhelakopa-1, Ambari, Bhelapeta, and Chhatoa located on the non-protected riverbank, showed higher vulnerability and lower sustainability. The inhabitants of Kaljani river basin face multiple natural threats, including bank erosion, floods, and avulsions. These hazards resulted damaged to houses, crops, public health infrastructure, water supply, transportation, and communications, have adversely affected the quality of life and well-being of the riparian people. In most of the villages climate change poses a risk to the sustainability of those living along the river, increasing their vulnerability to climate-related challenges. After analyzing the indicator-based methods, it can be inferred that all the LVI indices provided almost similar results in terms of livelihood vulnerability and sustainability, and these were properly validated through the LVI-IPCC framework. However, CVI provided more precise results than LVI and SLI. The livelihood vulnerability analysis can be of immense important to decision makers to adopt appropriate policies for river basin management and development. Proper environmentally friendly floodplain management in the lower catchment of the Kaljani river basin will minimize damage to infrastructure and crops, as well as reduce the potential for damage due to flooding and bank erosion without disrupting the natural flow of the river.

References

Alam GM, Alam K, Mushtaq S, Clarke ML (2017) Vulnerability to climatic change in riparian char and river-bank households in Bangladesh: implication for policy, livelihoods and social development. Ecol Indic 72:23–32. https://doi.org/10.1016/j.ecolind.2016.06.045

Arifah, Salman D, Yassi A, Demmallino EB (2022) Livelihood vulnerability of smallholder farmers to climate change: a comparative analysis based on irrigation access in South Sulawesi, Indonesia. Reg Sustain 3:244–253. https://doi.org/10.1016/j.regsus.2022.10.002

Chambers R, Conway GR (1991) Sustainable rural livelihoods: practical concepts for the 21st century. Discussion paper 296. Institute of Development Studies, Brighton, UK

Claude MJ, Martin NV, Abias M, Francoise M, Johnson U, Tonny K, Martine U (2020) Mapping landslide susceptibility and analyzing its impact on community livelihoods in Gakenke district, Northern Rwanda. J Geosci Environ Prot 8(5):41–55. https://doi.org/10.4236/gep.2020.85003

Department for International Development (1999) Sustainable livelihoods guidance sheets, numbers 1–8. Department for International Development, London. Available: www.livelihoods.org

Dey S, Mandal S (2022) Riverbank erosion hazards and channel morphodynamics. A perspective of fluvial geomorphology. Routledge

DoIW–GoWB: Department of Irrigation and Waterways, Govt. of West Bengal (1993) Annual flood report for the year 1993. Directorate of Advance Planning, Project Evaluation & Monitoring Cell, Kolkata, p 111

Dutta M, Mandal S (2022) Channel migration vulnerability in the Kaljani River Basin of eastern India. In: Mandal S, Maiti R, Nones M, Beckedahl HR (eds) Applied geomorphology and contemporary issues. Springer, pp 635–654. https://doi.org/10.1007/978-3-031-04532-5_33

Gansser A (1964) Geology of the Himalayas, 1st edn. Interscience Publishers, Zurich

Hahn MB, Riederer AM, Foster SO (2009) The livelihood vulnerability index: a pragmatic approach to assessing risks from climate variability and change—a case study in Mozambique. Glob Environ Change 19(1):74–88. https://doi.org/10.1016/j.gloenvcha.2008.11.002

Hoque MZ, Cui S, Xu L, Islam I, Tang J, Ding S (2019) Assessing agricultural livelihood vulnerability to climate change in coastal Bangladesh. Int J Environ Res Public Health 16(22):4552. https://doi.org/10.3390/ijerph16224552

Huong NTL, Yao S, Fahad S (2019) Assessing household livelihood vulnerability to climate change: the case of Northwest Vietnam. Hum Ecol Risk Assess Int J 25(5):1157–1175. https://doi.org/10.1080/10807039.2018.1460801

Intergovernmental Panel on Climate Change (2007) Summary for policy makers. In: Parry ML, Canziani OF, Palutikof JP, Linden PJ, Hanson CE (eds) Climate change 2007: impacts, adaptation and vulnerability. Contribution of working group II to the fourth assessment report of the Intergovernmental Panel on Climate Change. Cambridge University Press, Cambridge, UK

Jain SK, Sharma BR, Zahid A, Jin M, Shreshtha JL, Kumar V, Wang JA (2009) Comparative analysis of the hydrogeology of the Indus-Gangetic and Yellow River basins. In: Groundwater governance in the Indo-Gangetic and Yellow River basins: realities and challenges, pp 43–64. https://doi.org/10.1201/9780203874479.ch3

Jamshed A, Birkmann J, Rana IA, Feldmeyer D (2020) The effect of spatial proximity to cities on rural vulnerability against flooding: an indicator based approach. Ecol Indic 118. https://doi.org/10.1016/j.ecolind.2020.106704

Li T, Singh RK, Pandey R, Liu H, Cui L, Xu Z, Xia A, Wang F, Tang L, Wu W, Du J, Cui X, Wang Y (2023) Enhancing sustainable livelihoods in the Three Rivers Headwater Region: a geospatial and obstacles context. Ecol Indic 156. https://doi.org/10.1016/j.ecolind.2023.111134

Long S, McQuarrie N, Tobgay T, Grujic D, Hollister L (2011a) Geological map of Bhutan. J Maps 184–192. https://doi.org/10.4113/jom.2011.1159

Long S, McQuarrie N, Tobgay T, Rose C, Gehrels G, Grujic D (2011b) Tectonostratigraphy of the Lesser Himalaya of Bhutan: implications for the along-strike stratigraphic continuity of the northern Indian margin. Geol Soc Am GSA Bull 123(7):1406–1426. https://doi.org/10.1130/B30202.1

Mohamed SA, El-Raey ME (2020) Vulnerability assessment for flash floods using GIS spatial modeling and remotely sensed data in El-Arish City, North Sinai, Egypt. Nat Hazards 102:707–728. https://doi.org/10.1007/s11069-019-03571-x

Mondal J, Mandal S (2018) Monitoring changing course of the river Ganga and land-use dynamicity in Manikchak Diara of Malda district, West Bengal, India, using geospatial tools. Spat Inf Res 26(6):691–704. https://doi.org/10.1007/s41324-018-0210-2

Mukherjee N, Siddique G, Basak A, Roy A, Mandal MH (2019) Climate change and livelihood vulnerability of the local population on Sagar Island, India. Chin Geogr Sci 29(3):417–436. https://doi.org/10.1007/s11769-019-1042-2

NADMO (2010) Annual report of the National Disaster Management Organization. NADMO Regional Office, Tamale, Ghana

Norris FH, Steven SP, Pfefferbaum B, Wyche KF, Pfefferbaum RL (2008) Community resilience as a metaphor, theory, set of capacities, and strategy for disaster readiness. Am J Community Psychol 41:127–150. https://doi.org/10.1007/s10464-007-9156-6

Oven KJ, Petley DN, Rigg JR, Dunn CE, Rosser NJ (2008) Landscape, livelihoods and risk: a study of community vulnerability to landslide events in Central Nepal. In: Climate change disaster impact reduction, pp 94–102

Pandey R, Jha SK (2011) Climate vulnerability index—measure of climate change vulnerability to communities: a case of rural Lower Himalaya, India. Mitig Adapt Strateg Glob Change 17:487–506. https://doi.org/10.1007/s11027-011-9338-2

Pandey R, Jha S (2012) Climate vulnerability index—measure of climate change vulnerability to communities: a case of rural Lower Himalaya, India. Mitig Adapt Strateg Glob Change 17(5):487–506. https://doi.org/10.1007/s11027-011-9338-2

Pandey R, Jha SK, Alatalo JM, Archie KM, Gupta AK (2017) Sustainable livelihood framework-based indicators for assessing climate change vulnerability and adaptation for Himalayan communities. Ecol Indic 79:338–346. https://doi.org/10.1016/j.ecolind.2017.03.047

Pandey BW, Anand S, Negi VS, Pathak U, Prasad AS (2020) Ecological challenges and vulnerability assessment for exploring the adaptation-development nexus for sustainability in Alaknanda River Basin, Uttarakhand, India. In: Geoecology of landscape dynamics, pp 359–377. https://doi.org/10.1007/978-981-15-2097-6_22

Qin Z, Haili X, Xiao L, Luwei D, Bojie W, Fengqi C, Haiping T (2022) Livelihood vulnerability of pastoral households in the semiarid grasslands of northern China: measurement and determinants. Ecol Indic 140. https://doi.org/10.1016/j.ecolind.2022.109020

Sarker MNI, Wu M, Alam GM, Shouse RC (2019) Livelihood vulnerability of riverine-island dwellers in the face of natural disasters in Bangladesh. Sustainability. https://doi.org/10.3390/su11061623

Shreevastav BB, Tiwari KA, Mandal RA, Nepal A (2021) Assessing flood vulnerability on livelihood of the local community: a case from southern Bagmati corridor of Nepal. Prog Disaster Sci 12:1–15. https://doi.org/10.1016/j.pdisas.2021.100199

Starkel L, Soja R (2008) Geomorphic hazards by rain in the mountains of Eastern India. In: Singh S, Starkel L, Syiemlieh HJ (eds) Environmental changes and geomorphic hazards. Bookwell, New Delhi, pp 237–246

Tran PT, Vu BT, Ngo ST, Tran VD, Ho TDN (2022) Climate change and livelihood vulnerability of the rice farmers in the North Central Region of Vietnam: a case study in Nghe An province, Vietnam. Environ Chall 7. https://doi.org/10.1016/j.envc.2022.100460

Williams PA, Crespo O, Abu M (2018) Assessing vulnerability of horticultural smallholders to climate variability in Ghana: applying the livelihood vulnerability approach. Environ Dev Sustain 1–22. https://doi.org/10.1007/s10668-018-0292-y

Zhang C, Fang Y (2020) Application of capital-based approach in the measurement of livelihood sustainability: a case study from the Koshi River basin community in Nepal. Ecol Indic 116. https://doi.org/10.1016/j.ecolind.2020.106474

Chapter 9
Resilient Communities and Health Systems: Kerala's Lessons in Tackling Health Vulnerabilities

Vishnu N. Ramani

Abstract Kerala's health system resilience during the COVID-19 pandemic has been lauded globally. The state's robust public healthcare system has accelerated the state's ability to manage disasters effectively by containing the spread of infection. Additionally, a multisectoral action through decentralised healthcare and social participation in the health system preparedness and response has been the core of Kerala's unique legacy. Nonetheless, the social intermediation and health system responsiveness amidst a disaster can be considered a model for other states to learn from. This chapter will examine the COVID-19 response at the local level through an explorative study. The chapter's central argument will focus on how social brokerage among various stakeholders enabled COVID-19 and how it has facilitated the health system to become more resilient. Furthermore, the chapter delves into what are the ways in which different socio-political groups in an institutional set-up have enabled the response to COVID-19 at the grassroots level, and their constructive integration into the bottom-up strategies of the health system has paved the way for a better response. Kerala's resilient communities and health systems stand as exemplars forever and help prepare for future disasters.

Keywords Disaster response · People's participation · COVID-19 · Local self-government · Social intermediation · Health system resilience

9.1 Introduction

The world before and after Coronavirus disease 2019 (COVID-19) is critical in marking the evolving developments in global health and disaster response. The virus has devastated people's lives and challenged systems of various kinds. Furthermore, it has determined new norms of life and value systems for humankind. When world leaders, national governments, international development agencies, NGOs, and

V. N. Ramani (✉)
Independent Researcher, New Delhi, India
e-mail: vishnu.n.ramani@gmail.com

A. Roy et al. (eds.), *Environment and Public Health*, Sustainability Solutions,
https://doi.org/10.1007/978-3-031-99770-9_9

humanitarian organisations were in panic and uncertain about a concrete measure to curb the spread of COVID-19 across the countries, there was an emergence of unique voices and stories of success in tackling the virus from a few parts of the world. This chapter will delve deep into how such a diverse story came out of Kerala in its response to the pandemic, which has been a significant model in disaster response mechanisms across the globe. This implies the necessity of documenting and replicating such localised response strategies to prepare for future health shocks in a highly globalised world. Kerala's response was multisectoral compared to other Indian states since a robust primary healthcare system, a proactive civil society, and state machinery could act on an immediate call. Manifold bits and pieces of work on the COVID-19 response, ranging from social, economic, political and epidemiological aspects of the response activities, have already been done. The health system's response to any epidemic in pluralistic health systems is only possible with a people-centred and holistic approach to dealing with the problem. Consequently, the health and community system's response to the outbreak of COVID-19 infection is of high significance in studying the interconnectedness of various social systems and institutional structures that bring together diverse stakeholders in action to save people from suffering and uphold the motto of healthy, thriving communities.

Kerala's health system has always been in the headlines of global media and international development agencies, which helped to get immense attention for its exemplary performance in basic health indices (Balabanova et al. 2013). The state is far ahead of other Indian states regarding outstanding health outcomes, especially maternal and child health (Gandhi et al. 2021). However, the state was the first to be hit by COVID-19 in India because the people resided in nearly more than half of the countries in the world. Kerala's migration, especially to the gulf and other Western countries, and its influence on the state's economy through remittances is a well-studied area. Furthermore, student migration, especially professional migration, has accelerated in the recent past, which has shown a surge in the number of students pursuing higher education abroad. Kerala Model of Development, a tag assigned by development scholars across the globe for the state since the 1970s, has emphasised the need to examine the model's resilience amid the global crisis. The achievements in the social sectors of the state have a deep-rooted connection with history, along with strong socio-political movements for an egalitarian society. Its gains in reducing extreme poverty, building empowered communities, consistent efforts to invest in human development, favourable public financing of welfare-centric schemes and projects, higher literacy and a land of people with high literacy and hospitality all led to significant improvement in the social cohesion. Thus, this chapter will critically engage the social dynamics, how social brokerage among various stakeholders enabled COVID-19 and how it has facilitated the health system to become more resilient. Moreover, the chapter delves into what are the ways in which different socio-political groups in an institutional set-up have accelerated the response, and their constructive integration into the bottom-up strategies of the health system has resulted in a better response to the COVID-19 infection at the local level. The chapter has five sections, starting with the introduction and rationale of the study, methodology, findings, discussion, policy implications, and concluding remarks. Each section will have subsections to organise the arguments and findings of the study, along with in-depth details of the research topic being critically examined.

9.1.1 The Rationale of the Study

In a post-COVID era, the legacies and learning of the response mechanisms that nation-states worldwide have taken are a history and politics of knowledge, especially in the global health discourse. The present study is also centred on the public health underpinning of the disaster response amidst COVID-19, as it mainly deals with prevention and primary health care. The collective action at the local level, supported by an institutional structure, made it a unique strategy of intensive community mobilisation to prepare for the outbreak. Therefore, documenting and putting out stories of success along the lines of social intermediation in health during difficult times needs to be studied critically. The study was conducted in 2020 and 2022 to capture the localised health system response to the COVID-19 epidemic in Kerala's semi-urban areas. It was a time when this was novel to the world and caused social, economic, political, and global health threats to the international community and people worldwide. There were varied responses to the virus based on the nation-states' local conditions and domestic characteristics. Nonetheless, some countries and their local administrative regions have responded effectively to the virus, and much of this has been associated with a robust primary healthcare system and proactive governments with higher public spending on health. In the Kerala context, there were many factors responsible for the success of the virus's containment in the state during the initial phase of the pandemic. However, the later waves due to newer strains of the virus could shake all Indian states, including Kerala. This chapter looks at the COVID-19 response at the local level in the state's first wave of the pandemic.

9.1.2 Conceptual Framework

The study has been conceptualised on the foundations of social intermediation, which is a process that facilitates connections, enhances the capacity of individual and social groups, and builds relationships. It acts as a bridge, linking diverse community members and groups with the institutions, resources, knowledge, and social networks they need to advance their socioeconomic betterment (Kabir and Krishnan 1973). In this chapter, the diverse range of elements of social intermediation during COVID-19 includes resource mobilisation, community participation, social cohesion and trust, cooperation and coordination, social inclusion, information dissemination, spreading disaster awareness, and leveraging local knowledge. Similarly, the actors in the process of social intermediation during the pandemic in Kerala society were health and social sector actors, including community leaders, community-based organisations, political leaders, youth organisations, diverse representatives of the local self-government institutions, and the people's collectives. These factors could significantly positively affect people-centred COVID-19 preparedness and response in Kerala, which was later celebrated nationally and internationally. Therefore, community resilience was accelerated through the effective functioning of the elements of the social intermediation process, which has been a unique characteristic of Kerala society ever since the formation of the state (Kabir and Krishnan 1973).

9.2 Methodology

This explorative study on COVID-19 has two broad research objectives, primarily exploring the COVID-19 response and its preparedness by various stakeholders in the Manjeri municipality of Malappuram district in Kerala. Secondly, it examines the role of innovative community mobilisation and digital intervention in the strategies implemented at the grassroots levels of the study area. These research objectives will be discussed in detail in the upcoming section on the conceptual framework. These included the role of health and social sector actors, intersectoral coordination and social intermediation, decentralised governance and local health action, and lastly, building back better after the crises.

9.2.1 Research Design

The study adopted an exploratory qualitative research design because of the nature of this chapter's research questions, such as micro-level COVID-19 responses in the selected locality. A purposive sampling technique was employed to gather information, and it was conducted when most respondents were in the field fighting the virus. Meanwhile, various respondents from diverse backgrounds have been recruited to get a clear picture of the local dynamics in COVID-19 preparedness and response at the local level. Figure 9.1 shows the flowchart of the research methods used for the study.

The study location is Manjeri municipality of Malappuram district, Kerala, which was selected on the criteria of being one of the state's largest municipalities and the district's commercial capital. It has a medical college, a few primary health centres and sub-centres. The Municipality has 50 local self-government-level wards with three revenue villages: Payyanad, Manjeri, and Narukara. According to the 2011 census, the Municipality's total population is 97,102, with a sex ratio of 1058 and a population density of 1830 per square kilometre. The health system of Manjeri consists of primary, secondary, and tertiary-level healthcare institutions, both public and private, such as five for-profit Allopathy hospitals, two government

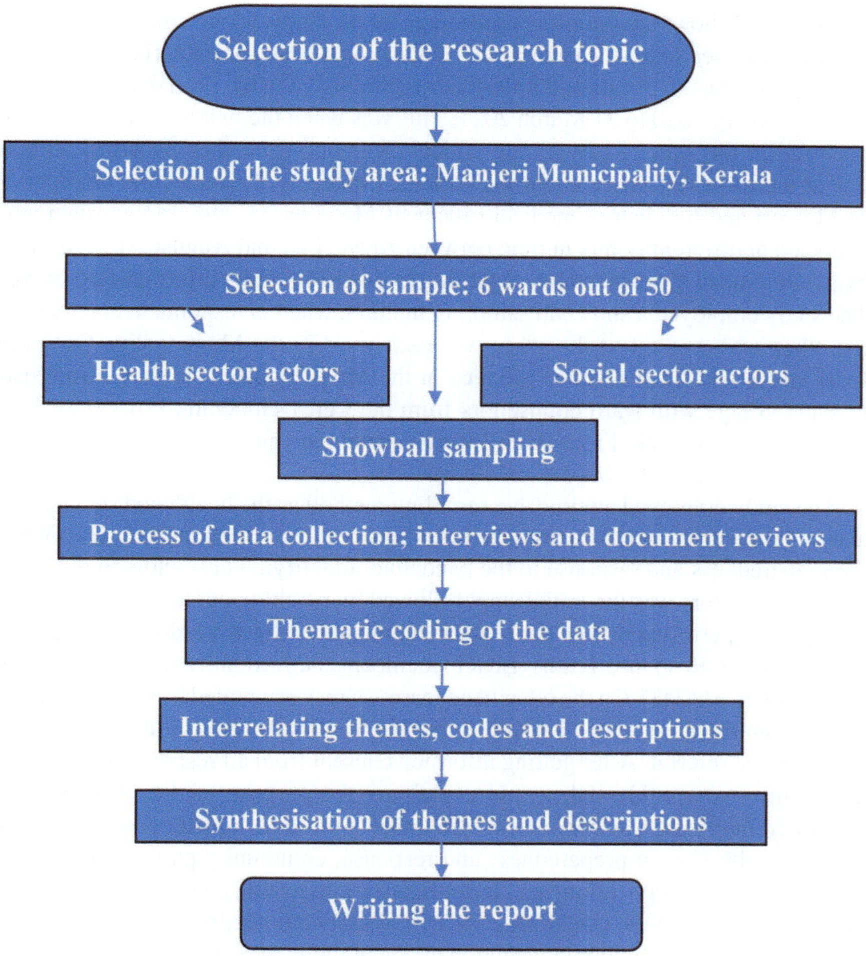

Fig. 9.1 Flowchart of the exploratory research methods

Ayurveda hospitals, one government Homoeopathy hospital, one Veterinary hospital, one hospital for Tuberculosis, two family health centres, and twelve sub-centres.

The study utilised primary and secondary data sources, including articles, reports, policy documents, advisories, and guidelines on COVID-19, minutes of meetings, key-informant interviews, and in-depth interviews with various stakeholders of the local health system. The respondents for the primary data were diverse actors in the COVID-19 response at the local level, which included medical officers, health inspectors, public health nurses, accredited social health activists from the health sector and chairperson for health at the Municipality, ward members, local rapid response team members, and members of Kudumbasree units were from the social sector. Since the study emphasised the collective action of health and social sector actors, the potential excellent data sources were these key stakeholders.

Figure 9.2 shows the sampling procedures for the study, which will be discussed in the upcoming sections. The primary data collection focused on gathering information on the public health events and milestones regarding COVID-19 in the Municipality from 1 January 2020 to 31 March 2021. This was when the first wave of COVID-19 occurred in the state. After getting ethical clearance from the concerned authorities at the local self-government institutions, especially the municipal chairperson of Manjeri and the district medical officer of Malappuram, the primary data were collected at different points in time between April 2021 and August 2021, followed by a follow-up of the respondents regarding cross-checking the information provided. The study employed a snowball sampling method, where a respondent connected to the other, and the unit of the study was each ward in the Municipality. Out of 50 local wards and three revenue villages in the Municipality, two wards from each revenue village with ward counsellors from the Left Democratic Front (LDF) and the United Democratic Front (UDF) were considered equally to get the same level of representation of the ward counsellors.

The study employed a purposive sampling method as the health and social sector actors are diverse, and to capture rich perspectives from those actively involved in the preparedness and response to the pandemic. Identifying key actors in the health and social sectors was the initial step, followed by reaching out to ward councillors after getting permission from the Manjeri municipality. Each ward member referred to Local Rapid Response Teams (LRRT) members and Kudumbasree Area Development Societies (ADS). On the other hand, permission was granted by the health sector actors from the district medical office, and the health inspector of the Municipality gave the first referral. After getting informed consent from all respondents, ensuring anonymity and confidentiality, and explaining the purpose of the study, a semi-structured interview was conducted. It included a pre-determined open-ended question on health system preparedness and response, community participation, social cohesion, local self-governments, decentralisation in health, and mental health. The particular themes of the questions consisted of early preparedness and awareness of the virus, lockdown and implementation of containment measures in terms of testing,

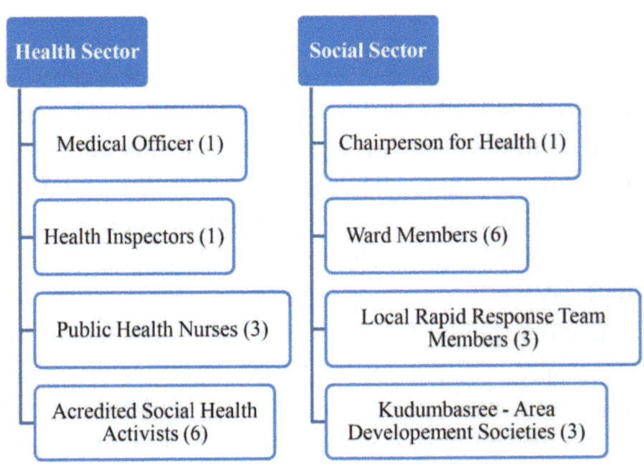

Fig. 9.2 Selection of the sample

tracing and isolation, healthcare capacity in terms of number of health workers recruited and deployed, essential medicines, public participation in the facilitation of first line, second line and reverse quarantine treatment centres, communication strategies with the public, intersectoral coordination and multisectoral stakeholder participation, social support systems and relief efforts, and challenges faced and lessons learned.

These questions were asked only after the respondents felt comfortable sharing lived experiences and trusted the researcher, and active listening was followed to capture verbal and non-verbal cues. Further, probing questions were also used when there was a need for deeper understanding and clarification. These interviews were recorded with informed consent, often on the phone, and an accurate translation was carried out for the analysis and the notes taken during the interview. Interview lengths varied, especially between frontline health workers, other health sector actors, and social sector actors. The recorded interviews were transcribed and translated from Malayalam to English, and a process of reading and re-reading the transcripts was done.

The qualitative primary data and the research objectives were analysed using thematic analysis. During the data analysis, text segments relevant to the research objectives were identified and labelled manually. The information has been collated and merged with appropriate context and factors, which helped further the accuracy and validity of the data. To add more, the transcripts' most consistent themes were picked up, and respective codes were given to summarise the information. Later, to identify patterns and themes in the data, grouping themes were generated, and an examination of identified themes was conducted to ensure that the data were accurate and meaningful. The coding themes arose through inductive reasoning aligned to each research objective and were under critical examination. Finally, findings from each were combined to test the content material overlap and divergence areas. Subsequently, the data were utilised to test the convergence of meanings and prominence of themes, which helped generate meaningful thematic information on the research aims.

9.3 Results and Discussion

This section brings about empirical evidence from the field on the discourse on COVID-19 response in Kerala, particularly from an urban locality with the characteristics of semi-urban areas. The exploration of COVID-19 preparedness and response through a wide range of stakeholders and social participation is the central theme of the chapter; the people, their culture, social relations, politics, caste, gender, social stratification, religion, discrimination, and all other social institutions have been critically taken into the analysis. Henceforth, it formulates the larger picture of the local dynamics of the COVID-19 response. Out of the selected six wards from the peripherals of the Manjeri municipality, there is heterogeneity within these localities in social relationships.

The two research objectives are the cornerstone of the chapter. However, exploring how health and development have been in progress in the state and the locality has a significant place in the discourse on the COVID-19 response. Additionally, the empirical analysis of the health system preparedness and response at the community level has been put at the centre of the analysis. Social participation and community engagement, along with the role of social cohesion, have emphasised significant narratives in the chapter and are a marker of the resilience of both health systems and communities, which was reflected in the collective and cooperative actions of both health and social sector actors. In addition, social intermediation through multisectoral action has taken a significant chunk of the description of how these processes have gained the upper hand in preventing the spread of the virus, which helped to curb the first wave of the pandemic in the state. Last, a section on building back better after the crisis has also been included to avoid missing the link between the previously mentioned themes.

9.3.1 Health and Development in Kerala

Kerala's health development has been determined by investment in health infrastructure in public, cooperative and private sectors, in addition to community members' health awareness (Centre for Development Studies 2005). The progress in the health status and development of healthcare services, which resulted in social development in Kerala, is backed by a historical legacy of consistent public investment in health and education since the eighteenth century (Franke and Chasin 1998a). There were interregional disparities in development, particularly in health and education, as these were caused by the social welfare measures implemented by the local rulers (Kabir and Krishnan 1973). The British introduced Western medicine to the state when vaccination was made compulsory for everyone in Travancore (Kutty 2000). Nonetheless, there were sharp systemic disparities in health and education among the marginalised communities of Kerala (Franke and Chasin 1998b; Kurien 1995). The state's public healthcare facilities were considered superior to those of other Indian states; however, longer waiting times due to overcrowding have become a crucial challenge (Peters et al. 2002). The neoliberal economic policies of the 1990s impacted Kerala's health sector, as there was a gradual decline in public spending on health, which led to a reduction in the number of beds and human resources and a surge in for-profit hospitals (Kutty 2000). Consequently, there were inequalities in the distribution of hospitals, as most hospitals were located in the cities and towns, leaving rural areas short of secondary and tertiary healthcare institutions (Narayana and Kurup 2000). The underinvestment in public health has also driven a faster growth of not-for-profit and for-profit hospitals and, later, a transition of not-for-profit hospitals to for-profit hospitals (Dilip 2010).

In the mid-1990s, people's campaign for decentralised planning also expanded to health, which led to the devolution of resources to local self-governments, accountability and transparency to health services provided through public facilities (Elamon et al. 2004). The 73rd and 74th amendments and the passing of the Kerala Panchayat Raj Nagara Palika Act, launched as part of the People's Planning Campaign in 1996, have been important milestones in democratic decentralisation (Thomas and Rajesh 2011). The same period witnessed the rapid expansion of for-profit healthcare institutions, ranging from clinics to multi-speciality hospitals, which was an outcome of the National Health Policy 1982, which eased regulations in the health sector (Kutty 2000; Thresia 2013). To be competitive in the market, the unregulated private sector began to provide quality health services, and the neoliberal economic policies of the 1990s impoverished the public healthcare system (Nair and Varma 2021).

The past decade has shown a government commitment to revamp public healthcare institutions to achieve universal health coverage; 2012 marked a turning point in the state's healthcare reforms, making a formal foray into UHC reforms (Perabathina and Ellangovan 2016). After a series of efforts by the state government to strengthen the public sector, the Aardram mission was introduced in 2017, aiming to counteract the market-driven, unregulated private sector by providing people-friendly outpatient services, transforming primary healthcare centres (PHC) into family health centres (FHCs), improving access to comprehensive services for the vulnerable, and standardising services (Nambiar et al. 2020). All these developments in Kerala's health system have influenced the COVID-19 response in the state and each unit at the grassroots level.

9.3.2 Health System Preparedness and Response

When the first case was reported in Kerala, all local self-governments were advised to prepare their local health system to respond to the outbreak, as there was immediate mobility of people from various parts of the country and abroad. Wearing masks in public places, washing hands with sanitiser, maintaining physical distancing in public spaces, quarantining people with travel history, ensuring food security for the vulnerable, providing contact tracing of the suspected infection, providing mental health support for those in need, monitoring of the health status of those in quarantine, etc., were the significant activities performed by the key stakeholders. In the first wave of the pandemic, the Municipality had its first case reported on 14 March 2020; later, when the expatriates had been allowed back to their native land in a phased manner, the number of cases had gone up.

Nonetheless, there were systemic constraints which posed a significant challenge to the preparedness and response to COVID-19 in the Municipality; some of them included a considerably large number of impoverished and lower-middle-class households, especially in the peripheries, their knowledge and attitude towards the preparedness activities at the local level, in the centre of the town, the well-off families show a resistance which came out of their class complex, a few expatriates were not ready to stay in quarantine facilities by calling it as substandard. A 31-year-old Rapid Response Team (RRT) member said that '*an expatriate who worked abroad returned during the lockdown was asked to stay in the public quarantine facilities*

Table 9.1 Distribution of workers

Category	Male	Female	Total
Total workers	22,142	3959	26,101
Main workers	19,090	3018	22,108
Main worker cultivators	665	17	682
Agriculture labourers	1089	134	1223
Household industries	214	38	252
Other workers	17,122	2829	19,951
Marginal workers	3052	941	3993
Non-working persons	25,029	45,972	71,001

Source Office of the Registrar General & Census Commissioner, India (2011)

as it was the rule, but he refused to stay at the facility by saying the facilities are not up to the mark of his standard of lifestyle' (July 2021). There were a few similar incidents where the resistance came from the affluent section of society, which was considered an obstacle to preparedness at the grassroots level. Since the study was carried out in the peripheries of the Municipality, especially in semi-urban contexts, the socioeconomic and political dynamics at the state and district levels also significantly impact the local level. Most households work in the informal sector, as shown in Table 9.1.

The data in Table 9.1 show that other workers represent the highest majority of the total workers compared to other categories. Since the study was conducted in the peripheral areas of the Municipality, most of the people's livelihoods were in the informal sector work, as seen in the table. This can also include male informal workers in Gulf countries whose families are left behind and headed by older women, especially in Muslim communities. A larger section of informal workers underwent severe distress during the lockdown period, and community-led relief mechanisms were the support systems.

In the initial period of the pandemic spread, there were fewer testing centres and hospitals equipped with treatment options, even though Manjeri municipality could handle the cases from the whole Malappuram district. One of the primary tasks of the stakeholders in terms of COVID-19 preparedness and response was to contain the spread of the virus, thereby dealing with the peak of the surge in the incidence of cases and fatalities. So, there was particular attention to people with comorbidities and older adults in the locality; there was an extensive distribution network of life-saving drugs and medicine by the government authorities and voluntary associations. A 28-year-old medical officer said, '*I was scared of the older people with the chronic disease during the lockdown, as they would not be able to come to our health centres due to the strict guidelines not to visit any hospital unless there were any symptomsof COVID-19. Later, the local authorities suggested an option to distribute essential medicines for those in need, and our community health workers were intermediaries in facilitating the need for medicines*' (May 2021). This highlights how a local health system put in place could foresee what should be done for those vulnerable sections of the population in difficult times. Furthermore, the DISHA helpline was set-up for those who need help in terms of counselling and assistance if they face any mental health issues from quarantine and isolation.

9.3.3 Social Participation as a Driving Force of Resilience

Community participation was the heart of the COVID-19 response all over the state, as there were systems in place to integrate diverse communities to come together for collective action. Manjeri municipality has also seen similar activities where political parties, civil society organisations, residence associations, religious organisations, and Kudumbasree units had a massive role in mobilising the community. Table 9.2 shows the religious composition of the population in the Municipality. According to the 2011 census, Muslims constitute the majority of the population, which accounts for 70% of the total population, and the second largest is Hindus and lastly, Christians with 2%. Additionally, the skewed distribution of religious groups has never distorted communal harmony because of civic sense among the members of the society; as a ward member noted, '*I have never seen people come together regardless of caste, class, political affiliation, religious beliefs and work collectively to fight the virus*' (July 2021). This denotes the social cohesion and communal harmony in the Municipality, which has a long history of social reform movements. The predominance of a community in a higher percentage has not negatively impacted communal harmony, as it is usually seen in many contexts. However, a caste-based stratification is present both among the Hindus and Muslims, which had an adverse effect on the social intermediation process, as neglect of marginalised sections during the distribution of essential food items during the crisis.

Table 9.2 Distribution of religious composition

Religion	Total population	Male population	Female population	Population share (%)
Hindu	26,820	13,196	13,624	27.62
Muslim	68,135	32,933	35,202	70.17
Christian	1977	958	1019	2.04
No religion	132	65	67	0.14
Other religions	19	8	11	0.02
Buddhist	11	4	7	0.01
Sikh	6	5	1	0.01
Jain	2	2	0	0.00

Source Office of the Registrar General & Census Commissioner, India (2011)

The heterogeneity in the population on the lines of caste, class, gender, and religion has influenced the social relationships and stratification in the Municipality. A 35-year-old RRT member said, '*I have noticed that when there was massive scale distribution of food and other material during the lockdown, there were some partiality practices by the socioreligious organisations, leaving a section of the society neglected*' (May 2021). This was partly due to the divisive mindsets of a few people during difficult times, which can hinder social integration and cohesion, which was not a general trait of Kerala society. Further, the discrimination against frontline health workers like Accredited Social Health Activists (ASHA) and Junior Public Health Nurses (JPHN) during their work in the community has also been reported as a 40-year-old ASHA said, '*I have been called a carrier of the infection and an older man yelled at me while visiting their home as part of field visits*' (June 2021). Most of the Dalits in the Municipality live in colonies, which are ghettoised in socially demarcated colonies. These colonies lack basic amenities, and most work in blue-collar jobs, making them vulnerable to health and other livelihood challenges. Most of the ASHA workers were from Dalit and other backward castes communities, and they had a clear-cut idea about the needs of these communities. Table 9.3 shows the population distribution of scheduled caste and scheduled tribe and their share of the total population in the Municipality. It shows how minute their share of the total population becomes a constraint in demanding their rights and entitlements during the pandemic, especially during the lockdown. Therefore, their living conditions have remained the same. They have worsened due to stringent lockdown rules.

Socio-political organisations such as the Muslim League, the Democratic Youth Federation of India (DYFI), the Youth Congress, the Youth League, and other minor voluntary associations have coordinated COVID-19 response activities. A Junior Public Health Nurse (JPHN) recalled, '*Our team struggled to get vehicles to visit the field. However, people from youth organisations have come forward to help us*' (June 2021). It is an example of how socio-political organisations have become an enabler in the COVID-19 response in Manjeri. Their voluntary and active involvement stems from facilitating the Break the Chain Campaign, which was basically to install facilities for washing hands in public spaces, sanitising the infected areas, provide masks to the community members, arrange food and other essentials for the people, help contact tracing, arranging emergency vehicle support for visiting hospitals, providing digital gadgets for those families in need when school education went online, and other favourable activities were some of the most crucial contributions during the time.

Table 9.3 Scheduled caste and scheduled tribe population in Manjeri municipality

Category	Population	Share of population (%)
Total population	97,102	Not applicable
Total scheduled caste population	8257	8.51
Total scheduled tribe population	264	0.27

Source Office of the Registrar General & Census Commissioner, India (2011)

On the other hand, Kudumbasree, a self-help group collective of women from poor backgrounds, their smallest unit, Ayalkkootams, which is a collective of 10–15 women members who thrive in finance and other social development activities in their locality. Area Development Societies (ADS) of Kudumbasree have engaged in RRTs and observed those who are in quarantine. Further, they extensively worked to ensure food security in their locality by providing free meals through community kitchens. A 46-year-old ADS said, '*I could manage to watch around 10 people who were in home quarantine, and a few of them were not cooperating with the RRT as they wanted to step out of the room. So, I had to call ward members to intervene and ensure they stayed home until the completion of their quarantine period*' (August 2021). This implies Kudumbasree ADS's role in ensuring people in home quarantine do not violate the rules and their power to inform authorities to ensure the Break the Chain Campaign is effective in its implementation. ADSs have identified the most vulnerable and poorest households to provide essential social safety nets for survival. Additionally, they have identified households to provide digital gadgets for continuing online education.

9.3.4 Actors in the Health and Social Sectors

There were two main actors in the COVID-19 preparedness and response in the Manjeri municipality, namely, health and social sector stakeholders. Given that the health system is a part of the larger socioeconomic systems, the analysis could not be possible without taking these two sectors in tandem. Figure 9.3 shows the hierarchy of actors in the health sector, where the Manjeri Medical College superintendent is at the top, and ASHA workers are at the bottom, as they are the ones who contact people at the ground level.

ASHAs are the frontline workers who link the community with health service systems in their localities; they perform reporting of cases from the field and help RRT members function in their localities. COVID-19 nodal officers at the medical college and medical officers of urban primary healthcare and family health centres coordinated the activities in the field and with higher authorities. Health supervisors and health inspectors were responsible for scrutinising the data to report to their reporting officers, and public health nurses and junior public health nurses have engaged with ASHAs to conduct fieldwork and COVID-19 response at the community level. This organisational structure has been effective in information flow from the field to the top, as well as any new guidelines or modifications for the activities, such as classifying containment zones to break the chain of the spread of the infection.

The social sector actors shown in Fig. 9.4 portray the municipal chairperson as the lead of COVID-19 response mechanisms at the Municipality, even though more responsibilities and duties were for the municipal standing committee chairperson for health. Regardless of their higher position in the hierarchy, the work burden was on the heads of ward members because the duties of the municipal chairperson and

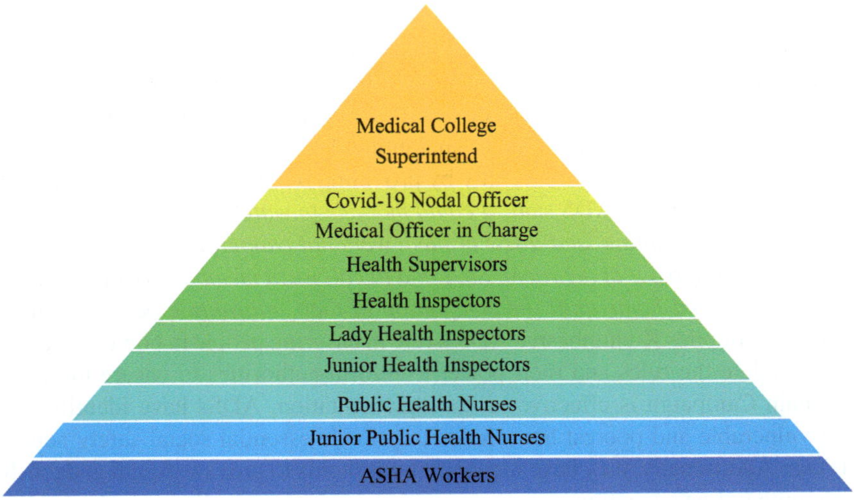

Fig. 9.3 Health sector actors. *Source* Primary data, Manjeri, June 2021

municipal standing committee chairperson for health are confined to overseeing the works carried out in the 50 municipal wards.

The collective and combined action of both social and health sector actors has accelerated an effective COVID-19 preparedness and response strategy in Kerala, which has been lauded globally, especially during the first wave of the pandemic in the state. It was done through a unique system where public trust was gained in

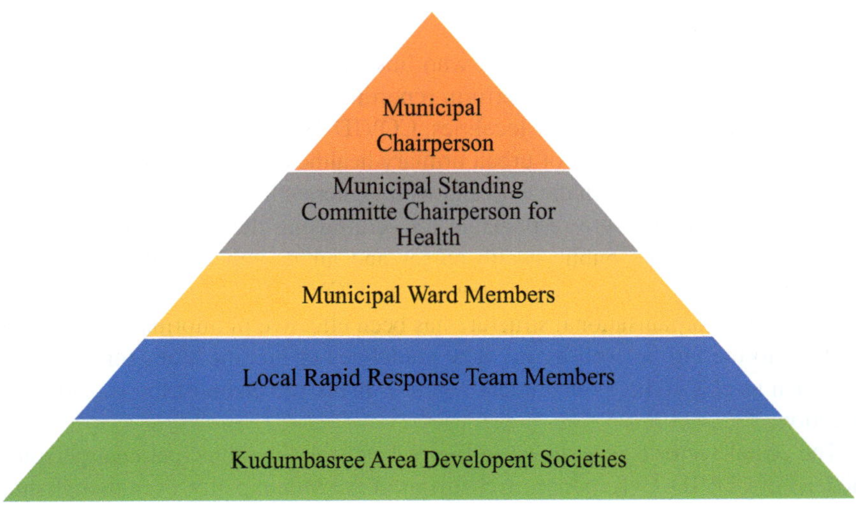

Fig. 9.4 Social sector actors. *Source* Primary data, Manjeri, May 2021

the system and state mechanism, and a digitally empowered disaster response was carried out all over the state. For instance, contact tracing of the infected was done by publishing the route map of the infected on social media, which helped those who came in contact or became secondary contacts to quarantine at home and prevent further spread of the virus. A 43-year-old health inspector said, '*In the initial phase of the crisis, we were asked to focus on contact tracing of the suspected and infected persons, which had helped us detect cases and reduce the incidents*' (July 2021). One of the core reasons behind the success of Kerala was the early detection and prevention of the spread of the virus in the community.

9.3.5 Multisectoral Coordination and Social Intermediation

One of the major attractions of the COVID-19 response in Kerala was multisectoral coordination, which, of course, implies health in all policies. Public health actions that address social determinants of health consider health a multidimensional state and need an integrated, primarily healthcare-oriented approach. Thus, the COVID-19 response in Kerala was not the responsibility of the Department of Health; instead, there was an innovative multisectoral action. It was a collaborative effort by a range of departments consisting of health and family welfare, local self-governments, revenue, labour, social justice, finance, general administration, civil supplies, police, information and media, education, electricity, tourism, water and sanitation, transport, and the public works department. One of the foremost important departments actively involved in the works at the grassroots level was local self-government institutions, where the elected representatives from respective localities were the best ones in identifying the challenges and prospects of their administrative areas. This has eased the collection and collation of information, spread awareness, and assessed the interventions' strengths and weaknesses. In other words, this was one of the primary departments that could be sensitive to the needs of community members.

Figure 9.5 shows the organisational structure of the local rapid response team at each ward in the Manjeri municipality. While this mechanism is the best example of community engagement, this system was more appropriate for discussing intersectoral coordination among various departments. RRTs function at the ward level and are headed by ward members and community volunteers at the bottom of the hierarchy. Apart from the usual health and social sector actors in the COVID-19 response, the government employees from each department involved in multisectoral coordination were also part of the structure. Therefore, the multisectoral coordination was effective in its functioning. The revenue and finance departments were actively engaged with potential departments in resource mobilisation and deployment of staff during the crisis.

Meanwhile, police and general administration looked after law and order, especially by demarcating containment zones and ensuring frontline health workers' safety. The civil supply department ensured food security through public distribution systems and prevented hoarding and black marketing of food items. The Department

Fig. 9.5 Rapid response teams. *Source* Primary data, Manjeri, April 2021

of Education's involvement was not by postponing classes but by deploying teaching staff to the COVID-19 activities on the ground. The Department of Electricity and Water Supply facilitates sustained services and waves of bill payments for three months. The Department of Transportation offered services to the various actors and COVID-19 warriors. Furthermore, the public work department has built and arranged facilities for quarantine for the expatriates and stranded migrant workers. The tourism department has enabled the COVID-19 response by closing down all tourist spots, and the social justice department has keenly observed and ensured the welfare of the marginalised sections of society. Similarly, the Department of Information has been providing authentic, reliable information on the daily updates of COVID-19 response outcomes. To add on, this was the period when the state machinery took social welfare of the whole people care with commitment, for example, provisioning of universal public distribution of food items, financial assistance for those in need, further strengthening of public systems, and a proactive role of bureaucracy in implanting social safety nets in the communities.

Multisectoral action for the COVID-19 response at Manjeri municipality has built a system of collective action by all sectors of society, which enabled positive public action in the state during the pandemic. Additionally, multisectoral action has accelerated social intermediation by building trust in the social and public systems, particularly in the institutional capacity of community-based organisations and collectives. On the other hand, social participation as an integrative part of multisectoral coordination has fostered community empowerment and trust building, resulting in resilience in health systems and social cohesion.

9.3.6 *Limitations of the Study*

This explorative study on resilient communities and health systems delved into the COVID-19 preparedness and response process at the Manjeri municipality, which is not complete without the epidemiological data of the locality. Since the authorities were not ready to share the epidemiological data regarding the incidence of cases, recovered cases, and fatality rates, validating the COVID-19 response based on just numbers is not the scope of the chapter. Instead, this chapter tried to include all the stories and incidents of COVID-19 preparedness and response at a locality in Kerala, where the world applauded Kerala for containing the spread of the virus in the first wave.

9.4 Conclusion

Kerala's response to COVID-19 was a testimony to the resilience of both health and social systems in tackling health vulnerabilities. Social participation and cohesion were the central points in the Kerala model of COVID-19 preparedness and response stories. Furthermore, an energetic state machinery of executives and administration, a strong public healthcare system, and civic engagement have been among the reasons for Kerala's success in containing the virus (Menon et al. 2020). Decentralised governance and institutions, and the legacy of the Kerala Model of Development, paved the way for success, even though there were adversarial politics which limited public action (Chathukulam and Tharamangalam 2021). Some enabling elements in the local self-governments' involvement in the COVID-19 response are organisational capacity, networking with various agencies, community engagements, resource mobilisation, and intersectoral coordination (Radha et al. 2020). On the other hand, social capital and active participation of socio-political groups through local self-government institutions have fostered the strategies relevant to the state's success in containing the virus in the first wave (Thomas Isaac and Sadanandan 2020; Prajitha et al. 2021). Additionally, the state had 75% digital literacy during the outbreak of COVID-19, which has helped the preventive measures through communication systems, surveillance, clinical treatment, and non-medical support (Ummer et al. 2021). The crucial role of frontline health workers, such as ASHAs and Anganwadi workers backed by community volunteers, averted the possibility of localised transmission of the virus (Choolayil and Putran 2021). Additionally, the strategies implemented at the local level resulted from synergies between the district and state health administrators and allied departments, rapid dissemination of reliable health information, civic engagement in community surveillance and disaster preparedness (Vaman et al. 2020). There were many criticisms that the Kerala Model of COVID-19 management faced during the second and third waves of the pandemic, precisely due to the unique health and epidemiological characteristics, even though the first wave of the pandemic was unquestionably effectively managed by the people of Kerala and the government. There are many lessons for other states and nations to learn from Kerala in managing COVID-19, especially in light of the first wave and public investment in the state's health sector.

Acknowledgements The author thanks the respondents who participated in the study, even amidst a deadly second wave of the virus that had just occurred, and conveys gratitude to the people who helped get to each respondent. Last but not least, the author thanks his M.Phil. supervisor, Prof. Rama V. Baru, a former faculty member at the Centre of Social Medicine and Community Health, Jawaharlal Nehru University, New Delhi, India, for her constant guidance and help in completing the dissertation.

References

Balabanova D, Mills A, Conteh L, Akkazieva B, Banteyerga H, Dash U, Gilson L, Harmer A, Ibraimova A, Islam Z, Kidanu A, Koehlmoos TP, Limwattananon S, Muraleedharan VR, Murzalieva G, Palafox B, Panichkriangkrai W, Patcharanarumol W, Penn-Kekana L, Powell-Jackson T et al (2013) Good health at low cost 25 years on: lessons for the future of health systems strengthening. Lancet (London, England) 381(9883):2118–2133. https://doi.org/10.1016/S0140-6736(12)62000-5

Centre for Development Studies (2005) Human development report 2005. Government of Kerala

Chathukulam J, Tharamangalam J (2021) The Kerala model in the time of COVID-19: rethinking state, society and democracy. World Dev 137:105207. https://doi.org/10.1016/j.worlddev.2020.105207

Choolayil AC, Putran L (2021) COVID-19, the local and the global: lessons from Kerala. South Asia Res 41(1):7–21. https://doi.org/10.1177/0262728020966102

Dilip TR (2010) Utilisation of inpatient care from private hospitals: trends emerging from Kerala, India. Health Policy Plan 25(5):437–446. https://doi.org/10.1093/heapol/czq012

Elamon J, Franke RW, Ekbal B (2004) Decentralisation of health services: the Kerala people's campaign. Int J Health Serv Plan Adm Eval 34(4):681–708. https://doi.org/10.2190/4L9M-8K7N-G6AC-WEHN

Franke RW, Chasin BH (1998a) The Kerala model of development: a debate (part 1). Bull Concern Asian Sch 30(3):25–36. https://doi.org/10.1080/14672715.1998.10411051

Franke RW, Chasin BH (1998b) The Kerala model of development: a debate (part 2). Bull Concern Asian Sch 30(4):35–47. https://doi.org/10.1080/14672715.1998.10411060

Gandhi S, Maharatha TM, Dash U (2021) Level of inequality and the role of governance indicators in the coverage of reproductive maternal and child healthcare services: findings from India. PLoS ONE 16(11):e0258244. https://doi.org/10.1371/journal.pone.0258244

Kabir M, Krishnan TN (1973) Social intermediation and health transition: lessons from Kerala. Working paper no. 251. Centre for Development Studies

Kurien J (1995) The Kerala model: its central tendency and the outlier. Soc Sci 23(1/3):70–90. https://doi.org/10.2307/3517892

Kutty VR (2000) Historical analysis of the development of health care facilities in Kerala State, India. Health Policy Plan 15(1):103–109. https://doi.org/10.1093/heapol/15.1.103

Menon JC, Rakesh PS, John D, Thachathodiyl R, Banerjee A (2020) What was right about Kerala's response to the COVID-19 pandemic? BMJ Glob Health 5(7):e003212. https://doi.org/10.1136/bmjgh-2020-003212

Nair MR, Varma RP (2021) Availability, distribution and utilisation of healthcare services in Kerala. Achutha Menon Centre for Health Science Studies, Sree Chitra Tirunal Institute for Medical Sciences and Technology, Trivandrum

Nambiar D, Sankar DH, Negi J, Nair A, Sadanandan R (2020) Monitoring universal health coverage reforms in primary health care facilities: creating a framework, selecting and field-testing indicators in Kerala, India. PLoS ONE 15(8):e0236169. https://doi.org/10.1371/journal.pone.0236169

Narayana D, Kurup KKH (2000) Decentralisation of the health care. Working paper no. 298. Centre for Development Studies

Office of the Registrar General & Census Commissioner, India (2011) District census handbook: Malappuram (part B). Directorate of Census Operations, Kerala. https://censusindia.gov.in/nada/index.php/catalog/659/download/2284/DH_2011_3205_PART_B_DCHB_MALAPPURAM.pdf

Perabathina S, Ellangovan K (2016) Universal health coverage in Kerala through a primary care pilot project. ACCESS. Accessed 17 Jan 2025. https://accessh.org/reports/universal-health-coverage-in-kerala-through-a-primary-care-pilot-project/

Peters DH, Yazbeck AS, Sharma RR, Ramana GNV, Pritchett LH, Wagstaff A (2002) Better health systems for India's poor: findings, analysis, and options. The World Bank. https://doi.org/10.1596/0-8213-5029-3

Prajitha KC, Rahul A, Chintha S, Soumya G, Maheswari Suresh M, Nalina Kumari Kesavan Nair A, Valamparampil MJ, Reghukumar A, Venkitaraman S, Anish TSN (2021) Strategies and challenges in Kerala's response to the initial phase of COVID-19 pandemic: a qualitative descriptive study. BMJ Open 11(7):e051410. https://doi.org/10.1136/bmjopen-2021-051410

Radha D, Kumaran JAB, Nair MTR (2020) Role of local self-governments in control of COVID-19 in Kerala: an exploratory study. Int J Community Med Public Health 7(12):5027–5034. https://doi.org/10.18203/2394-6040.ijcmph20205180

Thomas MB, Rajesh K (2011) Decentralisation and interventions in health sector: a critical inquiry into the experience of local self-governments in Kerala. Working paper no. 271. Institute for Social and Economic Change

Thomas Isaac TM, Sadanandan R (2020) COVID-19, public health system and local governance in Kerala. Econ Polit Wkly 55(21):35–40

Thresia CU (2013) Rising private sector and falling 'good health at low cost': health challenges in China, Sri Lanka, and Indian state of Kerala. Int J Health Serv Plan Adm Eval 43(1):31–48. https://doi.org/10.2190/HS.43.1.c

Ummer O, Scott K, Mohan D, Chakraborty A, Lefevre AE (2021) Connecting the dots: Kerala's use of digital technology during the COVID-19 response. BMJ Glob Health 6:1–14. https://doi.org/10.1136/bmjgh-2021-005355

Vaman RS, Valamparampil MJ, Augustine AE (2020) Using test positivity rate (TPR) as an indicator for strategic action in COVID-19: a situational analysis in Kerala, India. Indian J Clin Med 10(1–2):31–35. https://doi.org/10.1177/26339447211054234

Chapter 10
Climate Change and Public Health Preparedness in Lower-Middle-Income Countries

Irsa Fatima Makhdoom, Najma Iqbal Malik, Mohsin Atta, Danish Iqbal, Naeem Asim, and Sidra Zia

Abstract Climate change significantly impacts public health, particularly in lower-middle-income countries (LMICs), which face heightened vulnerability due to socioeconomic and geographic factors. LMICs often lack adaptive capacity, face extreme weather events, and depend heavily on climate-sensitive sectors like agriculture. Rising temperatures, altered precipitation patterns, and heatwaves owing to greenhouse gases, certain types of fuels, and urbanization increase the risks of vector-borne diseases like malaria and dengue, exacerbate heat-related illnesses and threaten food security, leading to malnutrition. Mental health challenges, such as anxiety and depression, are also emerging concerns, especially in disaster-affected regions. Limited resources, weak health systems, policy gaps, and low community awareness hinder public health preparedness in LMICs. However, innovative solutions and case studies offer hope. To strengthen resilience, LMICs must invest in climate-resilient infrastructure, integrate health and climate policies, engage communities, and foster international collaborations. Addressing these challenges through global solidarity is critical for protecting vulnerable populations and ensuring sustainable development.

I. F. Makhdoom · N. I. Malik · M. Atta · D. Iqbal
Department of Psychology, University of Sargodha, Sargodha, Pakistan
e-mail: irsamakhdoom@gmail.com

N. I. Malik
e-mail: najmamalik@gmail.com

M. Atta
e-mail: gotamabbasi@gmail.com

D. Iqbal
e-mail: danishiqbalkhattak@gmail.com

N. Asim
Department of Public Health, The Islamia University of Bahawalpur, Bahawalpur, Pakistan
e-mail: naeem.asim@iub.edu.pk

S. Zia (✉)
Independent Researcher, Islamabad, Pakistan
e-mail: sidrazia2010@hotmail.com

A. Roy et al. (eds.), *Environment and Public Health*, Sustainability Solutions,
https://doi.org/10.1007/978-3-031-99770-9_10

Keywords Climate change · Public health · Lower- and middle-income countries · Climate-related health issues

10.1 Introduction

Climate change refers to a long-term change in the overall weather patterns of an area. Environmental and climate scientists warn that if not treated, this climate change challenge can pose serious concerns to the health and well-being of human beings. It has become the most pressing challenge of our time, posing significant risks to ecosystems, economies, and societies worldwide. This climate change is manifesting itself in forms of rising temperature, increased amount of CO_2 in the atmosphere, heavy rainfalls, altered humidity levels, and heat and cold waves in different regions of the world (Hegerl et al. 2019). This climate change is causing several disastrous impacts on the ecosystem and the individuals living in these affected ecosystems through hotter temperatures, more severe storms, increased drought, rising oceans, loss of species, lack of food, poverty-related issues, and more health risks. Therefore, the researchers, the policy makers, and other stakeholders around the world are more concerned about such changes than ever before. Several factors are contributing to this climate change across the globe. Major contributing factors include greenhouse gases (including CO_2, methane, fluorinated gases, and nitrous oxide), natural forcing (such as solar and volcanic effects), global warming, tectonic shifts, changes in solar radiation, and probably a slight change in the orbit of the Earth (Turrentine 2022).

Once considered a future problem, climate change is no more a future problem; rather, it is happening now and affecting the world, particularly the global south. Many of the lower-middle-income countries fall in these areas, and unfortunately, they have limited resources to survive and combat this challenge. The problem of climate change affects LMICs at two levels. On the one hand, most of these countries are in regions which are at higher risk of climate change events and in these countries, almost 3.6 billion individuals are susceptible to being affected by the events of climate change. Thus, these countries are most vulnerable to be affected by the consequences of climate change events in terms of food, habitat, health, infrastructure, etc. (Bianco et al. 2024). The second, and even more severe issue is, these countries have the least resources to combat these effects. The health status of individuals living in these regions is already poor; they have fewer available resources to combat any climate change event; their climate readiness is lower, and their resilience (i.e., bouncing back from disasters) is even lower. Thus, they are more vulnerable to climate change events and the effects of these events; on the other hand, the health status and systems of these countries are even poorer, making them the most vulnerable group.

Among the myriad consequences of climate change, public health emerges as a domain profoundly affected by climate variability and extreme weather events. The interplay between climate change and health is particularly pronounced in lower-middle-income countries (LMICs), where socioeconomic vulnerabilities and underdeveloped health systems exacerbate the warming planet's challenges. This chapter explores the nexus between climate change and public health preparedness in LMICs. Drawing on empirical evidence from 2014 onward, it's the health impacts of the climate change, the challenges these countries face, and strategies to enhance resilience and preparedness.

Climate change refers to long-term temperature, precipitation, and weather changes due to natural processes and anthropogenic activities, primarily greenhouse gas emissions. While its impacts are global, LMICs face unique vulnerabilities due to their geographic location, dependence on climate-sensitive sectors, and limited adaptive capacity.

10.2 Methods

The following section describes the details of the methodology of the present scoping review.

10.2.1 Study Design

The researchers carried out a scoping review in order to answer the question of the present research. The following steps were followed in the present research: identifying the research question, identifying relevant studies, selecting studies, charting the data, summarizing and reporting the results. The details of these steps are as follows.

10.2.2 Identify the Research Question

Since the researchers were interested in exploring the current scenario on public health preparedness with relevance to climate change in lower- and middle-income countries, the research question was formulated as "What are the current trends, challenges, and future prospects for lower- and middle-income countries (LMICs) in addressing climate change impacts on public health preparedness?" and "Prospects faced by existing literature states about climate change and public health preparedness in lower- and middle-income countries?"

10.2.3 Identifying Relevant Studies

After identifying the research question, the researchers conducted literature search with the keywords including "public health", "public health preparedness", "low-middle-income countries" and "climate change" from various search engines, including "Google Scholar", "Web of Science", "ResearchGate", and "PubMed". For every search engine, the researcher specified the years between 2014 and 2024 so that the developing perspective on literature of climate change and its relationship with public health preparedness in LMICs can be reviewed. With these keywords, more than 2300 articles were selected.

10.2.4 Selection of Studies

Firstly, the research team searched articles for duplication from the four databases. After excluding duplicate papers, the research team reviewed the titles of the papers along with the abstracts to be included in the study. The inclusion and exclusion criteria were specified as only these tidies were included in the review, which were published after 2013, focused mainly on public health issues of only LMICs and not the high-income countries, and specifically included the public health outcome of climate change in these countries. Moreover, the book chapters, the qualitative research and conference papers were excluded from the pool. After excluding the duplicated, irrelevant studies and those which didn't qualify for inclusion criteria, a total of 26 studies were left in the pool and were reviewed for themes.

10.2.5 Charting the Data

The researchers further extracted data on the bases of main results of the studies, the current scenario of climate change in low- and middle-income countries with relevance to the author, year, and the broader categories including vulnerability factors of LMICs, direct and indirect effect of climate change on health-related outcomes in LMICs, challenges, limitations and future impact of these challenges.

10.2.6 Summarizing the Results

Finally, the researchers summarized and reported the data using thematic analysis. The major themes included vulnerabilities of LMICs, public health impacts in LMICs, challenges to public health, and strategies for enhancing public health

Table 10.1 Current trends and challenges and future prospects of climate change and public health preparedness in LMICs

No.	Theme	Sub-theme
1	Vulnerabilities of LMICs	• Geographic factors • Economic dependence • Infrastructure deficits • Urban heat island (UHI) effect
2	Climate change and public health impacts in LMICs	• Vector-borne diseases • Heat-related illnesses • Food security and nutrition • Mental health implications
3	Challenges to public health preparedness in LMICs	• Limited resources and weak health systems • Policy and governance gaps • Community awareness
4	Strategies for enhancing public health preparedness	• Strengthening health systems • Integrated policies • Community engagement • General to more specific strategies • International collaborations

preparedness. Table 10.1 summarizes the themes and their sub-themes are identified from the literature.

The detail of these themes and their sub-themes is provided in the following section.

10.3 Results and Discussion

The following section describes the themes of the extracted data.

10.3.1 Vulnerabilities of LMICs

The LMICs are more vulnerable to climate change and its effects; however, before moving forward, we must understand what the term *vulnerability* implies. "Vulnerable" refers to the liable to emotionally hurt or physically damaged. It also means (which is more appropriate in terms of climate change vulnerability) being exposed to or being susceptible to the influence of some destructive agent. Therefore, two major elements are the part of the term vulnerability i.e., an exposure to a threatening situation and a potential or risk of harm associated with it. Thus, vulnerability will depend upon the severity of the event (in this case, the climate change event) and the potential readiness or individual factors such as age, gender, socioeconomic status, health status, etc. (Clark and Preto 2018). Keeping in view the above definition, we define vulnerabilities in the domain of public health by identifying the individuals who are at greater risk of developing a health-related issue in terms of socioeconomic status, geographical region of the world, gender, age, ethnicity, and other related factors.

Though the climate change challenge is the biggest challenge of the present century, it is the low- and middle-income countries which can be classified as the most vulnerable groups to the events of climate change. Several factors, such as their geographical location, their economic dependence on factors which are most vulnerable to climate change and the infrastructure deficit, all pose serious concerns to the people living in these countries. We discuss these factors one by one in the following section.

10.3.1.1 Geographic Factors

Many LMICs are in regions prone to extreme weather events, such as tropical storms, floods, and droughts. These climate change-related events often do not come alone; rather often they are often in combination. For instance, storms appear along with erosions, floods, destruction of vegetation and infrastructure and other associated challenges. For instance, tropical storms in Philippines have largely affected the households of individuals from lower socioeconomic status (Schleypen et al. 2024). The same is the case with droughts in that the poor countries of the world are currently facing the most severe scarcity of water and thus facing droughts. The statistics are alarming in this regard, as 85% of the people who are currently affected by droughts are from LMICs (Zaveri et al. 2023).

10.3.1.2 Economic Dependence

Agriculture, a sector heavily affected by climate change, constitutes a significant portion of GDP and employment in LMICs. A report concluded that in Philippines, the tropical storms resulted in income loss in terms of income reduction in livestock and poultry, entrepreneurial activities, fishing and mining, 15%, 4.2%, 31.8%, and 3.1% respectively (Schleypen et al. 2024). In the countries with low to middle income other climate change events such as droughts are even more disastrous in such that in these countries even a moderate level drought reduces the GDP of the country by 0.39% and extreme droughts bringing even more disasters as it reduces the GDP of the area by 0.85% as compared to the high-income countries where this damage is almost half of the GDP reduction as experienced by lower- and middle-income countries (Zaveri et al. 2023). Such reasons contribute to the annual GDP loss of these countries, and thus, nine out of the top-ten countries losing their annual GDP due to climate-related disasters are low- and middle-income countries and only one country is included in this list (UNDRR 2018).

10.3.1.3 Infrastructure Deficits

Poor infrastructure limits the ability to respond effectively to climate-related health emergencies. Timely forecasting of a disastrous event may reduce the losses (both economic and non-economic); however, many of the LMICs do not have systems that could forecast a climate change event such as floods or heat and cold waves. Besides, these countries score very low on community resilience, which is highly required in times of crises and may help a community foster or at least rebuild itself after a climate disaster (Bhandari and Alonge 2020). A review on resilience of LMICs revealed that many of these countries were not able to deploy resilient efforts during the COVID-19 pandemic in such that only 8 out of 44 countries were able to allocate monetary resources for the resilience during COVID-19; and only 13 out of the studied 44 countries could utilize and extra healthcare workforce as frontline workers against COVID-19 and surveillance systems were maintained by almost 11 LMICs (Pradhan et al. 2023). Further, these countries already have weak healthcare systems, reduced access to clean water, disrupted food chains, and have little access to innovation and technological advancements. Further, these countries have too low resources to put on environmental safety measures and build a safe infrastructure (for instance, building flood defenses, drought-resistant crops, building irrigation systems, and more). Therefore, the LMICs not only are at greater risk of experiencing climate change-related disasters but also have weaker defense and resilience systems to combat these disasters, thus making them more vulnerable to the disastrous effects of climate change.

10.3.1.4 Urban Heat Island (UHI) Effect

Urban heat island effect refers to the higher temperature experienced by urban areas and the rural areas that surround these areas. This effect is due to human activities and heat producing infrastructure of the urban areas. In the cities, several materials absorb heat during the day and gradually emit this heat during the night, and this increases the overall temperature of the area. Other factors include transportation, industrial emissions, dense buildings, more population, reduced airflow, and less vegetation contribute even increasing the temperature and thus cause the urban heat island effect. Though the problem is faced by most of the developed and high-income countries, the densely populated cities of the low- and middle-income countries are also facing the issue. For instance, Karachi, a densely populated and the biggest city of Pakistan, is often susceptible to the heat waves with as many as 1000 casualties during the phase of heat wave. Moreover, unlike the high-income countries, the LMICs are less equipped with the resources to combat the effect and thus, the effect is more disastrous to such countries as compared to their counterparts (Khan and Shahid 2024).

10.3.2 Climate Change and Public Health Impacts in LMICs

Climate change is a significant phenomenon that influences individuals both directly and indirectly. The direct effects of climate change on people's lives can be witnessed in the forms of injuries and casualties in response to climate change events such as floods, heat waves, rainfalls, and hurricanes. However, the indirect effects are more subtle in nature and affect the public health of the individuals in LMICs through affecting the water and sanitation, creating malnutrition, affecting the health of people through forced displacements, and living under uncertain conditions (Myers and Bernstein 2011). The health impacts of climate change in LMICs are multifaceted, affecting physical and mental health outcomes and straining health systems.

Bianco and colleagues (2024) identified eight categories of how public health is impacted by climate change in LMICs. These included respiratory diseases, enteric infections, non-disease specified temperature-related mortality, neglected tropical diseases, cardiovascular diseases, malaria, other communicable diseases, and other health outcomes. The non-disease-specific temperature-related mortality was the most studied health outcome, and the least one was respiratory diseases. Progressively severe climate change scenarios were associated with worse health outcomes.

10.3.2.1 Vector-Borne Diseases

Vectors are the arthropods that transmit illnesses from one living being to another, such as fleas, mosquitoes, flies and ticks, etc. These agents carry pathogens from a living being, multiply these pathogens in their bodies, and then they transmit these to the hosts, thus communicating illness from one living being to others. Malaria, dengue, Lyme disease, West Nile Virus Infection, are some of the vector-borne diseases which are attributed to the changing climate across the globe. Climate change may affect the vector, the pathogens, the host, and the effects of the pathogens on hosts. For instance, increased heat affects the behavior, the life history and the physiological characteristics of both the vectors and pathogens carried by them. The interaction of vectors with pathogens and the temperature affects the risk of spread of disease across individuals. The exact role of climate change is manifested in the climate-mediated ecosystem changes where vectors are either no longer able to carry a pathogen or carry a new pathogen with them (Casadevall 2020; Thomson and Stanberry 2022). Rising temperatures and altered precipitation patterns expand the range of vectors such as mosquitoes, increasing the incidence of diseases like malaria and dengue.

10.3.2.2 Heat-Related Illnesses

Heat waves are defined as a prolonged period of high temperature compared to the usual temperature of that geographical area. Every city has a specific temperature threshold, and an average increase of 1 °C in this threshold can result in significant health-related outcomes. However, heat may produce both direct and indirect effects on the health outcomes of the individuals. These effects may include (but are not limited to) accidents and injuries, increased preterm births, food and water-related diseases, endocrine disorders, renal diseases and cardiorespiratory diseases and, in extreme cases, increased mortality. Further, extreme heat events can affect the health of the affected ones by altering their ability to control their body temperature. The illnesses that are related to inability to regulate body temperature include heat exhaustion, heat cramps, hyperthermia and heat stroke with several accompanying symptoms including dry lips, nosebleeds, arm and leg cramps, rashes, nausea, excessive thirst and sweating, fever, and dizziness. Further, extreme heat change can exaggerate the effects of several chronic disorders, including respiratory diseases, cardiovascular and cerebrovascular diseases, and diabetes-related illnesses (Ebi et al. 2021).

10.3.2.3 Food Security and Nutrition

Climate-induced disruptions to agriculture compromise food availability and quality, leading to malnutrition. A 2021 report by the World Food Program identified climate change as a key driver of food insecurity in LMICs, with over 40 million people affected. Tropical cyclones, rise in temperature, rainfall variability (either increase or decrease), floods and rise in the sea level, all the climate change events affect the food security and nutrition in the LMICs more specifically, both directly and indirectly. For instance, climate change can affect the river flow, which is (the high river flow as well as the low river flow), detrimental for agricultural growth as the former causes floods while the latter results in droughts. The glacier melts resulting from climate change can also result in flooding, affecting agriculture and thus reducing food production and security. Further, the droughts are significant events directly affecting the security of food in terms of both the quality and the quantity of the food. Finally, the climate change events affect the agro-ecological conditions of the geographical region of the country on one hand; on the other hand, it affects the distribution of income across the country, hence altering the supply–demand ratio of the individuals. Certain other outcomes are associated with this food insecurity, among which the top of the list is malnutrition. Climate change results in more undernutrition by the interaction of food security, health of the individuals (particularly that of marginalized groups such as women, children and the elderly) and care practices of the system (Krishnamurthy et al. 2012).

10.3.2.4 Mental Health Implications

The psychological toll of climate change, including anxiety, depression, and post-traumatic stress disorder, is gaining recognition. The extreme climate events result in stress and other clinical disorders, including post-traumatic stress disorders, anxiety, sleep disturbances, aggressive tendencies, mood disorders, substance use, self-harm, and suicidal ideations. The climate-related mental health outcomes can be categorized into three categories, including acute, sub-acute, and long-term effects. The acute outcomes are experienced by those who are directly affected by the climate change events such as fires, floods, and droughts. Often these are the most severe mental health injuries experienced shortly after a disaster; however, since they appear shortly after the disaster, they receive the highest attention from the policy makers and other stakeholders. The sub-acute outcomes are experienced by those individuals who themselves have not experienced the disaster but have closely witnessed such events. Such individuals report symptoms of anxiety and future uncertainty, but since they are not the direct victims of the incident, they receive less support in the rehabilitation efforts. Long-term effects are witnessed in terms of PTSD, violence originating from limited resources, and mental health issues arising because of forced migration, etc. (Cianconi et al. 2020).

10.3.3 Challenges to Public Health Preparedness in LMICs

Despite the issue's urgency, LMICs face numerous obstacles in building climate-resilient health systems.

10.3.3.1 Limited Resources and Weak Health Systems

Insufficient funding for health and climate adaptation initiatives constrains the ability of LMICs to invest in infrastructure, workforce training, and technology. Fragmented health systems with inadequate surveillance and response mechanisms hinder the effective management of climate-related health risks. Within the context of lower-middle-income countries, for example, the heat-related illnesses cause a burden on health system in many ways including increased burden on primary healthcare systems, increased hospitalization, increased load on ambulances, shortage of medicine, slower response time, and more burden on emergency and outpatient departments of the hospitals and emergency services (Sapari et al. 2023). Moreover, in these countries, factors such as lack of basic infrastructure, lower allocation of GDP to health, limited access to health care, at one hand and there is more vulnerable population in these areas due to low SES, lower adaptability to heat waves and lesser facilities to cooling the effects of heat such as air conditioning services in home and offices (Green et al. 2019; Guo et al. 2011; McCracken and Phillips 2016; Watts et al. 2018).

10.3.3.2 Policy and Governance Gaps

The challenge of policy and governance gap affects the low- and middle-income countries at many levels. Firstly, most of the policies in these countries are neither research-based nor data-driven. Most of these countries allocate little, if any portion of the GDP, to deriving the data on climate impacts, vulnerability, and health outcomes associated with the climate change events. The unavailability of recent, up-to-date, real-world data hinders the targeted, specific, and real-world policies for improvement. Further, in the case they are able to collect real-world data and generate the policies, often the internal conflicts of these governments, lack of available funds and other factors impede them from applying those policies successfully.

Moreover, the climate challenge and its consequences are often not a priority for such countries since they have very basic needs in their priority, including food and nutrition challenges, basic education, basic health facilities, and infectious diseases. Therefore, the LMICs often have to prioritize their resource allocation on these immediate and short-term needs over the long-term climate change challenge. Further, many of the low- and middle-income countries are facing internal conflicts and thus facing internal problems such as weak governance structures, fragile inter-sectoral coordination and corruption, which affects the ability of a country to develop and apply policies related to climate change. The internal conflict of these countries is often associated with another challenge, i.e., the coordination among different institutes of the state (for instance, the coordination between federal and provincial governments) and among different departments of the state (such as the coordination between health and disaster management departments, etc.). Owing to this lack of coordination, neither appropriate and targeted policies are developed, nor can they be implemented in a true sense.

10.3.3.3 Community Awareness

Community awareness about the interaction of climate change and public health is another challenge faced by many LMICs. The major reason behind the lower community awareness of the issue is the lower literacy rate of such countries. Besides the general low literacy rate, the rate of health literacy is even lower in these countries (Meherali et al. 2020). Most of the time, in case of the spread of epidemics and pandemics, the reason is a lower rate of health literacy. In a review of countries from South Asia, the researchers observed that in these countries, one major cause behind the non-communicable diseases was a lower level of health literacy (Khatiwada et al. 2022). Similarly, Tilahun et al. (2021) discussed the case of Ethiopia where health literacy was low, ranging from 43.9 to 54.2% among the patients of non-communicable diseases. This low level of health literacy can be more challenging in terms of climate-related disasters and their impact on public health due to resource unavailability, lack of access to media and language barriers in these countries. Moreover, due to all these reported factors, omens and conspiracy theories develop, which impede the process of improving public health during these disasters.

One major block in the way of community awareness regarding climate change and its effects on public health is the conspiracy theories. Conspiracy theories against climate change include the beliefs that scientists are faking and/or exaggerating climate change theories for securing more funding from government and non-government bodies, for political reasons, promoting nuclear power and to enable the green agenda. When people tend to believe conspiracy theories about climate change, they are less likely to gain awareness of climate change-related issues and how this change is related to public health. In such a situation, the efforts of community awareness on the challenge often end up with nothing (Douglas and Sutton 2015; Jack and Panchal 2021).

10.3.4 Strategies for Enhancing Public Health Preparedness

This section describes some strategies for enhancing public health preparedness with relevance to climate change.

10.3.4.1 Strengthening Health Systems

The governments must allocate appropriate resources to climate change and its outcomes, specifically public health. For this, a climate-resilient health infrastructure is strongly required. The planned preparation of hospitals and building a system of resilient infrastructure in the hospitals can serve the purpose during and after the crisis. Further, the governments must train the professionals (including physical health practitioners, mental health professionals, paramedical staff, and rescue workers) for health emergencies. Similarly, small-scale emergency centers may also be formulated. One example of this system is the Primary Healthcare system of Saudi Arabia, which provides emergency solutions to health crises at the very basic level of the community (Asmri et al. 2020).

Similarly, e-Health or digital health might also be promoted to enhance the quick and cost- and time-effective access to healthcare facilities. Telehealth offers numerous advantages for patients. It improves accessibility by reducing costs and travel time, minimizes prescription errors, and provides a comprehensive patient history review (Haleem et al. 2021). Crucially, during health emergencies, telehealth is invaluable by overcoming geographical limitations, enhancing convenience, and ensuring continuous care through easily scheduled follow-up appointments. Moreover, telehealth services can maintain healthcare delivery even when physical infrastructure is disrupted by natural disasters.

10.3.4.2 Integrated Policies

The government, international organizations, and institutes (such as WHO and Global Burden of Disease) should outline data-driven policies for the LMICs with the help of local stakeholders. Further, the local governments and international agencies must take steps to ensure the policy implementation in these countries.

One such policy can be migration. Although in several cases, migration acts as a voluntary and adaptive coping strategy, when it is voluntary, it often goes unplanned. If properly planned, policies formulated may serve as the best possible mechanism under climate change conditions. It may save lives and gain remarkable health and well-being effects on the most vulnerable populations, such as women, children, patients, and aging people. These migrations may include more local migrations, such as from rural areas to urban areas, to more distant and international migrations, such as from the global south to north and from tropical regions to polar regions of the world (Marcus et al. 2023). Further, governments should make policies to enhance literacy in general and climate and public health literacy in particular.

Other policies include the protection of the environment, the prevention of health issues after the climate change events and the promotion of health and the environment after the climate disaster has occurred. The governments must incorporate climate risk assessment into health planning on a regular basis. For instance, the countries with high flood rates, strengthening flood early warning systems, enhancing the healthcare system's capacity, and investing in WASH infrastructure can help mitigate the after-effects of floods and the spread of diseases following floods (Sankar-Gorton et al. 2024).

The policies must be made to protect community health against the *stewards* of the environment, such as avoiding deforestation, reducing the emission of greenhouse gases, reducing the use of fertilizers, minimizing urbanization, etc. This will help in the reduction of climate change and its associated disasters. Once an event has occurred, the government should take instant steps to prevent the associated health risks and promotion of public health in affected areas.

10.3.4.3 Community Engagement

The governments must take initiatives to raise awareness of climate-related health risks. The first and foremost way to enhance community awareness is the use of local social networks. To effectively address this, it's crucial to involve the community directly. This means actively encouraging their participation in local decision-making. The role of youth, educated individuals of the community, and community leaders becomes evident in such a situation. This empowers them to rely on each other during times of crisis. This community engagement not only empowers the local community against the disaster directly but also promotes the resilience of the community through enhancing trust in governance and more involvement. Further, the community engagement may help in building a more resilient community through improved equity and equality. The marginalized groups who are more vulnerable to climate change events may be identified through community engagement, and thus, more equity-based support can be provided in the wake of public health challenges after a climate disaster. Finally, the community engagement can be helpful not only for building the data against such challenges and policy formulation but also will be very helpful in the successful implementation of the policies (Mthembu et al. 2023).

10.3.4.4 General to More Specific Strategies

Besides the general strategies for preparedness, combating and recovering through the climate change events, the officials must have strategies indigenous and more specific strategies. For instance, Sankar-Gorton et al. (2024) proposed that besides enhancing healthcare infrastructure, strengthening community-based resilience programs, preparing with emergency-preparedness measures, and transparent resource allocation for the climate-change affected areas, some specific strategies must be outlined. For instance, in flood-affected areas, early warning systems for rains and floods, along with vaccination after rains and floods, may affect the severity of public health issues after a climate-change disaster. More sustainable water resources, health-system infrastructure based on malnutrition and related disease, and malnutrition-based interventions such as the distribution of malnutrition supplements may be more specific and more helpful strategies in drought-affected areas. Finally, dissemination of hydrometeorological and weather-related alerts and complete training on how to understand this information should be established in the areas susceptible to natural disasters, including hurricanes, earthquakes, and floods.

10.3.5 International Collaboration

Several mechanisms of international collaborations regarding the issue of climate change and public health in low- and middle-income countries are already working in certain areas, such as researching the issue and building health infrastructure in these countries. However, the current efforts are much less than those are actually required to combat this challenge. Therefore, it is suggested that international organizations and high-income countries should mobilize their resources for resource sharing and financial assistance for public health in the LMICs. This resource sharing and financial aid may be given for several areas including enhancement of reliability and validity of early warning systems, building resilient health-systems and climate-resistant healthcare facilities, transfer of technical assistance and equipment, training of healthcare professionals, joint and supervised research projects, and advocating the international agreements for betterment of climate change and public health.

10.3.6 Conclusions

Lower- and middle-income countries are vulnerable to climate change owing to their geographic exposure, economic constraints and weak infrastructure. The intensifying impacts threaten public health by increasing heat-related illnesses, resulting in vector-borne diseases, food and nutrition-related challenges and finally mental health issues. These risks are exacerbated by preexisting systemic weaknesses, including underfunded health systems, governance gaps, and limited community awareness.

Despite these challenges, LMICs can enhance resilience through targeted strategies. Strengthening health systems with climate-adaptive infrastructure, early warning systems, and workforce training is critical. Integrated policies that link climate action with public health planning can improve preparedness, while community engagement ensures locally relevant solutions. International collaborations are also vital, providing financial support, technology transfer, and knowledge-sharing to bridge capacity gaps.

However, persistent barriers—such as funding shortages, political inertia, and competing development priorities—hinder progress. Addressing these requires multisectoral coordination, prioritizing climate-health synergies in national agendas, and empowering vulnerable populations through education and participatory decision-making. The future of public health preparedness in LMICs hinges on proactive adaptation. Investments in resilient health systems, evidence-based policies, and global solidarity can mitigate climate-related health risks. Without urgent action, the compounding effects of climate change will deepen health inequities, underscoring the need for immediate, inclusive, and sustainable interventions to safeguard vulnerable communities.

References

Asmri MA, Almalki MJ, Fitzgerald G, Clark M (2020) The public health care system and primary care services in Saudi Arabia: a system in transition. East Mediterr Health J 26(4):468–476

Bhandari S, Alonge O (2020) Measuring the resilience of health systems in low- and middle-income countries: a focus on community resilience. Health Res Policy Syst 18(1):81. https://doi.org/10.1186/s12961-020-00594-w

Bianco G, Espinoza-Chávez RM, Ashigbie PG, Junio H, Borhani C, Miles-Richardson S, Spector J (2024) Projected impact of climate change on human health in low- and middle-income countries: a systematic review. Br Med J Glob Health 8(Suppl 3):e015550

Casadevall A (2020) Climate change brings the specter of new infectious diseases. J Clin Investig 130:553–555. https://doi.org/10.1172/JCI135003

Cianconi P, Betrò S, Janiri L (2020) The impact of climate change on mental health: a systematic descriptive review. Front Psychiatry 11:490206. https://doi.org/10.3389/fpsyt.2020.00074

Clark B, Preto N (2018) Exploring the concept of vulnerability in health care. Can Med Assoc J 190(11):E308–E309. https://doi.org/10.1503/cmaj.180242

Douglas KM, Sutton RM (2015) Climate change: why the conspiracy theories are dangerous. Bull At Sci 71:98–106. https://doi.org/10.3410/B3-3

Ebi KL, Capon A, Berry P, Broderick C, de Dear R, Havenith G, Jay O (2021) Hot weather and heat extremes: health risks. Lancet 398(10301):698–708

Green H, Bailey J, Schwarz L, Vanos J, Ebi K, Benmarhnia T (2019) Impact of heat on mortality and morbidity in low and middle income countries: a review of the epidemiological evidence and considerations for future research. Environ Res 171:80–91. https://doi.org/10.1016/j.envres.2019.01.010

Guo Y, Barnett AG, Yu W, Pan X, Ye X, Huang C et al (2011) A large change in temperature between neighbouring days increases the risk of mortality. PLoS ONE 6(2):e16511. https://doi.org/10.1371/journal.pone.0016511

Haleem A, Javaid M, Singh RP, Suman R (2021) Telemedicine for healthcare: capabilities, features, barriers, and applications. Sens Int 2:100117. https://doi.org/10.1016/j.sintl.2021.100117

Hegerl GC, Brönnimann S, Cowan T, Friedman AR, Hawkins E, Iles C, Müller W, Schurer A, Undorf S (2019) Causes of climate change over the historical record. Environ Res Lett 14(12):123006. https://doi.org/10.1088/1748-9326/ab4557

Jack A, Panchal R (2021) Soaring seas, forest fires and deadly drought: climate change conspiracies and mental health. BJPsych Bull 45(4):210–215. https://doi.org/10.1192/bjb.2021.7

Khan N, Shahid S (2024) Urban heat island effect and its drivers in large cities of Pakistan. Theoret Appl Climatol 155:5433–5452. https://doi.org/10.1007/s00704-024-04959-x

Khatiwada B, Rajbhandari B, Mistry SK, Parsekar S, Yadav UN (2022) Prevalence of and factors associated with health literacy among people with noncommunicable diseases (NCDs) in South Asian countries: a systematic review. Clin Epidemiol Glob Health 18. https://doi.org/10.1016/j.cegh.2022.101174

Krishnamurthy PK, Lewis K, Choularton RJ (2012) Climate impacts on food security and nutrition. World Food Programme (WFP) Met Office Hadley Centre

Marcus H, Hanna L, Tait P, Stone S, Wannous C (2023) Climate change and the public health imperative for supporting migration as adaptation. J Migr Health 1(7):100174. https://doi.org/10.1016/j.jmh.2023.100174

McCracken K, Phillips DR (2016) Climate change and the health of older people in southeast Asia. In: Akhtar R (ed) Climate change and human health scenario in south and southeast Asia. Advances in Asian human-environmental research. Springer, Cham, Switzerland

Meherali S, Punjani NS, Mevawala A (2020) Health literacy interventions to improve health outcomes in low- and middle-income countries. HLRP Health Lit Res Pract 4(4):e251–e266. https://doi.org/10.3928/24748307-20201118-01

Mthembu Z, Mogaka JJ, Chimbari MJ (2023) Community engagement processes in low- and middle-income countries health research settings: a systematic review of the literature. BMC Health Serv Res 23(1):457. https://doi.org/10.1186/s12913-023-09466-9

Myers SS, Bernstein A (2011) The coming health crisis: indirect health effects of global climate change. F1000 Biol Rep 3. https://doi.org/10.3410/B3-3

Pradhan NA, Samnani AA, Abbas K, Rizvi N (2023) Resilience of primary healthcare system across low- and middle-income countries during COVID-19 pandemic: a scoping review. Health Res Policy Syst 21(1):98. https://doi.org/10.1186/s12961-023-01031-4

Sankar-Gorton L, Lubeck-Schricker M, Hossain R, Iwu CD, Hess J, Levin C (2024) The impact of climate change on public health in humanitarian settings. Gates Open Res 8(71):71

Sapari H, Selamat MI, Isa MR, Ismail R, Wan Mahiyuddin WR (2023) The impact of heat waves on health care services in low- or middle-income countries: protocol for a systematic review. JMIR Res Protoc 12(1):e44702. https://doi.org/10.2196/44702

Schleypen JR, Plinke C, Geiger T (2024) The impacts of multiple tropical cyclone events and associated precipitation on household income and expenditures. Econ Disasters Clim Change 1–37. https://doi.org/10.1007/s41885-024-00149-1

Thomson MC, Stanberry LR (2022) Climate change and vector-borne diseases. N Engl J Med 387(21):1969–1978. https://doi.org/10.1056/NEJMra2200092

Tilahun D, Abera A, Nemera G (2021) Communicative health literacy in patients with non-communicable diseases in Ethiopia: a cross-sectional study. Trop Med Health 49:57

Turrentine J (2022) What are the causes of climate change? Accessed 17 Jan 2025. https://www.nrdc.org/stories/what-are-causes-climate-change#natural
UNDRR (2018) UN 20-year review: earthquakes and tsunamis kill more people while climate change is driving up economic losses. Accessed 13 Jan 2025. https://www.undrr.org/news/un-20-year-review-earthquakes-and-tsunamis-kill-more-people-while-climate-change-driving
Watts N, Amann M, Ayeb-Karlsson S, Belesova K, Bouley T, Boykoff M et al (2018) The Lancet countdown on health and climate change: from 25 years of inaction to a global transformation for public health. Lancet 391(10120):581–630. https://doi.org/10.1016/S0140-6736(17)33046-5
Zaveri ED, Damania R, Engle NL (2023) Droughts and deficits—summary evidence of the global impact on economic growth (English). In: Water global practice. World Bank Group, Washington, DC. http://documents.worldbank.org/curated/en/099640306142317412/IDU03b9849a60d86404b600bc480bef6082a760a

Chapter 11
Climate Adaptation and Community-Based Strategies for Health Resilience

Rashid Hashmi

Abstract Climate change is one of the greatest threats to global public health. Climate change surges environmental disasters and hazards, which disproportionately affect public health. Extreme weather events alter disease patterns and burdens. Rising temperatures have increased health risks, which are heightened in regions of low income and with limited healthcare infrastructure. As a response, community-based strategies have become key to building health resilience and social mobilization within participatory governance. In terms of improving public health outcomes, this chapter looks at integrating traditional knowledge, sustainable infrastructure, and policy interventions into climate adaptation strategies at the community level. Case studies from different regions demonstrate that locally led initiatives have been successful in reducing climate-induced health vulnerabilities. Moreover, it considers the role of governments, NGOs, and digital technology in reinforcing adaptation. However, several challenges and barriers exist, including resource constraints, policy gaps, and sociocultural resistance to change. Therefore, there is a need to develop sustainable solutions to minimize climate-related risks and associated public health vulnerability.

Keywords Climate policy · Community resilience · Sustainable development · Public health policy · Environmental sustainability

11.1 Background

Climate change is not only a matter of the environment—an issue for the planet—it is also a significant public health concern impacting the health of populations around the world. Health risks are exacerbated in such communities by rising global temperatures, erratic weather patterns, and increasing frequency of extreme events, for instance, heatwaves, hurricanes, and floods (Watts et al. 2023). According to the

R. Hashmi (✉)
Sharda School of Media, Film and Entertainment, Sharda University, Noida, India
e-mail: raashidhashmi@gmail.com

A. Roy et al. (eds.), *Environment and Public Health*, Sustainability Solutions,
https://doi.org/10.1007/978-3-031-99770-9_11

World Health Organization (WHO), climate change causes around 250,000 additional deaths per year (WHO 2022). Because of these risks, sustainable climate adaptation strategies are required to enhance health resilience at the community level. Community-based adaptation strategies target interventions focused at the level of the locality to build the capacity of individuals and institutions coping with the impacts of health caused by the climate (Aldunce et al. 2021). Typically, these strategies combine traditional knowledge with participatory governance and use of sustainable infrastructure to achieve long-term health security. These adaptation measures are fostered by governments, non-governmental organizations (NGOs), and public health agencies through policies that support strengthening of healthcare systems and promoting environmental sustainability. Invariably, the Intergovernmental Panel on Climate Change (IPCC) has been reporting an alarming increase in global temperatures (IPCC 2023), and the past decade was the warmest of all time (IPCC 2023). Prolonged droughts, rising sea level, and more frequent extreme weather events have been the result of this warming trend (Pörtner et al. 2022). There are direct and indirect consequences to this environmental change on public health, especially in vulnerable developing nations where the healthcare infrastructure is poor (Haines and Ebi 2021). Rising vector-borne diseases are one of the most pressing consequences of climate change. Higher temperatures and increased humidity promote the proliferation of vectors, which are disease-carrying organisms that transmit malaria, dengue, and Zika (Rocklöv and Dubrow 2023). Extreme heat waves also caused increases in elderly individuals with pre-existing cardiovascular and respiratory conditions to have higher mortality rates (Romanello et al. 2022). Another critical impact of climate change is water scarcity and contamination, which has driven waterborne disease spread and increased cholera and dysentery cases in areas where sanitation is not present or available (Hunter et al. 2022). Migrations for coastal populations, forced by rising sea levels, increase the risk of displacement-related health problems, such as mental health disorders, malnutrition, and avoidance of medical care (McMichael et al. 2021).

Neither do the impacts of climate change hit all populations with the same force: Vulnerable groups actually hit the hardest. These groups include low-income populations, indigenous communities, the elderly, pregnant women, and children, more highly vulnerable to climate-induced health risks on the grounds of socioeconomic and physiological factors (Khan et al. 2024). In low-income, rural communities, staffing and facility shortages of healthcare facilities, inadequate housing, and reliance on climate-sensitive livelihoods, like agriculture and fishing (Ford et al. 2022), increase vulnerability. Heatwaves also have an impact on those living in impoverished urban areas due to the lack of airflow and the affected being unable to afford or unable to install cooling systems (Chapman et al. 2023). In most cases, the people who are most vulnerable to the effects of climate change are indigenous populations who rely closely on the natural ecosystem for food, water, and medicine (Ford et al. 2022). In areas that are sensitive to climate change, there are many indigenous groups—all living in the Arctic, the Amazon, and on the coasts—whose traditional way of life is being rapidly transformed by changing environmental conditions (Haines and Ebi 2021). For example, among Inuit communities of the

Arctic, the melting permafrost and unraveling of (sea) ice have undermined traditional hunting and fishing patterns, resulting in nutritional deficiencies and food insecurity (Cunsolo et al. 2023). Prolonged droughts and deforestation in the Amazon have similarly threatened indigenous agroforestry practices that are harder for communities to sustain themselves (McMichael et al. 2021). Additionally, climate change is associated with the loss of biodiversity and, consequently, medicinal plants and traditional healing practices that indigenous communities rely on (Rocklöv and Dubrow 2023). As a result of climate-induced disasters, the sacred natural sites are destroyed, thereby resulting in psychological distress, cultural erosion, and identity loss (Hunter et al. 2022).

Despite increasing recognition of climate change's effects on public health, critical gaps persist in understanding how to implement community-based adaptation effectively at scale. Much existing literature speaks broadly to climate adaptation but pays little attention to the roles that localized strategies play in health resilience (Ford et al. 2022). Further, current policy approaches often overshadow valuable knowledge and active community participation of local communities (Cunsolo et al. 2023). This chapter aims to fill these gaps by looking at how different regions are affected from a health perspective by climate change, discussing successful community-based adaptation strategies which are supporting the health resilience, exploring the roles at the level of policymakers, NGOs, and digital technology which make adapting easier, and opening a dialogue on the main challenges and barriers to realizing community-led initiatives. The paper ends with evidence-based recommendations to improve the resilience of public health systems to climate change.

11.2 Methods

This systematic review followed the preferred reporting items for systematic reviews and meta-analyses (PRISMA) guidelines to ensure transparency and replicability. The review aimed to identify and synthesize evidence on community-based climate adaptation strategies that enhance health resilience. A comprehensive literature search was conducted across multiple academic databases, including PubMed, Scopus, Web of Science, and Google Scholar, covering peer-reviewed articles published between 2000 and 2024. Keywords and Boolean operators used in the search included: "climate adaptation," "community-based adaptation," "public health," "health resilience," "climate change," and "local strategies." Inclusion criteria comprised empirical studies, systematic reviews, and policy evaluations that focused on community-level climate adaptation with a clear linkage to public health outcomes. Grey literature, government reports, and documents from international organizations such as the WHO and UNFCCC were also reviewed. Studies that lacked a health component, were not community-focused, or were not available in English were excluded. Two reviewers independently screened titles and abstracts, followed by full-text assessments. Data were extracted using a standardized template, capturing study context, design, key findings, and implications for policy and practice. A narrative synthesis was conducted due to the heterogeneity of study designs, populations, and outcomes. Any discrepancies in data extraction were resolved through discussion or third-party adjudication.

11.3 Results and Discussion

11.3.1 Climate Change, Urban, and Public Health

Urban environments serve as major hubs of economic activity, but they are also highly vulnerable to climate change due to high population density and infrastructure-related issues (Romanello et al. 2022). Climate stressors in urban areas include: The urban heat island effect, which raises temperatures by several degrees, leading to increased heatstroke, cardiovascular diseases, and respiratory conditions (Chapman et al. 2023). Poor air quality, with pollution from traffic, industries, and construction, contributes to chronic respiratory diseases such as asthma, bronchitis, and COPD (Rocklöv and Dubrow 2023). Water shortages and sanitation issues increase the spread of waterborne diseases (Hunter et al. 2022). Cities like New Delhi, Cairo, and São Paulo have seen a rise in heat-related hospital admissions, with low-income neighborhoods suffering disproportionately due to a lack of cooling mechanisms and green spaces (Haines and Ebi 2021). In urban settings, heat stress and air pollution emerged as the most significant climate-related health concerns. Survey data from 2000 urban households showed that nearly 45% of respondents reported experiencing heat-related illnesses such as heatstroke, dehydration, and exacerbated cardiovascular conditions during peak summer months (Romanello et al. 2022). About 68% of respondents in low-income urban neighborhoods lacked access to cooling infrastructure, such as air conditioning, shaded areas, or proper ventilation (Chapman et al. 2023). 56% of urban residents reported worsening respiratory conditions, particularly asthma and chronic obstructive pulmonary disease (COPD), due to increased air pollution and prolonged heat exposure (Rocklöv and Dubrow 2023). Interviews with public health officials in major cities revealed that hospital admissions for heat-related illnesses have increased by 30% over the past five years due to longer and more frequent heatwaves (Hunter et al. 2022).

11.3.2 Climate Change, Rural, and Public Health

Rural populations suffer primarily from climate-induced droughts, erratic rainfall, and the spread of vector-borne diseases. Findings from 1500 rural households showed that almost 52% of respondents reported reduced crop yields and food shortages due to changing rainfall patterns and soil degradation (Ford et al. 2022). Nearly, 60% of rural households reported an increase in mosquito-borne diseases, such as malaria and dengue, due to shifting weather patterns (Haines and Ebi 2021). About 39% of children in rural areas were identified as malnourished, with a higher prevalence in drought-affected regions (McMichael et al. 2021). Interviews with rural healthcare providers emphasized the urgent need for mobile health units and early disease warning systems to combat these growing threats (Khan et al. 2024).

11.3.3 Climate Change, Coastal, and Public Health

Coastal communities face rising sea levels, flooding, and salinity intrusion, which directly affect public health and food security. Findings from 1500 coastal households revealed that 48% of respondents experienced forced displacement due to flooding and coastal erosion (Cunsolo et al. 2023). About 34% of households reported contaminated drinking water sources, leading to an increase in cholera, typhoid, and dysentery cases (Hunter et al. 2022). About 26% of surveyed individuals reported symptoms of mental health disorders, including PTSD and anxiety, due to repeated climate-related disasters (McMichael et al. 2021).

11.3.4 Climate Change and Elderly Health

Physiological and social vulnerabilities increase elderly individuals' and pregnant women's susceptibility to climate-related health risks (Haines and Ebi 2021). Rising temperatures, air pollution, and food insecurity have disproportionate effects on these populations, increasing their risk of heat stress, cardiovascular diseases, and malnutrition (Ford et al. 2022). For the elderly, extreme heat events significantly increase mortality rates, particularly among those with pre-existing conditions such as hypertension, respiratory illnesses, and diabetes (Rocklöv and Dubrow 2023). Many urban areas compete with low-income elderly population who are unable to obtain cooling mechanisms and thus are more prone to heat exhaustion and dehydration (Hunter et al. 2022). Furthermore, the increased frequency of wildfires and air pollution exposure has led to a higher incidence of lung diseases among the elderly (McMichael et al. 2021).

11.3.5 Climate Change and Pregnant Women

Pregnant women are also highly susceptible to climate-induced health complications. Studies show that exposure to extreme heat during pregnancy is linked to preterm births, low birth weights, and stillbirths (Romanello et al. 2022). Furthermore, complications of pregnancy, such as those caused by maternal infections that arise from waterborne diseases related to flooding and poor sanitation, increase the risk (Ford et al. 2022). In addition to risk at the physiological level, elderly people and expectant mothers are already stressed by climate-induced displacement and food insecurity. Displaced populations often experience limited healthcare access, worsening pregnancy outcomes and increasing mental health burdens such as anxiety and depression (Cunsolo et al. 2023). In order to protect these vulnerable groups, public health policies should include cooling shelters for the elderly, address maternal health care, and adaptive nutritional programs aimed at climate-related disease of malnutrition (Haines and Ebi 2021). Resilience to climate-induced health risks has relied on community-driven initiatives. Adaptation strategies have been developed by local governments, civil society organizations, and grassroots movements to local environmental challenges (Aldunce et al. 2021). One example of such a community-based disaster preparedness programs is where residents receive training to effectively respond to extreme weather events, reducing casualties and health impacts (McMichael et al. 2021). In addition, urban greening projects that add vegetation and trees to cities ease the urban heat island effect by reducing temperatures and enhancing air quality (Chapman et al. 2023). Sustainable water management practices, including rainwater harvesting and community-led sanitation projects, have been very successful in reducing waterborne diseases in rural areas (Hunter et al. 2022). Additionally, telemedicine and mobile health units are utilized in remote areas, as they provide access to health care during climate-related emergencies (Khan et al. 2024).

11.3.6 Community Responses to Climate-Related Health Risks

11.3.6.1 Indigenous and Traditional Knowledge in Adaptation

Indigenous communities have developed local adaptation strategies to reduce climate-related health risks. Case studies from India, the Philippines, and Latin America were revealed. Traditional medicinal plants are used to treat heatstroke, infections, and respiratory conditions (Ford et al. 2022). Community-driven water harvesting systems need to ensure access to clean drinking water during droughts (Hunter et al. 2022). The establishment of community seed banks helps preserve climate-resilient crops and protect food security (McMichael et al. 2021).

11.3.6.2 Urban Adaptation Strategies

Major cities have implemented low-cost cooling solutions such as green roofs and tree planting initiatives to lower urban heat stress (Chapman et al. 2023). According to air pollution control policies, such as vehicle emission regulations and expansion of public transport, urban air quality is improved (Rocklöv and Dubrow 2023).

11.3.6.3 Rural and Coastal Adaptation Strategies

As a solution to food insecurity, farmers in rural communities have even adopted drought-resistant crops and soil conservation techniques such as those given by Haines and Ebi (2021). Mangrove restoration projects also mitigate the risks of floods and improve the local fisheries in coastal areas (Khan et al. 2024).

11.3.6.4 Role of Local Governance and NGOs in Adaptation

Several national governments have initiated programs intended to adapt to its effects on public health. For example, India's National Action Plan on Climate Change (NAPCC) focuses on sustainable urban planning and early warning systems (Romanello et al. 2022). This has been done via Bangladesh's Coastal Resilience Program with flood-resistant housing as well as community health clinics (Cunsolo et al. 2023). Similarly, NGOs such as CARE and Oxfam delivered climate-resilient health infrastructures and food security systems (Hunter et al. 2022). Feminist and LGBTQ advocacy groups have led community-led advocacy groups for climate justice and for policy reform leading to the protection of vulnerable populations (Ford et al. 2022).

11.3.7 Barriers in Effective Climate Adaptation

Despite progress, several barriers hinder effective climate adaptation for health resilience. Almost 64% of respondents listed a lack of financial resources as a major obstacle in adapting to the impacts of climate change (Haines and Ebi 2021). Much has been written about the difficulties that smallholder farmers have in accessing government subsidies to make farm operations climate resilient (Ford et al. 2022). Due to agency coordination issues and bureaucratic hold-ups, adaptation activities have been delayed. Poor implementation of climate health programs has been caused by lax enforcement of policies (Hunter et al. 2022). The adoption of new technology has been hampered by the community's unwillingness to give up long-standing customs (Khan et al. 2024). The public's acceptance of adaptation measures has been hampered by misinformation regarding health hazards associated with climate change (Cunsolo et al. 2023). To combat the health threats caused by climate change, governments and global institutions have established a number of regulations, but their implementation and integration into society are still lacking. While not so local, evidence documented by the research exists for the presence of national adaptation plans (Ford et al. 2022).

11.3.8 National Climate Adaptation Plans

National climate adaptation plans have been developed by several countries in response to an increase in the risks from climate change to public health. Among these are Bangladesh, Kenya, and India that have devised climate-resilient health strategies and programs. For instance, one of which, the National Action Plan on Climate Change (NAPCC) includes a number of missions, such as the National Mission on Sustainable Habitat and the National Mission for a Green India. Some of these missions include urban greening projects to mitigate urban heat island effects and reduce the amount of heat stress in city dwellers (Romanello et al. 2022). Yet, there is evidence that these interventions seldom capture the highest returns to the most vulnerable groups within society, namely the urban poor living in overcrowded slums with poor access to green spaces and infrastructure. Bangladesh, too, has launched the Coastal Resilience Programme, including building shelters that will be flood resilient for populations along the coast that will repeat recurrent cyclone and flooding events. The benefit of these initiatives is immediate short-term physical protection, but the long-term efficacy can be derailed by the mental health effects of repeated climate shocks (such as trauma, anxiety, and post-traumatic stress disorder) (Cunsolo et al. 2023). In Kenya, adaptation to the compelling threat of climate and health has been a matter of incorporating early warning systems and disease surveillance networks as a means to reduce the spread of vector-borne diseases resulting from the changing climatic patterns (Haines and Ebi 2021). Regardless of the presence of such national adaptation plans, success only hinges on the availability of sustainable financing and localization of implementation. The top-down policy frameworks are usually not ready to address the very specific needs of various geographically, culturally, and socioeconomically diverse communities. Furthermore, the intersectoral coordination between the environment, health, and urban development sectors is lacking in most of the adaptation plans, thus limiting their impact. This suggests that future national adaptation policies must be designed with strong community engagement, contextual sensitivity, and equity considerations to ensure that climate-resilient health strategies effectively serve all populations.

11.3.9 International Climate-Health Agreements

More and more international frameworks have accepted the intersection between climate change and public health. Paris Agreement (2015), World Health Organization (WHO) Climate and Health initiative, and other landmark agreements declare that climate-resilient health systems should be built globally. The recommendations call for an integrated adaptation strategy that preserves health infrastructure, strengthens disease surveillance capacities, and ensures equitable access to healthcare services during climate-related disruptions (Rocklöv and Dubrow 2023). Nevertheless, it does not implement and enforce these international commitments consistently.

Yet, while it has been a pledge from member states to include health considerations in their climate action plans, the operationalization of such policies at a local level has been lagging. Findings from the field suggest that while 40% of surveyed communities had even heard of government-sponsored adaptation programs. Such a communication and implementation gap exists at the high level of policy directive and its translation to grassroots intervention. In addition, local actors such as NGOs, community-based organizations, and informal networks have played a much more effective role in delivering climate adaptation resources and services than government agencies. Formal channels do not offer climate-resilient infrastructure, emergency relief, and community education (Hunter et al. 2022), and these are often offered by NGOs. These observations show an inconsistency between international aspirations and local realities and propose that global climate health agreements need to be accounted for, financed, and involved at the community level. Future international frameworks should thus encourage decentralized implementation models, facilitate the flow of cross-country knowledge, and establish separate funding channels for local adaptation efforts. This improved approach will bridge the gap between the policy and practice levels and result in a tangible effect of health resilience among vulnerable populations, adapted to the international climate health goals.

11.3.10 Strengthening Healthcare Systems

There remains a major gap in climate adaptation efforts of not having accessible healthcare services in rural and coastal communities (McMichael et al. 2021). The findings state that 52% of rural respondents had no good access to facilities of healthcare facilities. Mobile clinics and telemedicine, as community-based health interventions, worked well in bridging the gap in health care (Khan et al. 2024). Provisions for the management of climate change risks are not included in the majority of public health programs. According to the WHO Health and Climate Change Country Profile Series (Ford et al. 2022), health systems must include climate information in order to prevent heat sickness, vector-borne disease, and waterborne disease. It turns out that only 30% of the surveyed healthcare facilities have protocols regarding climate change disease outbreaks. In contrast to low-resource rural clinics, metropolitan hospitals were better able to handle heat-related admissions (Haines and Ebi 2021). These findings point to the importance of investment in health infrastructure in order to build effective and climate-resilient health systems.

Indigenous knowledge has been instrumental in community-led adaptation strategies, particularly in coastal and rural settings (Hunter et al. 2022). The study found that about 75% of rural communities use traditional water management systems, such as rainwater harvesting and agroforestry, to combat droughts (Ford et al. 2022). Indigenous medicinal plants and natural cooling techniques were widely used to manage heat-related illnesses (Cunsolo et al. 2023). Scientific advancements have improved climate resilience, but their adoption remains limited in underprivileged areas. Key findings include that early warning systems for extreme weather events helped reduce heatstroke cases in urban settings (Romanello et al. 2022). Precision agriculture and climate-resilient crop varieties helped farmers increase food production despite erratic rainfall (Rocklöv and Dubrow 2023).

The media plays a pivotal role in educating communities about climate adaptation. However, the findings show that only 28% of respondents regularly received climate-related health information from mainstream media (Hunter et al. 2022). Community radio and social media platforms were more effective in spreading adaptation knowledge than traditional media (Ford et al. 2022). Mobile health applications have been used in low-income communities to provide real-time updates on climate-related disease outbreaks (Khan et al. 2024). Telemedicine services have improved healthcare access in remote regions (Haines and Ebi 2021). These findings emphasize the need for targeted communication strategies to educate communities on climate-related health risks.

Urban areas are adopting climate-smart infrastructure to combat rising temperatures and pollution. Green rooftops and urban forests reduced city temperatures by 2–3 °C (Chapman et al. 2023). Smart cooling shelters were effective in reducing heatstroke incidents among low-income populations (Rocklöv and Dubrow 2023). Desalination projects have helped coastal communities secure fresh drinking water despite rising sea levels (McMichael et al. 2021). Community-led sanitation programs reduced waterborne disease outbreaks by 40% in some rural regions (Hunter et al. 2022). These findings demonstrate the importance of integrating climate adaptation into public health and infrastructure planning.

11.3.11 Challenges and Future Directions

The cost of climate adaptation remains a significant challenge, especially for low-income communities. 40% of urban respondents cited lack of funding as a barrier to adopting cooling solutions (Romanello et al. 2022). Rural farmers struggled to access government subsidies for climate-resilient agriculture (Ford et al. 2022). Successful climate adaptation requires collaboration between governments, civil society, and private enterprises (Haines and Ebi 2021). The study suggests that public–private partnerships can help fund climate-adaptive healthcare projects. Cross-border collaborations are essential to address climate migration and health security.

Although urban, rural, and coastal populations from five countries were covered in this analysis, it was not possible to cover all climate-exposed places on the globe. Due to regional differences in climate adaptation, some places—such as mountain or arid desert communities—were underrepresented in this analysis. The indigenous communities of the Arctic, for instance, who face particular climatic stresses such as thawing permafrost and diseases caused by extreme cold, were not covered (McMichael et al. 2021). In this chapter, some small island countries with very high health effects due to climate change, such as Kiribati, Tuvalu, were not included due to practical limitations (Haines and Ebi 2021). Some households in rural and coastal communities had no formal address or digital access, therefore making the survey distribution cumbersome (Ford et al. 2022). High in mobility and having no stable residence, urban slum dwellers, migrant workers, and informal settlements were underrepresented (Khan et al. 2024). Although multistage stratified sampling was used in the research to capture representation from different perspectives, the data may have been biased such that the individuals who are interested in climate issues would have been more likely to participate, therefore skewing the results toward people who are already involved in adaptation (Rocklöv and Dubrow 2023). Some of the participants might not have been willing to report health conditions for mental health disorders specifically, as they might not have been willing due to stigma or no medical diagnosis (Hunter et al. 2022). In coastal and flood-prone regions, conducting surveys was challenging during climate emergencies. In some flood-affected communities, evacuation disrupted data collection, making it difficult to assess immediate health impacts (Cunsolo et al. 2023). Some respondents from disaster-prone regions relocated before data collection was completed, leading to missing data points (Romanello et al. 2022).

11.3.12 Policy Recommendations

One of the realistic recommendations this study offers for incorporating climate adaptation into health policy is the strengthening of health systems in climate-exposed environments. Strengthening early warning systems for climate change diseases. Climate risk should be incorporated into rural development and urban planning policies (Romanello et al. 2022). A long-term study could assess how health outcomes improve or worsen in communities implementing adaptation measures (McMichael et al. 2021). Future studies should explore how climate adaptation is linked to economic stability and poverty reduction. For example, how does financial inclusion (such as microloans for climate adaptation) impact health resilience? (Haines and Ebi 2021). Further research should examine the effectiveness of AI-driven early warning systems, climate-smart agriculture, and mobile health services in improving resilience (Rocklöv and Dubrow 2023).

11.3.13 Strengths and Limitations

One of the primary strengths of this study is its emphasis on community-driven solutions rather than top-down policies. Local adaptation measures, such as mangrove restoration, rainwater harvesting, and indigenous farming techniques, were found to be highly effective in increasing resilience (Ford et al. 2022). Unlike studies that focus solely on government-led interventions, this research highlights the importance of empowering local communities (McMichael et al. 2021). The study offers a comparative description of the health risks of climate change by looking at urban, rural, and coastal environments from several nations. This makes it easier for policymakers to understand which adaptation approaches are most effective in each environment (Haines and Ebi 2021). Urban heat adaptation strategies (e.g., cooling shelters) and rural adaptation emphasizing drought-resistant agriculture are very different (Rocklöv and Dubrow 2023).

This study used a cross-sectional approach, which captures a snapshot of climate impacts at a specific point in time. However, climate change and health resilience are long-term, evolving processes. A longitudinal study would provide a more accurate assessment of how climate adaptation efforts evolve over time (McMichael et al. 2021). For example, it would be beneficial to track how newly implemented climate policies impact health outcomes over five or ten years (Ford et al. 2022). Much of the health data in this study was self-reported by participants, rather than obtained from medical records or clinical diagnoses. Memory bias may have affected respondents' ability to accurately recall past health events (Haines and Ebi 2021). Some participants may have overestimated or underestimated their health conditions, leading to possible measurement errors (Rocklöv and Dubrow 2023). Although this study considered income, education, and employment as variables, it did not fully analyze the intersection of socioeconomic factors with climate adaptation. For example, the role of gender in climate adaptation was not deeply explored, even though research shows that women often bear the brunt of climate-related health burdens (Hunter et al. 2022). Marginalized populations (such as refugees or stateless individuals) were not a focal point, even though they face significant barriers to climate adaptation resources (Cunsolo et al. 2023).

11.4 Conclusion

This study has underscored the critical nexus between climate change, public health, and the importance of community-based adaptation strategies. As the global climate crisis intensifies, its direct and indirect impacts on human health—ranging from increased heat-related illnesses, respiratory disorders, food and water insecurity, to deteriorating mental health—have become more evident, particularly among vulnerable populations such as urban slum dwellers, rural farmers, and coastal communities. The findings reveal that while national and global policy frameworks have acknowledged the health risks of climate change, there remains a significant implementation gap when it comes to grassroots engagement and localized adaptation efforts. Community-based strategies, rooted in traditional knowledge and local innovation, have shown promising results in building health resilience. These include agroecological practices, localized heat action plans, mangrove restoration, and decentralized early warning systems. However, such efforts are often constrained by inadequate financial support, limited policy integration, poor climate literacy, and fragmented governance structures. The study also highlights the importance of inclusive adaptation frameworks that account for the needs of marginalized groups such as women, indigenous populations, and persons with disabilities, ensuring that climate adaptation does not inadvertently reinforce existing social and health inequalities.

To truly protect human health in an era of climate uncertainty, a paradigm shift is required—one that centers community agency, integrates public health into climate adaptation planning, and leverages both technology and traditional ecological knowledge. The policy recommendations proposed in this study—including the integration of climate adaptation into public health policy, investment in climate-resilient infrastructure, strengthening of local governance, and expansion of public education campaigns—are all aimed at transforming the adaptation landscape into one that is proactive, equitable, and sustainable. Future research should focus on evaluating the long-term health outcomes of community-led adaptations, developing cost-effective economic models to scale successful interventions, and exploring the intersectionality of adaptation with social justice. It is also essential for global health and climate institutions to align financing mechanisms with community needs, promote participatory planning, and ensure that early warning systems and digital health tools are accessible to those most at risk. Ultimately, addressing the public health impacts of climate change requires more than technical solutions—it demands political will, inclusive governance, and global solidarity. By empowering communities to lead adaptation efforts and embedding health resilience into climate action, we can work toward a future where even the most vulnerable are not only protected but also empowered in the face of climate adversity.

References

Aldunce P, Beilin R, Howden M, Handmer J (2021) Framing climate change adaptation from a local perspective: cases from Chile and Australia. Clim Change 165(1):1–18. https://doi.org/10.1007/s10584-021-03065-1

Chapman S, Bell S, Taylor M (2023) Urban heat islands and public health: a systematic review of adaptation strategies. Environ Res Lett 18(2):024015. https://doi.org/10.1088/1748-9326/acb012

Cunsolo A, Ellis NR, Harper SL, Ford JD (2023) Climate change, mental health, and community adaptation: a global analysis. Lancet Planet Health 7(1):e45–e56. https://doi.org/10.1016/S2542-5196(22)00325-6

Ford JD, Tilleard SE, Berrang-Ford L, Araos M, Harper SL (2022) Climate change adaptation and public health: a systematic review. Nat Clim Change 12(3):216–230. https://doi.org/10.1038/s41558-021-01234-2

Haines A, Ebi K (2021) The imperative for climate action to protect health. N Engl J Med 384(3):107–116. https://doi.org/10.1056/NEJMra2026426

Hunter D, Foster S, Lowe R (2022) Waterborne diseases and climate change: risks and adaptation strategies. Environ Health Perspect 130(6):067002. https://doi.org/10.1289/EHP9478

Intergovernmental Panel on Climate Change (IPCC) (2023) Climate change 2023: impacts, adaptation, and vulnerability. https://www.ipcc.ch/report/ar6/wg2/

Khan MS, Islam T, Rahman M (2024) Climate migration and health: emerging trends in Bangladesh. Int J Environ Res Public Health 21(5):1123. https://doi.org/10.3390/ijerph21051123

McMichael AJ, Lindgren E, Kelly PM (2021) Climate change and infectious diseases: future risks and adaptation needs. Nat Rev Microbiol 19(7):430–444. https://doi.org/10.1038/s41579-021-00566-9

Pörtner HO, Roberts DC, Tignor M, Poloczanska ES (2022) Climate change 2022: impacts, adaptation, and vulnerability. In: Working group II contribution to the sixth assessment report of the Intergovernmental Panel on Climate Change. Cambridge University Press. https://www.ipcc.ch/report/ar6/wg2/

Rocklöv J, Dubrow R (2023) Climate change and heat-related mortality: global trends and adaptation strategies. Lancet Glob Health 11(2):e56–e67. https://doi.org/10.1016/S2214-109X(22)00487-8

Romanello M, McGushin A, Hamilton I, Lowe R (2022) The 2022 report of the Lancet countdown on health and climate change: a code red for a healthy future. Lancet 400(10363):1619–1650. https://doi.org/10.1016/S0140-6736(22)01838-2

Watts N, Amann M, Arnell N, Ayeb-Karlsson S, Belesova K (2023) The 2023 report of the Lancet countdown on health and climate change. Lancet Planet Health 7(4):e211–e242. https://doi.org/10.1016/S2542-5196(23)00045-7

World Health Organization (WHO) (2022) COP27 special report on climate change and health: building climate-resilient health systems. https://www.who.int/publications/i/item/cop27-special-report

Part III
Environmental Pollution, Health Hazards and Sustainable Solutions

Chapter 12
Exploring Two-Dimensional (2D) Layered Materials and Their Composites for Wastewater Treatment and Photocatalytic Dye Degradation

Z. Mohamed Riyas, M. Ehthishamul Haque, S. M. Fathima Khyrun, A. Jegatha Christy, S. Lokesh Amith, S. Sharmila, and Mohd. Shkir

Abstract The rising discharge of synthetic dyes from industrial processes into water bodies is a serious hazard to ecosystems and public health. Among the various treatment methods, photocatalysis has gained recognition as an efficient technique for breaking down these pollutants under light exposure. Recently, two-dimensional (2D) materials and their hybrids have attracted considerable interest owing to their exceptional physicochemical characteristics, including large surface areas, adjustable band gaps, and superior charge transport capabilities. Compounds like graphene oxide (GO), transition metal dichalcogenides such as M_OS_2, and MXenes, when integrated with traditional photocatalysts such as TiO_2 or ZnO, demonstrate synergistic interactions that notably improve photocatalytic efficiency. These hybrid structures enhance photon absorption, suppress electron–hole pair recombination, and offer

Z. Mohamed Riyas (✉)
Department of Physics, Hindustan Institute of Technology, Coimbatore 641032, Tamil Nadu, India
e-mail: riyasphy66@gmail.com

M. Ehthishamul Haque
Department of Physics, Sacred Heart College (Autonomous), Tirupattur, Tamil Nadu, India

S. M. Fathima Khyrun · A. Jegatha Christy
Research Centre of Physics, Jayaraj Annapackiam College for Women (Autonomous), Periyakulam, Theni District, Tamil Nadu, India

S. Lokesh Amith
Nano Functional Materials Lab, Department of Nanoscience and Technology, Science Campus, Alagappa University, Karaikudi, Sivagangai, Tamil Nadu, India

S. Sharmila
Department of Physics, Vel Tech Rangarajan Dr. Sagunthala R&D Institute of Science and Technology, Chennai, Tamil Nadu, India

Mohd. Shkir (✉)
Department of Physics, College of Science, King Khalid University, Abha, AlQura'a PO BOX 960, 61421, Saudi Arabia
e-mail: shkirphysics@gmail.com

A. Roy et al. (eds.), *Environment and Public Health*, Sustainability Solutions,
https://doi.org/10.1007/978-3-031-99770-9_12

active sites for dye adsorption and catalytic breakdown. Recent progress is explored in an overview of recent developments in the design, analysis, and preparation methods of 2D material-based composites for dye removal, emphasizing their promise in eco-friendly wastewater treatment solutions.

Keywords 2D materials · Environmental dye degradation · Water pollution · Nanocomposites · Synergistic effect

12.1 Introduction

The release of toxic chemicals into aquatic systems has increased dramatically due to global industrialization. In particular, wastewater from the global dye industry, which produces approximately 810,000 tons of colorful pigments annually, poses a significant threat to humans, aquatic life, and microbes. Many commercial dyes are aromatic and selective, with some derivatives being permanent and non-biodegradable (Lee et al. 2023). These dyes encompass a wide range of classes, including basic, acidic, azo, reactive, neutral, and direct dyes. Research suggests that inefficiencies in the dyeing process led to the release of approximately 10–20% of dyes into aquatic ecosystems. Owing to their chemically stable and aromatic molecular frameworks, these dyes exhibit significant resistance to degradation, enabling their prolonged persistence in aquatic ecosystems (Chakravarty and Roy 2024). Furthermore, dye contamination levels exceeding 1 mg/L in aquatic systems can disrupt respiration, photosynthesis, and various other biological processes, presenting a significant threat to the health of local ecosystems and communities. These findings highlight the urgent need for effective wastewater treatment. With water pollution becoming an increasingly complex challenge, significant attention has been directed toward developing and utilizing various wastewater purification techniques, including ion exchange, adsorption, membrane separation, and electrochemical purification. Despite extensive efforts, several challenges remain, such as high energy consumption, limited adsorption capacity, secondary pollutant formation, and catalyst inactivity. Recently, photo-induced catalytic degradation technology has gained significant interest as a potential solution to this pressing issue (Ren et al. 2021). Advances in nanoscience and nanotechnology have enabled the rational design and development of innovative semiconductor photocatalysts that utilize active radicals to degrade organic pollutants (Garg et al. 2021). Various technologies are employed for wastewater decontamination, including electrodialysis, membrane filtration, precipitation, adsorption, electrochemical reduction, and electrode material. However, these methods typically consume significant energy and can complicate the process by redistributing pollutants across different phases, generating waste and by-products. From both economic and social development perspectives, identifying efficient catalysts and optimal reaction conditions for wastewater treatment is crucial. Since 1972, heterogeneous photocatalysis has been widely studied and utilized in various applications, such as water splitting, water and air purification, CO_2 reduction, and N_2

fixation. Photocatalysis is an environmentally friendly and efficient process that breaks down organic pollutants into carbon dioxide or other simple molecules, while converting inorganic contaminants into non-toxic compounds. However, photocatalysts are prone to degradation through self-etching as a result of their structural instability. Extensive research has been conducted to improve photocatalytic performance by designing heterojunction structures, nanocomposites, and oxidation–reduction strategies for catalyst regeneration and recycling. Nanoparticles (NPs) with semiconducting properties, such as the absorption of light and the transfer of charge carriers, are widely used as photocatalysts. These photoactive nanoparticles play a vital role in environmental remediation and the generation of sustainable energy. Compared to bulk materials, photocatalysts exhibit enhanced activity owing to their distinctive structures and increased surface area-to-volume ratio. Through precise manipulation of the size and morphology of photocatalytic materials at the nanoscale, novel functional materials can be designed for a diverse array of applications (Naikoo et al. 2022). Two-dimensional (2D) materials have been extensively employed to tackle environmental issues such as air pollution and water scarcity by either developing new solutions or enhancing existing technologies. Due to their exceptional electrical properties and adsorption characteristics, various 2D materials are employed as electrocatalysts in carbon, nitrogen, and water cycles. These advanced materials are categorized based on their nanoscale dimensions: (a) Zero-dimensional materials, where all three dimensions are in the nanometer range, (b) One-dimensional materials, such as nanotubes or nanowires, where two dimensions are in the nanometer range, and the third is significantly larger, (c) Two-dimensional materials, with one dimension in the nanometer range, and (d) Bulk materials, which do not exhibit nanoscale dimensions (Kittel 2004). The shift from conventional wastewater treatment methods to an emphasis on photocatalysis utilizing 2D materials should be more seamlessly articulated to adequately explain the preference for this approach. Techniques such as adsorption, electrodialysis, and electrochemical reduction are proficient in removing pollutants; however, they often encounter drawbacks like high operational expenses, potential secondary pollution, restricted reusability, and limited efficiency in completely breaking down complex organic dyes. In contrast, photocatalysis, particularly when powered by visible light, provides a sustainable, eco-friendly, and effective means of decomposing dyes into harmless end products such as CO_2 and H_2O. The integration of 2D materials further boosts the efficacy of this technique due to their expansive surface area, adjustable band gaps, superior charge carrier mobility, and capability for structural alterations. These characteristics facilitate enhanced light absorption, improved charge separation, and an increase in active sites, rendering 2D material-based photocatalysts more effective than conventional methods for dye degradation in wastewater treatment. A well-defined comparison and transition would delineate the reasons for concentrating on 2D photocatalytic systems in this research. Notably, the thickness of 2D materials can often be reduced to a single atomic layer, resulting in drastic changes in their properties. These materials, which include single-layer carbons, chalcogenides, and layered silicates, exhibit distinct physical and chemical properties when compared to their bulk counterparts. Layered materials can be used to create 2D nanostructures with high surface areas, characterized by strong in-plane bonding and weak van der Waals interactions between the layers. Owing to their distinct structural and electronic properties, 2D materials are highly effective photocatalysts. Their tunable electronic band structures enable efficient light absorption and energy alignment for various photocatalytic processes, and extensive surface area offers numerous active sites for reactions. The atomic-scale thickness of 2D materials improves charge separation and transport, minimizing the recombination losses of photogenerated electron–hole pairs. Through defect engineering, additional active sites can be introduced to enhance catalytic performance, while their intrinsic chemical stability ensures durability under harsh reaction conditions (Wang et al. 2023). The application of 2D-based nanocomposites in photocatalytic dye degradation represents a promising strategy for wastewater treatment; however, several research gaps still exist. Although numerous studies highlight the photocatalytic capabilities of individual 2D materials like graphene oxide and MoS_2, improving their efficiency, stability, and selectivity under visible light remains a challenge. One significant gap involves the limited understanding of charge transfer mechanisms within these nanocomposites and the role of structural modifications in enhancing their performance. This study addresses these challenges by designing and synthesizing advanced 2D-based nanocomposites with enhanced light absorption, improved charge separation, and greater surface reactivity. Through the combination of complementary materials and precise interface engineering, the work aims to boost photocatalytic activity specifically for dye removal in wastewater. The primary objective is to develop a photocatalyst that is not only efficient and stable under visible light but also scalable for real-world use, while offering deeper insight into the mechanisms responsible for its enhanced performance. This focused approach contributes to bridging the gap between experimental research and practical environmental solutions.

12.2 2D Materials for Photocatalysis

Rapid advancements in structural cognition and synthetic techniques have enabled the efficient synthesis of diverse 2D nanostructures for photocatalytic applications. Generally, materials that can form two-dimensional (2D) structures are categorized into two main types: layered and non-layered systems. Layered materials consist of stacked layers with weak interlayer interactions, allowing them to be easily exfoliated into 2D nanosheets using top-down techniques. This category includes black phosphorus (BP), hexagonal boron nitride, bismuthine, transition metal carbides/nitrides (MXenes), graphitic carbon nitride, transition metal dichalcogenides (TMDs), and certain transition metal oxides. In contrast, non-layered substances have an isotropic atomic configuration across all three spatial dimensions, as observed in certain metal oxides and metal chalcogenides. Despite their structural differences, “2D materials” refer to those that exhibit photocatalytic activity or can function as co-catalysts in photochemical or photo-induced reactions. This section highlights photocatalysts

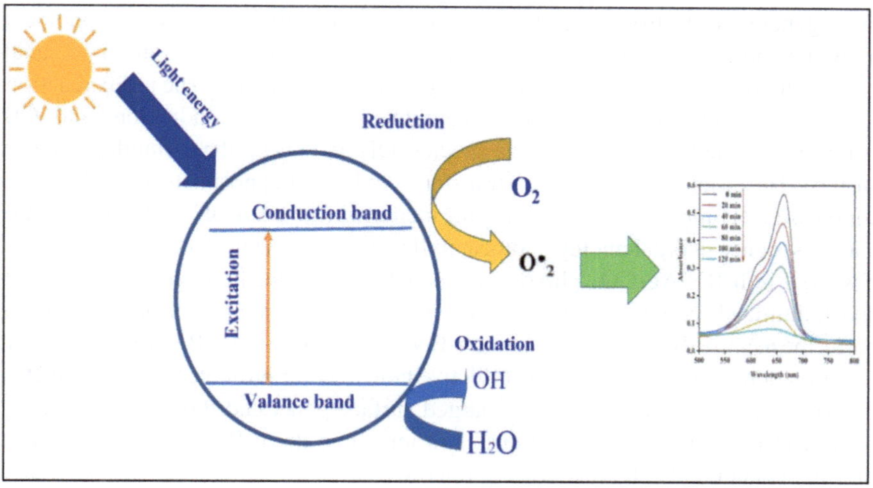

Fig. 12.1 Schematic illustration of photocatalysis mechanism

derived from two-dimensional (2D) materials, emphasizing their varied chemical compositions and structural configurations (Fig. 12.1) (Balarabe and Atabaev 2024).

12.2.1 Hexagonal Boron Nitride (h-BN) Composite for Photocatalysis

The broadband gap of hexagonal boron nitride (h-BN), approximately 5.5 eV, limits its photocatalytic activity to the UV spectrum. However, despite this limitation, h-BN nanostructures are widely employed in photocatalytic nanocomposites due to their exceptional thermal and chemical stability, as well as their capacity to facilitate hole transfer. Incorporating h-BN into nanocomposites can significantly improve photocatalytic performance by minimizing the recombination of photogenerated electron–hole (e^-/h^+) pairs, a key factor in improving overall photocatalytic performance. Despite the favorable characteristics of h-BN and its composites, including remarkable thermal stability, chemical inertness, and strong mechanical properties, their photocatalytic efficiency is impeded by a fundamental issue with light absorption. To enhance performance, methods such as doping, creating nanocomposites with heterostructures made from narrow-bandgap semiconductors, or tuning the band gap are commonly explored to improve sensitivity to visible light and broaden photocatalytic applications. Chen et al. reported the enhancement of bismuth phosphate through integration with hexagonal boron nitride (h-BN) nanosheets, which resulted in lower photoluminescence (PL) intensity compared to pure bismuth phosphate (Zhou et al. 2022). This reduction was attributed to the reduced recombination of photogenerated electron–hole (h^+/e^-) pairs. Additionally, the improved photocatalytic performance of the modified composites was validated through kinetic analysis of photodegradation and reduced charge transfer resistance. The redox reactions $OH + h^+ \rightarrow \cdot OH$ (valence band, VB) and $O_2 + e^- \rightarrow O_2^{\cdot}$ (conduction band, CB) contribute to the formation of active species ·OH and $O_2^{\cdot}$ on the bismuth phosphate surface. The efficient carrier (h^+/e^-) transfer from bismuth phosphate to h-BN further suppressed recombination, facilitating the increased formation of highly reactive ·OH and $O_2^{\cdot}$ species, thus enhancing photocatalytic efficiency. A similar effect may also be observed in h-BN/Ag_2CO_2 photocatalysts, where enhanced charge carrier transfer and minimized recombination rates contribute to improved performance. Furthermore, the negatively charged surface of h-BN nanosheets in the h-BN/TiO_2 composite has been demonstrated to act as an effective hole transfer promoter (Gad-Allah et al. 2023). By utilizing the negatively charged surface generated through ball milling, h-BN can also serve as a hole transfer promoter in the h-BN/ZnO photocatalyst, further enhancing its photocatalytic properties.

12.2.2 Graphitic Carbon Nitride (gC_3N_4)-Based Composite for Photocatalysis

Graphitic carbon nitride (g-C_3N_4) and its modified composites have proven effective in removing organic pollutants. These composites demonstrate superior performance in degrading dyes such as Rhodamine B (RhB), methyl orange (MO), basic fuchsin (BF), and methylene blue (MB) compared to TiO_2 and other semiconductors. For example, while TiO_2 eliminated only 10% of MO in 20 min, pure (g-C_3N_4) achieved 15% degradation under the same conditions. Moreover, CdS-doped g-C_3N_4 exhibited an impressive 90% degradation of MO under similar circumstances (Fang et al. 2022). In contrast, Bi_2S_3 nanorods removed only about 10% of MO, whereas the ZnO/SnO_2 catalyst achieved approximately 50% degradation within 20 min. Dyes, commonly utilized in the printing, textile, and plastic industries, present substantial environmental challenges due to their complex chemical structures, high color intensity, and water degradation resistance. As a result, these dyes are difficult to eliminate from aquatic systems. Extensive research has been conducted on the applications of in g-C_3N_4photocatalysis, including water splitting, hydrogen (H_2) production, and CO_2 reduction. This study focuses on evaluating the photocatalytic performance and underlying mechanisms of g-C_3N_4 composites for organic pollutant removal under visible light irradiation, comparing their effectiveness with that of widely used TiO_2 nanoparticles. Furthermore, it provides a comprehensive overview of modifications to g-C_3N_4 nanomaterials, including element or compound doping, morphological control, and immobilization. These modifications aim to improve the photodegradation capabilities of g-C_3N_4 for organic pollutants. The practical applications of synthesized g-C_3N_4 catalysts are also investigated, focusing on their reusability, the design of photocatalytic reactors, the influence of environmental factors on degradation processes, and the development of predictive models for the removal of organic contaminants (Ismael 2020).

12.2.3 Bismuthine-Based Composite for Photocatalysis

Group VA includes bismuthine (Bi), which possesses a tunable energy band gap that can range from nearly 0–0.55 eV, making it highly responsive to solar energy and suitable for various applications. As a result, bismuthine has been employed in gas sensing, biosensing, biomedicine, and catalysis. It demonstrates outstanding topological and metallic characteristics, such as high carrier mobility, extended light-material interaction times, a significant atomic weight, and robust spin–orbit coupling. Research suggests that bismuthine's 2D structure functions as a topological insulator; however, this remains a subject of debate. Notably, bismuthine was first discovered on a Bi (111) surface in 1997. It can adopt either a planar or sp^2 honeycomb structure. Photocatalysis is a process that enhances chemical reactions in semiconducting materials when exposed to light irradiation, which is an energy-efficient, cost-effective, and operable under ambient conditions. When a semiconductor absorbs light, electrons in the valence band (VB) are excited to the conduction band (CB), generating electron–hole pairs. These charge carriers then move to the surface of the photocatalyst, where they interact with surrounding species such as water and oxygen molecules, facilitating redox reactions. However, not all photogenerated charge carriers contribute to these surface reactions, as many undergo recombination through radiative and non-radiative pathways, reducing the photocatalytic efficiency. To enhance photocatalytic performance, several strategies have been explored, including bandgap tuning, morphological control, surface defect engineering, co-catalyst integration, and the formation of heterojunctions. The development of 2D materials with sheet-like structures and atomic-scale thickness has shown great potential in minimizing recombination losses. When materials are scaled down to their 2D form, they display distinct properties that differ from their bulk versions. For example, their increased surface-to-volume ratios offer more active surface sites, while the reduced thickness shortens the distance for charge carrier transport, which helps reduce recombination rates. Additionally, adjusting the number of layers can optimize light absorption properties. Recent studies have shown that forming a close interface with a co-catalyst (heterojunction) can produce a synergistic effect, enhancing light absorption, promoting charge transfer, and ensuring the spatial separation of photogenerated carriers, thereby enhancing overall photocatalytic efficiency.

12.2.4 Borophene-Based Composite for Photocatalysis

Borophene, a recent member of the two-dimensional (2D) material family, has ignited considerable scientific excitement since its first successful experimental creation. Borophene nanosheets and their nanocomposites, based on the principles of condensed matter physics, signify an innovative class of materials aimed at enhancing photocatalysis (Kaneti et al. 2021). Due to its highly anisotropic lattice structure and complex electronic band structure, which can be adjusted between metallic and semiconducting phases, influenced by dopants and synthesis conditions, this flexibility enables borophene to effectively absorb light over a wide range, facilitating the generation of electron–hole pairs essential for photocatalytic processes. Its thickness promotes quantum confinement effects, which elevate the density of electronic states near the Fermi level. This enhancement, in turn, boosts charge separation and light absorption efficiency. Furthermore, remarkable in-plane electrical conductivity facilitates the quick movement of photogenerated charge carriers, thereby reducing recombination losses. The surface plasmon resonance effect significantly amplifies local electromagnetic fields at designated wavelengths, substantially enhancing light-harvesting capabilities. Within nanocomposites, borophene collaborates synergistically with metals, semiconductors, and carbon-based frameworks to improve photocatalytic performance. Serving as an electron sink in hybrid structures, it creates Schottky barriers or heterojunctions that inhibit charge recombination and prolong carrier lifetimes. These interfaces encourage effective transfer of charges, optimizing the system for targeted photocatalytic applications like hydrogen evolution, CO_2 reduction, and organic pollutant breakdown. The strong interactions within borophene and its composite partners often alter band alignment, further enhancing photocatalytic efficiency. These localized states function as active sites, which assist in the adsorption and activation of reactant molecules. Additionally, phonon-mediated interactions help stabilize the material during reactions, ensuring effective energy transfer without compromising structural integrity, making them promising candidates for future catalytic systems (Zhang et al. 2020).

12.2.5 MXene-Based Composite for Photocatalysis

MXenes represent a new class of 2D materials composed of transition metal nitrides, carbides, and carbonitrides, which were initially discovered in 2011. These materials feature a layered structure defined by the general formula $M_n + 1X_nT_x$, where M signifies a transition metal, X stands for either carbon or nitrogen, n (ranging from 1 to 4) indicates the number of metal layers, and T_x represents various surface terminations such as –OH, –O, F (Deshmukh et al. 2022). MXenes and their derivatives have attracted considerable attention for water purification applications owing to their outstanding features, such as adjustable band gaps that can be tailored through modifications in surface chemistry. They exhibit high electrical conductivity, unique adsorptive and reductive characteristics, and remarkable photocatalytic performance in degrading a wide range of organic pollutants. For instance, the incorporation of dimethyl sulfoxide into MXene enables the synthesis of AgNP-loaded MXene/polymer/Fe_3O_4 nanocomposite exhibits excellent photocatalytic efficiency in the degradation of 2-nitroaniline and 4-nitrophenol, highlighting its potential for effective removal of hazardous organic pollutants. Additionally, MXenes serve as promising co-catalysts for photocatalytic applications, particularly under sunlight irradiation (Chinnasamy et al. 2023). Following the wet chemical etching process, MXenes exhibit strong catalytic potential due to the formation of multiple bonds with different reactants. One recent study developed $Ti_3C_2T_x$ nanosheets combined with Cu_2O to form composite photocatalysts featuring a Schottky heterojunction structure that facilitates the efficient degradation of tetracycline (TC) antibiotics under visible light irradiation, enhancing charge separation and promoting photocatalytic activity. This nanocomposite achieved an impressive 97.6% degradation of tetracycline (TC) within 50 min, primarily attributed to the generation of superoxide anion radicals ($O_2^{\cdot}$) and photogenerated holes (h^+). The formation of a Schottky heterojunction significantly contributed to the enhanced photocatalytic performance by promoting effective charge carrier separation and reducing recombination. Another study involved the addition of Ti_3C_2 nanosheets to a Bi/Bi_OCl system, forming a Ti_3C_2-Bi/Bi_OCl composite. This composite exhibited remarkable photocatalytic efficiency in degrading ciprofloxacin (CIP) residues in aqueous environments, highlighting its potential for antibiotic pollutant remediation. The efficiency of CIP removal was found to be 89%, with Bi/Bi_OCl contributing 71% and Bi_OCl accounting for only 10%. The improved photocatalytic performance was ascribed to the heterostructure-forming capability of Ti_3C_2, which significantly enhanced charge carrier separation and thereby boosted the overall degradation efficiency.

12.2.6 2D Transition Metal Dichalcogenides Nanocomposite for Photocatalysis

Two-dimensional (2D) transition metal dichalcogenides (TMDs), often regarded as the rising stars of the post-graphene era, have gained significant interest as promising materials for photocatalysis (Thakur et al. 2024). Their exceptional electrical, optical, and chemical properties make them highly efficient for light-harvesting and redox reactions in photocatalytic systems. Unlike conventional metal oxides, metal chalcogenides possess narrower band gaps, rendering them more responsive to visible light and making them ideal candidates for visible-light-active photocatalysis. Additionally, their large surface area, tunable morphology, and variable bandgap energies further enhance their photocatalytic potential (Rahman et al. 2022). MoS_2 has been extensively investigated for its potential in hydrogen evolution reactions, oxidative desulfurization, and the breakdown of organic contaminants. Its structure, characterized by a hexagonal lattice of molybdenum and sulfur atoms, contributes significantly to its catalytic performance. MoS_2 is often compared to grapheme; however, it features a narrower band gap, allowing the generation of electron–hole pairs (e^-/h^+) when exposed to visible light, initiating photocatalytic activity. This property renders MoS_2 an excellent candidate for photocatalytic applications driven by visible light. Moreover, its superior light absorption capacity and strong chemical stability have further solidified its role in photocatalytic applications. While metal oxides typically exhibit wide band gaps that limit their use under visible light, metal chalcogenides, due to their narrower band gaps, offer a more viable alternative for visible-light-driven photocatalysis. Despite extensive research, several challenges remain in the photocatalytic application of chalcogenides. One major challenge is the synthesis of these materials. Although considerable progress has been achieved in the development of visible-light-active chalcogenide materials, most synthesis approaches demand an inert atmosphere to prevent oxidation and preserve the structural and electronic integrity of the resulting compounds. A key concern with chalcogenide-based photocatalysts is their limited structural and chemical stability during photocatalytic reactions, impacting long-term efficiency. Photo corrosion and short excited-state lifetimes often hinder their overall performance. However, MoS_2 demonstrates remarkable photostability in solution, attributed to the antibonding interaction between molybdenum dz^2 and sulfur pz orbitals at the valence band maximum, which inhibits substantial photocorrosion. Beyond stability concerns, the recovery of photocatalysts after photocatalytic reactions presents another challenge. Additionally, a primary drawback of most metal chalcogenides lies in their limited efficiency in separating and transporting photogenerated charge carriers to the active catalytic sites. This limitation is due to the short lifetimes of charge carriers and their low mobility, which significantly affect photocatalytic performance. Addressing these issues through material modifications and hybridization strategies remains a key focus in the field of chalcogenide-based photocatalysis.

12.3 Preparation of 2D Material

12.3.1 Mechanical Exfoliation Method

Mechanical exfoliation is a simple and commonly used technique for isolating two-dimensional (2D) materials from their bulk forms. Examples include graphene, boron nitride, and transition metal dichalcogenides (Gao et al. 2018). Because it can produce high-quality samples with minimal defects, this technique has remained a cornerstone of 2D materials research. It operates by exploiting the weak van der Waals forces that hold the layers of these materials together, allowing for easy separation. The low yield quality associated with mechanical exfoliation arises from the manual and non-scalable nature of the technique, which involves extracting layers from bulk

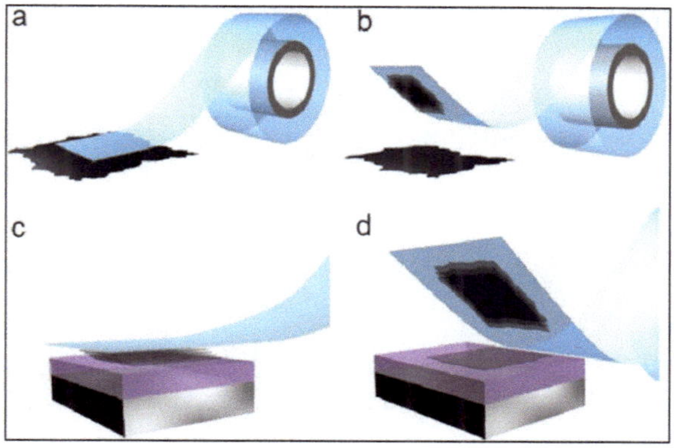

Fig. 12.2 Schematic of the mechanical exfoliation method of 2D material using **a, b** an adhesive tape is applied to a 2D crystal, allowing the upper layers to stick to the tape. **c** The tape holding the layered material crystals is applied to a selected surface. **d** When the tape is removed, the bottom layer remains on the substrate (Novoselov and Neto 2012)

materials using tools such as adhesive tape. This method results in a limited quantity of usable material, as significant portions are lost or remain in thicker, less desirable forms. Moreover, achieving large-area, uniform monolayer poses challenges and inconsistencies. Consequently, even though the exfoliated samples exhibit high quality, the overall amount is insufficient for extensive applications, rendering the yield inadequate for industrial use (Fig. 12.2).

12.3.2 Intercalation-Assisted Expansion and Exfoliation

Layered materials like graphite and certain transition metal dichalcogenides can be effectively divided into thinner sheets, sometimes even achieving single-layer isolation through the strategic application of intercalation-assisted expansion and exfoliation techniques. This technique has garnered significant interest among researchers in the field. The process begins with intercalation, where guest species such as ions, molecules, or chemicals are introduced between the material's layers. These intercalants weaken interlayer interactions by increasing spacing or disrupting bonding forces, thereby facilitating separation. Subsequent expansion is typically achieved by applying external stimuli such as heat, chemical reactions, or electrochemical techniques, which generate forces like gas release or electrostatic repulsion to push the layers apart. Finally, exfoliation is performed to isolate thin or single-layered materials, often using mechanical or ultrasonic agitation. Owing to its efficiency and scalability, this method is extensively used in the production of advanced nanomaterials, including grapheme. The schematic illustration is shown in Fig. 12.3.

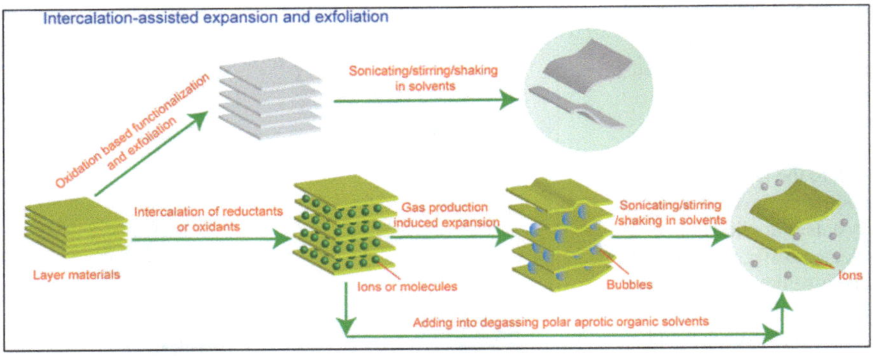

Fig. 12.3 Intercalation-assisted expansion and exfoliation of 2D material (Cai et al. 2018)

12.3.3 Ultrasonics-Supported Liquid-Phase Exfoliation Technique

In the ultrasonication-assisted liquid-phase exfoliation process, ultrasonic waves applied to the bulk material create cavitation bubbles, which generate high surface and kinetic energy within the liquid. Cavitation involves the formation, expansion, and rapid collapse of micron-sized voids or bubbles in a liquid caused by pressure fluctuations. This phenomenon affects the bulk material, facilitating exfoliation and enhancing the production of nanosheets. The yield and properties of exfoliated nanosheets are influenced by several sonication parameters, such as the type, duration, and intensity of sonication. Sonication can be categorized into two main types: direct and indirect. Indirect sonication is performed using a bath sonicator, whereas direct sonication involves a probe sonicator. Exfoliation is commonly achieved by dispersing the bulk material in an appropriate solvent, followed by ultrasonic treatment using a probe sonicator, a sonication bath, or a combination of both. Although both approaches utilize ultrasonic waves to improve the distribution and engagement of the active materials, their fundamental processes and effectiveness vary considerably (Telkhozhayeva et al. 2021). Nevertheless, this method has several constraints. A key issue is its significant dependence on solvents; the effectiveness of the exfoliation process largely hinges on selecting a solvent whose surface energy closely aligns with that of the intended material. Using inappropriate solvents can lead to inadequate dispersion and reduced exfoliation effectiveness. Another major concern is the creation of defects caused by the high-energy ultrasonic waves. These waves can harm the structure of the exfoliated layers, resulting in broken sheets, vacancies, or other structural imperfections that adversely affect the properties of the material. When comparing the two methods, mechanical exfoliation produces high-quality, defect-free monolayers; however, it is hindered by very low efficiency and limited scalability. On the other hand, ultrasonication-assisted exfoliation provides higher yields and improved scalability, making it more appropriate for large-scale applications, although it compromises structural integrity and uniformity. Consequently,

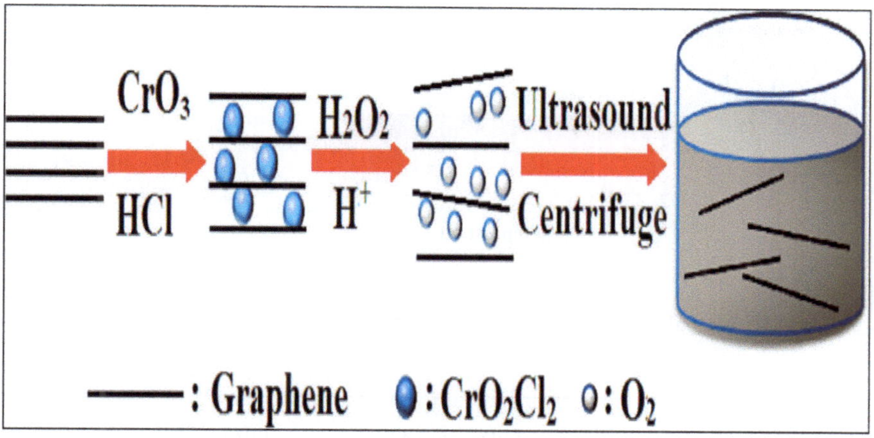

Fig. 12.4 Ultrasonics-assisted liquid-phase exfoliation method (Gu et al. 2019)

both methods play important roles but demonstrate a compromise between sample quality and production efficiency, emphasizing the necessity of choosing a method based on the specific application. The schematic illustration is shown in Fig. 12.4.

12.3.4 Chemical Vapor Deposition of 2D Material

Chemical vapor deposition (CVD) is a commonly employed technique for the synthesis of high-quality two-dimensional materials such as graphene, MoS_2, and WS_2. In this process, a film substrate (like copper or silicon) is placed inside a furnace that reaches high temperatures, and gases containing the necessary elements are introduced into the environment. Under the influence of heat, these gases decompose and chemically react on the substrate's surface, leading to the formation of an extremely thin layer, often just a single atomic layer or a few atoms thick. This layer constitutes the 2D material, which is shown in Fig. 12.5. CVD enables precise management of the film thickness and uniformity, making it an excellent choice for creating large-area, high-purity materials. Although the associated equipment and procedures can be costly and necessitate high temperatures, CVD continues to be one of the most effective methods for producing 2D materials that are suitable for use in electronics, sensors, and research purposes.

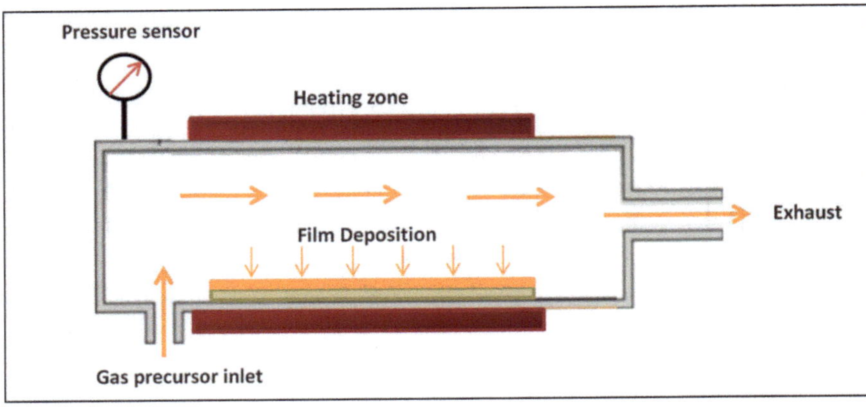

Fig. 12.5 Chemical vapor deposition of 2D material (Zhang et al. 2016)

12.4 Degradation Rates, Quantum Yields, and Stability Assessments of 2D Materials for Photocatalysis

In the realm of photocatalysis, two-dimensional (2D) materials such as transition metal dichalcogenides (TMDs), graphitic carbon nitride (g-C_3N_4), and MXenes have emerged as promising candidates owing to their expansive surface areas, adjustable electronic structures, and efficient interaction with light. To determine their real-world effectiveness, it is important to evaluate their degradation behavior, quantum efficiency, and long-term stability under working conditions. The degradation rate typically describes the decline in photocatalytic performance over time, which may result from physical deterioration, surface contamination, or instability under light exposure. For example, g-C_3N_4 is known to undergo photocorrosion or nitrogen depletion during extended light irradiation, which negatively affects its catalytic ability. Quantum yield (QY) plays a key role by indicating how efficiently a material can convert incoming photons into desired chemical products, such as hydrogen or degraded pollutants. Although the QY of pristine 2D materials is generally modest, enhancements can be achieved through methods like elemental doping or creating heterojunctions (Li et al. 2016). Stability evaluations are commonly performed by exposing the material to prolonged photocatalytic activity under artificial sunlight or UV light, followed by multiple usage cycles to assess performance consistency. Physicochemical characterization is frequently employed to detect any structural or compositional alterations. These analyses are crucial for the development of robust and high-performance 2D photocatalysts for environmental cleanup and sustainable energy applications (Table 12.1).

Table 12.1 2D material-based composite photocatalysts for various photocatalytic reactions

Sl. No	2D material-based composite	Preparation technique	Pollutant	Previous studies
1	2D Bismuthene	Liquid-phase exfoliation	CO_2 reduction	Ozer et al. (2022)
2	g-C_3N_4/CuS	Wet impregnation process	Rhodamine	Muthuganesh et al. (2024)
3	g-C_3N_4/based composite	Solid state reaction	CO_2 reduction	Prasad et al. (2023)
4	rGO/TiO_2	Hydrothermal	Methylene blue	Akyüz (2021)
5	rGO/CdS	Surface modification	Methylene blue	Bera et al. (2015)
6	h-BN	Liquid-phase exfoliation	CO_2 reduction	Laghaei et al. (2023)
7	h-BN/composite	Ball milling method	Rhodamine	Wang et al. (2018)
8	MoS_2/ZnO	Hydrothermal	Methylene blue	Krishnan et al. (2016)
9	MXene–$BiFeO_3$–ZnO	Hydrothermal	Rhodamine	Nag et al. (2023)
10	CuS/MXene	Liquid-phase exfoliation	H_2 evaluation	Xie et al. (2020)

12.5 Conclusions

The escalating demand for advanced wastewater treatment strategies has catalyzed significant research into two-dimensional (2D) layered materials and their heterostructure composites, owing to their exceptional physicochemical properties and promising performance in the photocatalytic degradation of dye pollutants. Industrial wastewater, particularly from the textile, pharmaceutical, and dyeing industries, contains persistent organic pollutants that pose severe environmental and health hazards. Conventional treatment methods often fail to fully degrade dyes, leading to secondary pollution and excessive waste generation. Photocatalysis has gained recognition as an effective and environmentally sustainable strategy for the degradation of organic pollutants in aqueous environments. Among the various photocatalysts explored, 2D materials offer unique advantages, including high surface area, tunable band gaps, superior charge carrier mobility, and efficient light absorption. Materials like graphitic carbon nitride (g-C_3N_4), MXenes, borophene, transition metal dichalcogenides (TMDs), and hexagonal boron nitride (h-BN) have exhibited remarkable capabilities in visible-light-induced photocatalytic applications due to their unique electronic structures and tunable physicochemical properties. Advances in heterojunction formation, doping, and bandgap engineering have further enhanced their efficiency. The fabrication of two-dimensional (2D) materials is pivotal for enhancing their photocatalytic efficiency. Various synthesis strategies, such as mechanical exfoliation, intercalation-assisted expansion, ultrasonic-assisted liquid-phase exfoliation, and chemical vapor deposition (CVD), have been systematically investigated to tailor their structural and functional properties for optimal performance. Each method offers distinct advantages, from high-quality nanosheet production to large-scale synthesis for industrial applications. However, challenges remain in terms of stability, charge separation efficiency, and large-scale reproducibility. Future research should focus on enhancing the durability, scalability, and real-world applicability of 2D photocatalysts.

Acknowledgements The corresponding author, Dr. Z. Mohamed Riyas, thanks Dr. Esakkimuthu Shanmugasundaram and Chakravarthy Kumaraesan for the School of Industrial Chemistry and the School of Management, Alagappa University, Karaikudi, for their aid.

Data Availability There is no data generated in this research.
Declaration of Competing Interest There is no conflict of interest.

References

Akyüz D (2021) RGO-TiO_2-CdO-ZnO-Ag photocatalyst for enhancing photocatalytic degradation of methylene blue. Opt Mater 116:111090

Balarabe BY, Atabaev TS (2024) Advancing photocatalysis: insights from 2D materials and operational parameters for organic pollutants removal. Adv Sustain Syst 8(12):2400483

Bera R, Kundu S, Patra A (2015) 2D hybrid nanostructure of reduced graphene oxide–CdS nanosheet for enhanced photocatalysis. ACS Appl Mater Interfaces 7(24):13251–13259

Cai X, Luo Y, Liu B, Cheng HM (2018) Preparation of 2D material dispersions and their applications. Chem Soc Rev 47(16):6224–6266

Chakravarty A, Roy S (2024) A Review of photocatalysis, basic principles, processes, and materials. Sustain Chem Environ 8:100155

Chinnasamy C, Perumal N, Choubey A, Rajendran S (2023) Recent advancements in MXene-based nanocomposites as photocatalysts for hazardous pollutant degradation—a review. Environ Res 233:116459

Deshmukh K, Muzaffar A, Kovářík T, Ahamed MB, Pasha SK (2022). Introduction to 2D MXenes: fundamental aspects, MAX phases and MXene derivatives, current challenges, and future prospects. In: Mxenes and their composites, pp 1–47

Fang J, Xie K, Kang Q, Gou Y (2022) Facile fabrication of g-C_3N_4/CdS heterojunctions with enhanced visible-light photocatalytic degradation performances. J Sci Adv Mater Dev 7(1):100409

Gad-Allah TA, Zhang R, Wang Y, Zhou Y (2023) Facile one-pot synthesis of defective (001)-TiO_2-x/h-BN photocatalyst for environmental applications. J Alloy Compd 954:170187

Gao E, Lin SZ, Qin Z, Buehler MJ, Feng XQ, Xu Z (2018) Mechanical exfoliation of two-dimensional materials. J Mech Phys Solids 115:248–262

Garg A, Basu S, Shetti NP, Reddy KR (2021) 2D materials and their heterostructures photocatalysts: synthesis, properties, functionalization, and applications in environmental remediation. J Environ Chem Eng 9(6):106408

Gu X, Zhao Y, Sun K, Vieira CL, Jia Z, Cui C, Wang Z, Walsh A, Huang S (2019) Method of ultrasound-assisted liquid-phase exfoliation to prepare graphene. Ultrason Sonochem 58:104630

Ismael M (2020) A review on graphitic carbon nitride ($g\text{-}C_3N_4$) based nanocomposites: synthesis, categories, and their application in photocatalysis. J Alloy Compd 846:156446

Kaneti YV, Benu DP, Xu X, Yuliarto B, Yamauchi Y, Golberg D (2021) Borophene: two-dimensional boron monolayer: synthesis, properties, and potential applications. Chem Rev 122(1):1000–1051

Kittel C (2004) Introduction to solid state physics, 8th edn. Wiley, New York, NY

Krishnan U, Kaur M, Kaur G, Singh K, Dogra AR, Kumar M, Kumar A (2019) MoS_2/ZnO nanocomposites for efficient photocatalytic degradation of industrial pollutants. Mater Res Bull 111:212–221

Laghaei M, Ghasemian M, Lei W, Kong L, Chao Q (2023) A review of boron nitride-based photocatalysts for carbon dioxide reduction. J Mater Chem A 11(23):11925–11963

Lee DE, Kim MK, Danish M, Jo WK (2023) State-of-the-art review on photocatalysis for efficient wastewater treatment: Attractive approach in photocatalyst design and parameters affecting the photocatalytic degradation. Catal Commun 183:106764

Li X, Yu J, Jaroniec M, Chen X (2016) Cocatalysts for selective photoreduction of CO_2 into solar fuels. Chem Rev 119(6):3962–4179

Muthuganesh A, Jacob ID, Soundranayagam JP, Surender S, Elangovan P, Flora XH (2024) Fabrication of $g\text{-}C_3N_4$/CuS heterostructures for efficient visible-light-driven photocatalysts. Inorg Chem Commun 159:111813

Nag R, Das S, Das D, Venimadhav A, Bera A (2023) A MXene–$BiFeO_3$–ZnO nanocomposite photocatalyst served as a high-performance supercapacitor electrode. Phys Chem Chem Phys 25(34):23125–23132

Naikoo GA, Arshad F, Almas M, Hassan IU, Pedram MZ, Aljabali AA, Mishra V, Serrano-Aroca Á, Birkett M, Charbe NB, Goyal R (2022) 2D materials, synthesis, characterization, and toxicity: a critical review. Chem Biol Interact 365:110081

Novoselov KS, Neto AC (2012) Two-dimensional crystals-based heterostructures: materials with tailored properties. Phys Scr 2012(T146):014006

Ozer MS, Eroglu Z, Yalin AS, Kılıç M, Rothlisberger U, Metin O (2022) Bismuthene is a versatile photocatalyst operating under variable conditions for the photoredox CH bond functionalization. Appl Catal B 304:120957

Prasad C, Madkhali N, Govinda V, Choi HY, Bahadur I, Sangaraju S (2023) Recent progress on developing gC_3N_4-based composite material and their photocatalytic application of CO_2 reductions. J Environ Chem Eng 11(3):109727

Rahman A, Jennings JR, Tan AL, Khan MM (2022) Molybdenum disulfide-based nanomaterials for visible-light-induced photocatalysis. ACS Omega 7(26):22089–22110

Ren G, Han H, Wang Y, Liu S, Zhao J, Meng X, Li Z (2021) Recent advances of photocatalytic application in water treatment: a review. Nanomaterials 11(7):1804

Telkhozhayeva M, Teblum E, Konar R, Girshevitz O, Perelshtein I, Aviv H, Tischler YR, Nessim GD (2021) Higher ultrasonic frequency liquid phase exfoliation leads to larger and monolayer to few-layer flakes of 2D layered materials. Langmuir 37(15):4504–4514

Thakur D, Porwal C, Chauhan VS, Balakrishnan V, Vaish R (2024) 2D transition metal dichalcogenides: synthesis methods and their pivotal role in photo, piezo, and photo-photocatalytic processes. Sep Purif Technol 337:126462

Wang N, Yang G, Wang H, Sun R, Wong CP (2018) Visible light-responsive photocatalytic activity of boron nitride incorporated composites. Front Chem 6:440

Wang H, Duan Y, Fei G, Yan T, Kang YM, Dionysiou DD (2023) Design, synthesis, and modification of 2D nanomaterials-based photocatalysts for pollutant degradation and photodegradation experiments from lab-scale to grand-scale. Chem Eng J 477:147219

Xie Y, Rahman MM, Kareem S, Dong H, Qiao F, Xiong W, Liu X, Li N, Zhao X (2020) Facile synthesis of CuS/MXene nanocomposites for efficient photocatalytic hydrogen generation. CrystEngComm 22(11):2060–2066

Zhang Q, Sando D, Nagarajan V (2016) Chemical route-derived bismuth ferrite thin films and nanomaterials. Journal of Materials Chemistry C 4(19):4092–4124

Zhang S, Li B, Wang X, Zhao G, Hu B, Lu Z, Wen T, Chen J, Wang X (2020) Recent developments of two-dimensional graphene-based composites in visible-light photocatalysis for eliminating persistent organic pollutants from wastewater. Chem Eng J 390:124642

Zhou J, Duo F, Wang C, Chu L, Zhang M, Yan D (2022) Robust photocatalytic activity of two-dimensional h-BN/Bi_2O_3 heterostructure quantum sheets. RSC Adv 12(21):13535–13547

Chapter 13
Water Contamination Risk During Flood and Ensuring Safe Water for Health: A Case Study from Bangladesh, 2024

Fahim Abrar Abid and Md. Tanzirul Alam

Abstract Bangladesh witnessed the worst flood crisis of recent times in 2024. The severity of the flood was striking not only during the flood but also at the time of the rescuing period, when people died due to, *inter alia*, heavily contaminated water. Overflow of pollutants into water supplies further escalated widespread outbreaks of waterborne diseases. As a climate-vulnerable country, this flood exposed Bangladesh's infrastructural weakness in tackling such crisis circumstances, just like any other Global South country. This weakness includes public health, environmental legal policy framework, and inconsistency in implementing international standards and sustainable development goals. In these circumstances of growing risk, this research employs a qualitative methodology incorporating case studies, policy review, and literature analysis. It aims to identify the loopholes by analysing the existing safeguard mechanisms to prevent water contamination and propose future guidelines of sustainable water management for a socioeconomically vulnerable country like Bangladesh to ensure safe water. The scope of this research is limited to public health and environmental policies by the government, NGOs, and community initiatives only, but not the scientific and technological aspects of the subject matter.

Keywords Water contamination · Safe drinking water · Public health · Environmental policy and legal framework · Sustainable water management

F. A. Abid (✉)
Bangladesh Campaign, Global Human Rights Defence, Den Haag, The Netherlands
e-mail: fahim.abrar@ghrd.org

University of Glasgow, Glasgow, UK

IBEI, Barcelona, Spain

University of Tartu, Tartu, Estonia

Bangladesh Society of International Law, Dhaka, Bangladesh

Md. T. Alam
University of Gothenburg, Gothenburg, Sweden
e-mail: tanzirul.alam@bracu.ac.bd

BRAC James P Grant School of Public Health, BRAC University, Dhaka, Bangladesh

A. Roy et al. (eds.), *Environment and Public Health*, Sustainability Solutions,
https://doi.org/10.1007/978-3-031-99770-9_13

13.1 Introduction

Bangladesh is highly vulnerable to floods due to its low-lying topography and unique geographical location (Huq and Moinuddin 2004). The country experiences regular flooding, devastatingly affecting communities and infrastructure (World Bank 2021). However, Bangladesh continues to suffer every time during the floods despite having the prior knowledge that the country will frequently confront such natural calamities due to its geography. The flood crises have not only been plaguing the country more frequently than they used to, but they will also continue to worsen with time (Teirstein 2023). Still not being able to deal with this issue properly can be because of the country's infrastructural weakness in tackling such crisis circumstances and the "inability" to utilise the existing infrastructure. While the former is a genuine limitation as a Global South country, the latter is an issue of serious concern. In an overpopulated country like Bangladesh, resource management stands at the core of the country's resistance and development in dealing with challenging issues like flood control. With the addition of these factors to the unalterable geography, the flood crisis became a national crisis for Bangladesh with multifaceted effects.

In Bangladesh, floods have become a part of life, shaping the land and the lives of millions living near the sea and river shores. For generations, these communities have learned to adapt, finding ways to live with the rising waters. However, floods have become more frequent and severe in recent years, largely due to climate change (World Bank 2021). The most recent one in 2024 was bigger and more detrimental than anything the nation had witnessed before (IPCC 2022). Because it not only panicked the general people but also questioned existing climate change-related implications and improved preparedness strategies, as the whole system failed (IPCC 2022).

The 2024 flood in Bangladesh was a clear example of the growing challenges that the country faces in handling flood crises. In addition to damaging houses and displacing families, the flood affected the country's existing water supply systems. A large number of drinking water sources were damaged and became contaminated (Araf 2024). As a result, waterborne diseases and other associated illnesses increased in the flooded areas (Araf 2024). The severity of the flood was striking not only during the flood but also at the time of the rescuing period, when people died due to, *inter alia*, heavily contaminated water. As communities worked collectively to recover, the emergency highlighted the need for stronger infrastructure and improved disaster preparedness.

Exploring the flood issue in Bangladesh, this chapter will particularly examine the water contamination issue during floods. The main focus will remain on the 2024 Bangladesh flood, but the discussions of previous floods will also be brought up when necessary and equally relevant. Then, this chapter will critically analyse the existing policy and legal framework to evaluate the strengths and weaknesses of existing safe drinking water management practices in Bangladesh *vis-a-vis* its effectiveness during floods *inter alia* other natural disasters. Next, this chapter will provide multi-faceted recommendations to tackle increasing difficulties caused by climate change and socioeconomic vulnerabilities. Finally, this chapter will conclude by emphasising the need to integrate public health policies and environmental legal frameworks to implement international standards and achieve sustainable development goals for implementing ideal water management practices in socioeconomically challenged countries with high vulnerability to climate change.

This research employs a qualitative methodology incorporating case studies, literature analysis, legislation, and policy review. To strengthen the in-depth understanding of the water management discourse during the 2024 flood, we conducted five interviews with the lived experiences of individuals from Feni, Noakhali, and Cumilla—the most affected districts in that flood. These insights will give us a meaningful perspective on both the specific challenges residents face regarding water access, contamination, and sanitation, and how current policies and practices are perceived and experienced by those directly impacted, thereby revealing gaps between policy intent and practical outcomes. The scope of this research is limited to public health and environmental policies by the government, NGOs, and community initiatives only, but not the scientific and technological aspects of the subject matter.

13.2 Conceptual Framework

In Bangladesh, floodwater-induced contamination typically occurs due to the intersection of industrial and agricultural chemical pollutants, infrastructural gaps, and a majorly lack of public awareness. During floods, people face severe vulnerability due to inadequate drainage structures, limited access to flood-resistant water sources, and weak healthcare systems (World Bank 2021). Flood water is often mixed with various harmful substances, agricultural runoffs, and human waste and carries detrimental microbes, which most likely create many life-threatening contagious diseases (WHO 2014). Children, pregnant women, and the elderly are the most vulnerable during the flood events and are susceptible to major health crises, like increased risks of diarrhoea, cholera, skin diseases, respiratory infections, and malnutrition due to the structural gaps that arise during the emergencies. Additionally, socioeconomic factors such as food security, total family income, access to health care, and drinking water sources are exacerbated by the associated health risk, which is connected with the flooding incident (Teirstein 2023).

Figure 13.1 indicates "flooding events" leading to "infrastructural vulnerabilities", which enables "contamination agents" to pollute water. Moreover, "contaminated water" causes "exposure pathways" resulting in "health outcomes." The "socio-economic factors" are a bidirectional influence, worsening exposure, and outcomes while being impacted by them. Additionally, dashed feedback arrows from "health outcomes" back to "infrastructural vulnerabilities" and "socio-economic factors," illustrating the cyclical process. This framework conceptualises a complex relationship between flood and its impact on public health. Understanding this interplay is

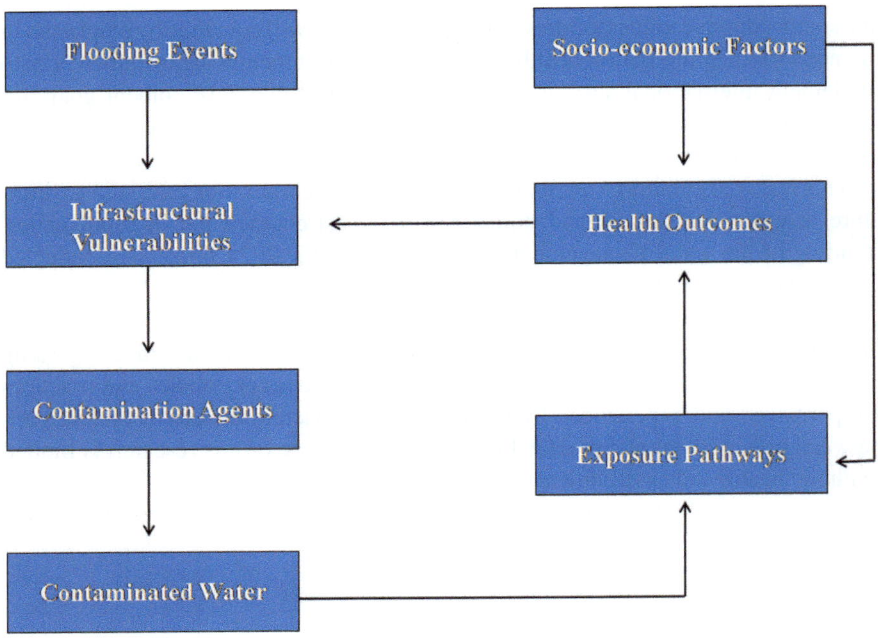

Fig. 13.1 How flooding contributes to the contamination of water sources and its impact on health

essential to identifying the gaps and challenges in flood preparedness and initiating sustainable solutions to tackle future climate emergencies.

13.3 Flood Issues in Bangladesh

Around 60% of the population of Bangladesh is exposed to a high flood risk, only second to the Netherlands in terms of proportion (Letsch et al. 2023). Bangladesh's coastal and riverine areas are highly flood-prone due to climatic and geographical features (Adnan et al. 2023). The country's dense system of rivers, coupled with its weather patterns and low-to-medium relief, forms the basis for recurring flooding (Bernard et al. 2022). Due to these factors, the region is at a much higher risk of being affected by challenging issues like increasing water levels. In this segment, the chapter will analyse the major causes and impacts of floods in Bangladesh while re-exploring the previous floods.

13.3.1 Major Causes of Flood

Flood types in Bangladesh can be categorised into four major groups, defined by the source of the water and the mechanisms causing the event. Firstly, flash floods are characterised by a sudden onset due to rainfall intensity. Secondly, localised rain floods are caused by heavy rainfall within a specific area. Thirdly, seasonal river floods are prolonged monsoonal floods resulting from large volumes of rain. Lastly, the storm-surge floods carry the seawater inland due to tropical cyclones. It is not uncommon for such floods to happen frequently since Bangladesh is situated in the lowest-lying area in the Hindu-Kush-Himalayan (HKH) region (Uddin et al. 2021).

Furthermore, rising sea levels caused by global warming increase the base level of rivers and lower the flow gradient, causing water to slow down. This backwater effect restricts river water from moving inland. As a result of this slow movement, the water fails to drain, leading to accumulation and a significant risk of floods in vulnerable regions. This geographical positioning creates a natural basin for the country to take in water flowing from the upstream area, thus increasing flood vulnerability. In addition, heavy rainfall contributes to the flood situation. Moreover, man-made factors like large-scale deforestation, rapid urbanisation, and massive irrigation also play a considerable role in flooding in Bangladesh (BBC 2014).

13.3.2 Impacts of Flood

Floods in Bangladesh cause widespread devastation, destroying homes, crops, and infrastructure. It also results in millions being displaced every year. Thus, it brought along with it the long-term loss of many aspects, threats of food shortages, and vulnerability to communities residing mainly in low-lying areas. However, the health impacts of floods are severe, with drowning and injuries being common occurrences in affected regions. Waterborne diseases, vector-borne diseases, and other associated illnesses also surge in the aftermath of flooding (Araf 2024). Following the flooding incident in 2017, the flood-prone areas reported 13,000 cases of diarrhoeal disease (IFRC 2017). In addition, prolonged stagnant floodwaters can promote the breeding of mosquitoes, which increases the risk of dengue and malaria, among other vector-borne diseases. The dengue outbreak in 2019, in which over 100,000 cases were reported in Bangladesh, was also partly linked to flooding (Hasan et al. 2024). Also, floods result in crop failures that directly contribute to malnutrition due to food shortages in the affected areas. Other significant issues include mental health problems like anxiety, depression, an increase in domestic violence, and post-traumatic stress disorder (PTSD). For example, according to a study, about 67% of women suffered from severe or extremely severe depression, and 65% portrayed a high level of anxiety, where 37% faced cases of severe stress (Rahman et al. 2023). Likewise, about 89% of women reported severe or extremely severe depression, 88%

intense anxiety, and 58% significant stress due to family violence during flood events (Rahman et al. 2023).

A big challenge is seen in the disruption of healthcare services during floods due to the damage to healthcare facilities and frequent disruption of supply chains for medicines, limiting access to essential medical services. Pregnant women, the elderly, children, and individuals with chronic illnesses are particularly vulnerable in this situation since it becomes hard for them to access emergency healthcare services during flood events. For example, a large number of health facilities were damaged in the 2022 flash flood, which caused long-term healthcare service gaps among several vulnerable communities in the northern and eastern parts of Bangladesh (Chakraborty et al. 2024). All health-related risks caused by flooding are aggravated by socioeconomic factors, including food security, financial capacity, vaccination history, access to medical care, and drinking water treatment facilities (Teirstein 2023).

13.3.3 Previous Floods in Bangladesh

Bangladesh has a long-lasting and enduring history of devastating floods. As mentioned in the chapeau, each flood has significantly made the country suffer. According to the National Adaptation Programme of Action (NAPA), the earliest "documented" flood in Bangladesh occurred in 1954 (then East Pakistan) due to vigorous monsoon rains and the broadening of the Brahmaputra River. It flooded 55% of the country's land, crops, and houses whilst claiming innumerable lives (NAPA 2009). Next, in 1974, after the independence of Bangladesh, another flood took place, which was triggered by incessant rainfall and swelling of the Ganges and the Brahmaputra. This flood drowned 36% of the nation's area, rendering crops and food, further impoverishing the weak economy (NAPA 2009). Then, the 1987 flood ruined cultivated land and covered over 50,000 square kilometres (NAPA 2009). It was caused by excessive rains in the Brahmaputra basin and impacted 20 million citizens throughout the country (NAPA 2009). A year later, the 1988 flood managed to outdo its predecessor by drowning 60% of the population and thereby bankrupting the economy (NAPA 2009).

In 1998, the country had almost 70% of its area submerged following record monsoon rains, overflowing rivers, and Himalayan meltwater (NAPA 2009). Over 30 million people were impacted, and billions of dollars' worth of damage was suffered, laying bare the country's infrastructure and weakness (NAPA 2009). The same was the pattern for the 2004 flood, which hit 36 million people with heavy rains and overflowing rivers, badly damaging agriculture and the economy (NAPA 2009). More recently, the 2007 flood caused another episode of devastation. Heavy monsoon rains and overflowing rivers submerged large areas, impacting millions and damaging homes, crops, and infrastructure (NAPA 2009). These recurring disasters have strongly underscored Bangladesh's urgent need for improved flood management and disaster preparedness. Although the data shows that the monsoon was part of a larger trend, South Asia's rainy season has become more intense and inconsistent due to climate change (Letsch et al. 2023). The growing concern was also visible in the 2022 flood during the summer, resulting in at least 141 deaths and millions affected (Teirstein 2023).

13.4 Bangladesh 2024 Flood: A Case Study

Bangladesh experienced one of the deadliest floods in August 2024 due to the prolonged heavy rainfall in northeast India and southeast Bangladesh. The rainfall caused more overflowing water in the cross-border rivers than in regular times. With no options left, the sluice gate of the hydroelectric plants in India was forced to open. Specifically, the heavy water flow from the Dambur Dam of Tripura State of India severely affected the flooding in 73 sub-districts and 528 municipalities across 11 districts in northeastern and southeastern Bangladesh (Bangladesh Red Crescent Society 2024).

The "danger level", a measurement indicating extreme flood risk and potential for widespread damage, crossed above 14 points in seven interconnected rivers in the southeastern part of Bangladesh. Millions from Feni (most affected), Noakhali, Cumilla, and other neighbouring districts in that region were flooded suddenly and waterlogged for weeks after that event. During this natural disaster, a major safe water supply crisis was observed. This case study reflects the scarcity of drinking water and alternative water sources during the flood in Feni and other neighbouring districts.

The residents of Feni, like any other district of Bangladesh, typically depend on two types of water sources, i.e. surface water and groundwater. However, due to heavy rainfall and flooding, the surface water sources overflowed and became unusable for drinking. As mentioned earlier, the water level rose above the danger mark, and the natural forceful flow of floodwater disrupted the electric supply across the district. Hence, people who relied on groundwater through electric pumps could not extract water. While some individuals managed to collect rainwater, they faced challenges with storage capacity, leaving many without access to drinking water for days. In desperation, some consumed floodwater, which led to severe cases of diarrhoea. Overall, ensuring access to safe drinking water was an immense challenge for the affected communities during this flood.

> Those who suffered the most were the ones who couldn't reach a centre or any shelter, and their situation was horrifying. One of my relatives had to eat tree leaves along with his little daughter for days. [35-year-old housewife, Feni]

During the first week of the event, many people did not receive any support from the government or NGOs due to the extreme weather conditions, making it difficult for the aid workers to reach affected areas. By the following week, when individuals were relocated to shelters, NGOs initially provided bottled water. However, transportation barriers continued due to the existence of the floodwater along the way. As a result, the shortage of supplies made it a significant challenge to ensure the availability of safe drinking water in the shelters. As the situation grew more critical, NGOs began distributing water purification tablets to families who could not afford alternative solutions for drinking water.

> I had to spend a large amount on the water because the large water bottles ran out quickly at first. I had to buy cases of small water bottles, which cost around 3,000 to 4,000 taka per case. Can you imagine people starting a syndicate business in this situation? Only Allah knows how people cried for drinking fresh water. Moreover, issues occurred while collecting them from the local market during heavy rainfall. [24-year-old male student, Feni]

After the floodwaters receded, layers of sediment and pollutants remained, making local water sources unsafe for months. Many families relied on tube wells and electric pumps for groundwater, but floodwater poured into wells, contaminating them with harmful bacteria and chemicals. The situation worsened as prolonged power outages left electric pumps unusable, cutting off access to clean water. Consequently, many had no choice but to use unsafe water, leading to widespread illnesses like diarrhoea and other waterborne diseases.

> During the flood, I spent 50,000 taka to buy water for my family and relatives. After the flood, I had to spend 1,50,000 taka on repairing the water storage and installing a new water pump, which was destroyed by the flood. [45-year-old businessman, Feni]

The 2024 flood has multifaceted effects. For example, excessive rain caused the mobile network to go down, which made the job difficult for the rescue team. At least 1.2 million families, counting over 5.8 million individuals, were affected in just two weeks (Bangladesh Red Crescent Society 2024). It also caused financial and economic difficulties. Twenty-six dams were breached due to the flood, causing over 300 million Taka (3 million USD) damage to infrastructure in the first round of flooding (Chowdhury and Lipi 2024). Moreover, over 15,000 hectares of different types of paddy seedlings, 50 hectares of seedbeds, and 350 hectares of vegetables have been submerged only in the Fatikchari sub-district, which provides an assumption of the damage to crops across the 73 affected sub-districts (Chowdhury and Lipi 2024).

Most families were unprepared for the flood's aftermath and had limited knowledge of emergency water storage in this region. The communities have no collective water storage initiatives or very few government initiatives to store large amounts of water that could be useful during natural disasters. The flood not only severely damaged Feni's water sources but also raised a question on the preparedness strategies for such long-term climate challenges, particularly for ensuring safe drinking water for the community.

13.5 Water Contamination During Floods

Water contamination is a major concern during floods as overflowing rivers and floodwaters mix with sewage, industrial waste, and agricultural runoff (WHO 2014). This polluted water often carries harmful bacteria, chemicals, and toxins, making it unsafe for drinking and household use. Contaminated water increases the risk of waterborne diseases like cholera, diarrhoea, and typhoid, posing serious health risks to affected communities (WHO 2023), especially in areas with limited access to clean water and sanitation facilities (Yeboah et al. 2024). In this section, the chapter will describe the sources and public health impacts of water contamination during floods.

13.5.1 Sources of Contamination

Water contamination can be caused during floods by multiple factors. In both urban and rural areas of Bangladesh, the sewage system becomes overloaded by floodwater. Consequently, untreated contamination gets mixed in with floodwater. Moreover, toxic industrial waste from factories operating around localisation is another reason for contamination. This waste carries harmful chemicals and hazardous substances (WHO 2014). These elements are highly hazardous to human health, and the effect multiplies many times if the factories are in nearby densely populated areas, which is the case all over Bangladesh. These substances easily mix with the water flow during the flood, further exacerbating circumstances. Similarly, another way of water contamination during floods is agricultural runoff, including pesticides, fertilisers, and animal waste. Moreover, household waste and garbage, especially in urban slums, also contribute to water pollution during flood time (Hasan et al. 2019). Furthermore, physical contamination results from debris, sediment, and solid waste carried by the floodwaters (Al-Jubouri et al. 2022).

This mixture of pollutants poses a significant threat to drinking water. In addition, due to the lack of a safe water supply, people had to use this contaminated water for other day-to-day activities. Any form of use of contaminated water leaves these affected people further vulnerable in the post-flood conditions. The situation worsens when potentially all possible contamination factors combine and create a bigger effect.

13.5.2 Public Health Impacts of Water Contamination

The public health impact of water contamination in Bangladesh is a serious public health issue with a significant impact on human health, including increased illness and mortality, especially among vulnerable populations. Floods in Bangladesh come with a "staggering death toll", not only in the immediate aftermath of the flooding itself but also in the months that follow (Rerolle et al. 2023).

Short-term health impacts of water contamination are immediate and severe, while long-term health consequences are equally concerning. A study found that a moderate elevation in risk of acute respiratory infection due to contaminated water occurs for up to two years after flooding (Milojevic et al. 2012). The cholera outbreak in 1998 flood in Bangladesh affected 4 million people in Dhaka city only, and the affected people were at a six times higher cholera risk compared to the people in areas unaffected by the flood (Lee et al. 2020). Another study showed that the short-term and long-term impact of the flood depends not only on the flood's characteristics but also on people's vulnerability (Alderman et al. 2012). Pregnant women, children, the elderly, and individuals with compromised immune systems particularly stand at a higher risk of experiencing severe health complications arising from this water contamination (Araf 2024).

Children with developing immune systems faced severe dehydration from diarrhoea, increasing mortality risks (Milojevic et al. 2012). Children under 5 years of age are found to be most vulnerable to waterborne diseases (Lee et al. 2020). According to Researchers of the University of California, 1,52,753 infants (children less than 11 months old) died in Bangladesh due to floods between 1998 and 2017 (Rerolle et al. 2023). Also, infant deaths per 1000 births were 5.3 times higher in flood-prone areas (Rerolle et al. 2023). These deaths are mostly linked to three flood aftermaths—diarrheal diseases, dengue, and massive crop loss (Teirstein 2023).

Similarly, the elderly are also severely impacted by water contamination during the flood because of their weakened physiological resilience. In the absence of adequate sanitation, they are prone to urinary tract infections and exacerbated chronic conditions. However, they are most vulnerable to "flood-related eye diseases, including focal retinal whitening, necrotising retinochoroiditis, and trachoma infection" (Lee et al. 2020). Additionally, individuals with pre-existing conditions, such as diabetes or hypertension, experienced uncontrolled symptoms when clean water and medications were unavailable. For example, disrupted insulin supplies during the flood may lead to critical health deterioration among diabetic patients.

Pregnant women faced heightened risks due to contaminated water exposure, compounding maternal health challenges. For example, contaminated water can infect the genital organs of women (Nasreen et al. 2017). In addition, the flood severely disrupts pregnancy healthcare services (Nasreen et al. 2017). It is further compounded by the increase in sexual violence during the flood time (Nasreen et al. 2017).

The 2024 flood particularly compromised safe drinking water and sanitation services. Like any other time, compromising these two important aspects of human health rendered "previously potable water sources lethal vehicles for pervasive waterborne diseases" (Araf 2024). As a result of the biological contamination, there were outbreaks of cholera, diarrhoea, and other waterborne diseases. The chances of these diseases increased since the floodwaters combined with various contaminants, including human excreta and agricultural and industrial waste (Araf 2024). Moreover, prolonged exposure to chemical pollutants in the water is associated with the development of serious chronic health conditions, including malignancies and renal impairment (UNDP 2011). Even the shelters made the risk of contagion in the absence of a clean water supply and adequate sanitation, as the shelter centres were densely populated (Araf 2024). According to the Directorate General of Health Services (DHGS) of Bangladesh's Ministry of Family and Health Welfare, mosquito-borne diseases like dengue were raised alarmingly in the flood-affected areas.

Along with the immediate and longstanding health consequences, the economic fallout from water contamination puts additional pressure on healthcare systems as well as on individuals' livelihoods. Moreover, the costs associated with treating waterborne illnesses, coupled with the necessity of ensuring access to safe drinking water, require additional financial resources among vulnerable groups. These shortages could further enhance poverty and social inequity, especially in communities battling the consequences of flooding.

13.6 Safe Water Policy in Bangladesh

According to the 2008 UNICEF "Handbook on Water Quality," insufficient water supplies coupled with poor sanitation cause 3.4 million deaths per year, which translates into someone dying every 10 s. As discussed throughout the four previous segments, the issue is even worse for Bangladesh due to its geography, economy, and poor infrastructure. Hence, it is imperative to have effective laws and policies in existence as guidelines to respond to emergencies. While Bangladesh has established essential laws and policies to manage and ensure water safety, the 2024 flood revealed significant gaps and weaknesses in these systems. While there have been a few studies on the issue from scientific, environmental, and public health perspectives, there is still a need for a comprehensive analysis of the policies and legal frameworks that govern water safety in flood situations. Therefore, this section will briefly outline the existing policies, international standards, and implementation gaps in the practical context.

13.6.1 Existing Framework

Despite being a flood-prone and climate-vulnerable country, Bangladesh does not have an explicit and dedicated law or policy to deal with public health issues, *inter alia*, ensuring safe water during floods. However, the country has different frameworks for public health, safe water, and flood prevention. The landscape of the framework comprises the legislations, policy, and the Constitution. It reflects a concerted effort to address its unique hydrological challenges as a flood-prone, lower-riparian nation.

Firstly, central to this framework is the Water Act of 2013, which establishes the right to potable water as a priority under section 3(2). This provision prioritises hygiene and sanitation as the highest priorities (Bangladesh Water Act 2013). Moreover, this Act establishes the National Water Resources Council (NWRC) as the apex decision-making body tasked with policy formulation, resource planning, and water conservation under section 4. The emphasis given by this Act is evident from the fact that NWRC will be chaired by the Prime Minister of Bangladesh, the most powerful position of the government (Bangladesh Water Act 2013). Furthermore, the Act also establishes an Executive Committee under section 9. This subordinate entity holds enforcement powers, including issuing protection orders to prevent water misuse on private lands according to section 3(3) and removal orders for illegal constructions obstructing water courses, exercising section 13. These provisions suggest a proactive intent to safeguard water resources, yet their focus on structural and administrative control raises questions about their adaptability to dynamic flood scenarios.

Secondly, the National Water Policy (1999), introduced by the Ministry of Water Resources, complements the Water Act 2013 by addressing Bangladesh's transboundary vulnerabilities. It acknowledges the country's lack of control over rivers entering from upstream nations like India and proposes international cooperation for data exchange, flood mitigation, and basin-wide planning (The National Water Policy 1999). Additionally, its emphasis on early warning systems and flood risk zoning aligns with disaster preparedness goals, but the policy's broad ambitions contrast with its limited enforcement mechanisms—a tension evident during the 2024 floods.

Thirdly, the National Policy for Safe Water Supply and Sanitation of 1998 shifts focus to public health, aiming to universalise access to safe water and sanitation through community participation and private-sector engagement. This policy aims to reduce the number of users per rural tube-well from 105 to 50 and ensure sanitary latrines for all households (National Policy for Safe Water Supply and Sanitation 1998). This vision reflects a pragmatic approach to health equity. However, its reliance on gradual cost-sharing and local government capacity assumes a stability disrupted by floods, undermining its efficacy in crisis contexts.

Fourthly, the National Plan for Disaster Management 2021–2025 integrates water safety into broader disaster risk governance, prioritising flood-proofing and early warning systems. This plan's multi-hazard approach is commendable, yet its implementation hinges on inter-agency coordination, a perennial challenge in Bangladesh's fragmented bureaucracy.

Finally and most importantly, Article 18 of the Constitution of Bangladesh upholds the right to public health among the State's primary duties. However, this provision falls under Part II of the Constitution—the Fundamental Principle of State Policy. While the provisions of Part III (Fundamental Rights) are judicially enforceable, the right enshrined in Part II has not been given the same status. Which means, although the right to public health of a citizen of Bangladesh exists, one cannot necessarily take the matter into a court of law in case of a violation of the right according to Article 8(2) of the Constitution—which, at this point, is a non-amendable part known as the basic structure of the Constitution (The Constitution of the People's Republic of Bangladesh 1972). Nevertheless, the Constitution Reform Commission formed under the Interim Government of Bangladesh has proposed implementing health rights based on the availability of resources, which is indeed a positive approach.

Analytically, the framework forms a robust theoretical scaffold, blending legal authority, international collaboration, and community involvement. However, their design reveals a disconnect between ambition and operational reality. The Water Act's top-down structure prioritises centralised control over grassroots adaptability, while the 1999 Policy's transboundary focus lacks binding mechanisms to compel upstream cooperation. The 1998 Policy's health-centric goals falter without resilient infrastructure, and the 2021–2025 Plan's success depends on overcoming bureaucratic inertia. Similarly, the framework that could implement these provisions, at least by containing the authority in pen and paper—the Constitution, also seems too ineffective to deal with the twofold issues of safe water during flood time. This framework, while progressive, struggles to translate intent into action during floods, as evidenced by the 2024 crisis.

13.6.2 International Standards and Bangladesh's Obligations

The World Health Organisation's (WHO) Guidelines for Drinking-Water Quality: Small Water Supplies (2024) offer a benchmark for assessing Bangladesh's policies. These guidelines advocate Water Safety Plans (WSPs)—a risk-management framework spanning the water supply chain from catchment to consumer (WHO 2024). WSPs prioritise proactive hazard identification and operational planning for emergencies like floods, drawing from global successes such as Cambodia's climate-resilient water kiosks operated by Teuk Saat 1001 (WHO 2024). This approach contrasts with Bangladesh's reactive, compliance-driven policies, highlighting a potential shift towards anticipatory governance.

WSPs emphasise adaptability by tailoring interventions to local socioeconomic conditions. It is a critical consideration for Bangladesh since the country has resource constraints, and the flood frequency amplifies the vulnerability. For instance, the WHO's focus on embedding WSPs into daily operations through standard operating procedures and preparing for contamination events offers a model for enhancing resilience. On the contrary, Bangladesh's policies lack such specificity. For instance, the Water Act's enforcement mechanisms address violations post-hoc rather than preempting risks, while the 1999 Policy's disaster provisions remain aspirational without operational detail. This gap suggests that adopting WSPs could bridge policy intent and practical outcomes, aligning with United Nations Sustainable Development Goal (SDG) 6 (clean water and sanitation) but requiring significant capacity-building.

On that note, Bangladesh has failed to comply with its multiple international obligations. As a UN member country, Bangladesh is committed to achieving UN SDG 6 as part of the 2030 Agenda for Sustainable Development. Although it is a "soft law" by nature, the SDGs foster commitment and collaboration among countries for global welfare. Fulfilling SDG goals, though not binding, is a reflection of performing a moral obligation, which increases international reputation and, most importantly, brings funding for emergency situations like floods and public health.

Bangladesh's failure to ensure public health during the 2024 flood violates the International Covenant on Economic, Social and Cultural Rights (ICESCR). Bangladesh is a State party to this core human rights treaty and is obliged to respect the provisions in every circumstance. Article 11(1) of ICESCR provides the right to "an adequate standard of living, including adequate food, clothing, and housing, as well as continuous improvement of living conditions", and Article 12 upholds the right to enjoy the highest attainable physical and mental health. Reading these provisions and looking back at the 2024 flood further reinforces that such emergency situations have multiple effects on human life. Another fundamental international instrument Bangladesh failed to comply with on this occasion was the UN Convention on the Rights of the Child (CRC). Article 24(2)(c) of CRC requires the signatory States to ensure the highest attainable public health standard for children, including safe water. Article 24(1) also obliges the States to diminish child mortality.

The Paris Agreement on Climate Change 2015, the International Watercourses Convention 1997, and the Sendai Framework for Disaster Risk Reduction (2015–2030) require Bangladesh to manage and utilise its water resources better, with the obligation to prevent significant harm. The obligations include enhancing adaptive capacity, strengthening resilience, and cooperation among the States to reduce vulnerability to climate change. Under these frameworks, Bangladesh is also committed to enhancing disaster preparedness for effective response and strengthening disaster risk governance.

13.6.3 The Gap in Implementation

The 2024 floods starkly exposed the chasm between policy design and execution. Reports from the Department of Public Health Engineering of the Bangladesh government documented 20,260 damaged water points and 1,34,269 damaged toilets, contributing to 1397 cases of acute watery diarrhoea (AWD) and widespread skin diseases (UNICEF 2024; UN Bangladesh 2024). Flooded roads and disrupted transport networks hampered healthcare access, while contaminated water sources fuelled disease outbreaks (ACAPS 2024). These outcomes reflect a failure to operationalise existing safeguards, raising critical questions about enforcement, infrastructure resilience, and emergency preparedness.

The Water Act's provisions for removal orders (section 13) and protection of flood control zones (section 25) were rendered ineffective by the scale and speed of the 2024 floods, suggesting a lack of preemptive enforcement. Similarly, the 1999 Policy's call for early warning systems faltered, with communities in Feni and Noakhali unprepared for sudden inundation (UN Bangladesh 2024). Submerged facilities undermined the 1998 Policy's sanitation targets, and the 2021–2025 Plan's governance framework struggled with coordination delays, as evidenced by shortages of medical supplies and midwives (UN Bangladesh 2024).

Also, as mentioned in the last segment, the country has failed to comply with a series of international obligations, reflecting another gap. While the country itself is committed to achieving those obligations simply because it has willingly signed and ratified those standards, its weak infrastructure is obstructing its progress towards its commitments.

These gaps stem from systemic issues, i.e. inadequate funding, weak local capacity, and a reactive rather than proactive stance. These challenges suggest that while policies exist, their enforcement and adaptation to flood scenarios are inadequate, necessitating a critical review of resource allocation and coordination.

13.7 Recommendations

Drawing on the complexities of the public health issue, particularly concerning safe water during the flood crisis in Bangladesh, this part of the chapter suggests a four-layer recommendation aligning with Bangladesh's socioeconomic structure.

13.7.1 Role of Government

To tackle such a national crisis, the government has to play the pivotal role of structural fixes to systemic resilience, prioritising flood-resistant infrastructure and robust enforcement. Investments in elevated tube wells and protected sanitation facilities, as piloted in coastal projects like the Char Development & Settlement Project (CDSP-4), could withstand flood impacts, reducing contamination risks. Moreover, the Bangladesh Water Development Board (BWDB)'s flood forecasting needs modernisation. It has provided five-day flash flood predictions since 2012, but should be expanded with real-time data integration and community dissemination, addressing the 2024 warning failures.

Furthermore, the government must ensure both immediate and long-term responses to the post-flood public health crisis (Rupa and Hossian 2024). This includes restoring clean drinking water through purification tablets and mobile purification units (Rupa and Hossian 2024). Moreover, public awareness can be created about how water can be purified with generally available elements, such as boiling, which purifies contaminated water. In addition, surveillance and rapid response units can be implemented to monitor waterborne and vector-borne diseases in the affected areas (Rupa and Hossian 2024). However, the most crucial thing the government should focus on is restoring the damaged healthcare system as soon as possible.

Critically, enforcing the Water Act's provisions must shift from punitive measures to preventive action. Regular water quality monitoring during monsoons, empowered by the Executive Committee's authority under Section 12(3) of the Act, could preempt outbreaks. At the same time, stricter penalties for industrial and agricultural pollution align with WHO standards. However, this requires overcoming bureaucratic silos and increasing budgetary allocations, a political challenge given competing developmental priorities.

13.7.2 Role of NGOs and INGOs

NGOs and INGOs offer agility and grassroots reach, filling gaps left by government capacity. For example, WaterAid Bangladesh's WASH programs in flood-prone areas can be proven as scalable solutions to deliver safe water through community-initiated systems. This will be particularly helpful when the aid workers from the government and NGOs cannot reach the affected communities on time. If the local people are adequately trained, the dependency on official aid will decrease. While floods cannot be prevented, such strategies can mitigate the situation. Another praiseworthy initiative that local and international organisations can follow was what was done by the UN Office for the Coordination and Humanitarian Affairs (OCHA). As early as 16 min of the early warning flood alert in Bangladesh, the OCHA's Central Emergency Response Fund released $6.2 million to humanitarian partners in Bangladesh, resulting in supporting 4,30,000 people with cash transfers and other necessary assistance (UN OCHA 2025). Such pre-disaster funding has proven to be successful, but not adequate.

Nevertheless, these initiatives underscore NGOs' potential to pilot innovations—e.g. portable purification units—before government adoption. Their advocacy role is equally critical in pressuring policymakers to align with international standards like WSPs. However, their impact hinges on coordination with government agencies, a challenge that has been historically persistent.

13.7.3 Role of Local Communities

Local communities, often sidelined in top-down frameworks, hold untapped potential for resilience-building. The 1999 Policy's mandate for stakeholder participation aligns with bottom-up approaches, as seen in tidal river management—a cost-effective, Indigenous practice reducing flood impacts. Similarly, as seen in the Bangladesh 2024 flood, the community members supported their communities, and they expanded their water storage. Such community-led water storage initiatives, if trained beforehand, will be helpful for people to rescue themselves in the future and enable them to support each other until the relief and aid team reaches them. Most importantly, empowering communities requires education on water safety and contamination risks, addressing the knowledge gaps evident in 2024 (e.g. consuming floodwater). However, the challenge lies in scaling these efforts amidst poverty and resource scarcity, necessitating government and NGO support.

13.7.4 Role of Regional and Transboundary Cooperation

Bangladesh's flood vulnerability is inseparable from its transboundary context, with 54 shared rivers amplifying downstream risks (Letsch et al. 2023). The 1996 Ganges Water Treaty with India mitigated dry-season shortages but faltered during the 2024 floods, with upstream dam releases exacerbating inundation (ACAPS 2024). Strengthening this treaty through real-time data sharing and joint flood management, as proposed in the 1999 Policy, could reduce uncertainty and mistrust.

Comparatively, the Indus Water Treaty between India and Pakistan offers lessons in benefit-sharing and balancing hydroelectric needs with downstream flows. Adopting someone like this to the Ganges–Brahmaputra-Meghna basin could yield mutual gains—flood control for Bangladesh, hydropower for India—but requires diplomatic will and regional frameworks beyond bilateral talks. The 2024 dam release controversy highlights the urgency of such cooperation, yet political tensions and upstream priorities pose formidable barriers. In such circumstances, regional bodies like the South Asian Association for Regional Cooperation (SAARC) may come into play for adapting approaches that will be sustainable for the broader benefit of the region.

13.8 Limitations of the Study

While numerous studies have investigated floods and other environmental parameters, relatively few studies correlate flood-related injuries with public health outcomes in the context of Bangladesh. Specifically, there is a lack of research quantifying the number of people contaminated due to specific flood-related public health issues. Further complicating matters, there is a dearth of numerical evidence and academic literature on the specific links between drinking water contamination and public health policy, particularly during flood crises. This gap in knowledge highlights a critical need for immediate attention, yet government initiatives to address this issue are currently lacking. Moreover, the study was conducted during a period of political transition in Bangladesh, which imposed significant constraints. Consequently, due to time limitations and the country's socio-political unrest, the researchers could not gather data from a broader range of individuals, limiting the inclusion of more diverse lived experience insights.

13.9 Conclusion

The 2024 floods in Bangladesh crystallised the complex interplay of water contamination, public health, and flood emergency. If there is any positive outcome to be taken from the disaster scenario, it would understand the need to strengthen and adapt holistic approaches. Public health concerns, such as social, economic, and cultural rights, have inherently been less prioritised in Bangladesh's broader legal and policy framework. Even the Constitution Reform Commission also focused more on civil and political rights. Despite the existence of a few laws, their implementation crumbled under flood pressure, as seen in the widespread AWD outbreaks and infrastructure collapse. While fulfilling Bangladesh's international commitments and adopting further international standards like WSPs offers a path to resilience, the country's socioeconomic constraints demand localised adaptation. Hence, since the policy efficiency exposed both fragilities and priorities of safe water framework implementation, the only solution seems to be the holistic adaptation of government-led infrastructure upgrades, NGO-driven innovation, community empowerment, and transboundary cooperation to bridge these gaps. The 2024 crisis reveals a deeper challenge of reconciling ambitious policies with climate vulnerability and resource limitation realities. Success hinges on shifting from reactive governance to proactive resilience, integrating local knowledge with global best practices. For a nation where floods are a perennial threat, ensuring safe water is not just a public health imperative but a cornerstone of sustainable development, demanding sustained commitment across stakeholders. The findings of this chapter serve as the starting point for policymakers, government agencies, NGOs, and community organisations to strongly consider public health concerns *inter alia* ensuring safe water, particularly during, before, and after flood emergencies in Bangladesh. If the due attention and acknowledgement come, further research can be driven towards finding detailed, sustainable, and more effective solutions specifically for countries with limited resources and infrastructure like Bangladesh.

Acknowledgements We would like to express our sincere gratitude to the interviewees for being strong and discussing the biggest traumatic event of their lives, strengthening this research. We also thank Md. Tahzib Ul Alam (Student, Peace and Conflict Studies, University of Dhaka) for assisting with the interviews, Muhammad Jabidul Islam Bhuiyan (Judge Md Wahidul Islam Research Assistant, Bangladesh Society of International Law & Final year undergraduate student, School of Law, BRAC University) for research support and Ms Mahbuba Kamal (LLM in Human Rights Candidate, University of Birmingham) for reviewing the piece. Without their generous help, this chapter would not have made it to life.

References

ACAPS (2024) Bangladesh: flooding. ACAPS. https://www.acaps.org/fileadmin/Data_Product/Main_media/20240904_ACAPS_Bangladesh_Flooding.pdf. Accessed 4 Sep 2024

Adnan MSG, Siam ZS, Kabir I, Kabir Z, Ahmed MR, Hassan QK, Rahman RM, Dewan A (2023) A novel framework for addressing uncertainties in machine learning-based geospatial approaches for flood prediction. J Environ Manage 326:116813. https://doi.org/10.1016/j.jenvman.2022.116813

Alderman K, Turner LR, Tong S (2012) Floods and human health: a systematic review. Environ Int 47:37–47. https://doi.org/10.1016/j.envint.2012.06.003

Al-Jubouri SM, Waisi BI, Eslamian S (2022) Debris and solid wastes in flood plain management. In: Flood handbook. CRC Press, Boca Raton, pp 421–434. https://doi.org/10.1201/9780429463327

Araf Y (2024) August 2024 Bangladesh floods: emerging health risks. Lancet 404(10461):1401–1402. https://doi.org/10.1016/S0140-6736(24)01856-7

Bangladesh Red Crescent Society (2024) Situation Report-3, Southeastern Flood in Bangladesh. BDRCS. https://bdrcs.org/wp-content/uploads/2024/08/BDRCS-Sitrep-3-Southeastern-Flood-August-2024.pdf. Accessed 28 Aug 2024

Bangladesh Water Act 2013 (Act No. XIV of 2013)

BBC (2014) Rivers and flooding—case study: Bangladesh. BBC. https://www.bbc.co.uk/bitesize/guides/zgycwmn/revision/4. Accessed 20 Aug 2014

Bernard A, Long N, Becker M, Khan J, Fanchette S (2022) Bangladesh's vulnerability to cyclonic coastal flooding. Nat Hazards Earth Syst Sci 22:729–751. https://doi.org/10.5194/nhess-22-729-2022

Chowdhury MH, Lipi S (2024) What made the August floods in Bangladesh so devastating? bdnews24. https://bdnews24.com/bangladesh/cffcb9fcc22d#:~:text=Sardar%20Uday%20Raihan%2C%20executive%20engineer,the%20water%20from%20fully%20receding. Accessed 23 Aug 2024

Chakraborty D, Mondal KP, Jubaer A, Islam ST, Talukder B (2024) Health impacts of rapid-onset event: 2022 flash flood in Bangladesh. In: Living with climate change. Elsevier, Amsterdam, pp 199–212. https://doi.org/10.1016/B978-0-443-18515-1.00004-6

Constitution of the People's Republic of Bangladesh (1972)

Hasan MK, Shahriar A, Jim KU (2019) Water pollution in Bangladesh and its impact on public health. Heliyon 5(8):e02145. https://doi.org/10.1016/j.heliyon.2019.e02145

Hasan MN, Khalil I, Chowdhury MAB, Rahman M, Asaduzzaman M, Billah M, Banj LA, Alam M, Ahsan A, Traore T, Uddin MJ, Galizi R, Russo I, Zumla A, Haider N (2024) Two decades of endemic dengue in Bangladesh (2000–2022): trends, seasonality, and impact of temperature and rainfall patterns on transmission dynamics. J Med Entomol 61(2):345–353. https://doi.org/10.1093/jme/tjae001

Huq S, Moinuddin K (2004) Climate Change, vulnerability, and adaptation in Bangladesh. In: Climate change. CRC Press, Boca Raton, pp 267–288. https://doi.org/10.1201/9781482294415

International Federation of Red Cross and Red Crescent Societies (IFRC) (2017) South Asia flood crisis: disease outbreaks, funding shortages compound suffering. ReliefWeb. https://reliefweb.int/report/bangladesh/south-asia-flood-crisis-disease-outbreaks-funding-shortages-compound-suffering. Accessed 5 Sep 2017

IPCC (2022) Climate change 2022: impacts, adaptation, and vulnerability. Contribution of working group II to the sixth assessment report of the intergovernmental panel on climate change. IPCC. https://www.ipcc.ch/report/ar6/wg2/. Accessed 28 Feb 2022

Lee J, Perera D, Glickman T, Taing L (2020) Water-related disasters and their health impacts: a global review. Prog Disaster Sci. https://doi.org/10.1016/j.pdisas.2020.100123

Letsch L, Dasgupta S, Robinson EJZ (2023) Tackling flooding in Bangladesh in a changing climate. LSE. https://www.lse.ac.uk/granthaminstitute/wp-content/uploads/2023/08/Tackling-flooding-in-Bangladesh-in-a-changing-climate.pdf. Accessed 23 Aug 2023

Milojevic A, Armstrong B, Hashizume M, McAllister K, Faruque A, Yunus M, Streatfield PK, Wilkinson KM (2012) Health effects of flooding in rural Bangladesh. Epidemiology 23(1):107–115. https://doi.org/10.1097/EDE.0b013e31823ac606

Nasreen M, Hossain KM, Azad AK, Hasan K (2017) Sexual and reproductive health during emergencies: situation analysis of disaster prone areas of Bangladesh. Institute of Disaster Management and Vulnerability Studies, Dhaka

National Adaptation Programme of Action (NAPA) (2009) Ministry of Environment and Forests, Government of the People's Republic of Bangladesh. https://unfccc.int/resource/docs/napa/ban02.pdf

National Policy for Safe Water Supply and Sanitation (1998) Ministry of Local Government, Rural Development and Cooperatives, Government of the People's Republic of Bangladesh. https://nda.codezonelab.com/en/c/publication/national-policy-for-safe-water-supply-and-sanitation-1998

Rahman MM, Shobuj IA, Hossain MT, Tasnim F (2023) Impact of disaster on mental health of women: a case study on 2022 flash flood in Bangladesh. Int J Disaster Risk Reduc 96:103935. https://doi.org/10.1016/j.ijdrr.2023.103935

Rerolle F, Arnold BF, Benmarhnia T (2023) Excess risk in infant mortality among populations living in flood-prone areas in Bangladesh: a cluster-matched cohort study over three decades, 1988 to 2017. Proc Natl Acad Sci 120(50):e2218789120. https://doi.org/10.1073/pnas.2218789120

Rupa FH, Hossian M (2024) Addressing public health risks: strategies to combat infectious diseases after the august 2024 floods in Bangladesh. J Prev Med Public Health 57(6):600–603. https://doi.org/10.3961/jpmph.24.495

Teirstein Z (2023) The hidden death toll of flooding in Bangladesh sends a grim signal about climate and health. Grist. https://grist.org/health/the-hidden-death-toll-of-flooding-in-bangladesh-sends-a-grim-signal-about-climate-and-health/. Accessed 11 Dec 2023

The National Water Policy (1999) Ministry of Water Resources, Government of the People's Republic of Bangladesh. https://nda.codezonelab.com/en/c/publication/national-water-policy-1999#:~:text=The%20National%20Water%20Policy%2C%20formulated,targeted%20objectives%20of%20the%20sector

Uddin K, Matin MA, Thapa RB (2021) Rapid flood mapping using multi-temporal SAR images: an example from Bangladesh. In: A decade of experience from SERVIR, p 201. https://doi.org/10.1007/978-3-030-73569-2_10

UN OCHA (2025) A race against floods in Bangladesh. UN OCHA. https://www.unocha.org/news/race-against-floods-bangladesh. Accessed 13 Feb 2025

UNDP (2011) Assessment of the investments and financial flows to adapt to climate change in the water sector. UNDP Climate Promise. https://climatepromise.undp.org/sites/default/files/research_report_document/undp-iff-bangladesh-assessment-water-en.pdf. Accessed 1 Jan 2012

UNICEF (2024) Situation report on cyclone REMAL and floods in Bangladesh No. 2. UNICEF. https://www.unicef.org/media/160426/file/Bangladesh-Cyclone-Remal-and-Floods-SitRep-21-August-2024.pdf. Accessed 21 Aug 2024

United Nations (2024) Bangladesh: eastern flash floods 2024 situation Report No. 02. United Nations in Bangladesh. https://bangladesh.un.org/en/277637-bangladesh-eastern-flash-floods-2024-situation-report-no-02#:~:text=02,-30%20August%202024&text=The%20recent%20flash%20floods%20in,were%20displaced%20in%20evacuation%20shelters. Accessed 30 Aug 2024

WHO (2014) Floods and health fact sheets for health professionals. WHO. https://www.who.int/europe/publications/i/item/WHO-EURO--2014-8519-48291-71705. Accessed 31 Dec 2014

WHO (2023) Drinking-water. https://www.who.int/news-room/fact-sheets/detail/drinking-water. Accessed 18 Aug 2025

WHO (2024) Guidelines for drinking-water quality: small water supplies. WHO. https://www.who.int/publications/i/item/9789240088740. Accessed 15 Feb 2024

World Bank (2021) Climate change knowledge portal: Bangladesh. World Bank. https://climateknowledgeportal.worldbank.org/country/bangladesh. Accessed 1 Jan 2022

Yeboah SIIK, Prince Antwi-Agyei P, Kabo-Bah AT, Ackerson NOB (2024) Water, environment, and health nexus: understanding the risk factors for waterborne diseases in communities along the Tano River Basin, Ghana. J Water & Health 22(8):1556–1577. https://doi.org/10.2166/wh.2024.186

Chapter 14
Sustainable Strategies for Access and Utilization of Antenatal Care Services in Flood-Prone Regions of India

Bikash Barman, Mithun Sarkar, Koyel Majumder, and Pradip Chouhan

Abstract Environmental hazard has a major effect on healthcare services particularly in antenatal care services, posing a number of difficulties that impact the accessibility and utilization of care for expectant mothers and new moms. Natural disasters like flooding and antenatal care are closely related, which is a serious problem that calls for all-encompassing disaster planning and response plans. The purpose of the current study is to examine the relationship between flood events and antenatal care services among Indian women. The current study utilized both the Vulnerability Atlas of India (2019) along the NFHS-5 (2019–21) data. More than 4 ANC visits have been taken as outcome variables, whereas flood-induced vulnerability and socio-demographic variables have been run as control variables. The bivariate analysis and multivariate analysis have been used for the statistical analysis. The bivariate results show that ANC utilization ($\geq$4 ANC visits) and the occurrence of floods are significantly correlated. Flood predominantly affects the access and utilization of ANC. The result of binary logistic regression shows that the likelihood of receiving at least 4 ANC visits during pregnancy is 12% higher among the women of non-flood-prone areas than the women of flood-prone areas of India. The significant association between flood and receiving $\geq$ 4 ANC visits stated the crucial strategies for disaster readiness and ANC, mainly in the flood-prone areas across India. This study provides valuable takeaways for promoting proper access and utilization of antenatal care services during the pregnancy period in the flood-prone areas, which may lead to a decrease of maternal as well as infant mortality across the nation.

Keywords Flood · Antenatal care · Maternal health vulnerability · NFHS

B. Barman (✉)
Department of Geography, Malda Women's College, Malda, India
e-mail: barmanbikash2013@gmail.com

M. Sarkar · K. Majumder · P. Chouhan
Department of Geography, University of Gour Banga, Malda, India

A. Roy et al. (eds.), *Environment and Public Health*, Sustainability Solutions,
https://doi.org/10.1007/978-3-031-99770-9_14

14.1 Introduction

Maternal mortality and infant mortality now act as a serious global health issue mainly in the middle-income countries and low-income countries, which act as a barrier to achieving the overall health and well-being as well as play a main constraint to achieve the Sustainable Goal-3 (Good Health and Well-being) and Millennium Development Goal-5 (Improved maternal health) globally (Ahern et al. 2005; WHO 2019; Afzal et al. 2024). Maternal mortality is unacceptably high in low and middle-income nations. In 2020, around 287,000 women died during pregnancy periods and delivery, and also this year these nations reported about 95% of all maternal deaths, and most of death could have been avoided (Alderman et al. 2012; Ali and McLean 2015). The high rate of maternal death in some parts of the world exacerbates the wealth divide and reproduces disparities in access to high-quality health care (Barman et al. 2023; Begum and Hamid 2023). The MMR in 2020 was 13/100,000 live births and 430/100,000 live births in high- and low-income nations, respectively. In India, due to maternal causes, 97 women per 100,000 live births die every year (Antipova and Curtis 2015; Biswas et al. 2024). This indicates that around 24,000 maternal deaths occur every year, which is very alarming. On the other hand, infant mortality rate is also still rising; IMR is 26/1000 live births each year, which suggests the seriousness of the issue in India (Mohan and Sharma 2016). The WHO stated that the incidence of maternal mortality and infant mortality can be checked through the appropriate use of child and MHC services by Indian women because the usage of MHC services directly affects the mother's health as well as newborn health (Patel and Tiwari 2017). Proper usage of ANC can reduce the chances of pregnancy complications, delivery complications, infants' birth weight, diseases like fever, jaundice, anaemia, etc. (Sahoo and Saha 2018). The access and utilization of antenatal care services depend on the different socio-demographic, economic, and environmental factors.

One of the most common natural calamities is flooding in the developed and developing nations, which has a significant effect on morbidity and mortality globally (Thomas and Coker 2012). Over the past 30 years, flooding has caused over 2.8 billion injuries and over 200,000 fatalities globally (Zhang and Li 2021; Pappas et al. 2024). Depending on how vulnerable a population is, floods might have different health effects. It significantly lowers everyone's quality of life in the impacted area (Callaghan et al. 2007). However, floods may have a number of harmful psychological and pathological effects on women's reproductive health, and women who are pregnant are especially vulnerable to the negative consequences of flood-related alterations (Baloch et al. 2012). Flood is known as a temporary geographic barrier which directly affects the degree of access and usage of MHC services during the wet season. It has a direct impact on the increase of morbidity as well as mortality, mainly in developing nations, but it also causes damage and injuries in developed countries (Dancause et al. 2011). During the flood situation, the transportation connectivity is lost, the healthcare centres are closed, restricting healthcare access and isolating the entire communities from each other (Datar et al. 2013; Gurley and Rohan 2019). In this situation, women who are pregnant face psychological stress, and they have to go through several health difficulties because there are insufficient facilities for adequate sanitation, hygiene, and drinking water, which may lead to several infectious diseases (Harville et al. 2010; Bukhari and Rizvi 2015). As a result, the morbidity and mortality are increased in the wet season compared to the dry season. Many previous studies show the flood's effects on access to health care and utilization, but this study will focus on the impact of flooding on the access and utilization of ANC (treatment-seeking behaviour) services among Indian reproductive women.

14.2 Materials and Methods

14.2.1 Data

The present study is cross-sectional, and the required data have been collected from two secondary data sources, i.e. NFHS-5 (2019–2021) and the Vulnerability Atlas of India (3rd edition), published in 2019. In the NFHS-5, with a 97% response rate of 724,115 women aged 15–49 and multi-stage stratified sampling technique was used to conduct the survey. The second source of data was published by the Building Materials and Technology Promotion Council (BMTPC), Ministry of Housing and Urban Affairs, Government of India. It was used to identify the flood-prone areas. This Atlas combination of diverse data sources delineates the risk zone across India and vulnerability degree, and such sources are the Survey of India (SOI), the Indian Meteorological Department (IMD), the Geological Survey of India (GSI), the Census of India, etc. To illustrate the flood hazards vulnerable region, a zonation map has been prepared and after that to classify the primary sampling units and households in various regions National Family Health Survey (NFHS) data has been used to know the impact of flood on the utilization of antenatal care services.

14.2.2 Study Sample

The present study used individual-level data from the IR file of NFHS-5. A total 724,115 numbers of ever-married women aged 15–49 were interviewed. In the five years prior to the survey, 232,920 of these mothers gave birth to at least one live child. However, regarding information about antenatal care services, mainly the number of ANC visits, 125,993 women's data were available. Therefore, 125,993 women (aged 15–49 years) have been selected for this study as a study sample.

14.2.3 Outcome Variables

In this study, antenatal care (ANC) services have been considered the outcome variable. Here highlighted that one component of ANC, i.e. four or more ANC visits, is the key outcome variable in this study. For the empirical analysis, variables have been divided into two binary variables, i.e. '1' indicating the ≥ 4 ANC visits and '0' indicating the < 4 ANC visits.

14.2.4 Explanatory Variables

In this study, the occurrence of floods has been taken as a key predictor variable. Occurrences of floods are divided into binary categories, i.e. '0' means no, and '1' means yes, for empirical analysis. Other covariates have been selected for this study based on the previous research. Other covariates are maternal age (15–24, 25–34, and 35–49 years), maternal education (no education, primary, secondary, and higher), birth order (first, second, third, and four or more), residence (rural and urban), social group (SC, ST, OBC, and others), religion (Hindu, Muslim, Christian, and others), family wealth index (poorest, poorer, middle, richer, and richest), mass media exposure (low, partial, and high), and geographical region (north, north-east, west, central, east, and south).

14.2.5 Statistical Analysis

The study investigates the association between floods and antenatal care services by using Pearson's chi-square statistics and multivariable binary logistic regression, and also descriptive statistics for showing the frequency and percentage distribution of study participants. Here we used sample weights to show the outcome variables and explanatory variables by percentage. The results of Pearson's chi-square statistic have been considered significant if the p-value is < 0.05. In the case of binary logistic regression, we estimate the crude and adjusted odds ratio with 95% CI (confidence interval), and the results are considered taken as significant if the p-value is < 0.05. All the statistics have been done using Stata version 12.8.

14.3 Results

14.3.1 Participant's Characteristics

Out of 125,993 participants, 62.72% of participants received 4 or more ANC visits. Among the total respondents, 60.68% live in flood-prone areas. In this study, 61.11% of participants are in the age group 25–34 years. High media exposure (radio, TV, and newspaper/magazine) respondents were 7.07%. About 15.13% of women had higher education. The majority of respondents reside in rural areas (78.18%), believe in the Hindu religion (74.70%), and fall into the OBC category (40.50%). Most of the women live in the poorest family (23.84%) and the central part of India (24.03%) (Table 14.1).

The analysis of flood-prone areas across Indian states reveals significant spatial disparities in vulnerability (Fig. 14.1), with certain regions facing substantially higher flood risk due to their physiographic and hydrological characteristics. Punjab exhibits the highest proportion of flood-prone land, with approximately 73.45% of its total geographical area susceptible to flooding. This is closely followed by Haryana (53.15%), Bihar (45.25%), and Assam (40.13%). These states lie predominantly in the Indo-Gangetic plains and are traversed by major rivers such as the Ganga, Yamuna, Brahmaputra, Sutlej, and Ghaghra. The flat terrain, coupled with monsoonal rainfall and poor drainage infrastructure, makes these areas particularly vulnerable to seasonal inundation. Bihar, for instance, is frequently affected by floods due to the north-flowing Himalayan rivers that swell during the monsoon. Other states with moderate flood vulnerability include West Bengal (29.86%), Tripura (31.49%), Kerala (22.40%), and the NCT Delhi (33.78%). These states experience flooding due to a combination of riverine overflow, tidal surges in coastal belts (particularly in Kerala and West Bengal), and urban flooding in densely populated and poorly drained urban centres like Delhi. In contrast, states such as Karnataka (0.10%), Madhya Pradesh (0.84%), Maharashtra (0.75%), and Meghalaya (0.90%) have relatively low percentages of flood-prone areas. These regions are typically located on elevated plateaus or in hilly terrains where natural drainage is more effective, thereby reducing the risk of widespread flooding. The spatial patterns suggest that northern and eastern India are considerably more flood-prone than the southern and central regions. This has important implications for disaster management and planning. States with higher flood exposure require targeted interventions, including the construction of embankments and levees, strengthening of early warning systems, sustainable land use planning, and community-based disaster preparedness. Moreover, as climate change is expected to exacerbate the frequency and intensity of extreme rainfall events, even states with historically low flood risk may need to reassess their vulnerability and enhance their adaptive capacity accordingly. Thus, understanding the percentage of flood-prone areas at the state level provides a critical foundation for formulating region-specific mitigation and resilience strategies.

Table 14.1 Descriptive statistics of the women aged 15–49 years included in the study, NFHS-5, 2019–2021

Variables	Frequency	Percentage
Independent variables		
Hazard		
Flood-prone areas	76,452	60.68
Non-flood-prone areas	49,541	39.32
Maternal age (years)		
15–24	34,950	27.74
25–34	76,996	61.11
35–49	14,047	11.15
Mass media exposure		
Low	32,590	25.87
Partial	84,495	67.06
High	8908	7.07
Maternal education		
No education	24,207	19.21
Primary	15,349	12.18
Secondary	67,376	53.48
Higher	19,061	15.13
Birth order		
1	41,986	33.32
2	45,973	36.49
3	21,334	16.93
4 or more	16,700	13.25
Residence		
Rural	98,497	78.18
Urban	27,496	21.82
Social group		
Scheduled caste	24,812	20.89
Scheduled tribe	24,506	20.63
Other backward class	48,108	40.50
Other	21,359	17.98
Religion		
Hindu	94,122	74.70
Muslim	17,570	13.95
Christian	9107	7.23
Others	5194	4.12
Wealth index		
Poorest	30,032	23.84
Poorer	28,872	22.92

(continued)

Table 14.1 (continued)

Variables	Frequency	Percentage
Middle	25,392	20.15
Richer	22,845	18.13
Richest	18,852	14.96
Region		
North	24,111	19.14
North-east	19,038	15.11
East	23,498	18.65
Central	30,274	24.03
West	11,949	9.48
South	17,123	13.59
Outcome variables		
ANC visits		
< 4	46,975	37.28
4 or more	79,018	62.72

Source NFHS-5 (2019–2021)

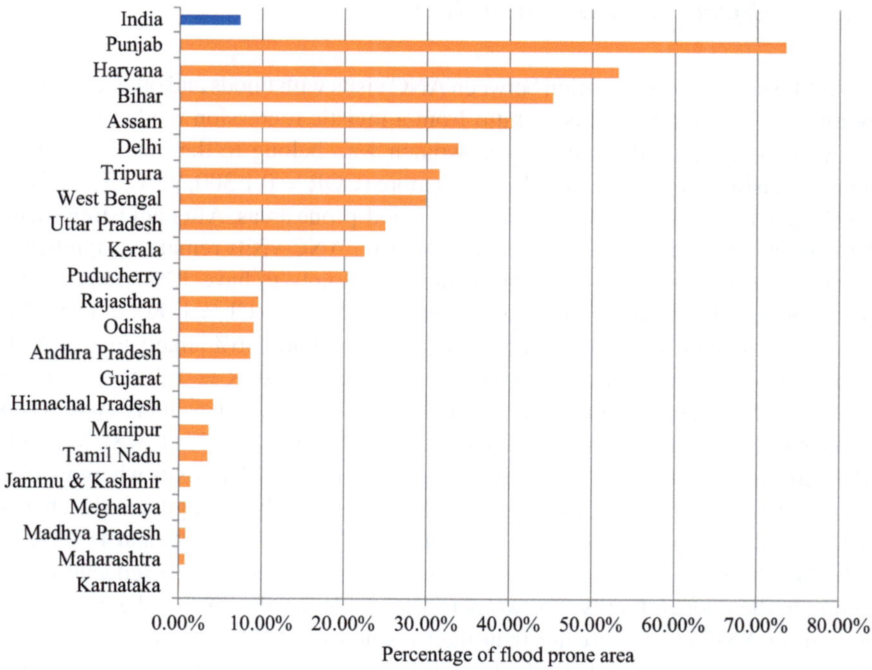

Fig. 14.1 State-wise percentage of flood-prone area in India. *Source* Vulnerability Atlas of India (3rd edition), 2019. *Note* These percentages are calculated based on available data on flood-prone areas and total geographical areas of the respective states

14.3.2 ANC Visits by Predictor Variables

Table 14.2 represents the percentage of women (15–49 years) with ANC visits by predictor variables and other covariates. Here is the significant difference in ANC visits by hazard and socio-demographic variables. Four or more ANC visits are higher (60.33%) among women who live in non-flood-prone areas compared to those who reside in flood-prone areas. ANC visits (≥4 times) are lower in the age group 15–24 years and low media exposure women (49.50%). Women who have no formal education among them ≥ 4 ANC visits have lower (49.75%) while higher educated women have higher (73.39%) ≥ 4 ANC visits. ANC visits (≥4 times) are reduced when birth order increases. It is found that ANC visits ≥ 4 times are lower (72.12%) among women who reside in rural areas and belong to the scheduled caste (59.75%). Christian women are visiting ≥ 4 ANC has lower (58.15%). The wealth index is positively associated with ANC visits, and women who belong to the poorest families have low ANC visits. Four or more ANC visits have a higher rate (80.91%) in the western part of India.

14.3.3 Effects of Floods on ANC Visits

Table 14.3 shows the association between ANC visits with floods and other covariates by unadjusted and adjusted odds ratio from a logistic regression model in India. It is found from the crude analysis that women who belong to the non-flood-prone areas are visiting ANC are likely 1.3 times more (COR = 0.1.300; 95% CI = 1.269–1.331) than those women who belong to the flood-prone areas. After adjusting socio-demographic variables, the net effect of floods on ANC visits remained significant. Women who reside in non-flood-prone areas are likely to have 12% greater ANC visits compared to their counterparts (AOR = 1.126; 95% CI = 1.095–1.159). The result reveals that women age group 35–49 years are more (36%) likely to visit ANC than women who fall into the 15–24 years age group (AOR = 1.365; 95% CI = 1.298–1.435). Among women who are highly exposure, 1.3 times more ANC visits compared to low exposure women (AOR = 1.301; 95% CI = 1.227–1.378). Maternal education and wealth index are positively associated with ANC visits where women who have higher education and belong to the richest family among them the chance of ANC visits are likely 33% (AOR = 1.334; 95% CI = 1.226–1.405) and 45% (AOR = 1.454; 95% CI = 1.373–1.539) more than those women who have no education and belong to the poorest family. For women who reside in urban areas, the probability of visiting ANC is 11% greater than their counterpart (AOR = 1.111; 95% CI = 1.072–1.151). It is found that scheduled tribe women visiting ANC is 26% higher than scheduled caste women (AOR = 1.263; 95% CI = 1.206–1.319). It was also observed that birth order is negatively associated with ANC visits. Four or more birth order women visiting health facilities for ANC is likely 40% lower than women who have first birth order (AOR = 0.607; 95% CI = 0.573–0.629). Muslim women had

Table 14.2 Association between floods and socio-demographic factors on ANC visits

Variables	ANC visits		p-value
	< 4	4 or more	
Hazard			
Flood-prone areas	39.67	60.33	< 0.001
Non-flood-prone areas	33.59	66.41	
Maternal age (years)			
15–24	38.39	61.61	< 0.001
25–34	36.52	63.48	
35–49	38.70	61.30	
Mass media exposure			
Low	50.50	49.50	< 0.001
Partial	32.98	67.02	
High	29.78	70.21	
Maternal education			
No education	50.25	49.75	< 0.001
Primary	42.61	57.39	
Secondary	34.43	65.57	
Higher	26.61	73.39	
Birth order			
1	32.00	68.00	< 0.001
2	34.88	65.12	
3	42.04	57.96	
4 or more	51.11	48.89	
Residence			
Rural	39.91	60.09	< 0.001
Urban	27.88	72.12	
Social group			
Scheduled caste	40.25	59.75	< 0.001
Scheduled tribe	39.49	60.51	
Other backward class	37.76	62.24	
Other	30.51	69.49	
Religion			
Hindu	36.91	63.09	< 0.001
Muslim	36.53	63.47	
Christian	41.85	58.15	
Others	38.54	61.46	
Wealth index			
Poorest	50.36	49.64	< 0.001
Poorer	41.46	58.54	

(continued)

Table 14.2 (continued)

Variables	ANC visits		*p*-value
	< 4	4 or more	
Middle	34.11	65.89	
Richer	28.59	71.41	
Richest	24.87	75.13	
Region			
North	31.27	68.73	< 0.001
North-east	43.73	56.27	
East	46.05	53.95	
Central	48.09	51.91	
West	19.09	80.91	
South	20.15	79.85	

P-value: Significance level of chi-square statistic
Source NFHS-5 (2019–21)

an 11% higher likelihood of ANC visits (AOR = 1.116; 95% CI = 1.069–1.164). The western part has 3.2 times more ANC visits than women who live in the northern part of India (AOR = 3.261; 95% CI = 3.089–3.442).

14.4 Discussion and Conclusion

India is prone to recurrent floods, especially in states like Bihar, Uttar Pradesh, Assam, and West Bengal. Floods can disrupt local healthcare infrastructure, causing significant challenges in accessing essential services, including maternal health care (Harville et al. 2010; Varughese and Purushothaman 2021; Iltaf 2023). **ANC** is critical for safeguarding the health of both mothers and newborns during pregnancy (Maheen and Hoban 2017; Sorensen et al. 2018; Lafarga Previdi et al. 2022). In rural and flood-prone areas, women may struggle to access these services due to devastated roads, healthcare centres being inundated, and a general lack of transportation during flood events (Mallett and Etzel 2018; Tariq and Shah 2020; Menon et al. 2022). The present study demonstrates the relationship between the incidence of floods and the usage of ANC (At least 4 ANC visits or more) among Indian women during pregnancy. Many previous studies showed the flood's effects on maternity and child health abroad (Begum and Hamid 2023; Afzal et al. 2024), but there is no such study in India that has shown the impact of floods on antenatal care services. The international studies showed there is a negative impact of flood on the access and utilization of treatment-seeking behaviour among women during the pregnancy period (Sharma and Sharma 2017; Orderud et al. 2022; Partash et al. 2022; Mroz et al. 2023), and similar findings were also found in the present study, where the access and the utilization of 4 ANC visits are higher among women belonging

Table 14.3 Unadjusted and adjusted odds ratio assess the relationship between flood events and socio-demographic factors on ANC visits

Variables	Model 1: Unadjusted		Model 2: Adjusted	
	Odds ratio	95% CI	Odds ratio	95% CI
Hazard				
Flood-prone areas ®				
Non-flood-prone areas	1.300***	1.269–1.331	1.126***	1.095–1.159
Maternal age (years)				
15–24 ®				
25–34	1.083***	1.055–1.11	1.181***	1.145–1.219
35–49	0.987	0.948–1.027	1.365***	1.298–1.435
Mass media exposure				
Low ®				
Partial	2.073***	2.020–2.128	1.341***	1.299–1.384
High	2.404***	2.286–2.528	1.301***	1.227–1.378
Maternal education				
No education ®				
Primary	1.360***	1.306–1.417	1.191***	1.139–1.245
Secondary	1.923***	1.867–1.981	1.271***	1.225–1.319
Higher	2.784***	2.673–2.900	1.334***	1.266–1.405
Birth order				
1 ®				
2	0.878***	0.854–0.903	0.834***	0.808–0.861
3	0.649***	0.627–0.671	0.709***	0.681–0.738
4 or more	0.450***	0.434–0.467	0.607***	0.573–0.629
Residence				
Rural ®				
Urban	1.718***	1.668–1.769	1.111***	1.072–1.151
Social group				
Scheduled caste ®				
Scheduled tribe	1.032*	0.995–1.070	1.263***	1.209–1.319
Other backward class	1.110***	1.076–1.146	0.965*	0.933–0.998
Other	1.534***	1.476–1.594	1.173***	1.124–1.224
Religion				
Hindu ®				
Muslim	1.016	0.983–1.051	1.116***	1.069–1.164
Christian	0.813***	0.778–0.849	0.726***	0.683–0.772
Others	0.933**	0.888–0.988	0.735***	0.689–0.783

(continued)

Table 14.3 (continued)

Variables	Model 1: Unadjusted		Model 2: Adjusted	
	Odds ratio	95% CI	Odds ratio	95% CI
Wealth index				
Poorest ®				
Poorer	1.433***	1.387–1.480	1.097***	1.057–1.138
Middle	1.959***	1.893–2.028	1.189***	1.140–1.239
Richer	2.534***	2.443–2.629	1.304***	1.244–1.367
Richest	3.065***	2.945–3.191	1.454***	1.373–1.539
Region				
Central ®				
North-east	1.191***	1.149–1.236	1261***	1.200–1.326
East	1.085***	1.049–1.123	1.134***	1.094–1.175
North	2.036***	1.966–2.109	1.539***	1.478–1.603
West	3.927***	3.732–4.132	3.261***	3.089–3.442
South	3.670***	3.514–3.834	2.970***	2.836–3.110

Note ® Reference category; *CI* Confidence Interval
* $p < 0.05$; ** $p < 0.01$; *** $p < 0.001$
Source NFHS-5 (2019–2021)

to the other areas than the flood-affected areas in India. The possible reason for that is the lack of transportation, lack of medical personnel, lack of communication with the doctors and the medical personnel, etc., during the flood situation. The present study also shows the relationship between the usage of ANC services and the different socio-demographic and economic characteristics that play a critical role in the utilization of ANC services during the flood situation. The impact of floods on ANC services could vary by socio-economic factors. Women from marginalized communities or lower-income groups who have low education and have not been exposed to mass media belonged to rural areas might face greater barriers in accessing healthcare services during floods. The present article is very relevant and climate change, in particular, makes natural calamities more frequent and severe. The analysis of flood-related disturbances to ANC services in India could contribute valuable knowledge to the global treatise on health systems flexibility to climate tremors. If successful, it could also inform national and international policies aimed at preserving maternal health in vulnerable regions. The findings could have important policy implications for improving maternal health care during natural disasters. The article might recommend increasing healthcare resilience in flood-prone areas, including improved infrastructure such as elevated healthcare centres and mobile healthcare units; community-based disaster preparedness to ensure pregnant women have access to emergency care and strengthening health policies to provide more equitable access to ANC services, especially during crises.

References

Afzal F, Das A, Chatterjee S (2024) Drawing the linkage between women's reproductive health, climate change, natural disaster, and climate-driven migration: focusing on low-and middle-income countries-a systematic overview. Indian J Commun Med 49(1):28–38

Ahern M, Kovats RS, Wilkinson P, Few R, Matthies F (2005) Global health impacts of floods: epidemiologic evidence. Epidemiol Rev 27(1):36–46

Alderman K, Turner LR, Tong S (2012) Floods and human health: a systematic review. Environ Int 47:37–47

Ali SA, McLean E (2015) The impact of natural disasters on maternal and child health: a systematic review of the literature. BMC Pregnancy Childbirth 15:206. https://doi.org/10.1186/s12884-015-0650-2

Antipova A, Curtis A (2015) The post-disaster negative health legacy: pregnancy outcomes in Louisiana after Hurricane Andrew. Disasters 39(4):665–686

Baloch S, Khaskheli MN, Sheeba A (2012) Screening of reproductive health problems in flood affected pregnant women. J Liaquat Univ Med Health Sci 11:101–104

Barman P, Sarif N, Saha A (2023) Association between natural hazards and postnatal care among the neonates in India: a step towards full coverage using geospatial approach. BMC Emerg Med 23(1):76

Begum A, Hamid SA (2023) Maternal healthcare utilization in rural Bangladesh: a comparative analysis between high and low disaster-prone areas. PLOS Global Public Health 3(7):e0001409

Biswas S, Mondal S, Banerjee A, Alam A, Satpati L (2024) Investigating the association between floods and low birth weight in India: using the geospatial approach. Sci Total Environ 912:169593

Bukhari SIA, Rizvi SH (2015) Impact of floods on women: with special reference to flooding experience of 2010 flood in Pakistan. J Geogr Nat Disast 5(2):1–5

Callaghan WM, Rasmussen SA, Jamieson DJ, Ventura SJ, Farr SL, Sutton PD et al (2007) Health concerns of women and infants in times of natural disasters: lessons learned from Hurricane Katrina. Matern Child Health J 11:307–311

Dancause KN, Laplante DP, Oremus C, Fraser S, Brunet A, King S (2011) Disaster-related prenatal maternal stress influences birth outcomes: project ice storm. Early Human Dev 87(12):813–820

Datar A, Liu J, Linnemayr S, Stecher C (2013) The impact of natural disasters on child health and investments in rural India. Soc Sci Med 76:83–91

Gurley ES, Rohan E (2019) Assessing the effects of flooding on maternal and child health in low-lying areas: a case study of Bangladesh. Int J Environ Res Public Health 16(12):2192. https://doi.org/10.3390/ijerph16122192

Harville E, Xiong X, Buekens P (2010) Disasters and perinatal health: a systematic review. Obstet Gynecol Surv 65(11):713–728

Iltaf K (2023) Impact of flood-triggering rainfall on adverse pregnancy outcomes in Pakistan: a quantitative analysis

LafargaPrevidi I, Welton M, Díaz Rivera J, Watkins DJ, Díaz Z, Torres HR et al (2022) The impact of natural disasters on maternal health: hurricanes Irma and María in Puerto Rico. Children 9(7):940

Maheen H, Hoban E (2017) Rural women's experience of living and giving birth in relief camps in Pakistan. PLoS Curr 9:ecurrents-dis

Mallett LH, Etzel RA (2018) Flooding: what is the impact on pregnancy and child health? Disasters 42(3):432–458

Menon AB, Rajeev K, Prasad G, Thomas GM (2022) Contextual analysis on antenatal and post-natal effects of 2018 flood in Ernakulam, Kerala. In: Advances in construction materials and sustainable environment: select proceedings of ICCME 2020. Springer Singapore, pp 635–645

Mohan P, Sharma K (2016) Access to antenatal care and the effect of floods in rural areas of Pakistan. Lancet Glob Health 4(2):e99–e100. https://doi.org/10.1016/S2214-109X(16)30021-0

Mroz EJ, Willis T, Thomas C, Janes C, Singini D, Njungu M, Smith M (2023) Impacts of seasonal flooding on geographical access to maternal healthcare in the Barotse Floodplain, Zambia. Int J Health Geogr 22(1):17

Orderud H, Härkönen J, Hårsaker CT, Bogren M (2022) Floods and maternal healthcare utilisation in Bangladesh. Popul Environ 44(3):193–225

Pappas A, Kovats S, Ranganathan M (2024) Extreme weather events and maternal health in low-income and middle-income countries: a scoping review. BMJ Open 14(6):e079361

Partash N, Naghipour B, Rahmani SH, Asl YP, Arjmand A, Ashegvatan A, Faridaalaee G (2022) The impact of flood on pregnancy outcomes: a review article. Taiwan J Obstet Gynecol 61(1):10–14

Patel RS, Tiwari R (2017) Impact of flood-induced displacement on maternal health services in flood-prone regions of South Asia. Glob Health Action 10:1308159. https://doi.org/10.1080/16549716.2017.1308159

Sahoo HR, Saha S (2018) Flood-induced disruption of healthcare services in India: case studies from rural areas. J Environ Stud 25(4):142–152. https://doi.org/10.1016/j.jes.2018.03.004

Sharma R, Sharma A (2017) Natural disasters and maternal healthcare services: a study of the Nepal earthquake. BMC Public Health 17:768. https://doi.org/10.1186/s12889-017-4857-0

Sorensen C, Saunik S, Sehgal M, Tewary A, Govindan M, Lemery J, Balbus J (2018) Climate change and women's health: Impacts and opportunities in India. GeoHealth 2(10):283–297

Tariq R, Shah S (2020) Effects of flood-induced stress on pregnant women: a study from flood-affected regions of Pakistan. J Matern Fetal Neonatal Med 33(6):1057–1063. https://doi.org/10.1080/14767058.2019.1679342

Thomas SM, Coker R (2012) Impact of environmental disasters on healthcare access and delivery: a global overview. Disaster Manag Response 10(2):34–42. https://doi.org/10.1016/j.dmr.2012.01.001

Varughese A, Purushothaman C (2021) Climate change and public health in India: the 2018 Kerala floods. World Med Health Policy 13(1):16–35

World Health Organization (WHO) (2019) Floods: health impact. https://www.who.int/news-room/fact-sheets/detail/floods

Zhang S, Li X (2021) Resilience of maternal healthcare systems during floods: a comparative analysis of affected regions in China. Int J Disaster Risk Reduc 60:102295. https://doi.org/10.1016/j.ijdrr.2021.102295

Chapter 15
Sustainable Strategies for Mitigating Indoor Air Pollution and Cardiovascular Health Risks Among Aging Populations: A Case Study of Bankura District

Ujjwal Das, Nishamani Kar, and Sailajanda Saika

Abstract Stroke is a major contributor to the global burden of disease, particularly in low- and middle-income countries, where air pollution is an emerging risk factor. This study explored the impact of indoor cooking fuel use on blood pressure and stroke among elderly individuals in the Bankura district of West Bengal. The present study was conducted among 332 participants aged 45 years and above. The analysis revealed a higher prevalence of stroke in households using solid fuel for cooking (17.2%) compared to those using clean fuel (12.5%). Individuals with systolic blood pressure (SBP) $\geq$ 140 mmHg or diastolic blood pressure (DBP) $\geq$ 90 mmHg had an increased stroke risk. Females experienced a higher prevalence of stroke than males (20.1 vs. 18.8%). Obese and underweight participants were at greater risk than those with a normal weight. Key risk factors for stroke included age 65–74 years (OR = 3.69, $p = 0.004$), high body mass index (OR = 3.05, $p = 0.005$), alcohol consumption (OR = 2.24, $p = 0.004$), and smoking (OR = 1.96, $p < 0.001$). The findings highlight high blood pressure as an independent risk factor for stroke, emphasizing the need for targeted awareness and intervention strategies among middle-aged and elderly populations to mitigate stroke prevalence and its associated complications.

Keywords Stroke · Body mass index · Obesity · Alcohol · Smoking

U. Das (✉)
Department of Geography, Fakir Mohan University, Balasore, Odisha, India
e-mail: ujjwaldas608@gmail.com

N. Kar · S. Saika
Department of Geography, Rajiv Gandhi University, Itanagar, Arunachal Pradesh, India
e-mail: nmkar7@yahoo.co.in

S. Saika
e-mail: sailajananda.saikia@rgu.ac.in

A. Roy et al. (eds.), *Environment and Public Health*, Sustainability Solutions,
https://doi.org/10.1007/978-3-031-99770-9_15

15.1 Introduction

Clean air is crucial for human health. The World Health Organization (WHO) reports that nearly 91% of people worldwide are exposed to polluted air. In low- and middle-income countries, indoor air pollution (IAP) caused by the use of solid fuels such as wood, dung, and crop residues for cooking poses a serious public health concern, as traditional cooking methods are still commonly used. India's $PM_{2.5}$ average in 2022 stood at 53.3 μg/m^3, slightly lower than 2021's 58.1 μg/m^3 (Bedi and Bhattacharya 2024). "Despite some improvements, 12 of the 15 most polluted cities in Central and South Asia remained in India during both 2021 and 2022" (Liu et al.2023). Alarmingly, nearly 60% of Indian cities surveyed had $PM_{2.5}$ levels exceeding the World Health Organization's (WHO) air quality guidelines by more than sevenfold. According to data from the Central Pollution Control Board (CPCB), numerous Indian cities consistently exceed the permissible limits for air quality. "Concentrations of particulate matter ($PM_{2.5}$ and PM_{10}) and nitrogen dioxide (NO_2) frequently breach safety thresholds. For example, annual average $PM_{2.5}$ levels in metropolitan cities such as Delhi, Mumbai, and Kolkata significantly surpass the WHO's recommended levels, posing a serious health threat to their residents" (Pandey et al. 2021). The combustion of these fuels emits hazardous pollutants like fine particulate matter ($PM_{2.5}$) and carbon monoxide, which have been linked to negative impacts on cardiovascular health. Exposure to $PM_{2.5}$ in outdoor air has been associated with deaths from ischemic heart disease (IHD), stroke, and respiratory diseases (Kim et al. 2016). Among the elderly, who are particularly vulnerable due to age-related declines in health and pre-existing conditions, prolonged exposure to IAP can contribute to elevated blood pressure and an increased risk of stroke. These conditions, already significant contributors to morbidity and mortality in aging populations, are further exacerbated by environmental factors like household air pollution. Air pollution-related strokes are most prevalent in low- and middle-income countries, largely because of their dependence on solid fuels for cooking. The International Energy Agency's report stated that household air pollution leads to approximately 2.6 million premature deaths globally each year, primarily due to the traditional use of biomass for cooking in developing countries. This number exceeds the deaths caused by many other diseases (International Energy Agency 2018). Globally, around 2.5 billion people are exposed to indoor air pollution from cooking (Bonjour et al. 2013). According to Cohen et al. (2017), indoor air pollution resulting from the use of solid fuels for cooking led to approximately 2.8 million deaths. "The Global Burden of Disease (2017) reported 11.9 million new stroke cases, 104.2 million people living with stroke, and 6.2 million stroke-related deaths worldwide, contributing to over 116 million disability-adjusted life years (DALYs)" (Cohen et al. 2017). Of these, around 16.9% were attributed to air pollution (Krishnamurthi et al. 2020; Feigin et al. 2016). Solid fuels such as coal, wood, and agricultural residues emit large quantities of harmful pollutants, including fine particulate matter ($PM_{2.5}$ and PM_{10}), carbon monoxide, nitrogen dioxide, sulfur dioxide, and volatile organic compounds. These pollutants are linked to increased blood pressure, hypertension, and stroke"

(Everson et al. 2019; Liu et al. 2017; Wang et al. 2017; Yan et al. 2020; Yang et al. 2019). In addition, exposure to biomass fuel is associated with other adverse health outcomes such as low birth weight, cataracts, asthma, lower respiratory tract infections, and chronic obstructive pulmonary disease (Haider et al. 2016; Ravilla et al. 2016; Thacher et al. 2013; Agrawal and Yamamoto 2015; Ndlovu and Nkeh-Chungag 2024).

High systolic and diastolic blood pressure (>140 mmHg and > 90 mmHg, respectively) is closely linked to cardiovascular disease risk, particularly in older individuals (Yadav et al. 2021). Epidemiological studies increasingly highlight the association between air pollution and stroke, as measured by hospital admissions and mortality rates. Arku et al. (2020) observed an inverse relationship between the use of solid fuels and blood pressure. In contrast, a study conducted in Bangladesh revealed that women relying on solid fuels were more likely to report elevated blood pressure levels (Khan et al. 2021). However, a cross-sectional study in Pakistan involving 850 women under the age of 40 found no significant link between indoor air pollution and hypertension (Fatmi et al. 2019).

In India, hypertension affects 33.8% of the population and is responsible for 57% of all stroke-related deaths and 24% of deaths due to coronary heart disease (Gupta et al. 2004). In aging populations, elevated systolic blood pressure (SBP) accounts for most cases of uncontrolled hypertension (Kumar et al. 2023). The increased prevalence of elevated SBP with age is significantly correlated with higher stroke mortality rates (Xu et al. 2018). Besides high blood pressure, several other risk factors—such as physical inactivity, low socioeconomic status, elevated body mass index, smoking, and alcohol use—are strongly linked to an increased risk of stroke (Kumar et al. 2015; Pandian et al. 2011). Previous research indicates that individuals from lower socioeconomic backgrounds who rely on solid cooking fuels face a higher risk of stroke. Notably, 89% of strokes occur after age 65 in low-income populations (Avendano and Glymour 2008). Bankura, one of the most underdeveloped districts in West Bengal, has seen its elderly population grow from 8.5% in 2001 to 9.7% in 2011. Over 70% of the population resides in rural areas and uses solid fuels for cooking.

Considering these contributing factors, it is essential to examine the association between the use of solid fuels indoors and blood pressure among the elderly. This study targets the elderly population in Bankura district, West Bengal—a largely rural area where traditional cooking with solid fuels remains widespread. By investigating the connections between indoor air pollution, blood pressure, and stroke, the research seeks to shed light on the health consequences of conventional cooking practices and emphasize the need for focused interventions to lessen the burden of non-communicable diseases in similar regions. Understanding the relationship between solid fuel use and stroke further highlights the importance of managing stroke risk among individuals with hypertension, especially in rural and socioeconomically disadvantaged communities.

15.2 Materials and Methods

15.2.1 Study Population

The present study was conducted in the rural Bankura district of West Bengal from May 2023 to September 2023. The study population comprised all elderly individuals aged 45 years and above who had resided in the study area for at least one year. A sample size of 480 was calculated for the district, with a 95% confidence interval (CI) and a 5% margin of error. The total sample size for this study was determined using the following formula.

$$\text{Sample size: } n = \frac{z^2pq}{d^2} * D_{\text{eff}}.$$

where n = Sample size required.

p = Prevalence of the chronic disease among the elderly 50% (0.50). Therefore, $q = 1 - p$ is 0.50.

z = z value (1.96 at 95% level of confidence).

d = non-response rate of 5% (0.05).

Design effect = 1.25

$$n = \frac{1.96^2 * (0.50)(0.50)}{(0.05)^2} * 1.25$$

Hence, $n = 480.2$, and it was around 480.

15.2.2 Sampling Method

A multi-stage random sampling strategy was employed to select the required number of elderly participants. In the first stage, blocks were randomly selected, followed by the selection of villages in the second stage, and finally, individuals from the target population in the third stage. A total of five blocks were chosen at random, and from each block, three villages were selected, yielding a total of 15 villages to reach the desired sample size.

15.2.3 Ethical Approval

Ethical clearance for the study was granted by the Institutional Ethical Committee of Rajiv Gandhi University. Prior to data collection, informed consent was obtained from all participants, and the study's objectives were clearly communicated to them. Data were gathered through door-to-door interviews using pre-tested, pre-designed, and semi-structured questionnaires. Information on stroke was based on self-reports provided by the participants, while hypertension data were collected through different means.

15.2.4 Diagnostic Criteria

Hypertension was defined as having a systolic blood pressure (SBP) of 140 mmHg or higher, a diastolic blood pressure (DBP) of 90 mmHg or higher, or current use of antihypertensive medication. Out of 480 participants, 332 were classified as hypertensive. To investigate possible physiological mechanisms influencing blood pressure, pulse pressure (PP) was calculated by subtracting DBP from SBP, and mean arterial pressure (MAP) was computed using the formula MAP = PP/3 + DBP (Sesso et al. 2000). BP, DBP, PP, and MAP were analyzed as continuous variables, while stroke prevalence was recorded as a binary variable.

Stroke prevalence was determined based on participants' self-reports. To identify whether a respondent had experienced a stroke, they were asked the question, "Have you ever been diagnosed with a stroke?".

15.2.5 Indoor Cooking Fuel Use Assessment

The present study determined the type of solid fuel used for cooking by asking the question, "What is the main source of cooking fuel in your family?" The cooking fuels were then categorized into two groups: (1) indoor solid fuel use, including coal, charcoal, and wood; and (2) clean fuel use, including solar energy, natural gas, induction cookers, and other electrical appliances.

15.2.6 Covariates

Demographic factors, such as sex, age, marital status, caste, and education level, as well as lifestyle factors like smoking, alcohol consumption, and physical activity, were considered as covariates. Body mass index (BMI) was calculated by dividing weight (in kilograms) by the square of height (in meters) and classified based on

Asian standards: underweight (BMI < 18.5 kg/m^2), normal weight (18.5–22.9 kg/m^2), and overweight or obese (BMI $\geq$ 23 kg/m^2).

Socioeconomic status was evaluated using a wealth index, derived from the ownership of durable goods within the household (e.g., furniture, vehicles, appliances, electronics, and access to electricity). These items were recorded as binary variables, with a value of 0 for non-ownership and 1 for ownership. Principal component analysis was used to rank the assets from lowest to highest, and the households were grouped into five equal wealth categories: poorest (quintile 1), poorer (quintile 2), middle (quintile 3), richer (quintile 4), and richest (quintile 5).

15.2.7 *Statistical Analysis*

Descriptive statistics were used to summarize the characteristics of the study population according to their types of cooking fuel, and the corresponding *p*-values were used to determine whether there were statistically significant differences among the socioeconomic and demographic variables. Pearson's chi-squared test was employed for dichotomous variables, while the *t*-test was used for continuous variables. Multivariate regression models were employed to identify the risk factors associated with stroke due to the use of cooking fuel, with 95% confidence intervals.

The first model was a crude, unadjusted model. The base model (Model 1) included adjustments for age, gender, and marital status, and additional stroke risk factors were gradually added in subsequent models. Model 2 adjusted for other risk factors, including smoking, alcohol consumption, physical activity, BMI, and dietary habits. The fully adjusted model (Model 3) included the covariates from Model 2 as well as socioeconomic factors such as education and wealth status.

15.3 Results

Table 15.1 presents the sample of the study population according to their cooking fuel used, with background characteristics. The mean age of the study population was 60.51 years. Those households that use solid fuel had a higher prevalence of stroke as compared to those that use clean fuel (17.2% vs.12.5%). In addition, those who use solid fuel for cooking were at a higher prevalence of elevated blood pressure. The participants from Sarenga and Khatra block (29.84% and 21.99%) had a higher percentage of solid fuel use for cooking, while Simlapal, Taldangra and Raipur (23.88%, 23.53%, and 20.42%) had higher percentage of use of clean fuel. The participants who use clean fuel had higher socioeconomic background, both in terms of the highest year of education and wealth status, than those using indoor solid fuel, respectively.

Figure 15.1 illustrates the use of solid fuels for cooking purposes across the blocks. Nearly one-third (39.79%) of respondents reported that solid fuel is used in

Table 15.1 Individual study population used of solid and clean fuel for cooking with selected variables

Background characteristics	N	Cooking fuel		χ^2 (df)	p
		Solid fuel	Clean fuel		
Individuals	480	191 (39.79)	289 (60.21)		
SBP (mmHg) (mean, SD)	135.08 (12.32)	132.75 (13.79)	136.61 (11.00)	56.34 (18)	*< 0.001*
DBP (mmHg) (mean, SD)	89.45 (11.92)	89.92 (12.93)	89.14 (11.23)	41.25 (18)	*< 0.001*
PP (mmHg), (mean, SD)	45.63 (9.85)	42.83 (10.64)	47.47 (8.77)	90.0 (23)	*< 0.001*
MAP (mmHg) (mean, SD)	104.66 (11.13)	104.2 (12.24)	104.96 (10.36)	81.68 (43)	*< 0.001*
Hypertension (*n*, %)				0.63 (1)	*0.008*
No	143 (29.79)	53 (27.75)	90 (31.14)		
Yes	337 (70.21)	138 (72.25)	199 (68.86)		
Stroke (*n*, %)				1.94 (1)	*0.164*
No	409 (85.74)	156 (82.98)	253 (87.54)		
Yes	68 (14.26)	32 (17.02)	36 (12.46)		
Age (years, mean, SD)	60.51 (9.63)	59.56 (10.38)	61.14 (9.06)		
Sex (*n*, %)				3.69 (1)	*0.055*
Male	237 (49.38)	84 (43.98)	153 (52.94)		
Female	243 (50.63)	107 (56.02)	136 (47.06)		
Marital status (*n*, %)				9.95 (3)	*0.019*
Currently married	354 (73.75)	128 (67.02)	226 (78.20)		
Widowed/widower	113 (23.54)	59 (30.89)	54 (18.69)		
Divorced/separated	13 (2.71)	4 (2.10)	9 (3.12)		
Highest education (*n*, %)				66.89 (3)	*< 0.001*
No education	110 (22.92)	73 (38.22)	37 (12.80)		
Primary	106 (22.08)	48 (25.13)	58 (20.07)		
Secondary	178 (37.08)	61 (31.94)	117 (40.48)		
Higher secondary	86 (17.92)	9 (4.71)	77 (26.64)		
Wealth status (*n*, %)				192.94 (4)	*< 0.001*
Poorest	54 (11.25)	53 (27.75)	1 (0.35)		
Poorer	84 (17.50)	54 (28.27)	30 (10.38)		
Middle	91 (18.96)	50 (26.18)	41 (14.19)		
Richer	72 (15.00)	23 (12.04)	49 (16.96)		
Richest	179 (37.29)	11 (5.76)	168 (58.13)		

(continued)

Table 15.1 (continued)

Background characteristics	*N*	Cooking fuel		χ^2 (df)	*p*
		Solid fuel	Clean fuel		
Smoking status (*n*, %)				4.13 (1)	*0.042*
No	321 (66.88)	138 (72.25)	183 (63.32)		
Yes	159 (33.13)	53 (27.75)	106 (36.68)		
Alcohol use (*n*, %)				9.05 (1)	*0.003*
No	274 (57.08)	125 (65.45)	149 (51.56)		
Yes	206 (42.92)	66 (34.55)	140 (48.44)		
Tobacco use (*n*, %)				35.70 (1)	*< 0.001*
No	221 (46.04)	56 (29.32)	165 (57.09)		
Yes	259 (53.96)	135 (70.68)	124 (42.91)		
Physical work (*n*, %)				0.71 (1)	*0.399*
No	293 (61.04)	121 (63.35)	172 (59.52)		
Yes	187 (38.96)	70 (36.65)	117 (40.48)		
BMI (*n*, %)				8.66 (3)	*0.034*
< 24 (kg/m^2)	30 (6.25)	17 (8.90)	13 (4.50)		
> 24 (kg/m^2)	450 (93.75)	174 (91.1)	278 (95.50)		
Dietary diversity (*n*, %)				0.76 (1)	*0.383*
Vegetarian	91 (19.83)	34 (17.89)	57 (21.19)		
Non-vegetarian	368 (80.17)	156 (82.11)	212 (78.81)		
Indoor ventilation (*n*, %)				8.86 (3)	*< 0.001*
None	115 (23.96)	50 (26.17)	65 (22.49)		
Natural	250 (52.08)	120 (62.82)	130 (44.98)		
Mechanical	115 (23.96)	21 (10.99)	94 (32.52)		
Region/Block (*n*, %)				26.03 (4)	*< 0.001*
Simlapal	96 (20.00)	27 (14.14)	69 (23.88)		
Taldangra	96 (20.00)	28 (14.56)	68 (23.53)		
Sarengra	96 (20.00)	57 (29.84)	39 (13.49)		
Khatra	96 (20.00)	42 (21.99)	54 (18.69)		
Raipur	96 (20.00)	37 (19.37)	59 (20.42)		

χ^2 *(df)* chi-square, (degree of freedom), *SBP* Systolic blood pressure, *DBP* Diastolic blood pressure, *PP* Pulse pressure, *MAP* Mean arterial pressure
Source Author's own calculations
The significance of italicized values is the P value with 95% CI

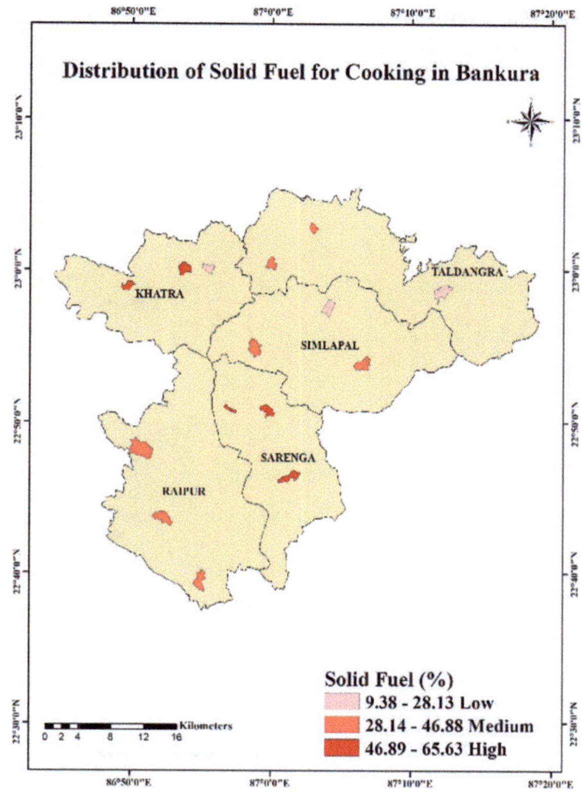

Fig. 15.1 Distribution of solid fuel

their village for cooking. The highest percentage of solid fuel usage was observed in Khatra, followed by Sarenga and Raipur blocks. The lowest percentage of solid fuel usage was recorded in Jaganathpur village in the Simlapal block.

Figure 15.2 shows that the prevalence of stroke was significantly associated with solid fuel use, with a higher prevalence observed in Raipur block. Bankata in Simlapal and Sarenga also reported a high percentage of solid fuel usage, along with a higher prevalence of stroke. In contrast, Dedua and Jibanpur in Khatra block, as well as Amjhor in Sarenga block, reported a lower prevalence of stroke.

Table 15.2 shows that the prevalence of stroke is associated with adjusted blood pressure. It was found that the prevalence of stroke was higher among individuals with systolic blood pressure (SBP) values of $\geq$ 160 mmHg compared to those with SBP values of < 120 mmHg (25.0% vs. 13.3%). Similarly, stroke prevalence significantly increased among participants with diastolic blood pressure (DBP) values of $\geq$ 110 mmHg compared to those with DBP values < 80 mmHg. The same trends were observed for both male and female participants, with stroke prevalence continuously increasing with higher blood pressure values. The prevalence of stroke was higher among female participants compared to male participants.

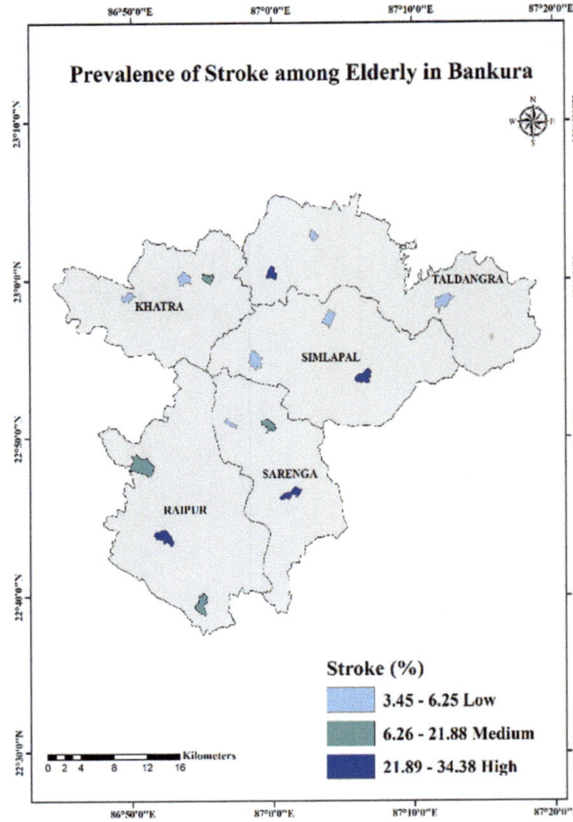

Fig. 15.2 Prevalence of stroke among elderly

Table 15.3 explores the association between cooking fuel type—solid versus clean—and various blood pressure (BP) parameters, as well as the odds of stroke among elderly individuals. In the **crude analysis**, using solid fuel for cooking was associated with increased BP levels and a higher stroke risk. Model 1 also showed that people who cooked primarily with indoor solid fuel had a higher risk of stroke (OR = 1.15; 95% CI: 0.72–2.16). Model 2, which further adjusted for smoking status, alcohol use, tobacco use, physical activity, BMI, and dietary patterns, demonstrated an increase in the association. In Model 3, which included all the previous adjustments plus educational attainment and income, SBP and DBP associations were slightly reduced. SBP was $\beta = 1.79$ (95% CI: 0.96–2.93, $p < 0.001$), and DBP was $\beta = 1.01$ (95% CI: 0.41–2.15, $p < 0.05$), indicating some attenuation of effects. Stroke risk was not significant in this final model (OR = 0.98, 95% CI: 0.71–1.16), suggesting that socioeconomic factors and education may partially mitigate the impact of solid fuel use on stroke risk.

Table 15.4 presents univariable and multivariable analyses assessing factors associated with stroke risk among elderly individuals exposed to solid fuel for cooking. Age was a significant predictor of stroke, with individuals aged 65–74 and 75 +

Table 15.2 Prevalence of stroke with adjusted blood pressure (BP) level

BP level	Number of participants	Number of cases	Prevalence (%)	χ^2	*p* value
Persons					
SBP level					
Normal < 120	15	2	13.33		
Elevated 130 ~	44	4	9.09	12.593	*0.006*
Hypertension Stage_1 140 ~	73	8	10.95		
Hypertension Stage_2 ≥ 160	200	50	25.0		
DBP level					
Normal < 80	3	1	33.33		
Elevated 90 ~	36	2	5.56	10.638	*0.014*
Hypertension Stage_1 100 ~	171	28	16.37		
Hypertension Stage_2 ≥ 110	120	33	27.5		
Male					
SBP level					
Normal < 120	8	3	12.5		
Elevated 130 ~	27	15	0.37	6.262	*0.1*
Hypertension Stage_1 140 ~	33	10	9.09		
Hypertension Stage_2 ≥ 160	95	22	23.15		
DBP level					
Normal < 80	1	1	100		
Elevated 90 ~	13	5	7.69	9.414	*0.024*
Hypertension Stage_1 100 ~	81	13	16.05		
Hypertension Stage _2 ≥ 110	59	15	25.42		
Female					
SBP level					
Normal < 120	7	1	14.28		
Elevated 130 ~	17	3	17.64	6.305	*0.098*
Hypertension Stage_1 140 ~	40	5	12.5		
Hypertension Stage_2 ≥ 160	105	28	26.67		

(continued)

Table 15.2 (continued)

BP level	Number of participants	Number of cases	Prevalence (%)	χ^2	*p* value
DBP level					
Normal < 80	2	0	0		
Elevated 90 ~	23	10	4.34	6.235	*0.101*
Hypertension Stage_1 100 ~	90	15	16.67		
Hypertension Stage_2 ≥ 110	61	18	29.51		

χ^2 *(df)* chi-square
Source Author's own calculated
The significance of italicized values is the P value with 95% CI

Table 15.3 Association between use of solid versus clean fuel for cooking, and BP parameters and odds of having a stroke

	SBP *β* (95% CI)	DBP *β* (95% CI)	PP *β* (95% CI)	MAP *β* (95% CI)	Stroke OR (95% CI)
Crude	2.56*** (1.42–3.56)	1.67*** (1.09–2.24)	1.12*** (0.15–2.26)	1.97*** (1.32–2.58)	1.15*** (0.72–2.16)
Model 1	1.75*** (0.75–2.45)	1.42*** (0.85–1.98)	0.56 (0.10–1.25)	1.53*** (0.82–2.16)	1.19*** (0.89–2.52)
Model 2	2.84*** (1.87–3.86)	1.65** (1.09–2.98)	1.06** (0.23–2.01)	2.01*** (1.42–2.65)	1.09** (0.86–2.97)
Model 3	1.79*** (0.96–2.93)	1.01* (0.41–2.15)	0.89 (0.06–1.83)	1.26*** (0.62–1.97)	0.98 (0.71–1.16)

Model 1: Age, Gender, and Marital status; Model 2: Model 1 + smoking Status, Alcohol use, Tobacco use, Physical work, BMI, and Dietary Pattern; Model 3: Model 2 + highest education and wealth status
OR odds ratio
Significance level: ***$p < 0.001$, **$p < 0.01$, *$p < 0.05$

showing increased odds of stroke compared to the reference group aged 45–54. The crude odds ratio (OR) for ages 65–74 was 3.62 (95% CI: 1.54–8.99), and for 75 + was 4.18 (95% CI: 3.57–7.57). Body mass index (BMI) over 24 kg/m^2 also correlated with higher stroke risk, with a crude OR of 3.11 times (95% CI: 2.20–4.14), and reduced slightly to 2.56 times (95% CI: 2.29–3.02) after adjustment of all the variables. Considering education, primary education had a higher odds (OR 2.06, 95% CI: 0.94–4.45) than those with no education. Marital status was also relevant, as widowed and divorced individuals showed higher risks of stroke compared to married counterparts, though these associations diminished in adjusted analyses. Wealth status indicated that those in poorer socioeconomic categories had a higher risk of stroke than the richest group (OR 1.42, 95% CI: 1.01–1.93). Engagement in physical activity was associated with reduced stroke risk; those who were physically active showed significantly lower risks of stroke (OR 0.19, 95% CI: 0.06–0.58). Behavioral factors such as smoking and alcohol use showed increased risks of stroke, with smokers having an adjusted OR of 1.35 (95% CI: 1.18–1.55) and alcohol users an adjusted OR of 3.10 (95% CI: 1.15–8.45). The place of residence (blocks) showed some association with stroke risk. Compared to Simlapal, blocks in Taldangra and Khatra had a lower risk of stroke due to solid fuel used (Taldangra OR 0.74, 95% CI: 0.2–0.89 and Khatra OR 0.51, 95% CI: 0.16–0.76) than the other blocks.

15.4 Discussion

The present study examined the impact of biomass cooking fuel on high blood pressure and stroke prevalence. It found that elderly people who used solid fuel for cooking had a significantly increased risk of stroke and high blood pressure (SBP, DBP, and MAP). Among participants who transitioned from clean to solid fuel, the risk of hypertension, including high blood pressure, was significantly higher than that of persistent clean fuel users. Solid fuel use was notably higher in Sarenga and Khatra blocks, reflecting the persistence of traditional cooking practices in these regions. In contrast, blocks such as Simlapal, Taldangra, and Raipur exhibited a higher adoption of clean fuel, suggesting progress toward cleaner energy use. These patterns were influenced by socioeconomic disparities, with households using clean fuel more likely to belong to higher wealth and education categories. These findings reinforce the role of socioeconomic status as a determinant of exposure to indoor air pollution and its associated health risks. The study confirms a higher prevalence of stroke and elevated blood pressure among individuals using solid fuel compared to those using clean fuel. The association was particularly pronounced in Raipur block and specific villages such as Bankata in Simlapal and Sarenga, which also reported high solid fuel usage. These geographic variations suggest localized impacts of indoor air pollution, influenced by cooking practices and other community-level factors.

The multivariable models demonstrated that solid fuel use was associated with increased systolic (SBP) and diastolic blood pressure (DBP), although the association with stroke risk became non-significant after adjusting for socioeconomic and lifestyle factors. These findings indicate that while solid fuel use contributes to elevated blood pressure, its impact on stroke risk is partly mediated by education, income, and behavioral factors such as smoking and alcohol use. This attenuation highlights the complex interplay of environmental, behavioral, and socioeconomic determinants of health in this population. Additionally, a higher risk was observed among participants aged 75 years and above compared to those aged 45–55 years. The results suggest that age is a significant factor mediating the association between solid fuel use and blood pressure and stroke among elderly residents in Bankura. These findings are consistent with an earlier study conducted in Peru, which included 1,004 elderly individuals and reported that solid cooking fuel was significantly associated with an increased risk of hypertension and stroke (adjusted OR: 3.5 and 4.1,

Table 15.4 Univariate and multivariable analyses for factors associated with stroke by use of solid fuel for cooking among elderly households

Background characteristics	Crude OR	95% CI	*p* value	Adjusted OR	95% CI	*p* value
Age						
45–54 ®						
55–64	0.23	0.08–0.73	*0.009*	0.16	0.04–0.61	*0.01*
65–74	3.62	1.54–8.99	*0.004*	2.76	0.93 -7.79	*0.07*
75 and above	4.18	3.57–7.57	*0.01*	3.53	2.55–5.89	*0.05*
Sex						
Male ®						
Female	1.09	0.63–1.88	*0.77*	1.89	0.79–4.56	*0.15*
BMI						
< 24 (kg/m^2) ®						
> 24(kg/m^2)	3.11	2.20–4.14	*0.005*	2.56	2.29–3.02	*0.1*
Education level						
No education ®						
Primary	2.06	0.94–4.45	*0.074*	1.07	0.33–3.27	*0.93*
Secondary	0.72	0.35–1.61	*0.446*	0.74	0.19–2.74	*0.65*
Higher secondary	1.14	0.48–2.66	*0.788*	1.18	0.49–2.62	*0.39*
Marital status						
Currently married ®						
Widowed/ widower	3.36	2.17–9.68	< *0.001*	1.22	0.49–3.03	*0.67*
Divorced/ separated	5.57	1.03–8.90	*0.047*	2.14	0.25–3.41	*0.21*
Wealth status						
Poorest ®						
Poorer	1.50	1.05–1.85	< *0.001*	1.42	1.01–1.93	< *0.001*
Middle	1.25	0.81–2.10	*0.02*	1.12	0.54–1.96	< *0.001*
Richer	0.85	0.15–1.40	*0.24*	0.78	0.21–1.20	*0.21*
Richest	0.56	0.16–1.87	*0.005*	0.31	0.15–0.98	*0.002*
Physical activity						
No ®						
Yes	0.16	0.07–0.42	< *0.001*	0.19	0.06–0.58	< *0.001*
Smoking status						
No ®						
Yes	1.96	1.80–2.13	< *0.001*	1.35	1.18–1.55	< *0.001*

(continued)

Table 15.4 (continued)

Background characteristics	Crude OR	95% CI	*p* value	Adjusted OR	95% CI	*p* value
Alcohol use						
No ®						
Yes	2.23	1.28–3.82	*0.005*	3.10	1.15–8.45	*0.03*
Region/block						
Simlapal ®						
Taldangra	0.87	0.37–1.02	*0.1*	0.74	0.26–0.89	< *0.001*
Sarengra	1.13	0.50–2.54	*0.750*	1.12	0.41–3.08	*0.822*
Khatra	0.48	0.18–1.27	*0.141*	0.51	0.16–0.76	< *0.001*
Raipur	1.72	0.80–3.68	*0.161*	1.46	0.57–3.73	*0.425*

OR odds ratio, ® Reference category
The level of statistical significance is set at $p < 0.05$
The significance of italicized values is the P value with 95% CI

respectively) (Burroughs Peña et al. 2015). Older people often have additional cardiovascular risk factors that may amplify the risk of air pollution-related strokes. On the contrary, middle-aged adults (45 years) may also be susceptible to air pollution. It is noteworthy that older individuals (above 60 years) spend more time indoors, which might lead to an overestimation of the effect of ambient air pollution (Yitshak Sade et al. 2015). However, another study in southern Nigeria reported no association between solid fuel use and the prevalence of stroke and high blood pressure. These inconsistencies may be attributed to differences in the types of solid fuels used, study locations, and study populations. The emission of pollutants and fine particulate matter (PM) varies with biomass fuel burning. In our study, biomass fuel mainly referred to charcoal, coal, firewood, straw, and wood, which differs from the sources examined in other studies. Approximately 75% of the population used solid biomass fuel for cooking purposes, with an almost equal proportion of men and women in the study population. Therefore, differences in livelihood status, dietary patterns, and activity levels related to cooking may influence the impact of solid fuel use on high blood pressure and stroke.

Various confounding factors significantly contributed to the elevation of blood pressure and the increased risk of stroke. The study found that women were at a higher risk of stroke due to solid fuel use compared to men. This can be explained by the fact that women are primarily responsible for cooking, meaning elderly women are exposed to air pollution for longer durations than men, resulting in higher exposure levels (Austin and Mejia 2017; Ofori et al. 2018). Another study noted that solid fuel use for heating had a more significant effect on systolic blood pressure in women than in men, while the impact on diastolic blood pressure was more pronounced in men than in women (Lin et al. 2021).

Women with a high body mass index (BMI > 24 kg/m^2) were at a greater risk of stroke compared to those with a lower BMI. Previous studies have indicated "that a high BMI exacerbates the impact of air pollution on blood pressure and stroke" (Dong et al. 2013; Yang et al. 2018). Physical activity was found to be protective against stroke, emphasizing its importance as a modifiable risk factor. These results suggest that promoting physical activity could mitigate some of the adverse effects associated with solid fuel use. Additionally, the findings on marital status and education highlight the social dimensions of health, where widowed or less-educated individuals may face greater health vulnerabilities due to limited access to resources and support systems. Again, the study found that elderly individuals from poorer socioeconomic backgrounds who used solid fuel had a higher prevalence of stroke compared to those from wealthier families. The study also observed an association between smoking status and the risk of stroke. Furthermore, the estimated association between indoor solid fuel use for heating and elevated systolic blood pressure (SBP) and stroke was relatively stronger among residents of Raipur and Sarenga compared to those living in other areas. A previous study mentioned that "indoor air pollution adversely impacts the respiratory and cardiovascular systems through mechanisms including systemic inflammation, oxidative stress, and altered cardiac autonomic and vascular function" (Mirowsky et al. 2016).

Although this study did not examine long-term exposure to $PM_{2.5}$, NO_2, and O_3, which are independently associated with stroke, previous research has established that hypertension and diabetes are significant risk factors for stroke and are also linked to air pollution (O'Donnell et al. 2010). Another study found that the risk of stroke increases with exposure to air pollution, especially in individuals with one or more cardiovascular risk factors, such as hypertension and diabetes (Ho et al. 2018). A Canadian study involving over 9000 patients reported a significantly higher stroke risk in individuals with a previous stroke compared to those without. The odds ratios were 2.31 (95% CI: 1.39–3.83) for individuals with a prior stroke and 1.25 (95% CI: 0.87–1.80) for those without, per 5.8 $\mu g/m^3$ (IQR) average increase in NO_2 levels during the three days prior to the stroke (Villeneuve et al. 2012).

In addition, our findings indicate that household use of solid fuel slightly increased SBP, DBP, PP, and MAP. A high level of PP suggests poor blood vessel elasticity (Deng et al. 2020). Some studies have shown that increases in SBP, DBP, and MAP are positively associated with cardiovascular risk factors (Arku et al. 2020). Previous research suggests that "the underlying mechanisms of a blood pressure (BP) increase with solid fuel use include increased inflammation and oxidative stress, which subsequently lead to elevated BP" (Dutta et al. 2012; Dinh et al. 2014). "Additionally, solid fuel in the combustion chamber produces toxic volatile organic compounds (VOCs), which can easily vaporize and participate in metabolic processes that elevate BP" (Sun et al. 2020). "Several studies have indicated that the incomplete combustion of biomass fuels produces black carbon, a component of particulate matter (PM), which is positively associated with elevated blood pressure" (Deepthi et al. 2019; Heffernan 2018; Honda et al. 2018). Furthermore, PM can disrupt autonomic nervous system balance, favoring increased sympathetic drive and enhancing the regulation of vasoconstriction by the nerves, thereby increasing the risk of stroke (Brook et al. 2010).

15.4.1 Policy Intervention to Reduce IAP

Given the significant health and mortality risks associated with indoor air pollution, various governments, non-governmental organizations (NGOs), and international bodies have implemented strategies aimed at reducing IAP in households. "One of the most crucial preventive measures is public education—raising awareness about the harmful effects of indoor air pollution and its serious impact on health and well-being" (Kankaria et al. 2014). Another key intervention involves promoting the use of cleaner alternatives, such as kerosene, in place of traditional biomass fuels. A third strategy includes the development, promotion, and subsidization of improved cooking stoves, which are designed to burn traditional fuels more efficiently and are thus affordable for low-income households. Additionally, enhancing ventilation is essential; during house construction, adequate airflow should be ensured. In poorly ventilated homes, improvements such as installing a window above the cooking area and enabling cross-ventilation through doors can significantly reduce exposure to indoor pollutants.

15.4.2 Limitations of the Study

The study had some limitations. First of all, the present study is based on self-reported information, and thus, we cannot make causal inference with the study population. Second, we lacked sufficient information about individual cooking practices, such as the specific types of cooking and the amount of time spent cooking with solid fuel. These factors may vary depending on familial and personal characteristics. Third, the study did not collect data on long-term exposure levels to pollutants such as $PM_{2.5}$, NO_2, and O_3, which are positively associated with hypertension and stroke prevalence. Fourth, for participants who switched cooking fuels during the follow-up period, the specific timing of the switch was not investigated. As a result, we cannot confirm the effect of the switch timing on blood pressure. Finally, the study is based on a cross-sectional design; therefore, it cannot establish a cause–effect relationship within the study population.

15.5 Conclusion

In conclusion, the study found that indoor solid fuel use was significantly associated with increases in systolic and diastolic blood pressure levels and an elevated risk of stroke prevalence. Age emerged as a critical factor influencing the relationship between biomass fuel use and SBP, PP, and MAP among elderly individuals. Older adults aged 65 years and above showed a higher prevalence of elevated SBP and increased stroke risk. Therefore, greater attention should be directed toward transitioning from biomass solid fuel to clean fuel to lower SBP and reduce the risk of stroke. Strategies such as promoting access to clean energy, improving education, and addressing socioeconomic inequalities could significantly reduce the health burden of indoor air pollution. Additionally, public health programs should prioritize integrated approaches that address both environmental and behavioral determinants of stroke and hypertension among the elderly. These efforts are crucial for achieving equitable health outcomes and improving the quality of life for aging populations in similar resource-limited settings.

Acknowledgements Not applicable

Authors' Contributions Conceptualization of study: UD, Analysis and/or interpretation of data: UD and NK participated in the writing of the manuscript; drafting the manuscript: UD and SS; revising the manuscript critically for important content: NK. The manuscript was prepared under the overall supervision of NK and SS. All the authors read and approved the final manuscript.

Funding The authors did not receive any funding to carry out this research.

Data Availability Statements The present study used the current primary sources of data in storage of statistical software tools which are available as per corresponding author request.

Ethics Approval and Consent to Participate The study was conducted in the Bankura district of West Bengal, India. The study aims to explore the familial setups, roles, health status, and expectations of the elderly. This study was approved by the Ethics Committee of the Rajiv Gandhi University, Arunachal Pradesh, India. Informed consent was obtained from participants involved in the study. Participation in this study was strictly voluntary, and all subjects were informed about the right to withdraw at any time without explanation. The study confirms that all methods were carried out according to relevant guidelines and regulations in the Declaration of Helsinki.

Consent for publication Not applicable.

Competing Interests The authors declare that they have no competing interests.

References

Agrawal S, Yamamoto S (2015) Effect of indoor air pollution from biomass and solid fuel combustion on symptoms of preeclampsia/eclampsia in Indian women. Indoor Air 25:341–352. https://doi.org/10.1111/ina.12144

Arku RE, Brauer M, Ahmed SH, AlHabib KF, Avezum Á, Bo J, Hystad P (2020) Long-term exposure to outdoor and household air pollution and blood pressure in the Prospective Urban and Rural Epidemiological (PURE) study. Environ Pollut 262:114197. https://doi.org/10.1016/j.envpol.2020.114197

Austin KF, Mejia MT (2017) Household air pollution as a silent killer: women's status and solid fuel use in developing nations. Popul Environ 39:1–25

Avendano M, Glymour MM (2008) Stroke disparities in older Americans: Is wealth a more powerful indicator of risk than income and education? Stroke 39:1533–1540. https://doi.org/10.1161/STROKEAHA.107.490383

Bonjour S, Adair-Rohani H, Wolf J, Bruce NG, Mehta S, Prüss-Ustün A, Smith KR (2013) Solid fuel use for household cooking: country and regional estimates for 1980–2010. Environ Health Perspect 121:784–790. https://doi.org/10.1289/ehp.1205987

Bedi TK, Bhattacharya SP (2024) Indoor air quality and health: an emerging challenge in Indian megacities. Dev Environ Sci 15:269–293. https://doi.org/10.1016/B978-0-443-21948-1.00013-3

Brook RD, Rajagopalan S, Pope CA III, Brook JR, Bhatnagar A, Diez-Roux AV, Kaufman JD (2010) Particulate matter air pollution and cardiovascular disease: an update to the scientific statement from the American Heart Association. Circulation 121:2331–2378. https://doi.org/10.1161/CIR.0b013e3181dbece1

Burroughs Peña M, Romero KM, Velazquez EJ, Davila-Roman VG, Gilman RH, Wise RA, Checkley W (2015) Relationship between daily exposure to biomass fuel smoke and blood pressure in high-altitude Peru. Hypertension 65:1134–1140. https://doi.org/10.1161/HYPERTENSIONAHA.114.04840

Cohen AJ, Brauer M, Burnett R, Anderson HR, Frostad J, Estep K, Forouzanfar MH (2017) Estimates and 25-year trends of the global burden of disease attributable to ambient air pollution: an analysis of data from the Global Burden of Diseases Study 2015. Lancet 389(10082):1907–1918. https://doi.org/10.1016/s0140-6736(17)30505-6

Deepthi Y, Nagendra SS, Gummadi SN (2019) Characteristics of indoor air pollution and estimation of respiratory dosage under varied fuel-type and kitchen-type in the rural areas of Telangana state in India. Sci Total Environ 650:616–625. https://doi.org/10.1016/j.scitotenv.2018.08.381

Deng Y, Gao Q, Yang D, Hua H, Wang N, Ou F, Liu R et al (2020) Association between biomass fuel use and risk of hypertension among Chinese older people: a cohort study. Environ Int 138:105620. https://doi.org/10.1016/j.envint.2020.105620

Dinh QN, Drummond GR, Sobey CG, Chrissobolis S (2014) Roles of inflammation, oxidative stress, and vascular dysfunction in hypertension. Biomed Res Int 2014:406960. https://doi.org/10.1155/2014/406960

Dong GH, Qian Z, Xaverius PK, Trevathan E, Maalouf S, Parker J, Simckes M (2013) Association between long-term air pollution and increased blood pressure and hypertension in China. Hypertension 61:578–584. https://doi.org/10.1161/HYPERTENSIONAHA.111.00003

Dutta A, Ray MR, Banerjee A (2012) Systemic inflammatory changes and increased oxidative stress in rural Indian women cooking with biomass fuels. Toxicol Appl Pharmacol 261:255–262. https://doi.org/10.1016/j.taap.2012.04.004

Everson F, De Boever P, Nawrot TS, Goswami N, Mthethwa M, Webster I, Strijdom H (2019) Personal NO_2 and volatile organic compounds exposure levels are associated with markers of cardiovascular risk in women in the Cape Town region of South Africa. Int J Environ Res Public Health 16:2284. https://doi.org/10.3390/ijerph16132284

Fatmi Z, Ntani G, Coggon D (2019) Coronary heart disease, hypertension and use of biomass fuel among women: comparative cross-sectional study. BMJ Open 9:e030881. https://doi.org/10.1136/bmjopen-2019-030881

Feigin VL, Roth GA, Naghavi M, Parmar P, Krishnamurthi R, Chugh S, Forouzanfar MH (2016) Global burden of stroke and risk factors in 188 countries, during 1990–2013: a systematic analysis for the Global Burden of Disease Study 2013. Lancet Neurol 15:913–924. https://doi.org/10.1016/S1474-4422(16)30073-4

Gupta AK, Patil RS, Gupta SK (2004) Influence of meteorological factors on air pollution concentration for a coastal region in India. Int J Environ Pollut 21 (3):253–262

Haider MR, Rahman MM, Islam F, Khan MM (2016) Association of low birthweight and indoor air pollution: biomass fuel use in Bangladesh. J Health Pollut 6:18–25. https://doi.org/10.5696/2156-9614-6-11.18

Heffernan KS (2018) What is the (particulate) matter with biomass fuel use? Indoor air toxicants and central haemodynamic burden. Heart 104:1478–1479. https://doi.org/10.1136/heartjnl-2018-313036

Ho AF, Zheng H, De Silva DA, Wah W, Earnest A, Pang YH, Ong ME (2018) The relationship between ambient air pollution and acute ischemic stroke: a time-stratified case-crossover study in a city-state with seasonal exposure to the Southeast Asian haze problem. Ann Emerg Med 72:591–601. https://doi.org/10.1016/j.annemergmed.2018.06.037

Honda T, Pun VC, Manjourides J, Suh H (2018) Associations of long-term fine particulate matter exposure with prevalent hypertension and increased blood pressure in older Americans. Environ Res 164:1–8. https://doi.org/10.1016/j.envres.2018.02.008

International Energy Agency (2018) World Energy Outlook 2018: highlights. International Energy Agency, Paris. https://doi.org/10.1787/weo-2018-2-en

Khan JR, Hossain MB, Gupta RD (2021) Household cooking fuels associated with elevated blood pressure among adult women: a national-wide assessment in Bangladesh. Environ Sci Pollut Res 28:67814–67821. https://doi.org/10.1007/s11356-021-15344-w

Kankaria A, Nongkynrih B, Gupta SK (2014) Indoor air pollution in India: implications on health and its control. Indian J Commun Med 39:203–207. https://doi.org/10.4103/0970-0218.143019

Kim C, Seow WJ, Shu XO, Bassig BA, Rothman N, Chen BE, Lan Q (2016) Cooking coal use and all-cause and cause-specific mortality in a prospective cohort study of women in Shanghai, China. Environ Health Perspect 124:1384–1389. https://doi.org/10.1289/EHP236

Krishnamurthi RV, Ikeda T, Feigin VL (2020) Global, regional and country-specific burden of ischaemic stroke, intracerebral haemorrhage and subarachnoid haemorrhage: a systematic analysis of the global burden of disease study 2017. Neuroepidemiology 54:171–179. https://doi.org/10.1159/000506396

Kumar A, Prasad M, Kathuria P, Nair P, Pandit AK, Sahu JK, Prasad K (2015) Low socioeconomic status is an independent risk factor for ischemic stroke: a case-control study in North Indian population. Neuroepidemiology 44:138–143. https://doi.org/10.1159/000374118

Kumar SS, Sasidharan A, Bagepally BS (2023) Air pollution and cardiovascular disease burden: changing patterns and implications for public health in India. Heart Lung Circul 32:90–94. https://doi.org/10.1016/j.hlc.2022.10.012

Lin L, Wang HH, Liu Y, Lu C, Chen W, Guo VY (2021) Indoor solid fuel use for heating and cooking with blood pressure and hypertension: a cross-sectional study among middle-aged and older adults in China. Indoor Air 31:2158–2166. https://doi.org/10.1111/ina.12872

Liu C, Chen R, Zhao Y, Ma Z, Bi J, Liu Y, Kan H (2017) Associations between ambient fine particulate air pollution and hypertension: a nationwide cross-sectional study in China. Sci Total Environ 584:869–874. https://doi.org/10.1016/j.scitotenv.2017.01.133

Liu Z, Fang C, Sun B, Liao X (2023) Governance matters: urban expansion, environmental regulation, and PM2.5 pollution. Sci Total Environ 876:162788. https://doi.org/10.1016/j.scitotenv.2023.162788

Mirowsky JE, Dailey LA, Devlin RB (2016) Differential expression of pro-inflammatory and oxidative stress mediators induced by nitrogen dioxide and ozone in primary human bronchial epithelial cells. Inhalation Toxicol 28:374–382. https://doi.org/10.1080/08958378.2016.1185199

Ndlovu N, Nkeh-Chungag BN (2024) Impact of indoor air pollutants on the cardiovascular health outcomes of older adults: systematic review. Clin Interv Aging 19:1629–1639. https://doi.org/10.2147/CIA.S480054

O'Donnell MJ, Xavier D, Liu L, Zhang H, Chin SL, Rao-Melacini P, Yusuf S (2010) Risk factors for ischaemic and intracerebral haemorrhagic stroke in 22 countries (the INTERSTROKE study): a case-control study. Lancet 376:112–123. https://doi.org/10.1016/S0140-6736(10)60834-3

Ofori SN, Fobil JN, Odia OJ (2018) Household biomass fuel use, blood pressure and carotid intima media thickness; a cross-sectional study of rural dwelling women in Southern Nigeria. Environ Pollut 242:390–397. https://doi.org/10.1016/j.envpol.2018.06.102

Pandian JD, Jyotsna R, Singh R, Sylaja PN, Vijaya P, Padma MV, Singh Y (2011) Premorbid nutrition and short-term outcome of stroke: a multicentre study from India. J Neurol Neurosurg Psychiatry. https://doi.org/10.1136/jnnp.2010.233429

Pandey A, Brauer M, Cropper ML, Balakrishnan K, Mathur P, Dey S et al (2021) Health and economic impact of air pollution in the states of India: the Global Burden of Disease Study 2019. Lancet Planet Health 5(1):e25–e38. https://doi.org/10.1016/S2542-5196(20)30298-9

Ravilla TD, Gupta S, Ravindran RD, Vashist P, Krishnan T, Maraini G, Fletcher AE (2016) Use of cooking fuels and cataract in a population-based study: the India Eye Disease study. Environ Health Perspect 124:1857–1862. https://doi.org/10.1289/EHP193

Sesso HD, Stampfer MJ, Rosner B, Hennekens CH, Gaziano JM, Manson JE, Glynn RJ (2000) Systolic and diastolic blood pressure, pulse pressure, and mean arterial pressure as predictors of cardiovascular disease risk in men. Hypertension 36(5):801–807

Sun Y, He Y, Kuang Y, Xu W, Song S, Ma N, Worsnop DR et al (2020) Chemical differences between PM1 and PM2. 5 in highly polluted environment and implications in air pollution studies. Geophys Res Lett 47(5):e2019GL086288

Thacher JD, Emmelin A, Madaki AJ, Thacher TD (2013) Biomass fuel use and the risk of asthma in Nigerian children. Respir Med 107:1845–1851. https://doi.org/10.1016/j.rmed.2013.09.009

Villeneuve PJ, Johnson JY, Pasichnyk D, Lowes J, Kirkland S, Rowe BH (2012) Short-term effects of ambient air pollution on stroke: Who is most vulnerable? Sci Total Environ 430:193–201. https://doi.org/10.1016/j.scitotenv.2012.05.002

Wang L, Xiang Z, Stevanovic S, Ristovski Z, Salimi F, Gao J, Li L (2017) Role of Chinese cooking emissions on ambient air quality and human health. Sci Total Environ 589:173–181. https://doi.org/10.1016/j.scitotenv.2017.02.124

Xu X, Mishra GD, Dobson AJ, Jones M (2018) Progression of diabetes, heart disease, and stroke multimorbidity in middle-aged women: a 20-year cohort study. PLoS Med 15:e1002516. https://doi.org/10.1371/journal.pmed.1002516

Yan M, Li C, Zhang L, Chen X, Yang X, Shan A, Tang NJ (2020) Association between long-term exposure to sulfur dioxide pollution and hypertension incidence in northern China: a 12-year cohort study. Environ Sci Pollut Res 27:21826–21835. https://doi.org/10.1007/s11356-020-08572-z

Yang BY, Guo Y, Morawska L, Bloom MS, Markevych I, Heinrich J, Dong GH (2019) Ambient PM1 air pollution and cardiovascular disease prevalence: insights from the 33 Communities Chinese Health Study. Environ Int 123:310–317. https://doi.org/10.1016/j.envint.2018.12.012

Yang BY, Qian Z, Howard SW, Vaughn MG, Fan SJ, Liu KK, Dong GH (2018) Global association between ambient air pollution and blood pressure: a systematic review and meta-analysis. Environ Pollut 235:576–588. https://doi.org/10.1016/j.envpol.2018.01.001

Yitshak Sade M, Novack V, Ifergane G, Horev A, Kloog I (2015) Air pollution and ischemic stroke among young adults. Stroke 46:3348–3353

Yadav R, Deora S, Yadav G (2021) Air pollution and its impact on cardiovascular health—It's time to act fast! Indian Heart J 73(1):1. https://doi.org/10.1016/j.ihj.2021.01.021

Chapter 16
Vegetation Cover for Sustainable Urban Planning: The Multifaceted Dimension

Nilpaban Mali, Pritimani Bharali, Ritima Mishra, and Nabanita Bhattacharyya

Abstract Sustainable urban planning involving vegetation cover is a transformative approach for creating cleaner and healthier cities. Incorporating green spaces into urban landscapes helps balance environmental, social, and economic well-being. Urban air, water, and soil get polluted from sources like automobile and industrial discharges, as well as municipal waste dumping. Control of such pollution is a critical issue that requires prioritized attention, as it poses serious risks to human health. Botanical solutions such as phytoremediation have tremendous importance in recent times in the sustainable management of pollutants, where plants are utilized to remove contaminants, including heavy metals, from soil and water. Similarly, air pollution-tolerant plants can be used to clean volatile toxic compounds released from automobiles and industries into the atmosphere of urban areas. This chapter has been prepared as a comprehensive and precise document by pulling together published information on some potential pollution-tolerant plants to create vegetation cover and their pollution mitigation mechanisms to help in the sustainable planning of urban infrastructures like bioswales as a pollution management strategy. The possibility of prospecting the naturally occurring native wasteland plants as potential and resilient candidates for sustainable remediation of urban pollutants by the process of phytoremediation is especially emphasized here. Scientific suggestions on potential plants included in this chapter will aid in developing sustainable urban planning for a better and healthier future. Researchers, environmentalists, urban planners, as well as policymakers, will benefit from the information cited in this chapter.

Keywords Air pollution · Bioswales · Phytoremediation · Soil pollution · Wasteland plants · Water pollution

N. Mali · P. Bharali · R. Mishra · N. Bhattacharyya (✉)
Department of Botany, Gauhati University, Guwahati, Assam, India
e-mail: nbh_17@gauhati.ac.in

A. Roy et al. (eds.), *Environment and Public Health*, Sustainability Solutions,
https://doi.org/10.1007/978-3-031-99770-9_16

16.1 Introduction

Global urbanization is making cities more crowded and polluted. To achieve social, economic, and environmental sustainability, urban areas worldwide have been implementing more programs, practices, and experiments in the past few decades (Tan et al. 2016). Continued urbanization increases pollution and the risk to human health. Urban green space offers a variety of ecosystem services which might assist in tackling many urban problems and enhance city dwellers' quality of life, particularly their health, in addition to preserving the environmental quality of urban areas. Apart from providing food, vegetation may purify air, eliminate pollution, reduce noise, maintain temperatures, absorb wastewater from storms, and replenish groundwater. For instance, trees in urban areas can help lower air pollution by removing some airborne contaminants from the atmosphere. By offering shade as well as cooling an area, urban forests and vegetation cover can also help regulate temperatures and lower the chance of illness caused by heat for urban residents. As a result of continuous industrialization and urbanization, soils are becoming more contaminated, losing their fertility, and turning into wasteland. It is necessary to reduce or restore this kind of contaminated land for sustainable life processes. Vegetation cover will be a good and effective solution to mitigate this type of pollution and health risk. The term "benefits people receive from an ecosystem" refers to the variety of ecosystem services that urban vegetation offers. For instance, city dwellers are less likely to suffer from heat-related illnesses when trees are present in urban areas because they provide shade and cool the air through transpiration. Furthermore, through photosynthetic activity and generation of biomass, trees serve as atmospheric CO_2 sinks (Ferrini et al. 2020). By increasing physical activity, lowering stress levels, and improving air quality, vegetation cover has a major positive influence on health. Urban greenery improves respiratory health by filtering pollutants like ozone and particulate matter, which helps reduce air pollution. Urban areas with higher concentrations of green vegetation have lower mortality and cardiovascular disease rates. Reports suggest that people who live in greener environments are less likely to have high blood pressure and heart disease. Vegetation cover is also important in public health strategies because it can reduce the incidence of respiratory ailments like asthma by lowering heat and providing cleaner air (James et al. 2015). As vegetation facilitates dry deposition processes, which remove pollutants from the atmosphere as they land on plant surfaces, it improves the quality of the air. Trees and other vegetation serve as physical barriers, keeping human receptors and pollution sources (such as roads) apart. Because of its physical presence, people are exposed to fewer toxic pollutants (Barwise and Kumar 2020). In areas where wastelands occupy a significant proportion of the landscape, the implementation of vegetation cover strategies is essential for improving ecological conditions. Vegetation acts as a natural barrier against land pollution by absorbing pollutants and improving soil quality. Through processes such as phytoremediation, plants can uptake heavy metals and other contaminants from the soil, effectively reducing their concentration and toxicity (Reeves et al. 2018). Because of their special capacity to accumulate heavy metals, some plant species offer a sustainable method of cleaning up contaminated land by drawing toxic substances like lead, cadmium, and arsenic from the soil and storing them in their tissues. We can successfully reduce the levels of heavy metals from contaminated areas while restoring vegetation cover and fostering ecological health by putting phytoremediation techniques into practice. Vegetation also plays a vital role in improving soil structure and nutrient content, which is essential for re-establishing ecological balance. The choice of suitable plant species has a direct impact on how well vegetation reduces pollution. Some species have characteristics that improve their capacity to filter pollutants; for example, trees with intricate leaf structures are good at capturing particulate matter. Several previous observational and experimental studies have also revealed that the establishment and maintenance of planned vegetation cover in urban setups have tremendous health benefits, including mental stress reduction, increased physical activity, and lowered risks of respiratory ailments (Browning et al. 2022). Overall vegetation cover can play multifaceted role to alleviate pollution and environment degradation and thereby to lessen health risk in urban areas (Fig. 16.1). However, there is lack of comprehensive document on selection of potential plants for creating vegetation cover in various levels to alleviate pollution and thereby to restore healthy environment in urban areas. Therefore, this chapter has been prepared by pulling together the published information on various environment-friendly aspects of vegetation cover in mitigating pollution and related health issues for better urban planning in a sustainable way. Emphasis has been given to certain eco-friendly recommendations for infrastructural planning in urban areas. Moreover, future potential research prospects in this area are also highlighted in light of available literature. This document will be helpful for environmentalists, urban planners, and policymakers to execute plans and policies in a scientific and nature-friendly way for a healthy future.

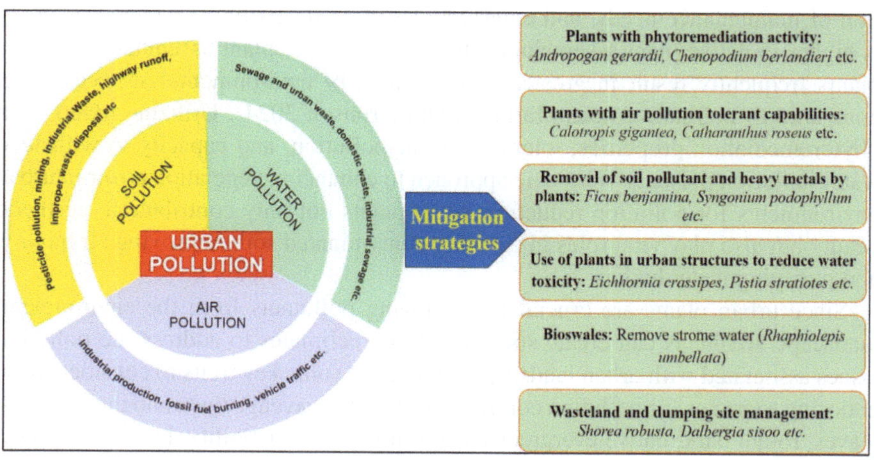

Fig. 16.1 Vegetation-based strategies for reducing urban pollution

16.2 Selection of Plants for Vegetation Cover

Once known for unique and rich biodiversity, the countries, especially in the Global South, are facing multidimensional challenges of globalization-induced, but poorly planned, rapid urbanization processes in recent decades (Verma and Das 2024). However, with tremendous plant diversity and favorable growth conditions in this region, including the North Eastern parts of India, there is ample scope to mitigate such undesirable challenges by implementing plant-based approaches such as the creation of vegetation cover in a green, sustainable way. The term "vegetation cover" describes the plant matter that covers the ground in a particular location. It can be divided into different types according to the predominant flora of a particular area. For ecological research, land management, and environmental preservation, an understanding of these kinds is crucial. Vegetation cover has a significant impact on soil erosion as well as the physical and chemical characteristics of soil (Pan et al. 2017). It also reduces runoff and restores ecosystems in ecologically sensitive regions. Vegetation can raise soil organic matter over time, enhance the physical characteristics and anti-erodibility of the soil, and safely lower runoff and erosion. As reported, several ground flora species, diversity index, population of bacteria, fungi, nematodes, micro arthropods, and vascular-arbuscular mycorrhizal fungi spores are more under plantation than that of unplanted degraded land. The soil under the plantation has significantly better fertility status as well as higher N, P, K content. Under the canopy of some plants like *Albizia procera* and *Leucanea leucocephala*, higher availability of N, P_2O_5, and K_2O as well as higher organic carbon percentage were reported (Kiran et al. 2009). However, choosing the right plants for vegetation is crucial. Native plants offer several advantages over imported species, especially in terms of vegetation cover and overall ecosystem health. These benefits include better adaptability to local environmental conditions, support for wildlife, reduction of maintenance needs, and enhancement of biodiversity. The faunal diversity is higher in native plant assemblages than in non-native plant assemblages. Native plants frequently result in greater establishment of environmental services when compared to non-native ones (Tartaglia and Aronson 2024). Utilizing plants with phytoremediation properties, tolerance to air pollution, and capacity to withstand water pollution presents a strategic approach to enhancing vegetation cover in urban environments for pollution reduction. Such plants not only contribute to aesthetic beauty but also play vital roles in cleaning contaminated soil, improving air quality, and purifying water bodies.

Since urban plants are crucial for removing pollutants from the environment, landscape architects can create specialized microclimates to address the pollution issues associated with urban landscapes. There are some key traits to consider while choosing a species for pollution control as well as to scavenge dust, and those are big leaves, rough bark, native evergreens, environment compatibility, low water needs, little maintenance, high pollutant absorption capacity, pollution resistance, agro-climatic compatibility, height, biomass productivity, etc. (Sandhya 2018). These morphological features can play a significant role for dust and other pollutants from the environment. Smaller compound leaves, like those of tamarind trees (*Tamarindus indicus*), are typically better at collecting particles than bigger leaves. Meanwhile, the previous studies reveal that the leaves of *Derris indica* (Pongamia), *Mangifera indica* (Mango), *Polyalthea longifolia* (Ashoka), and *Thespepsia populnea* (Umbrella trees) could collect more dust compared to other neighboring trees. This morphological diversification can be one of the pollution-mitigating factors in urban areas. Besides, trees including Java *Azadirchta indica* (Neem), *Bombax ceiba* (Silk cotton), *Cassia fistula* and *C. siamea* (Indian laburnum), *Dalbergia sissoo* (Rosewood), *Delonix regia* (Gulmohar), *Ficus religiosa* (Pipal), *Holoptelea integrifolia* (Indian Elm), *Jacaranda mimosifolia* (Jacaranda), *Lagerstroemia indica* (Indian lilac), *Plumeria rubra* and *P. alba* (temple or pagoda trees), and *Syzygium cumini* (plum) are reported to have high air pollution tolerance index (APTI) which can be planted in urban areas to reduce the air pollution load from automobiles and industries (Pokhriyal and Subba Rao 1986).

16.2.1 Plants with Phytoremediation Activity

Rapid urbanization and industrialization lead to depletion of soil fertility and soil quality, which causes soil pollution. Land resources are essential to human existence and the advancement of society. The fact that about 60% of the planet's land area is thought to be degraded makes it one of the biggest obstacles to its sustainable use (Jinger et al. 2023). According to the NITI Aayog, as of February 2024, approximately 16.96% of India's total geographical area is classified as wasteland. This classification includes various types of land, such as grassland, shrub land, and waterlogged areas, which are seen as underutilized or non-productive. The accumulation of potentially hazardous metals (PTMs) in soils has become a global issue over the past few decades because of ongoing industrialization and urbanization (Su et al. 2022). Metal soil pollution is increasing steadily because of industrial production activities constant emissions. Heavy metals in the polluted soil make their way into the food chain and cause serious health issues for humans (O'Connor et al. 2018). Wasteland reclamation is crucial, which aims to restore lands that are unproductive or degraded, which may result from natural processes or anthropogenic activities (Singh et al. 2023). Techniques such as planting trees and crops in these areas are critical for revitalizing the ecosystem and enhancing the ecological value of the land (Jinger et al. 2023). Wasteland reclamation through the utilization of plants, particularly via techniques such as phytoremediation, serves as an effective approach to restore degraded environments (Yan et al. 2020). As reported previously, several plant species have the unique capabilities to remove various pollutants from soil (Table 16.1).

Industrial waste dumping and leaching of chemicals from the industries pollute soil and degrade the soil fertility for plant growth (Bhattacharyya 2022). However, some plants can grow on polluted soil of wastelands and are able to mitigate stress as well as accumulate pollutants, including heavy metals. Some plants help in fixing

Table 16.1 Removal of soil pollutants and heavy metals by plants

Plant species	Removed pollutant	References
Azolla pinnata	Cd	Rai (2008)
Zea mays	Atrazine (pesticide)	Ibrahim et al. (2013)
Helianthus annuus	Cobalt	Lotfy and Mostafa (2014)
Pteris vittaia	As	Dong et al. (2005)
Pteris vittaia	Cr	Kalve et al. (2011)
Medicago sativa	Polychlorinated biphenyls	Xu et al. (2010)
Brassica juncea	Cu, Ni, Pb, Zn, Cr	Bluskov (2005)
Symphytum Officinale linn	Pb, Zn, Cd	Xiao et al. (2018)
Calatropis procera	Cd, Ni	Pathak and Shah (2024)
Parthenium hysterophorous	Pb	Pathak and Shah (2024)
Datura innoxia	Cr	Pathak and Shah (2024)

the nitrogen also which is helpful in the restoration of soil fertility for reclaiming the wasteland soils. Moreover, phytoremediation is an environmentally safe, economically feasible, and sustainable approach. Native plants should be chosen for phytoremediation as they are better in terms of survival, growth, and reproduction under environmental stress than plants introduced from other habitats. Each plant species has different mechanisms for removing soil pollutants, but most of them happen in the rhizosphere, where microorganisms and plants work together symbiotically to support soil recovery (Table 16.2). By eliminating pollutants, microorganisms lower soil toxicity and enhance plant growth and metabolism, while plants give microorganisms a growing environment and enhance soil structure. Different plant species are utilized as environmentally beneficial and financially viable methods of remediating land contaminated by harmful metals, for example, by increasing the diversity and abundance of soil microorganisms, and the rate of remediation varies species to species (Table 16.2). For instance, *Brassica napus* lowered the detrimental effects of Cd on soil. The exceptional ability of *Brassica juncea* to accumulate high concentrations of heavy metals, especially Ni, Zn, Cu, and Pb, has led to its widespread utilization in phytoremediation. *B. juncea* is regarded as a Cr hyperaccumulator as well. Its rapid growth is another advantage of employing this species in phytoremediation (Bluskov 2005). Numerous studies have confirmed that sunflowers (*Helianthus annuus* L.) have powerful stress tolerance, a higher metal bioconcentration factor (including U and Cd), and translocation factor (Chen et al. 2020). Similarly, *Calatropis procera*, *Datura innoxia*, *Parthenium hysterophorous*, and *Symphytum officinale* are good accumulators of heavy metals like Cd, Ni, Cr, Zn, and Pb (Xiao et al. 2018). Above-mentioned plants can be used as natural colonizer along with the application of organic supplements like cow dung to alleviate soil pollution in urban wastelands (Pathak and Shah 2024). On metalliferous and sulfide-bearing mine wasteland, direct seeding or planting of native plant species combined with lime and manure amelioration was proved to be a suitable method for the first establishment of a self-sustaining vegetation cover (Yang et al. 2009). Nanomaterials aid in raising the protein content of plants and interacting with microorganisms in the plant's rhizosphere to encourage the absorption of heavy metals and enhance soil fertility (Zhao et al. 2019). Again, biochar can also be used for reclaiming degraded soil as it improves the physical condition of soil and helps the hyperaccumulator plants in the absorption of heavy metals and organic contaminants from polluted soil (Sahu et al. 2020). Some other plant species with considerable phytoremediation properties worth to be used to alleviate urban soil and water pollution are wheat grass (*Agropyron cristatum*), big bluestem (*Andropogan gerardii*), musk thistle (*Carduus nutans*), periwinkle (*Catheranthus roseus*), lambsquarters (*Chenopodium berlandieri*), rhodes grass (*Chloris guyana*), barmuda grass (*Cynodon dactylon*), carrot (*Daucus carota*), Canada wild rye (*Elymus candensis*), water hyacinth (*Eichhornia cressipes*), tall fescue (*Festuca anundinacea*), sunflower oil plant (*Helianthus annus*), foxtail barley (*Hordeum jubatum*), kochia (*Kochia scoparia*), flatpea (*Lathyrus sylvestris*), duck weed (*Lemna minor*), alfa-alfa (*Medicago sativa*), red mulberry (*Morus rubra*), catnip (*Nepeta cataria*), switch grass (*Panicum variegatum*), bahia grass (*Paspalaum notatum*), reed grass (*Phragmites australis*), poplar tree (*Populus* spp.), white raddish (*Raphanus sativus*), nightshade (*Solanum nigrum*), Indian grass (*Sorghastrum nutans*), tall wheat grass (*Thynopyron elongatum*), cumbungi (*Typha angustifolia*), and vetiver grass (*Vetiveria zizanioides*) (Sinha et al. 2007). Transpiration loss is the basic principle of bio-drainage, which is one of the vegetation-based systems used widely for treating waterlogged areas. Fast-growing tree species like *Salix*, *Eucalypt*, *Acacia*, *Albezia*, *Terminalia*, *Prospis*, and *Papulus* are suitable plant species to use in bio-drainage method for the reclamation of water logged wastelands of urban areas (Sarvade et al. 2017).

16.2.2 Plants with Air Pollution Tolerance

Air pollution levels are always higher and alarming in crowded urban areas due to an excess of automobiles and industries. Nevertheless, there are certain plant species which have been reported as air pollution tolerant and capable of reducing the pollution level in the air by absorbing the polluting toxic elements both in outdoor and indoor environments of urban setups. As reported previously, plants like *A. lebbek*, *B. spectabilis*, *C. gigantea*, *C. roseus*, and *P. dulce* are tolerant to air pollution (Singh et al. 1991). Native trees, shrubs, and grasses are more helpful in reducing air pollution, as they are often better adapted to local climates, need fewer resources, and effectively capture particulate matter. Green spaces and vegetation have shown a reduction in the concentrations of airborne pollutants, particularly particulate matter. Exposure to air pollutants raises the risk of respiratory and cardiovascular disorders both immediately and over time. Usually, air pollutants gather on the leaf surface or enter the plant's tissues, some of which are used by the plant for its own metabolism (Gawronski and Gawronska 2017). Tree coverage and its density help to reduce different rates of air pollution accordingly. Conifers

Table 16.2 Removal of soil pollutants by plant species

Plants species	Types of pollutants	Removal rate	References
Syngonium podophyllum	Benzene	18.8–220.2 ng/m^3/cm^2/h	Yoo et al. (2006)
Scindapsus aureus	Benzene	43–77% in 72 h	Treesubsuntorn and Thiravetyan (2012)
Anthurium andraeanum	Benzene	0–13.28 μg/m^3/m^2/h	Yang et al. (2009)
Ficus benjamina	Formaldehyde	0.14–1.36 mg/m^3/cm^2 in 5 h	Kim and Lee (2008)
Ardisia crenata	Toluene	2.5–40 μg/m3 /h/m^2	Kim et al. (2012)
Ficus elastica	Benzene	28–91% in 12 h	Chun et al. (2010)
Epipremnum aureum	Propionaldehyde	0.8–2.4 nmol/m^2/s	Tan et al. (2016)
Azalea indica	Toluene	7–76 h to remove 95% of the initial dose	De Kempeneer et al. (2004)
Nephrolepis obliterata	Formaldehyde	90–100% in 48 h	Teiri et al. (2018)
Ardisia japonica	Formaldehyde	0.14–0.88 μg/m^3/cm^2 in 5 h	Kim and Lee (2008)

like *Pinus* show the best absorption capacity of particulate matter (PM), especially to eliminate PM 2.5, and *Magnolia* has a tolerance for NO_2, which is crucial for its metabolic processes (Gourdji 2018). Larger-leafed trees are more effective at removing PM from the atmosphere. Similarly, because of the unique structures in their leaves, stems, and roots, certain houseplants can enhance indoor air quality by absorbing and eliminating a variety of air pollutants. As a result of routine activities and materials, these plant structures aid in the absorption, capture, and degradation of hazardous chemicals, including airborne particles and volatile organic compounds (VOCs), which are frequently found indoors. In this process, pollutants are taken up through stomata, broken down chemically in the roots, and physically absorbed by the leaf surfaces. Formaldehyde, benzene, toluene, xylene, and ethylbenzene are among the most significant environmental contaminants that plants can eliminate. Houseplants provide a natural and affordable way to improve the air quality in homes, workplaces, and other enclosed spaces by using their natural biological processes to lower the concentration of these dangerous substances and promote healthier indoor environments. Indoor plants raise humidity and oxygen levels as well. Increased indoor humidity from plant leaf evapotranspiration can lessen static electricity, alleviate drying out of skin, and ease respiratory problems. Plants use their stomata to absorb volatile organic compounds (VOCs) during regular gas exchange, and then Calvin cycle transforms them into amino acids (Yoo et al. 2006). Microbes that reside in plant buds and the rhizosphere may also eliminate volatile organic compounds (Khaksar et al. 2016). The morphology (leaf shape, size, and hairiness) and anatomy (e.g., mesophyll and epidermis composition, stomatal density, and size) of the plants should be taken into consideration when developing scientific selection criteria for indoor plants. Common indoor air pollutants primarily consist of nitrogen oxides, carbon monoxide, carbon dioxide, volatile organic compounds (VOCs), toxic gases (formaldehyde and benzene), and various aromatics (Davardoost and Kahforoushan 2019). Houseplants have physiological advantages, but they also have positive psychological effects by lowering stress, elevating mood, and increasing productivity. They produce an atmosphere that is peaceful and beautiful and encourages a bond with the natural world. With their unique capacities to eliminate indoor pollutants, aloe vera, Sansevieria, pothos, and peace lily are all good houseplants for air purification (Talaie and Ponraj 2024). Hence, proper and scientific plan and policy to create vegetation cover with the above-mentioned plant species can prove effective in managing outdoor and indoor pollution of urban environments.

16.3 Plants to Reduce Water Toxicity in Urban Structures

Wetlands, which can be classified as either natural or artificial, are primarily defined as the area where terrestrial and aquatic ecosystems meet. Wetlands are called the kidneys of an ecosystem, as they recycle nutrients, treat wastewater, enhance plants' absorption capacity, and filter out toxic chemicals, heavy metals, and pollutants from the water. Wetland's physical and chemical characteristics determine unique interactions between plants and rhizospheric microbes and offer favorable conditions for phytoremediation of water pollutants. As reported previously, the natural colonization of algae was significantly reduced because of the nutrient depletion brought on by plants' absorption, thereby improving the water quality, including the pH and turbidity. Plants are used to extract nutrients like phosphorus and nitrogen from the water, thereby helping to help in reducing eutrophication in aquatic ecosystems (Yu et al. 2019). The efficacy of plants for heavy metal removal from water varies from species to species. For instance, *Pistia stratiotes* and *Eichhornia crassipes* are good accumulators of Hg; *Vallisneria spiralis* and *Salvinia natans* are both excellent Cr accumulators; *Ceratophyllum demersum* and *Myriphyllum triphyllum* are useful for extracting Cd from water; *Salvinia auriculata* and *S. minima* have the best removal effect on Pb. Likewise, several other aquatic weeds such as alligator weed, azolla spp., duckweed, water lettuce, and water hyacinth have been successfully used for remediation of several toxicants from water in previous works (Cho-Ruk et al. 2006; Nazir et al. 2020). Therefore, giving adequate emphasis to cover the natural water bodies and polluted wetlands (that are created due to various anthropogenic abuses) with previously reported potential aquatic plant species can improve the water quality by alleviating toxic contamination to restore healthy environments in urban areas.

16.4 Bioswales

Bioswales are a useful way to reduce contaminants associated with toxicity in urban stormwater. Contamination and toxicity associated with urban stormwater runoff are a growing concern because of the potential impacts on receiving systems. Engineered green areas known as bioswales gather runoff and filter out dangerous contaminants from stormwater before it is released into various surface water sources. To improve infiltration and treat stormwater, California water regulators required new urban development projects to incorporate green infrastructure. Low-impact development techniques, identified as parking lot bioswales, encourage runoff to be filtered through soil and vegetation. Previous research indicated that bioswales lowered levels of metals, hydrocarbons, and suspended sediments. Bioswales can reduce pollutants from water and reduce stormwater runoff in urban areas. Some properly engineered bioswales can reduce water runoff than normal bioswales, which makes them worth being used for sustainable urban development. In Davis, California, bioswales with engineered soil mix containing 75% native lava rock and 25% loam soil and planted with Blueberry Muffin Hawthorn [*Rhaphiolepis umbellata* (Thunb.) Makino] and four Red Tip Photinia [*Photinia* × *fraseri* Dress] could reduce surface runoff by 99.4% and reduce nitrogen, phosphate, and total organic carbon loading up to 99.1%, 99.5%, and 99.4%, respectively, in comparison with non-engineered control bioswales (Xiao et al. 2017). The addition of appropriate plants in bioswales is a novel method to reduce pollutants, including heavy metals from sewage water and to release stored water back into the atmosphere by the process of transpiration (Table 16.3). According to the previous reports, plants in bioswales reduced runoff by 46–72% of the total water output through transpiration. Species most likely to contribute to the function of the bioswales are those with large mature forms and high stomatal conductance (Scharenbroch et al. 2016). Hence, construction of properly engineered bioswales with selected potential plant species should be included in modern urban planning to alleviate pollution levels and related health risks faced by the urban population.

Table 16.3 Accumulation of heavy metals from sewage by plant species

Plant	Heavy metals	References
Pistia stratiotes	Hg	Molisani et al. (2006)
Salvinia natans	Cr	Dhir et al. (2009)
Eichhornia crassipes	Hg	Molisani et al. (2006)
Myriphyllum triphyllum	Cd	Sivaci et al. (2004)
Vallisneria spiralis	Cr	Verma et al. (2008)
Salvinia minima	Pb	Olguin et al. (2002)
Ceratophyllum demersum	Cd	Mishra et al. (2008)
Salvinia auriculata	Pb	Molisani et al. (2006)
Lemna minor	Cu	Kanoun-Boulé et al. (2009)

16.5 Wasteland and Dumping Site Management

Wasteland typically refers to a barren or desolate area, often characterized by the absence of vegetation and wildlife. In technical terms, it can be described as ecologically degraded areas that are suffering from soil erosion, contamination, or desertification and lands that have been abandoned or left unproductive due to industrial activities, urban sprawl, or natural disaster. There are 11 major types of wastelands or unutilized degraded landmasses, and they are gullied or ravenous lands, scrublands, water logged or marshy lands, lands with salinity/alkalinity, scrub forests, degraded pasture/grazing lands, deserts, mining areas, snowy glacial fields, degraded lands after *jhum*, and barren rocky fields (Krishnamoorthi and Ganguly 2022). The fact that about 60% of the planet's land area is thought to be degraded makes it one of the biggest obstacles to its sustainable use (Jinger et al. 2023). According to the NITI Aayog, Govt. of India, approximately 16.96% of India's total geographical area is classified as wasteland as of February 2024. The accumulation of potentially hazardous metals (PTMs) in soil has become a global issue over the past few decades because of ongoing industrialization and urbanization. Also, industrial waste dumping and leaching of chemicals from the industries pollute the soil, resulting in loss of soil fertility to grow plants (Das et al. 2021). Land degradation often leads to soil erosion, nutrient loss, and reduced agricultural productivity, further heightening the challenges posed by climate variability and changes. Wasteland environments often cause a decline in biodiversity as habitats are altered or destroyed (Singh et al. 2023). When lands are abandoned and turn into wastelands, the alteration and subsequent destruction of soil often lead to the release of stored carbon. This release contributes to increasing the atmospheric CO_2 level, thereby further intensifying climate change dynamics (Campos et al. 2022). Reclamation strategies can be thought of as environmental control systems that work to create the ideal conditions for the formation and development of favorable biotic assemblages. Reclamation of land contributes to the maintenance and replenishment of soil fertility, increases crop productivity, improves farming sustainability, and lessens the influence of weather and climate fluctuations on output capacity. Wasteland reclamation through the utilization of plants, particularly via techniques such as phytoremediation, serves as an effective approach to restore degraded environments (Yan et al. 2020).

Increased salinity of the soil is another concern in wastelands. Previous studies reveal that saline wastelands can be reclaimed by using specific plant species including *Acacia auriculiformis, Albinia guachepele, Albizia caribaea, Albizia lebbeck, Areca catechu, Callistemon lanceolatus, Camarilla glauca, Eucalyptus tereticornis, Pithecellobium dulce, Pongamia pinnata, Prosopis juliflora, Salix fruticosa, Tamarindus indica, Tamarix articulate, Tamarix indica, Vachellia nilotica*, and *Vachellia tortilis* (Tomar 1997). Likewise, alkaline wastelands can be reclaimed by using species like *Aegle marmelos, Albizia lebbeck, Azadirachta indica, Butea monosperma, Dalburgia sissoo, Emblica officinalis, Grevillia robusta, Hardwickea binnata, Kajellea pinnata, Morus alba, Populus deltoides, Punica granatum, Tectona grandis*, *Pongamia pinnata*, *Sesbania sesban*, and *Zizyphus mauritiana*, which can be used for dumping site management (Luna 2006). The native plant improves dump stability, reduces soil erosion, and has better binding capacity. Both the soil quality and the organic matter of the wastelands affect the growth of native plants. Leaf debris is crucial for enhancing the organic matter of the wastelands for the revegetation program to be implemented successfully (Ranjan et al. 2015). Compared to the traditional plantation method, the Miyawaki method is more effective for revegetating dumping areas. In a small pit or dump slope, a variety of native plants are planted close to each other in the Miyawaki tree-planting technique. Close planting enhances the green cover and strengthens the land's richness by planting numerous trees in a small area. In fact, each plant benefits from the coexisting plants in terms of essential nutrients, and they grow stronger and healthier. The region where such plantations have been established has its own ecology (Daou et al. 2024). Barren and polluted urban wastelands and dumping sites can be covered with native stress-resilient wasteland plant species by implementing the Miyawaki technique to recover the degraded land masses and to check the further extension of soil degradation.

16.6 Conclusion

With increasing urbanization in the course of development, maintaining a good environment is crucial for the future generation to sustain a healthy life with minimum risk of climate change. Maintenance of adequate vegetation cover helps to keep the urban ecosystem in balance by conserving different species of plants along with the soil and water sources. Adoption of specific vegetation cover, particularly involving potential stress-resilient native species of plants that are explicitly adapted to a particular climate and soil condition, can help in minimizing urban pollution and mitigate different types of health hazards caused by pollution. Besides cleaning soil, air, and water, vegetation cover also aids in reducing the urban heating that helps in the management of health-related issues caused due to increasing temperature. However, an extensive research program should be undertaken to understand the stress mitigation strategies and to quantify the pollution removal capacity of adaptive native species of plants for specific climatic regions with unique edaphic and ecological dynamics. In this context, explorative genetic diversity studies to find out more potential plant species for cleaning up major toxic pollutants from soil, water, and air should be encouraged. Novel plant–microbe interaction is another area of investigation for microbial-assisted removal of pollutants worth looking for. Mass propagation of potential species by in vitro propagation methods, including tissue culture technology, can be undertaken for the ease of availability of planting materials is another aspect to be taken into consideration during policy preparation.

Creating adequate and scientifically supported vegetation cover can offer scalable and sustainable solutions to modern-day challenges, aligning with the Sustainable Development Goals (SDGs) set by the United Nations (UN). Strategic vegetation cover involving plants with scientifically studied, based proven potential can certainly meet the challenges posed by unplanned and unsustainable urbanization, especially in the Global South, to fulfill the objective of "SDG-11", i.e., "Sustainable Cities and Communities". As highlighted in this chapter, several plants are potentially efficient in restoring "good health and well-being" of the urban population by minimizing pollutants and contaminants in all the spheres, including soil, water, and air, which is the primary objective of "SDG-3". Moreover, urban greenery can mitigate the harmful consequences of climate change, including rising temperature, artificial floods, and related health hazards, which come under the objective of "SDG-13", i.e., "Climate Action". Information in this chapter will be helpful for the researchers, environmentalists, urban planners, as well as policymakers, who are involved in sustainable urban planning programs for a better and healthier future.

References

Barwise Y, Kumar P (2020) Designing vegetation barriers for urban air pollution abatement: a practical review for appropriate plant species selection. Npj Clim Atmos Sci 3(1):12. https://doi.org/10.1038/s41612-020-0115-3

Bhattacharyya N (2022) Naturally growing native plants of wastelands: their stress management strategies and prospects in changing climate. In: Roy S, Mathur P, Chakraborty AP, Saha SP (eds) Plant stress: challenges and management in the new decade. Advances in science, technology & innovation (IEREK interdisciplinary series for sustainable development). Springer, Cham, pp 149–168. https://doi.org/10.1007/978-3-030-95365-2_10

Bluskov S (2005) Distribution of chromium and its effect on morphology and anatomy of *Brassica juncea* (Indian mustard) and soil microorganisms. University of Northern British Columbia.https://doi.org/10.24124/2005/bpgub334

Browning MHEM, Rigolon A, McAnirlin O, Yoon HV (2022) Where greenspace matters most: a systematic review of urbanicity, greenspace, and physical health. Landsc Urban Plan 217:104233. https://doi.org/10.1016/j.landurbplan.2021.104233

Campos MO, Cerri CEP, Scala NL (2022) Atmospheric CO_2, soil carbon stock and control variables in managed and degraded pastures in central Brazil. Remote Sens Appl: Soc Environ 28:100848. https://doi.org/10.1016/j.rsase.2022.100848

Chen L, Yang JY, Wang D (2020) Phytoremediation of uranium and cadmium contaminated soils by sunflower (*Helianthus annuus* L.) enhanced with biodegradable chelating agents. J Clean Prod 263:121491. https://doi.org/10.1016/j.jclepro.2020.121491

Cho-Ruk K, Kurukote J, Supprung P, Vetayasuporn S (2006) Perennial plants in the phytoremediation of lead contaminated soils. Biotechnology 5(1):1–4. https://doi.org/10.3923/biotech.2006.1.4

Chun SC, Yoo MH, Moon YS, Shin MH, Son KC, Chung IM, Kays SJ (2010) Effect of bacterial population from rhizosphere of various foliage plants on removal of indoor volatile organic compounds. Korean J Hortic Sci Technol 28(3):476–483

Daou A, Saliba M, Kallab A (2024) A review of the Miyawaki method. Available at SSRN 4728239. https://doi.org/10.2139/ssrn.4728239

Das P, Bora P, Paul N, Bhattacharyya N(2021)Vegetation composition and assessment of phytotoxicity in a paper mill dumpsite. Plant Sci Today 8(1):140–147. https://doi.org/10.14719/pst.2021.8.1.947

Davardoost F, Kahforoushan D (2019) Evaluation and investigation of the effects of ventilation layout, rate, and room temperature on pollution dispersion across a laboratory indoor environment. Environ Sci Pollut Res 26:5410–5421. https://doi.org/10.1007/s11356-018-3977-8

De Kempeneer L, Sercu B, Vanbrabant W, Van Langenhove H, Verstraete W (2004) Bioaugmentation of the phyllosphere for the removal of toluene from indoor air. Appl Microbiol Biotechnol 64(2):284–288. https://doi.org/10.1007/s00253-003-1415-3

Dhir B, Sharmila P, Saradhi PP, Nasim SA (2009) Physiological and antioxidant responses of *Salvinia natans* exposed to chromium-rich wastewater. Ecotoxicol Environ Saf 72(6):1790–1797. https://doi.org/10.1016/j.ecoenv.2009.03.015

Dong R, Formentin E, Losseso C, Carimi F, Benedetti P, Terzi M, Schiavo FL (2005) Molecular cloning and characterization of a phytochelatin synthase gene, PvPCS1, from *Pteris vittata* L. J Ind Microbiol Biotechnol 32(11–12):527–533. https://doi.org/10.1007/s10295-005-0234-1

Ferrini F, Fini A, Mori J, Gori A (2020) Role of vegetation as a mitigating factor in the urban context. Sustainability 12(10):4247. https://doi.org/10.3390/su12104247

Gawronski SW, Gawronska H (2017) Air Phytoremediation. In: Ansari A, Gill S, Gill R, Lanza RG, Newman L (eds) Phytoremediation. Springer, Cham. https://doi.org/10.1007/978-3-319-52381-1_19

Gourdji S (2018) Review of plants to mitigate particulate matter, ozone as well as nitrogen dioxide air pollutants and applicable recommendations for green roofs in Montreal, Quebec. Environ Pollut 241:378–387. https://doi.org/10.1016/j.envpol.2018.05.053

Ibrahim SI, Lateef MA, Khalifa HMS, Monem AA (2013) Phytoremediation of atrazine-contaminated soil using *Zea mays* (maize). Ann Agric Sci 58(1):69–75. https://doi.org/10.1016/j.aoas.2013.01.010

James P, Banay RF, Hart JE, Laden F (2015) A review of the health benefits of greenness. Curr Epidemiol Rep 2:131–142. https://doi.org/10.1007/s40471-015-0043-7

Jinger D, Kaushal R, Kumar R, Paramesh V, Verma A et al (2023) Degraded land rehabilitation through agroforestry in India: achievements, current understanding, and future prospectives. Front Ecol Evolut 11:1088796. https://doi.org/10.3389/fevo.2023.1088796

Kalve S, Sarangi BK, Pandey RA, Chakrabarti T (2011) Arsenic and chromium hyperaccumulation by an ecotype of *Pteris vittata*-prospective for phytoextraction from contaminated water and soil. Curr Sci 100(6):888–894

Kanoun-Boulé M, Vicente JAF, Nabais C, Prasad MNV, Freitas H (2009) Ecophysiological tolerance of duckweeds exposed to copper. Aquat Toxicol 91(1):1–9. https://doi.org/10.1016/j.aquatox.2008.09.009

Khaksar G, Treesubsuntorn Č, Thiravetyan P (2016) Effect of endophytic *Bacillus cereus* ERBP inoculation into non-native host: potentials and challenges for airborne formaldehyde removal. Plant Physiol Biochem 107:326–336. https://doi.org/10.1016/j.plaphy.2016.06.020

Kim K, Lee DW (2008) Efficiency of volatile formaldehyde removal of orchids as affected by species and crassulacean acid metabolism (CAM) Nature. Hortic Environ Biotechnol 49(2):132–137

Kim KJ, Yoo EH, Kays SJ (2012) Decay kinetics of toluene phytoremediation stimulation. Hortscience 47(8):1195–1198. https://doi.org/10.21273/hortsci.47.8.1195

Kiran KR, Rani M, Pal A (2009) Reclaiming degraded land in India through the cultivation of medicinal plants. Botany Res Int 2:174–181

Krishnamoorthi S, Ganguly S (2022) A short review on systematic approach for successful wasteland restoration. In: Ecological environment: a new perspective. LULU Publication, Raleigh, NC, pp 133–154

Lotfy SM, Mostafa AZ (2014) Phytoremediation of contaminated soil with cobalt and chromium. J Geochem Explor 144:367–373. https://doi.org/10.1016/j.gexplo.2013.07.003

Luna RK (2006) Plantation forestry in India. International book distributors, Dehradun, pp 361–41

Mishra VK, Upadhyaya AR, Pandey SK, Tripathi BD (2008) Heavy metal pollution induced due to coal mining effluent on surrounding aquatic ecosystem and its management through naturally occurring aquatic macrophytes. Bioresource Technol 99:930–936. https://doi.org/10.1016/j.biortech.2007.03.010

Molisani MM, Rocha R, Machado W, Barreto RC, Lacerda LD (2006) Mercury contents in aquatic macrophytes from two reservoirs in the Paraiba do Sul: Guandu river system. SE Brazil. Brazilian J Biol 66(1a):101–107. https://doi.org/10.1590/s1519-69842006000100013

Nazir MI, Idrees I, Danish P, Ahmad S, Ali Q, Malik A (2020) Potential of water hyacinth (*Eichhornia crassipes* L.) for phytoremidation of heavy metals from wastewater. Biol Clin Sci Res J 6. https://doi.org/10.54112/bcsrj.v2020i1.6

O'Connor D, Peng T, Zhang J, Tsang DCW, Alessi DS, Shen Z, Bolan NS, Hou D (2018) Biochar application for the remediation of heavy metal polluted land: a review of in situ field trials. Sci Total Environ 619–620:815–826. https://doi.org/10.1016/j.scitotenv.2017.11.132

Olguin EJ, Hernandez E, Ramos I (2002) The effect of both different light conditions and the pH value on the capacity of *Salvinia minima* BAKER for removing cadmium, lead and chromium. Acta Biotechonlogica 22(1–2):121–131. https://doi.org/10.1002/1521-3846(200205)22:1/2%3c121::aid-abio121%3e3.0.co;2-f

Pan J, Bai Z, Cao Y, Zhou W, Wang J (2017) Influence of soil physical properties and vegetation coverage at different slope aspects in a reclaimed dump. Environ Sci Pollut Res 24:23953–23965. https://doi.org/10.1007/s11356-017-9974-5

Pathak L, Shah K (2024) Natural colonizers effectively restore heavy metal polluted wasteland. Int J Phytorem 26(11):1873–1884. https://doi.org/10.1080/15226514.2024.2358380

Pokhriyal TC, Subba Rao BK (1986) Role of forests in mitigating air pollution. Indian Forester 112:573–582

Rai PK (2008) Phytoremediation of Hg and Cd from industrial effluents using an aquatic free floating macrophyte *Azolla pinnata*. Int J Phytorem 10(5):430–439. https://doi.org/10.1080/15226510802100606

Ranjan V, Sen P, Kumar D, Sarsawat A (2015) A review on dump slope stabilization by revegetation with reference to indigenous plant. Ecol Process 4(1):14. https://doi.org/10.1186/s13717-015-0041-1

Reeves RD, Baker AJ, Jaffré T, Erskine PD, Echevarria G, van Der Ent A (2018) A global database for plants that hyperaccumulate metal and metalloid trace elements. New Phytol 218(2):407–411. https://doi.org/10.1111/nph.14907

Sahu A, Sen S, Mishra SC (2020) Processing and properties of *Calotropis gigantea* bio-char: a wasteland weed. Mater Today: Proc 33:5334–5340. https://doi.org/10.1016/j.matpr.2020.03.024

Sandhya RD (2018) The role of native plants in restoring degraded ecosystems. J Emerg Technol Innov Res 5:12

Sarvade S, Gautam DS, Kathal D, Tiwari P (2017) Waterlogged wasteland treatment through agroforestry: a review. J Appl Nat Sci 9(1). https://doi.org/10.31018/jans.v9i1.1147

Scharenbroch BC, Morgenroth J, Maule B (2016) Tree species suitability to bioswales and impact on the urban water budget. J Environ Qual 45(1):199–206. https://doi.org/10.2134/jeq2015.01.0060

Singh R, Verma A, Prakash S (2023) The web of life: role of pollution in biodiversity decline. Int J Fauna Biol Stud 10(3):49–52. https://doi.org/10.22271/23940522.2023.v10.i3a.1003

Singh SK, Rao DN, Agrawal M, Pandey J, Naryan D (1991) Air pollution tolerance index of plants. J Environ Manag 32(1):45–55. https://doi.org/10.1016/s0301-4797(05)80080-5

Sinha RK, Herat S, Tandon PK (2007) Phytoremediation: role of plants in contaminated site management. Environmental bioremediation technologies, pp 315–330. https://doi.org/10.1007/978-3-540-34793-4_14

Sivaci ER, Sivaci A, Sokmen M (2004) Biosorption of cadmium by *Myriophyllum spicatum* L. and *Myriophyllum triphyllum* orchard. Chemosphere 56(11):1043–1048. https://doi.org/10.1016/j.chemosphere.2004.05.032

Su C, Meng J, Zhou Y, Bi R, Chen Z, Diao J, Huang Z, Kan Z, Wang T (2022) Heavy metals in soils from intense industrial areas in south China: spatial distribution, source apportionment, and risk assessment. Front Environ Sci 10:820536. https://doi.org/10.3389/fenvs.2022.820536

Talaie A, Ponraj M (2024) Impact of houseplants on reducing indoor air pollution. J Environ Treat Tech 12(3):1–13

Tan Y, Xu H, Zhang X (2016) Sustainable urbanization in China: a comprehensive literature review. Cities 55:82–93. https://doi.org/10.1016/j.cities.2016.04.002

Tartaglia ES, Aronson MF (2024) Plant native: comparing biodiversity benefits, ecosystem services provisioning, and plant performance of native and non-native plants in urban horticulture. Urban Ecosyst 27(6):2587–2611. https://doi.org/10.1007/s11252-024-01610-5

Teiri H, Pourzamzni H, Hajizadeh Y (2018) Phytoremediation of VOCs from indoor air by ornamental potted plants: a pilot study using a palm species under the controlled environment. Chemosphere 197:375–381. https://doi.org/10.1016/j.chemosphere.2018.01.078

Tomar OS (1997) Technologies of afforestation of salt-affected soils. Int Tree Crops J 9(2):131–158

Treesubsuntorn C, Thiravetyan P (2012) Removal of benzene from indoor air by *Dracaena sanderiana*: effect of wax and stomata. Atmos Environ 57:317–321. https://doi.org/10.1016/j.atmosenv.2012.04.016

Verma A, Das P (2024) Urbanization in the global south: perspectives and challenges. Vikalpa 49(1):110–112. https://doi.org/10.1177/02560909241234200

Verma VK, Tewari S, Rai Jpn (2008) Ion exchange during heavy metal bio-sorption from aqueous solution by dried biomass of macrophyte. Bioresource Technol 99:1932–1938. https://doi.org/10.1016/j.biortech.2007.03.042

Xiao Q, McPherson EG, Zhang Q, Ge X, Dahlgren R (2017) Performance of two bioswales on urban runoff management. Infrastructures 2(4):12. https://doi.org/10.3390/infrastructures2040012

Xiao R, Shen F, Du J, Li R, Lahori AH, Zhang Z (2018) Screening of native plants from wasteland surrounding a Zn smelter in Feng County China, for phytoremediation. Ecotoxicol Environ Saf 162:178–183. https://doi.org/10.1016/j.ecoenv.2018.06.095

Xu L, Teng Y, Li ZG, Norton JM, Luo YM (2010) Enhanced removal of polychlorinated biphenyls from alfalfa rhizosphere soil in a field study: the impact of a rhizobial inoculum. Sci Total Environ 408(5):1007–1013. https://doi.org/10.1016/j.scitotenv.2009.11.031

Yan A, Wang Y, Tan SN, Mohd Yusof ML, Ghosh S, Chen Z (2020) Phytoremediation: a promising approach for revegetation of heavy metal-polluted land. Front Plant Sci 11:359. https://doi.org/10.3389/fpls.2020.00359

Yang SZ, Jin HJ, Wei Z, He RX, Ji YJ et al (2009) Bioremediation of oil spills in cold environments: a review. Pedosphere 19(3):371–381. https://doi.org/10.1016/s1002-0160(09)60128-4

Yoo MH, Kwon YJ, Son KC, Kays SJ (2006) Efficacy of indoor plants for the removal of single and mixed volatile organic pollutants and physiological effects of the volatiles on the plants. J Am Soc Hortic Sci 131(4):452–458. https://doi.org/10.21273/JASHS.131.4.452

Yu S, Miao CP, Song H, Huang YQ, Chen W, He XY (2019) Efficiency of nitrogen and phosphorus removal by six macrophytes from eutrophic water. Int J Phytorem 21(7):643–651. https://doi.org/10.1080/15226514.2018.1556582

Zhao A, Gao L, Chen B, Feng L (2019) Phytoremediation potential of *Miscanthus sinensis* for mercury-polluted sites and its impacts on soil microbial community. Environ Sci Pollut Res 26:34818–34829

Part IV
Climate Extremes, Food Security, Nutritional Impacts and Sustainability

Chapter 17
Climate-Resilient Food Systems in Fiji: A Sustainable Path Forward

Mumtaz Alam and Kunal Avishek Gounder

Abstract Climate change is a global issue posing various implications for humans and the environment. Food security is one of the significant factors that need to be discussed in relation to climate change. Changing climate brings many impacts, such as rising sea levels, saltwater intrusions, coastal erosion, coastal flooding, and climate events, that affect food security. Hence, this chapter focuses on climate food-resilient systems in Fiji and the strategies that have been adopted to deal with improving food security. Fiji, being a SIDS, is vulnerable to the impacts of climate change, causing depletion of local produce. This results in the country becoming more reliant on imported foods, leading to various health-related problems, thus reducing the life expectancy rate overall. The chapter also elaborates on how improving climate-resilient food systems can help achieve various SDGS, such as 1, 2, 3, 11, 12, and 13. Nature-based solutions (NbS), community-based approaches, ecosystem-based solutions, and sustainable agricultural practices are just some of the ways Fiji is working to improve the food system, supported by the policies of the Fiji Government. The chapter findings suggested that the use of available land increases food production and reduces dependence on imported products.

Keywords Smart agriculture · Climate-resilient food systems · Food security · SDGs · Fiji

17.1 Introduction

Like many other Small Island Developing States (SIDS), Fiji is very vulnerable to climate change impacts. Impacts such as rising sea levels, coastal erosion, saltwater intrusion, changing precipitation patterns, and frequent extreme weather events

M. Alam (✉)
Department of Social Sciences, College of Humanities, Education and Law, Fiji National University, Lautoka, Fiji
e-mail: mumtaz.alam@fnu.ac.fj

K. A. Gounder
College of Humanities, Education and Law, Fiji National University, Suva, Fiji

A. Roy et al. (eds.), *Environment and Public Health*, Sustainability Solutions,
https://doi.org/10.1007/978-3-031-99770-9_17

are just the tip of the iceberg. All these impacts have a catastrophic impact on the country's food systems, which is significant for the survival of livelihood and their health. This chapter aims to provide a comprehensive analysis of the challenges of the food system posed by climate change and the opportunities for a climate-resilient food system in Fiji. This will mostly focus on sustainable agricultural practices, community-based adaptation strategies, and government interventions. This chapter aims to elaborate in-depth on how Fiji, being a SIDS, is leading the climate-resilient food system despite being vulnerable to the impacts of climate change. This study addresses a significant gap in the literature on sustainable agricultural adaptations in Small Island Developing States by uniquely combining conventional ecological knowledge with modern scientific methods to create a thorough framework for climate-resilient food systems in Fiji. Moreover, this work provides a novel examination of how ecosystem-based adaptations and nature-based solutions may simultaneously promote food security, enhance public health outcomes, and support several Sustainable Development Goals in the framework of Fiji's particular geographical and socioeconomic constraints.

17.1.1 Background and Context

Fiji's geographical characteristic makes it particularly vulnerable to several climate-related hazards, such as rising sea levels, increased frequency and intensity of tropical cyclones, and changes in precipitation pattern that affects agricultural production and water security. Fiji is a leading Pacific Island Country in the region that has been comprehending traditional ecological knowledge with contemporary scientific techniques in climate adaptation, mitigation, and resilience. Fiji has developed legislations like the National Climate Change Policy, Climate Change Act, and National Adaptation Plan, which provide a comprehensive guideline for tackling short-term and long-term climate concerns. The resilience strategy includes ecosystem-based adaptation techniques, like coral reef conservation and mangrove reforestation, which act as a natural barrier to protect the coastal area from storm surges and coastal erosion.

Fiji depends on its agricultural sector to meet domestic demand and for export purposes. Due to climate change it has affected agricultural production in many ways. Fiji has pioneered new methods to achieve resilience in the agriculture sector, adapting climate-smart farming methods with the integration of traditional knowledge to improve food security and economic stability. Community-based adaptation measures are essential to Fiji's resilience since traditional knowledge is actively integrated into disaster risk reduction initiatives and early warning systems. Traditional knowledge is a skill inherited from the ancestors on environmental issues that can help in resilience with the integration of modern techniques for a vast impact. Fiji is also dedicated to renewable energy as it has targeted to replace the current 45% of fossil fuel-generated electricity and achieve 100% renewable energy generation by

2030 (Long et al. 2024). To achieve this target, the Fiji government has initiated "Climate Finance" and policies such as the provision of green low-interest loans, asset purchase programs, and liberalization of collateral towards green assets. Climate resilience in Fiji has improved over the years due to changes in the regulations and coastal protection strategies. Fiji National Adaptation Plan addresses components of infrastructure for transport, water and sanitation, energy, and hazard management to achieve climate resilience (Ministry of Economy, Republic of Fiji 2021).

Notwithstanding these proactive initiatives, Fiji persists in encountering substantial problems, such as constrained resources, limitations in technical competence, and a continual need for foreign assistance. The nation's holistic strategy for climate resilience, which harmonizes environmental conservation with economic advancement and social fairness, exemplifies a model for other Small Island Developing States with analogous difficulties. The amalgamation of physical and soft adaptation strategies, with robust policy frameworks and international collaboration, establishes Fiji as a frontrunner in climate resilience among Pacific Island Countries. The nation's history illustrates that successful climate resilience requires solid institutional structures and technological solutions, substantial community involvement, and the preservation of traditional knowledge systems. As global climate issues escalate, Fiji's strategy for enhancing resilience provides significant insights for other susceptible countries, emphasizing the need for comprehensive, contextually relevant approaches that tackle both immediate climate effects and long-term sustainability objectives. The challenge of boosting agricultural production is the availability of land to the interested farmers and the limited support provided. 87% of the land is under communal tenure, which is given out for lease by the i-Taukei Land Trust Board of Fiji (Scheyvens and Russell 2012). Agricultural land leases are given out for 30 years, which makes it difficult for farmers to invest in agricultural production and obtain financial loans (Government of Fiji 1978). However, the food and agriculture sectors are increasingly embracing green economic principles to create more sustainable agricultural systems. These principles focus on balancing economic growth with environmental stewardship and social equity (Agan and Bayrak 2025).

17.1.2 Importance of Climate-Resilient Food Systems

Climate-resilient food systems are essential to food security, improve public health, and support the economic stability of Fiji. By implementing innovative approaches such as policies and practices, Fiji has the potential to reduce the dependence on imported foods, boost local production, and utilize the abundant vacant lands. According to Trading Economics (2025), Fiji's food imports accounted for 18.99% in 2023. This figure is high for a developing country like Fiji with a background in agriculture, land availability, and labour availability and indicates a notable vulnerability within its food security framework. The significant reliance on food imports in Fiji is troubling, especially given its status as a developing island nation with an agricultural background, available land, and labour resources, which suggests that it could feasibly enhance domestic food production. This statistic carries significant implications for food security.

Firstly, this highlights an excessive dependence on external food sources, rendering Fiji susceptible to global market fluctuations, supply chain disruptions, and international price volatility. Global crises, such as pandemics or conflicts, can result in food shortages and price increases that disproportionately impact vulnerable populations. Secondly, the extent of food importation signifies a lost opportunity to enhance local agricultural systems and bolster rural economies. Resources that could be allocated to enhancing domestic food production and generating local employment are instead being diverted to the acquisition of foreign goods. Thirdly, imported foods frequently comprise processed items that possess lower nutritional value than fresh local produce, thereby contributing to the diet-related health issues outlined in your chapter, particularly the elevated rates of diabetes and decreasing life expectancy in Fiji.

Local food production, rooted in communal traditions and climate-adaptive farming methods, has historically emphasized a plant-based diet. Dietary colonialism has replaced small-scale agriculture with non-nutritive commercial crops, intensive animal husbandry, and the exporting of high-quality food products. This has resulted in a diabetogenic environment, diminishing dietary quality and increasing the risk of type 2 diabetes. Nonetheless, locally sourced foods such as starchy tubers, legumes, fruits, and shellfish are vital for autonomous local economic production and nutritional well-being. Studies indicate that local food production, supported by local and Indigenous knowledge, cultural preservation, and collective action, may facilitate the reclamation of healthy and climate-resilient food systems in tiny islands (Marrero and Mattei 2022).

17.1.3 Food Security Challenge

The demographic situation of Fiji is challenging, with a population of 924,145 in 2023 projected to rise by 8.2% by 2050 (WHO n.d.). Figure 17.1 shows the demographic trend of Fiji over time. This reflects broader systematic health challenges to food security. The local food supply will be unable to satisfy the demands of a growing population, necessitating an increased reliance on imported and processed foods for sustenance. This dilemma also affects the coastal communities, where there is significant dependence on fishing and root crops for food security. In Pacific coastal communities, fish consumption is often two to four times higher than the world norm, with 50–90% of this fish sourced from subsistence fishing (Bell et al. 2009). The significant reliance on subsistence agriculture and fishing in rural regions arises from the restricted income opportunities attributable to the isolated locations of numerous islands, the elevated coastline-to-land area ratio, the deficiency of arable land resulting from the steep topography of many islands, and the prevalence of coral atolls characterized by poor, sandy soils (Bell and Taylor 2015). Addressing public health issues linked to lifestyle and dietary changes among rapidly expanding urban populations resulting from the influx of energy-dense, nutritionally deficient foods. These imports are supplanting indigenous meals due to the limited availability of arable land for agricultural production by migrants to urban areas, rising disposable income among city residents, and insufficient knowledge of the repercussions of inadequate nutrition. The impact on urban populations in Pacific Island nations has been substantial. Nine of the ten countries with the greatest prevalence of overweight and obesity and seven of the ten countries with the highest incidence of diabetes are Pacific Island nations (Bell and Taylor 2015). Figure 17.2 shows the decline in life expectancy in Fiji, and this will continue to decline if the local food insecurity continues, with diabetes being the major cause of death in Fiji (Fig. 17.3).

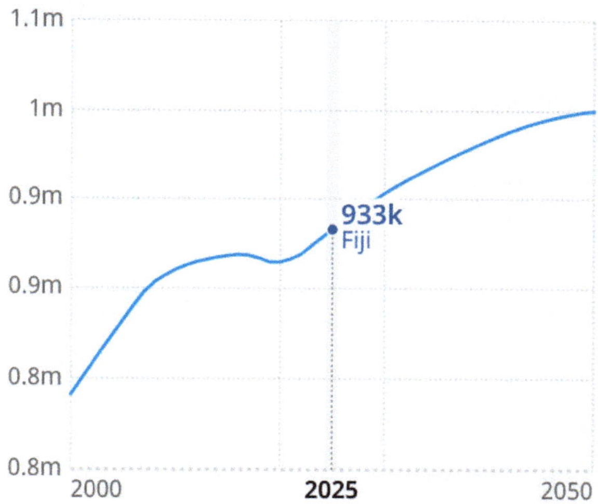

Fig. 17.1 Population trend and prediction of Fiji (*source* World Health Organization)

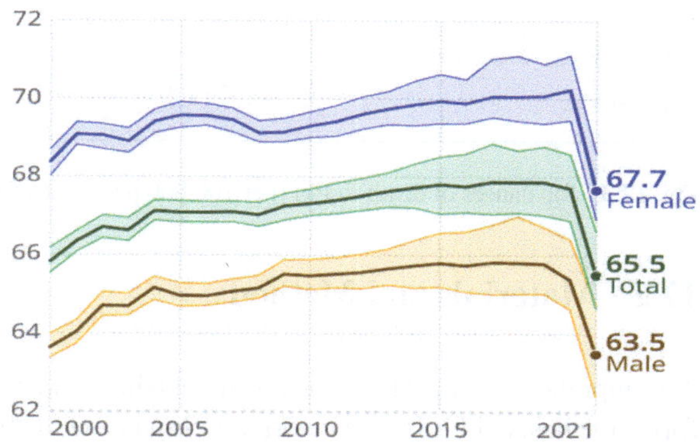

Fig. 17.2 Life expectancy of Fiji (2000–2021) (*source* World Health Organization)

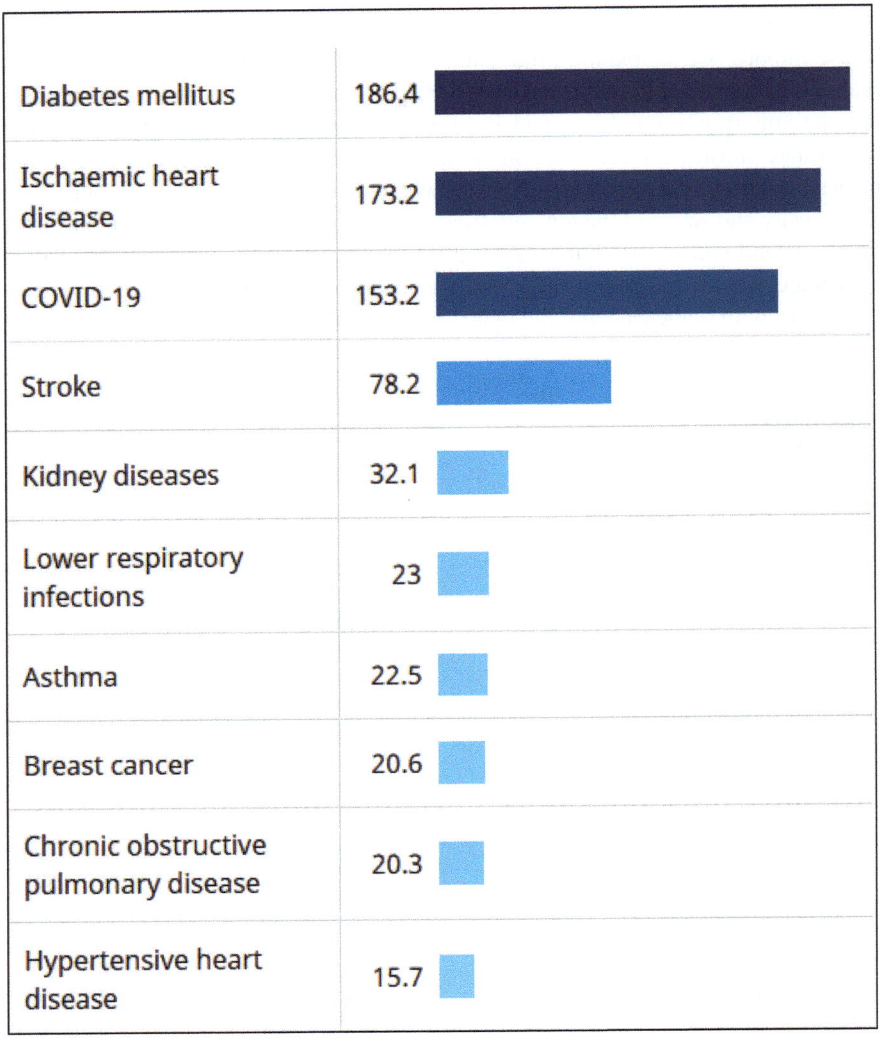

Fig. 17.3 Top causes of death in Fiji (per 100,000 people) (*source* World Health Organization)

17.2 Materials and Methods

A comprehensive desktop study synthesizing data from many secondary sources forms the basis of Fiji's framework for climate-resilient food systems. This approach systematically reviews and analyses government reports, scholarly literature, policy papers, and case studies with an eye towards Fiji's agriculture sector and climate adaptation methods. Important sources include peer-reviewed journal articles on Pacific Island food systems, climate vulnerability assessments particular to Fiji, publications from the Fijian Ministry of Agriculture, reports from international organizations, including the Food and Agriculture Organisation (FAO) and the United Nations Development Program (UNDP), and peer-reviewed journal entries. Four main pillars—traditional agriculture knowledge, climate-smart agriculture, value chain development, and policy and governance—are integrated into the conceptual framework shown in the diagram (Gennari et al. 2019). These pillars, which emphasize recurring characteristics vital to strengthening resilience in Fiji's food systems, were found through the study of the literature. The sources were selected purposively and interpreted. By means of effective implementation strategies recorded in the secondary sources, supporting mechanisms, including research and innovation, capacity building, and financing mechanisms, were similarly derived. The technique prioritized a comprehensive approach that respects current agricultural technologies and indigenous knowledge systems since they complement each other in generating sustainable solutions. Particularly focused on the connections between several food system components—shown by the diagram's connecting lines—was their identification. These relationships support the nature of climate resilience building in Fiji's food systems setting by showing how actions in one area can favourably affect outcomes in others (Fig. 17.4).

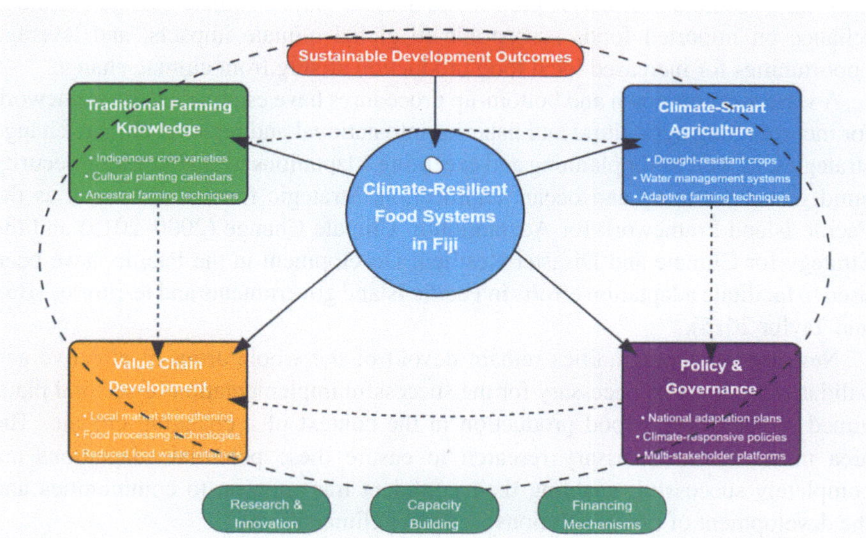

Fig. 17.4 Climate-resilient food system in Fiji

17.3 Results and Discussion

17.3.1 Climate Change and Food Security

Fiji recognizes that climate change is expected to impact its strategies for ensuring food security for expanding rural and urban populations. This insight arises from the endeavours of development partners and the extensive experience of Pacific Island communities in managing climate unpredictability in fisheries and agriculture. The region's reactions to climatic variability are influenced by its susceptibility to the fluctuations of the El Nino Southern Oscillation (ENSO), which impacts the location of the western Pacific warm pool, the South Pacific convergence zone (SPCZ), the intertropical convergence zone (ITCZ), and the incursion of the western Pacific monsoon, potentially leading to acute localized droughts in certain years and persistent flooding.

Although Fiji's population has shown resilience in managing climatic unpredictability, there is a broad acknowledgement that the area will need support to adapt to the anticipated severe climatic changes resulting from ongoing greenhouse gas emissions. It requires support to ensure that strategies for sustaining and enhancing local agricultural and fisheries production are resilient to climate change, diminish reliance on imported foods susceptible to global climate impacts, and leverage opportunities for increased local food production arising from climate change.

A variety of top-down and bottom-up procedures have established the framework for incorporating agriculture and fisheries into national and regional climate change strategies, as well as for planning and executing adaptations to enhance food security amid global warming and ocean acidification. Strategic frameworks, such as the Pacific Island Framework for Adaptation to Climate Change (2006–2015) and the Strategy for Climate and Disaster-Resilient Development in the Pacific, have been used to facilitate adaptation efforts in Pacific Island governments and territories (Bell and Taylor 2015).

Nevertheless, communities remain devoid of the whole array of effective and validated instruments necessary for the successful implementation of sectoral plans aimed at augmenting food production in the context of a changing climate. The area must do the necessary research to ensure these potential adaptations are completely successful, enabling their confident transmission to communities and the development of useful solutions for future climates.

17.3.2 Nature-Based Solutions

Nature-based solutions (NbS) provide an effective approach to tackling the interrelated issues of climate change, biodiversity decline, and stresses unrelated to climate in coastal communities. Nature-based solutions (NbS) are acts inspired by and derived from nature that provide many advantages, including the protection of biodiversity, environmental services, and economic benefits. They are economically efficient and may enhance synergies among the United Nations' Sustainable Development Goals for environmental, social, and economic advancement. Nature-based solutions (NbS) include sustainable management approaches, including protection and both active and passive restoration, fostering a comprehensive viewpoint. Nonetheless, apprehensions about possible misapplications and difficulties in execution remain. The IUCN established the Global Standard for NbS to delineate, articulate, and direct the execution and dissemination of NbS. Frameworks such as the social-ecological vulnerability framework have been established to assess the efficacy of nature-based solutions (NbS), taking into account ecological and socioeconomic vulnerabilities as well as adaptive capability.

The ocean links all coastal countries and encompasses the most extensive liveable area, abundant in different coastal and marine ecosystems. Coastal habitats, including mangrove forests and coral reefs, offer vital ecosystem services, including storm protection, carbon sequestration, water filtering, and habitat for diverse species. Marine ecosystems located in the open ocean include a diverse array of conditions and are essential for controlling the Earth's temperature, sustaining global fisheries, and preserving biodiversity, the carbon cycle, and nutrient distribution. Nonetheless, these services are becoming endangered by environmental stresses such as climate change and biodiversity decline. Natural biodiversity for marine and coastal ecosystems has the potential to leverage local and global socio-ecological advantages, including climate mitigation and adaptation, as well as a sustainable blue economy. Nonetheless, several hurdles exist in the implementation of nature-based solutions (NbS) across marine and coastal environments as opposed to terrestrial ecosystems. These include a substantial deficiency in comprehending how marine biodiversity and ecosystems provide ecosystem services, the intricacy of marine and coastal ecosystems, and the scarcity of long-term data for marine systems.

Marine and coastal National Biodiversity Sites must be conceived and executed at a seascape scale, taking into account the adjacent terrain and the social context of local populations. Coastal systems have jurisdictional constraints and complexities in regulating various public and private operations owing to the absence of clearly defined borders for the land–water coastal zone. Current data indicates that measures such as marine protected areas and coastal wetland restoration may be economically efficient. A transition to comprehensive spatial planning that incorporates coastal and marine nature-based solutions while departing from exclusively conventional grey infrastructure and exploring hybrid alternatives may result in more ecological and economically feasible coastal protection strategies.

To address challenges and enhance the implementation of nature-based solutions (NbS) in marine and coastal ecosystems, O'Leary et al. (2023) recommend three research priorities: elucidating the relationship between marine and coastal biodiversity and ecosystem services, offering scientific guidance on the optimal implementation of marine and coastal NbS, and fostering improved communication, collaboration, ocean literacy, and stewardship.

These objectives must be evaluated together with scientific insights from the following priorities:

- Informing the implementation instructions developed as a priority
- Initiatives to enhance awareness, promote cooperation, and jointly develop solutions under priority
- Continuous involvement is needed and must be included across all research objectives.

Diverse stakeholders, including local communities, policymakers, industry representatives, and multidisciplinary scholars, will engage in all research priorities, with involvement varying according to the activities implemented. By identifying the obstacles to implementation and enhancing the knowledge foundation for design, monitoring, and evaluation via research, we may bridge the disparity between anticipated and actual results of nature-based solutions in marine and coastal environments.

17.3.3 Ecosystem-Based Adaptation

Fiji, like other Pacific Island Countries, depends on the ecosystem for its survival and is quite attached to it as well. Ecosystem-based adaptation (EbA) projects have positive impacts on the people as they increase food production at the community level, allowing food security and building resilience to climate change. EbA allows the use of traditional knowledge to be put into practice, which is cost-effective and involves the community. The Fijian Government's National Adaptation Plan (NAP) emphasizes food security, land use, development, and agricultural growth in Fiji (Ministry of Economy, Republic of Fiji 2021). Ecosystem-based adaptation (EbA) strategies have been delineated for Taveuni, including ecological restoration and conservation efforts (SPREP 2018). The over-exploitation of soil resources for commercial agriculture has resulted in the degradation of soil and poor land utilization practices, increasing community vulnerability to climate change. Investing in ecosystem-based adaptation (EbA) strategies may enhance land management, expand livelihood opportunities, empower people, and safeguard land and marine ecosystems. Strategies like green manure, extended fallow times, and natural fertilizers have been tested to enhance soil mineral management in Taveuni. Macuata Province, the first province in Fiji to implement a provincial Natural Resource Management Strategy (NRMS) assisted by the World Wildlife Fund, has an active assistance program for traditional fishing areas along the coastline (SPREP 2018). The biggest qoliqoli, Qoliqoli Cokovata, is regarded as a critical access point for the whole Great Sea Reef system, designated as a Ramsar Wetlands Site (Prasad et al. 2022).

Another example of the EbA project is using Vetiver Grass (Chrysopogon zizanioides or "Sunshine") as a cost-effective traditional knowledge to protect the erosion of riverbanks during flooding and was first introduced in Fiji in the 1950s for soil and water conservation in the sugarcane plantations (Truong and Creighton 1994). When the impact of climate change and risk erosion became a serious problem, the need for Vetiver Grass was given importance again. Since 2015, the Fiji Government has started supporting the planting of Vetiver Grass for coastal and riverside communities to become resilient to erosion (Dekens et al. 2024). Ministry of Waterways was the implementation arm of the government to deliver support on vetiver planting; some of the communities that were assisted are Naveiveiwali, Malabe, and Nabouva in the Tailevu Province (Dekens et al. 2024).

17.3.4 Sustainable Agriculture Practices

Fiji's First Fiji 2020 Agriculture Sector Policy Agenda emphasizes the enhancement of climate-smart food production systems with the objective of modernizing Fijian agriculture. The strategy, developed with UN supervision, seeks to provide security against uncertainties created by climate change via import substitution. The strategy underscores the need to transition from subsistence agriculture, which constitutes around 80% of Fiji's agricultural output, to market-oriented production to enhance the economic viability and climatic resilience of Fiji's food security. The strategy seeks to completely industrialize Fiji's agricultural industry to provide resilience against uncertainties generated by climate change.

The Fijian government is prioritizing sustainable modernization while also promoting traditional agricultural techniques, often associated with organic farming. The development of traditional agriculture is attracting public interest as a crucial area for adaptation to climate change. Nevertheless, the official organic agricultural industry in Fiji developed slowly until the beginning of the twenty-first century. In 2007, efforts financed by the International Fund for Agricultural Development and the Secretariat of the Pacific Community (SPC) sought to mitigate the financial and administrative obstacles to certification. These initiatives resulted in the establishment of Pacific Regional Standards for Organic Agriculture Production and the formation of the Pacific Organic and Ethical Trade Community (POETCom), which is presently leading the promotion of organic agriculture in Fiji and the broader South Pacific region. Initiatives to equate traditional agriculture with organic agriculture are often advocated as an indigenous or distinctly Fijian approach to foster future resilience against climate change. The narratives arising from this development attempt are diverse and characterized by conflicting themes, including the tangible demands of climate change, nostalgic traditionalism, and economic advancement. All are dedicated to creating a disaster-free future for Fiji that embodies agricultural excellence rooted in traditional resilience via the revival of an ambiguous concept of Fijian agrarian heritage (Griswold 2021).

The Pacific Adaptation to Climate Change (PACC) initiative sought to rectify the inadequate management and rampant proliferation of invasive plants in low-lying catchments, which adversely impact agricultural productivity. National Institute of Water and Atmospheric Research's (New Zealand) proposals, including various strategies for maintaining drainage systems, enhanced the sustainability of the drainage system and project outcomes. The initiative highlighted the significance of drain maintenance to communities in pilot locations, and farmers acknowledged their responsibility in maintaining the drains via physical clearance and spraying. The initiative enhanced understanding of crop diversification and the use of technologies such as the extreme rainfall calculator, conveyance estimation system, and hydraulic models to support the Land and Water Resources Management Division in forthcoming drainage evaluations. The creation of a nursery by farmers in Nakelo reduced dependence on the Ministry of Agriculture for planting materials, hence enhancing farmers' self-sufficiency. In Visama, Nakelo, and Rovadrau, Deuba farmer groups will enhance the sustainability of operations after the project's conclusion. The initiative successfully enabled farmers to recover 10 ha of land that was previously uncultivable owing to waterlogging. Sixty-five farmers are using the crop varieties provided by the initiative and integrating the farming methods acquired during the training into their agricultural programs (Taylor et al. 2015). Thus, sustainable agriculture practices allow Fiji to be on track with Sustainable Development Goal 2: Zero Hunger.

17.3.5 Community-Based Approach

Medina Hidalgo et al. (2020) carried out extensive research on how the rural communities in Fiji are sustaining their food security despite climatic impacts. In order to reduce the risks posed by climate change, the community is implementing separate adaptation strategies. The intercropping of cassava with taro and yaqona is the most common method for providing shade under these conditions. Yaqona plants are being cultivated by farmers instead of taro and cassava, which is causing the output of these crops to decrease. The planting dates are being adjusted in order to maximize the use of the rainy season for crop establishment and to lengthen the duration of the planting process. The cultivation of yaqona in nurseries is yet another method that may be used to protect seedlings from high temperatures. In reaction to the tremendous rains, some farmers have resorted to using even the most basic irrigation techniques. The evaluation and assessment of the quality of each child's school meals is an example of a project that has had a positive impact on the dietary habits of students (Medina Hidalgo et al. 2020).

McGuigan et al. (2022) research to evaluate the resistance of a food system in Fiji built on agroforests to Cyclone Winston, with a particular emphasis on the scale distribution of starch crops. The agroforests that were found in the communities of Fiji were resilient. They became more abundant over the course of time, which contributed to their robustness and redundancy. The system's responsiveness and resourcefulness were brought to light by the findings of new starch crop species detected in 2017 and 2019. Through the introduction of new starch crop species and crop cultivars, the Fijian agroforests have shown their high level of dynamism. This has made it possible for substitutes and alterations to be made, which in turn enables adaptation and recovery. For this particular reason, the most often mentioned emergency food crops that farmers indicated were sweet potato, tannia, and spiny yam.

According to the findings of the research by McGuigan et al. (2022), 49,100 sweet potato cuttings were dispersed across Fiji after Cyclone Winston. This was most likely accomplished with the assistance of the government and several social networks. About 79% of farmers who farmed sweet potatoes planted new cultivars one year after the storm, according to the records that were kept before the cyclone (McGuigan et al. 2022). In the aftermath of the hurricane, the Fijian agroforest food systems demonstrated a high degree of responsiveness, as seen by the rapid replacement of taro with sweet potato. Taro was temporarily replaced by sweet potato. It is possible that Fiji's dependence on sweet potatoes has a reduced potential for climate change adaptation and mitigation techniques compared to other Pacific Island nations where sweet potatoes have played a significant role.

The introduction of new crop cultivars was followed by the loss of previous cultivars, particularly traditional cultivars, which village elders often described as being more resilient and resistant to environmental perturbation. This occurred despite the fact that crop cultivars had a high degree of dynamism. The efforts that are made to locate and disseminate these plants that are significant in terms of their cultural, gastronomic, and post-disaster recovery characteristics may help to strengthen resilience to natural disasters and contribute to the transmission of cultural information about these plants and the significance they have.

The robustness of Fiji's agroforest systems to severe weather events, especially energy-dense staple crops, is the primary emphasis of this research. The study suggests that Fiji agroforest systems still offer significant levels of food system resilience to severe weather events, particularly for energy-dense staple crops. This is the case even though there have been changes in biodiversity and a historical loss in total cultivar variety for the majority of crops throughout the Pacific. Despite this, agroforests in the South Pacific continue to face a persistent threat from agrodeforestation and socioeconomic development, both of which are driving changes in land ownership and usage.

17.3.6 Food Security Policy Initiative

The Agriculture Strategic Development Plan and Green Growth Framework for Fiji are essential planning tools, fundamentally grounded in national development planning mandates. Both plans emphasize initiatives concerning nutrition, health, agriculture, and climate change. The Plan of Action for Nutrition and the Non-Communicable Diseases Strategic Plan are the only papers that do not expressly reference or relate to other policies and strategies (Republic of Fiji 2017). The green growth framework highlights that Fiji does not possess a comprehensive national strategy on food security that encompasses agriculture, fisheries, biosecurity, health, nutrition, and education (Ministry of Strategic Planning 2014). The reviewed papers span from 2010 to 2050, with the majority of climate change planning documents created in 2018, likely as a result of Fiji's aggressive reaction to the adoption of the Paris Agreement. The Health Ministry's Climate Change and Health Strategic Action

Plan predates the majority of significant climate change publications (Ministry of Health 2016). The green growth framework predates several other planning documents, elucidating its role in producing mandates and defining priorities for various sectoral planning instruments.

The national development and adaptation planning documents of Fiji emphasize health, nutrition, and agriculture as key areas for intervention. The National Action Plan (NAP) comprises 160 adaptation measures, of which 33 pertain to these sectors (Medina Hidalgo et al. 2022). The majority of actions emphasize management and planning activities, followed by capacity-building. Most proposed actions seek to incorporate a response to climate change through the alignment of sectoral or national development goals with climate change adaptation objectives. The predominant integration method employed in the NAP and the green growth framework is a normative approach, comprising 50% and 45% of integration actions, respectively (Medina Hidalgo et al. 2022). The NAP delineates a series of measures not associated with specific sectors, referred to as "system components". Horizontal and vertical integration are encompassed within these system components, linked to a total of 21 specific actions. The majority of integration actions aim for coordination (81%), harmonization (14%), and prioritization (5%) (Medina Hidalgo et al. 2022).

Climate change adaptation is incorporated through targeted measures in Fiji's sectoral planning frameworks for agriculture and health. In the agricultural sector, initiatives emphasize the implementation or enhancement of practices and behaviours, including conservation agriculture, agroforestry, soil and water management techniques, and diversification of farming systems. The health sector has formulated the Climate Change and Health Strategic Action Plan, emphasizing the significance of climate change. The actions advocated in these plans primarily emphasize management, planning, and capacity-building initiatives aimed at enhancing the ability of organizations and stakeholders within the health sector to integrate climate change adaptation considerations into their services.

17.3.7 Sustainable Development Goals (SDGs) and Food Security in Fiji

The Sustainable Development Goals (SDGs) seek to establish a sustainable and secure food system through an emphasis on agriculture and nutrition. SDG 2 specifically seeks to eradicate hunger, attain food security and enhanced nutrition, and foster sustainable agriculture. Additional objectives pertaining to the food system encompass poverty alleviation (SDG 1), sustainable consumption and production (SDG 12), climate change adaptation and mitigation (SDG 13), land utilization and ecosystems, and sustainable fisheries management (Brooks 2016).

SDG 2 is based on "Good Health and Well-being", and this can be achieved if there is enough food security to meet the demand of all individuals, hence reducing hunger, malnutrition, and diet-related diseases, which improves the health of the general public. Food security also strengthens SDG 11, "Sustainable Cities and Communities", by making the food supply resilient and equitable food distribution, making the cities and communities sustainable rather than depending on imported foods.

Moreover, SDG 12 emphasizes sustainable consumption and development practices, encompassing eight distinct targets for food security. The objectives encompass production efficiency, reduction of food losses, waste management, sustainable business practices, and public procurement (Bizikova et al. 2020). The objectives for minimizing waste generation and the rationale for subsidizing fossil fuels encompass both production and consumption aspects. SDG 12 encompasses eight distinct targets aimed at promoting sustainable consumption and development patterns (Rabbi et al. 2021; Gasper et al. 2019). In addition, food security also contributed to SDG 13 "Climate Action", by promoting and adapting to climate-resilient farming methods such as NbS, ecological-based solutions, and agroforestry, which reduced greenhouse gas emissions and boosted the food system to adapt to climate change.

Rural regions constitute the residence of a significant portion of the impoverished global population, with agriculture serving as the primary economic endeavour. Significant enhancements in agricultural investment are essential to elevate incomes and augment the availability of sustainably produced food. Governments and public investment, bolstered by development assistance, can enhance and draw private investment. Key focus areas for public expenditure encompass research, innovation, rural infrastructure, social protection, and support mechanisms to enhance nutrition.

The growth of agricultural productivity will enhance food availability and advantage consumers, resulting in reduced domestic prices. Nevertheless, the reduction in prices undermines some of these benefits, necessitating comprehensive development to facilitate the transition of less competitive farmers into more lucrative endeavours. Trade will assume a progressively vital role in safeguarding global food security, necessitating that developed and significant emerging economies refrain from implementing policies that disrupt world markets.

Coordinated action is essential to prevent national policies from producing spillover effects that undermine food security in impoverished nations. Climate change and the degradation of land, water, and biodiversity necessitate modifications in production systems. National policies must align with sustainable productivity goals, necessitating increased efforts in agricultural research and development, technological advancement, and skill enhancement. Environmental policies are necessary to establish clearly defined property rights for natural resources and address comprehensive environmental challenges across the economy. Targeted agri-environmental policies are essential for effectively mitigating adverse environmental effects and enhancing resource management.

17.4 Conclusion

In conclusion, having a sustainable food system in Fiji plays a significant role as the catastrophic impacts of climate change are putting immense pressure on food security, as productivity is declining, and the population is increasing. This chapter provided a brief background of Fiji, highlighting how vulnerable Fiji is to climate change and the importance of having a sustainable, food-resilient system in Fiji. This chapter has also highlighted the current measures in place to secure food security, such as NbS, ecosystem-based solutions, sustainable agriculture practices, community-based approaches, and government interventions, such as policies introduced to support the growth of the agriculture sector in Fiji.

This chapter also discovered that there is an abundance of land available that can be utilized to increase food production to reduce the dependence on imported produce, but a lack of land security and support from the government is a challenge for interested farmers to invest in farming. There is a serious need to increase food production in Fiji as the country is importing a large number of goods that can be produced locally with less investment and support from the government. Fiji, an island community with strong relations with nature, land, and ocean, can invest more in food production and meet domestic demand and export.

Several important policy suggestions stand out as one looks forward to strengthening Fiji's climate-resilient food systems. First, the government could think about changing land tenure rules to offer longer-term agricultural leases past the present 30-year restriction, motivating farmers to adopt sustainable techniques. Second, more financial assistance and technical training for small-scale farmers using climate-smart agriculture practices would help to speed the shift from subsistence to market-oriented production. Future studies should aim to measure the economic advantages of ecosystem-based adaptations and nature-based solutions to support the argument for their more general use. Examining creative funding sources, including green bonds and climate finance, can help to address limited resources. How well climate-resilient food systems are included in national development planning will influence the attainment of SDGS 1, 2, 3, 11, 12, and 13 in Fiji. Fiji can build a thorough framework that concurrently tackles food security, economic growth, environmental preservation, and climate action by matching agriculture policy with these SDGS, thereby setting an example for other Small Island Developing States confronting comparable difficulties.

References

Agan B, Bayrak R (2025) The future of the food and agriculture sector in terms of green economics and sustainable development. In: SMEs in agriculture and food industry during the health crises: experience from the COVID-19. Springer, Cham, pp 119–146

Bell JD, Kronen M, Vunisea A, Nash WJ, Keeble G, Demmke A, Andréfouët S et al (2009) Planning the use of fish for food security in the Pacific. Mar Policy 33(1):64–76

Bell J, Taylor M (2015) Building climate-resilient food systems for Pacific Islands. WorldFish

Bizikova L, Jungcurt S, McDougal K, Tyler S (2020) How can agricultural interventions enhance contribution to food security and SDG 2.1? Glob Food Secur 26:100450

Brooks J (2016) Food security and the sustainable development goal

Dekens J, Bujold P, Mannle K (2024) Behavioural science for climate change adaptation: a case of ecosystem-based adaptation in Fiji. International Institute for Sustainable Development & the Center for Behavior & the Environment, Rare

Gasper D, Shah A, Tankha S (2019) The framing of sustainable consumption and production in SDG 12. Global Pol 10:83–95

Gennari P, Rosero-Moncayo J, Tubiello FN (2019) The FAO contribution to monitoring SDGs for food and agriculture. Nat Plants 5(12):1196–1197

Government of Fiji (1978) Agricultural Landlord and Tenant Act (Ed. 1978). https://www.agriculture.gov.fj/documents/legis/AgriculturalLandlord&TenantAct.pdf

Griswold D (2021) Growing hope: island agriculture and refusing catastrophe in climate change adaptation in Fiji. Island Stud J 16(2)

Long H, Prasad B, Krishna V, Tang K, Chang CP (2024) Understanding the key determinants of Fiji's renewable energy. Econ Anal Policy 82:1144–1157

Marrero A, Mattei J (2022) Reclaiming traditional, plant-based, climate-resilient food systems in small islands. Lancet Planet Health 6(2):e171–e179

McGuigan A, Ticktin T, Tora M, Tikonavuli V, Quazi SA, Dacks R (2022) Post-cyclone resilience of an agroforest-based food system in the Pacific Islands. Reg Environ Change 22(2):57

Medina Hidalgo D, Nunn PD, Beazley H, Burkhart S, Rantes J (2022) Adaptation, sustainable food systems and healthy diets: an analysis of climate policy integration in Fiji and Vanuatu. Climate Policy 22(9–10):1130–1145

Medina Hidalgo D, Witten I, Nunn PD, Burkhart S, Bogard JR, Beazley H, Herrero M (2020) Sustaining healthy diets in times of change: linking climate hazards, food systems and nutrition security in rural communities of the Fiji Islands. Reg Environ Change 20:1–13

Ministry of Economy, Republic of Fiji (2021) Fiji National Adaptation Plan. In: Fiji climate change portal. Retrieved 6 Oct 2024, from: https://fijiclimatechangeportal.gov.fj/wp-content/uploads/2022/01/Brief__Fiji_National_Adaptation_Plan_Policy.pdf

Ministry of Health. (2016). Climate change and health strategic action plan 2016–2020. Fiji.

Ministry of Strategic Planning NDAS (2014) A green growth framework for Fiji: restoring the balance in development that is sustainable for our future

O'Leary BC, Fonseca C, Cornet CC, de Vries MB, Degia AK, Failler P, ... Roberts CM (2023) Embracing nature-based solutions to promote resilient marine and coastal ecosystems. Nat Based Solutions 3:100044. https://doi.org/10.1016/j.nbsj.2022.100044

Prasad JM, Waqa-Sakiti H, Viliamu Iese L (2022) An analysis of ecosystem-based adaptations in Pacific Island countries

Rabbi MF, Hasan M, Kovács S (2021) Food security and transition towards sustainability. Sustainability 13(22):12433

Republic of Fiji (2017) 5-year & 20-year national development plan: transforming Fiji

Scheyvens R, Russell M (2012) Tourism, land tenure and poverty alleviation in Fiji. Tour Geogr 14(1):1–25

SPREP (2018) Planning for ecosystem-based adaptation in Taveuni, Fiji. A synthesis report by the Secretariat of the Pacific Regional Environment Programme, Apia, Samoa, 16 pp.

Taylor M, Ramsay D, Win K, Waqanidrola J (2015) PACC Demonstration Guide: building resilience to climate change in lowland farming communities in Fiji. In: Leavai P, Timmermans H, Pelesikoti N, Sweeney D (eds) PACC technical report. SPREP, p 36. https://www.sprep.org/attachments/Publications/CC/PACCTechRep17.pdf#page=7.06 (Original work published 2015)

Trading Economics. Fiji—Food imports (% of merchandise imports)—2025 data 2026 forecast 1967–2023 historical. Trading Economics. https://tradingeconomics.com/fiji/food-imports-percent-of-merchandise-imports-wb-data.html

Truong P, Creighton C (1994) Report on the potential weed problem of Vetiver grass and its effectiveness in soil erosion control in Fiji. Division of Land Management, Queensland Department of Primary Industries, Brisbane, Australia, pp 7–8

World Health Organization (n.d.) Health data overview for the Republic of Fiji. Datadot. Retrieved 8 Mar 2025 from: https://data.who.int/countries/242#:~:text=Overall%20Population%20over%20time&text=In%20Fiji%2C%20the%20current%20population,8.2%25%20to%201%2C000%2C261%20by%202050

Chapter 18
Ecosystem-Based Sustainable Agriculture in the Current Climate Change Context of Bangladesh: Legislative and Institutional Responses

Zelina Sultana and Md. Mofazzal Hossain

Abstract With the increasing population, the demand for agriculture in Bangladesh is growing, which tends to an unwritten competition to produce more crops from fewer pieces of land, ignoring ecological sustainability. Some common practices of this: the excessive application of chemical fertilisers and pesticides, leading to soil, air, and water pollution due to phosphorus, nitrate, and ammonia, disrupting soil nutrients and harming biodiversity, including birds, fish, and other organisms, shifting cultivation, conversion of the forest into agricultural land, changing agricultural and land use patterns, and producing cash crops like tobacco have dangerous impacts on the environment and ecology. However, the current climate change has worsened this situation, including food production, ecology/environment, and people's livelihood. Recent research has proved that only the agricultural sector could lose around USD 7.7 billion annually due to climate change, and rice production could decline by 33% within two decades. Depending on secondary data review and adopting a qualitative and policy analysis, this chapter will focus on how far the legal regime and institutions of Bangladesh are supportive of ensuring sustainable and eco-friendly agriculture in the current climate change context. This chapter finds that traditionally, the laws and policies of Bangladesh have concentrated on the primary issues of producing food, and there is no synergy among the associated institutions. Therefore, we argue that Bangladesh must develop comprehensive policies, strategic planning, and an integrated institutional approach to establish a sustainable and environmentally friendly agricultural system to meet the future demand in the context of climate change.

Z. Sultana (✉)
Faculty of Business and Law, Anglia Ruskin University, Cambridge, UK
e-mail: Zelina.Sultana@aru.ac.uk; zelina2011@gmail.com

Department of Law, Jagannath University, Dhaka, Bangladesh

Md. M. Hossain
Department of Law, Jagannath University, Dhaka, Bangladesh
e-mail: md.mofazzal@law.jnu.ac.bd

A. Roy et al. (eds.), *Environment and Public Health*, Sustainability Solutions,
https://doi.org/10.1007/978-3-031-99770-9_18

Keywords Climate change · Ecosystem · Sustainable agriculture · Sustainable development · Bangladesh

18.1 Introduction

Bangladesh is a densely populated country in the world facing resource constraints, land crises, water crises, and food crises. Because of these crises, it has emphasised producing more food by exaggerating the use of land, water, chemical pesticides, and fertilisers (Rasul and Thapa 2004). According to them, the increased use of chemical pesticides and fertilisers adversely impacts water and land nutrients and causes a decline in soil fertility and increases dependency on limited groundwater resources. As a result, the crop yields are decreasing or even contaminated, which is responsible for many public health safety issues (Garvey 2019). While dealing with these extant challenges, Bangladesh, as a low-lying, riverine, flat country, faces disasters and some events like salinity, sea level rise, and warming due to current climate change impacts. Hence, the climate change impacts worsen its existing issues, like food and water crises. For example, changes in rainfall patterns reduce the yields, which are deepened by weather and natural rain (Thomas et al. 2013) fall short of essential food for millions.

Additionally, the increasing demand for food and fodder is bound to shift agriculture mainly to the eastern part of Bangladesh. Until recently, the economy of the hilly area in Bangladesh has been based on shifting cultivation, locally called *Jhum*. For this cultivation, hilly people clear the forest; consequently, forest and land degradations contribute negatively to both the environment/climate and crop production (Rahman et al. 2012). The eastern region of the country is experiencing difficulties due to *Jhum* cultivation, while the northern region faces environmental challenges, including soil degradation, water and air pollution, and the degradation of natural resources, contributing to a loss of habitat diversity and increased mortality among domestic animals, primarily because of ongoing tobacco farming practices (Ali et al. 2023). Additionally, the cultivation of tobacco demands a substantial quantity of water, fertiliser, and pesticides to grow profitably. Again, the Southern regions adversely affected by climate change, salinity intrusion, and severe disasters due to climate change. Nonetheless, this area is associated with environmental harm and human health challenges (Rahman et al. 2019).

Besides, Bangladesh faces a drastic change in its land use pattern due to increased demand for housing and expanding urbanisation (Dewan et al. 2012). This rapid urbanisation is now accounting for the loss of a major portion of agricultural land, including wetlands and vegetation, causing degradation of the natural environment of the agriculture in Bangladesh. Moreover, the traditional agricultural pattern is changing. For instance, pure agricultural land is used for shrimp farming (Ali 2006), fish farming, fruit farming, or three- or four-times land use for more crop products and more profits. However, all these increase profits and sometimes food also temporarily, but excessive use of land for crop production ultimately negatively impacts soil health and causes significant environmental damage due to the widespread use of pesticides, herbicides, and fertilisers. Besides, this causes biodiversity loss, development of pesticide resistance and new pests, contamination and depletion of freshwater resources, soil degradation and erosion, and adverse health effects (Pretty 2018). The situation has worsened in Bangladesh because of the ongoing consequences of climate change on food production, ecosystems, and the environment, impacting people's livelihoods.

Moreover, Bangladesh is the most climate-affected country due to its geographical location and becomes vulnerable easily due to its overpopulation, limited resources, and inadequate infrastructural and technological development. The agro-based economy of Bangladesh is significantly impacted, contributing 12.1% to the nation's GDP and employing 39.71% of the workforce (World Bank 2020). The agricultural sector alone could face annual losses of approximately USD 7.7 billion, with rice production potentially decreasing by 33% over the next 20 years (Chowdhury et al. 2022). The entire agrosystem is under threat of frequent and severe climate change disasters and slow-onset events, including floods, cyclones, tropical storms, droughts, landslides, riverbank erosion and sea level rise, warming, changing precipitation, and salinity (Chowdhury et al. 2022). The production of crop yields is reduced due to temperature increases and changing rainfall in Bangladesh. And a sudden disaster caused a huge loss of agro-products within a few seconds (CRFS Alliance 2023; Chowdhury et al. 2022). For instance, Cyclone Remal visited Bangladesh in May 2024, damaging the crop, livestock, and fisheries sectors, totalling approximately USD 596 million, with 1.7 million farming households affected, standing crops destroyed, and vital agricultural assets lost (Anon 2024).

Considering the impact of climate change and the vulnerability in the agricultural sector of the country, eco-friendly and sustainable agriculture (hereafter SA), supportive of the environment, food supply, land, and soil, would be a great alternative to Bangladesh, which could also help to reduce climate change impacts. However, depending on secondary data and literature, this chapter first focuses on the necessity of sustainable and eco-based agricultural development in Bangladesh, with its meaning, as well as highlights the legal regime and institutional affiliation to promote sustainable and eco-friendly agriculture. After reviewing law and policy and secondary literature, this research demonstrates that conventional laws and policies have limited implications in ensuring food security by concentrating eco-based SA in a climate change context. Besides, there is no synergy among the associated institutions like Government Ministries, departments, institutions, etc., (Akter and Sultana 2022) in implementing the existing legal regime. This lack of policy implementation impacts on ensuring food safety and security in the near future and hinders the lives and livelihoods of millions of people in the climate crisis of Bangladesh. Therefore, we contend that Bangladesh requires strong planning, effective policy implementation, and a coordinated approach to establish a sustainable and environmentally friendly agricultural system that safeguards both people and the planet. The insights from this section hold global relevance for nations committed to promoting SA to address future food crises and develop appropriate policies.

The entire chapter is divided into five parts, of which the first part is related to the general introduction of the chapter, which is followed by chapter two to demonstrate the meaning of SA and the necessity of it. The third chapter deals with the Bangladesh approach to maintaining ecosystem-based farming evidenced by the literature. The next part will be based on a policy analysis of Bangladesh to ensure ecosystem-based SA. Consequently, the subsequent chapter will highlight the institutional affiliation for sustainable and ecosystem-based agriculture in the current climate change context. Part five of this chapter will contribute findings and suggestions, and the last part will demonstrate a general concluding remark.

18.2 Sustainable Agricultural Practice and Its Necessity

Sustainable agriculture promotes stable, continuous production with sufficient resources for the future, adhering to five principles outlined by FAO: boosting food chain productivity, improving people's well-being, protecting environmental resources and economic growth, fostering ecosystem resilience, and supporting governmental initiatives and regulations (FAO 2014). However, the concept of SA is not new, and in 1992, Pretty et al. conceptualised the SA, which should incorporate natural processes, reduce the use of toxic chemicals and non-renewable resources, and increase the participation of farmers, environmentalists, businesses, and consumers in decision-making (Pretty et al. 1992). They also place a strong emphasis on using local knowledge, assisting farmers and rural communities in becoming more self-sufficient and ensuring equitable access to resources. In the same decade, Harremoes also defined it in terms of the three pillars of sustainable development: social and economic equality, profitability, and environmental health. These include making sure agriculture does not harm the environment (soil, water, air, and other natural resources are preserved), increasing food chain productivity, and fostering economic growth, all the while ensuring equitable compensation for all parties involved, securing livelihoods, and promoting diversity, inclusion, and equity (Harremoes 1996).

Additionally, SA is a rapidly evolving field that aims to protect the environment and meet the world's food, fibre, and aesthetic needs. It utilises environmentally friendly technologies, such as managed intensive grazing and precision farming, to support sustainable farming practices. According to Horiuchi and Tsubota (1998), SA requires a balance between the environment and the people who work on the land. Farmers and landowners must be involved in the process, and research institutions, businesses, and government agencies must collaborate as learning organisations. Developing sustainable farming systems requires teamwork and support from various fields, including science, social studies, policymakers, and landowners (Horiuchi and Tsubota 1998). From the above, all the meanings of SA, the three main elements can be summed up as (i) to meet the present food needs, enhancing future well-being, (ii) to use environmentally friendly/ecosystem-based approaches, and (iii) to collaborate with stakeholders. However, for this chapter, SA means the food production system, based on an ecosystem-friendly way, collaborating with all associated stakeholders, assembling the present needs of the people, and considering the well-being of future generations.

18.2.1 Necessity of Sustainable Agriculture for Bangladesh

Sustainable agriculture is crucial for protecting the environment, ensuring food security, mitigating climate change, promoting human health, and building economic resilience. Conventional agricultural practices lead to environmental degradation, water pollution, soil erosion, and biodiversity loss (Faroque et al. 2011). SA prioritises reducing adverse effects, safeguarding ecosystems, and conserving natural resources (Anon 2021). Sustainable farming is essential to building a sustainable future for this globe, including Bangladesh. However, for Bangladesh, SA will help its economic growth, environmental protection, and increased food safety and security by lowering malnutrition as well as benefiting human health. Additionally, it helps to protect water, soil, and aquatic resources and thus helps to succeed in global goals like the Sustainable Development Goals, Aichi biodiversity targets, etc.

18.2.1.1 Economic Growth

As Bangladesh's economy is mainly agro-based, SA is one step forward to ensure economic growth, including the rural economy, by creating work opportunities and lowering poverty (Rahman 2017). However, increased land productivity, yield strength, and profitability from main crops are considered the indicators of economic growth (Rasul and Thapa 2004). It contributes to the GDP of Bangladesh (World Bank 2020). Thus, sustainable economic growth will enhance this contribution to the economy by sustainably increasing the productivity of crops.

18.2.1.2 Environmental Protection

Reducing the usage of chemical pesticides and fertilisers from the system is the most important advantage since it improves the protection of environmental elements like soil, water, and air. It can improve air quality by regulating smoke from agricultural burning, dust from tillage, pesticide drift from spraying, etc. (Muhie 2022). He also pointed out that this could prevent soil erosion by eliminating tillage and planting more trees. Sustainable farming's purpose is to achieve a healthy equilibrium between agricultural production and environmental conservation by minimising environmental degradation, boosting the health and resilience of the ecosystem, and safeguarding natural resources (Islam et al. 2024). Thus, balancing environmental equilibrium also assists in controlling rapid climate change and its impacts.

18.2.1.3 Increased Food Security and Safety

Sustainable agriculture increases crops and livestock, which help to maintain food security and proper nutrition, and food production helps people from malnutrition and provides safety. For instance, expanding crop variety and lowering post-harvest loss increased food availability (Muhie 2022). Additionally, crop breeding, microbiome management, carbon sequestration, and climate-smart irrigation are innovative methods that can guarantee sustainable food production to meet the needs of an increased population and lessen the adverse environmental effects of the agriculture sector (Kodaparthi et al. 2024).

18.2.1.4 Protecting Human Health

Crops produced by using SA practice reduce contamination, for instance, with arsenic. In conventional agricultural practice, more groundwater use tends to lead to arsenic contamination, which easily enters the food chain and affects food safety and human health (Heikens 2006). Furthermore, because of the addition of nitrogen, phosphorus, and persistent pesticides, the rising use of agrochemicals and pesticides is degrading the quality of groundwater and soil. This also triggers the contamination of groundwater and soil, travels to the food cycle of humans, and causes several human health issues (Srivastav 2020). Sustainable farming practices can reduce the use of pesticides and agrochemicals, improve the water and soil quality, and assist in protecting humans from many health issues.

18.2.1.5 Achieving Global Goals

Sustainable agriculture contributes to agricultural products and helps to eliminate hunger and poverty, and ensure clean water. The term 'SA' refers to a variety of methods and objectives that are intended to satisfy present and future demands for food and other agricultural goods while connecting to other international objectives and those about food safety. Although not one of the 17 official SDGs, SA is considered an important component for achieving many of the targets and goals (Atapattu et al. 2024). SA is responsible for achieving several SDGs, including poverty eradication (SDG 1), zero hunger (SDG 2), good health and well-being (SDG 3), gender equality (SDG 5), clean water and sanitation (SDG 6), affordable and clean energy (SDG 7), decent work and economic growth (SDG 8), industry, innovation, and infrastructure (SDG 9), life on land (SDG 15), etc. (Datta et al. 2024).

18.3 Sustainable Agriculture and Bangladesh Stands

Sustainable agriculture benefits the economy, environment, and health. Understanding this fact, Bangladesh is increasingly adopting an ecosystem-based approach for SA practices, though there are challenges to integrating it. Some process of SA is connected to pest management, organic farming, crop diversification, and climate-resilient agriculture. Currently, Bangladesh is facing obstruction in its agricultural sector due to the current climate change disasters and slow-onset events. These challenges bound Bangladesh to implement some SA, although they are undergoing some limitations. The following four ecosystem-based approaches for sustainable agricultural practices are ongoing in Bangladesh with some limitations.

18.3.1 Pest Management

Bangladesh loses about one-fourth of its agro-products each year due to pests (Kabir and Rainis 2013). Conventional chemical pesticides are not feasible in social, economic, and environmental aspects, leading to the need for an alternative strategy. Integrated pest management (hereafter IPM) is another alternative that ensures both productivity and environmental safety. IPM is widely used to ensure sufficient, equitable, steady and secure food and ecosystem services, as well as increased agriculture cost-effectiveness (Kabir and Rainis 2013). To ensure a sustainable future in agriculture, the government is disseminating IPM information to better adopt these practices among farmers. However, there are limitations to IPM, and to avoid those, future work should focus on increasing training for extension officers, government assistance, and integrating IoT with current pest management techniques. IPM recommendations ought to encourage sustainable farming practices that strike a balance between consumer affordability, food security, farmer profitability, and environmental preservation (Angon et al. 2023).

18.3.2 Organic Farming

In food production systems, conventional fertilisers are necessary to meet the gap between supply and demand. However, they can cause food safety issues, release harmful chemicals, and cause soil acidification. Overuse of chemical fertilisers can lead to air and soil pollution, soil acidification, and nutrient imbalance. On the other hand, organic fertilisers, by-products of animal and plant waste, can deliver macronutrients and micronutrients in a balanced manner, promoting plant root development. Because they decompose slowly and provide a steady supply of nutrients for longer periods, organic fertilisers reduce soil acidity, heavy metal availability, pest and disease infestation, and nutrient leaching. They also do not release harmful gases and chemicals into the environment, which ensures food safety (Khan et al. 2024). Organic farming may be able to fulfil the dual purposes of protecting the environment and ensuring food security for nations. More work needs to be done despite the nation's expanding organic agriculture trend, especially in the areas of research, extension, and raising awareness among those who are either directly or indirectly involved in organic farming (Patle et al. 2014).

18.3.3 Crop Diversification

Crop diversification is a strategy to reduce soil fertility reduction, pests and disease outbreaks, water level decline, and climate change, particularly temperature and rainfall pattern changes by using limited land and valuable water resources and making agriculture more environmentally friendly and sustainable. For instance, rice monoculture reduces non-rice crop production, erodes biodiversity, and creates a nutritional imbalance, which can be removed by employing crop diversity. Diversifying crop production allows families to create diverse livelihoods using a range of asset and resource combinations (Rahman et al. 2024). Additionally, crop diversification is crucial in addressing issues like quality seeds, efficient irrigation management, balanced fertiliser use, and credit availability. However, crop diversification is used now as a common strategy to enhance soil fertility on a countrywide scale. This needs more effort from the farmers' perspective, including funds. It also impacts the national seed policy, as it influences the production and distribution of seeds for different crops (Moniruzzaman 2019). Considering Bangladesh's large agriculture sector and its necessity for an ever-growing population, the government should provide support in cash/price and types for farmers, including the effect of crop diversification at the household level in the national agriculture policy.

18.3.4 Climate-Resilient Agriculture

For climate resilience agricultural extension, the Department of Agriculture recommends various agricultural practices to farmers to mitigate greenhouse gas (GHG) emissions, such as using dome digesters, conservation agriculture techniques, bio fuels, high-efficiency fertiliser application, artificial and participatory woodlot plantations, midseason drainage, off-season incorporation of rice straw, replacing urea with ammonium sulphate, using deep-placed briquette urea for rice, and replacing roughage with feed concentrates (Ali and Meisner 2017). Climate-smart agriculture (CSA) practices are currently essential in Bangladesh, where a large number of smallholders live in flood-prone deltas. Traditional methods such as the Sorjan system (the Sorjan system is a farming technique that uses raised beds and deep sinks to grow crops and raise fish), floating bed agriculture, vertical gardens, homestead production, and rooftop gardening can be incorporated into these tactics (O'Neill et al. 2014). Additionally, enhancing and simplifying climate information services would help farmers better adjust their farming methods (Ali and Meisner 2017). The lack of this poses a significant challenge to scaling out climate change practices. For instance, farmers may choose crops and prepare for suitable reaction and mitigation measures if they could use basic salinity metres to forecast salt incursion into irrigation canals. But for many Bangladeshi farmers, a lack of funding for CSA investments continues to be a barrier. In this respect, the government should recognise indigenous and local agrarian adaptation practices in its policies and improve new strategies for enhanced implementation of these adaptation techniques (Rahman et al. 2024).

Bangladesh, as a developing agro-based economic country, tries to initiate the above steps to support the ecosystem-based SA. This fact is already by employing some crucial steps like encouraging organic farming, crop diversification, control of pesticide uses, and adopting climate-resilient agriculture. However, the following section will highlight the laws and policies in support.

18.4 Legal Regime to Promote Sustainable Agriculture

Bangladesh's agriculture always plays an important role in building the economy and remains always in a priority to promote and enhance it in the current climate change context. The laws and policies also provide priority to the agriculture sector and enact laws and policies directly related to agriculture, land, and some natural resources like water, soil, etc., essential for agriculture. The first and most important Act is the State Acquisition and Tenancy Act of 1950, which aims for land reform, protection of agricultural land, prevention of land fragmentation, protection of wetlands, strengthening climate resilience in rural communities, and legal provisions for sustainable land management. The latest law, the Plant Varieties Protection Act, 2019, objects to protect plant breeders' rights and foster sustainable agricultural practices. It encourages the development of climate-resilient crop varieties, protects local and indigenous plant varieties, and promotes low-input farming. Thus, Bangladesh is enacting and updating policies for implementing SA in Bangladesh.

18.4.1 Laws Dealing Directly Focus on SA

As mentioned above, Bangladesh has enacted ten direct laws and policies to ensure SA, which started by adopting the State Acquisition and Tenancy Act, 1950, after its independence in 1971. Therefore, from its birth to now, the legal regime has been very supportive of shaping SA. Bangladesh tries to protect its agriculture by promoting SA and acknowledging its climate change vulnerability. Under the auspices of the United Nations Framework Convention on Climate Change (henceforth UNFCCC), Bangladesh has launched the Bangladesh Climate Change and Strategy Action Plan 2009 (henceforth BCCSAP). Though this policy is the first blueprint in Bangladesh relating to climate change, it focuses a little on SA issues. Hence, the country starts its journey towards achieving the SDGs by 2030 and has found out that SA is essential to reach the goals. Consequently, a tremendous development of laws is found highlighting the climate-resilient agriculture and SA after the year 2015. Six comprehensive laws have been enacted between 2015 and to present, concentrating on SA and climate change (Table 18.1).

18.4.2 Policies Dealing Directly Focus on SA

Bangladesh enacted some policies for climate change, development, and the agricultural sectors. The nation can adapt to climate change thanks to several agricultural and climate change-related policies, plans, and initiatives. By referring to the creation, formalisation, and active implementation of policies, the policy cycle classification finds opportunities and weaknesses in the policymaking process. Over the previous five years, allocations to the agriculture sector have not changed. However, private sector actors or some companies such as ACI, PRAN, and Lal Teer are increasingly participating in agricultural technology diffusion, mechanisation, and seed development. This has consequences for implementing climate change adaptation strategies and technology on a larger scale, which could come with greater and longer-term costs (Ali and Meisner 2017). Here, policy implications are essential for sustainable agro-business practice by the private sector to enhance climate change adaptation practices and technologies (Table 18.2).

18.4.3 Institutional Activities for SA

The implementation of law and policies for sustainable agricultural practice in the current climate change context demands a competent, knowledgeable, and inclusive as there will be no second chance to remedy this 'crisis multiplier'. The ministries and departments are overburdened with their respective activities, including the development of agriculture and now SA. Thus, various ministries are responsible for implementing the goal of SA in Bangladesh. For example, the Ministry of Agriculture (hereafter MoA) works with the Ministry of Water Resources, the Ministry of Disaster Management and Relief, and the Ministry of Environment, Forest and Climate Change to align disaster risk reduction policies and sustainable agricultural development. It also collaborates with international organisations to secure funding and technical assistance for climate-resilient agriculture.

However, the MoA is the main administrative body primarily focused on SA in Bangladesh. It is responsible for developing climate-smart agricultural policies and frameworks, supporting research on climate-resilient crops, drought-resistant rice varieties, and organic farming, and building farmer capacity through training programs. It also promotes climate-resilient infrastructure development, such as

Table 18.1 Laws focusing directly on sustainable agriculture

	Important laws	Provisions focusing SA
1	The Plant Varieties Protection Act, 2019	• Supports climate adaptation and research support • Facilitating the development of saline-resistant, drought-tolerant, and flood-resistant crops
2	The Seeds Act, 2018	• Regulate the production, certification, quality control, and distribution of seeds to ensure seed security, enhance agricultural productivity, and promote sustainable farming to address climate change challenges, including ensuring high-quality, climate-resilient seeds • Promoting indigenous and agroecological seed varieties, community seed banks and farmer-led seed networks, local adaptation strategies, and reducing dependence on synthetic fertilisers and pesticides
3	The Fertilizer (Management) (Amendment) Act, 2018	• Regulates the production, import, distribution, quality control, and use of fertilisers, including organic fertilisers, by promoting fertiliser management to mitigate the adverse impacts of chemical inputs on soil health, biodiversity, and climate change • Supporting climate-smart fertilisation practices, such as integrated soil fertility management and sustainable agricultural techniques
4	The Agricultural Marketing Act, 2018	• Aims to reduce post-harvest losses, promote local markets, and encourage sustainable practices
5	The Pesticide Act, 2018	• Regulates the import, production, sale, and use of pesticides to ensure safety and sustainability in agricultural practices • Promoting safe pesticide use, reducing chemical dependency, and encouraging environmentally friendly alternatives to mitigate climate change, and eco-friendly pest control methods • Reduction of pesticide pollution and climate impact, consumer and environmental protection • Monitoring and legal enforcement for sustainable practices, and support for climate-resilient agriculture
6	The Groundwater Management for Agricultural Purposes Act, 2018	• Regulates the installation and use of tube wells for irrigation • Assesses water levels, distance from other wells, agricultural productivity, and household water sources by promoting water-efficient agricultural practices • Encouraging farmers to adopt climate-smart irrigation methods, and climate resilience in agriculture by maintaining groundwater balance • Reducing reliance on deep aquifers and supporting sustainable irrigation planning
7	The Food Safety Act, 2013	• promotes SA by regulating agricultural inputs, monitoring food quality, and promoting safe farming practices • Ensures the use of environmentally friendly fertilisers, pesticides, and food additives, promoting food safety and health
8	The Plant Quarantine Act, 2011	• Protect crop diversity, soil health, and biodiversity, • Supports organic and agroecological farming systems, reducing agriculture's carbon footprint

(continued)

Table 18.1 (continued)

	Important laws	Provisions focusing SA
9	The Bangladesh Environment Conservation Act, 1995	• Requiring Environmental Impact Assessments (EIAs) for agricultural projects • Regulating agricultural pollution through waste, pesticide use, and water management
10	The State Acquisition and Tenancy Act, 1950	• Promote sustainable land management, equitable land distribution, and tenant rights, including land reform, protection of agricultural land • Prevention of land fragmentation, protection of wetlands, strengthening climate resilience in rural communities, promoting smallholder farming and sustainable land use planning

Source Authors after studying laws in Bangladesh

irrigation facilities and drainage systems, to mitigate climate impacts. It provides financial support for climate-smart technologies, organic fertilisers, and drought-resistant seeds. In order to execute climate-resilient initiatives, it also interacts with international organisations such as the International Rice Research Institute (IRRI) and the Food and Agriculture Organisation (FAO). It is promoting climate-smart agriculture (CSA) techniques, including agroforestry, conservation agriculture, and stress-tolerant crop varieties. It is also promoting integrated pest and soil fertility management, promoting organic fertilisers and crop rotation. The MoA has expanded irrigation coverage and promoted alternate wetting and drying techniques to reduce water consumption. It has launched digital agriculture and extension services, such as mobile-based advisory services and precision agriculture tools, to improve decision-making and resource efficiency. Additionally, the MoA has improved early warning systems to assist farmers in getting ready for severe weather conditions (Table 18.3).

18.5 Findings and Suggestions

A wide variety of elements and methods are needed for sustainable agriculture, which is a complicated and multidimensional strategy. The skills of extension workers are not being updated to increase their technical competency. Due to a lack of knowledge among policymakers regarding the advantages of sustainable practices and possible opposition from the pesticide industry and traditional players in the supply chain for intensive agriculture, the legislative environment is not favourable for the broad adoption of SA. There are several obstacles to climate change adaptation in Bangladesh's agriculture sector, such as a lack of suitable technology, the availability of substitute production technologies, a lack of cooperation and coordination, and the choice of suitable dissemination methods. These elements restrict formal institutions' ability to provide agriculture with climate change adaptation. However, there are no sole and independent laws, policies, and institutions to target the SA in Bangladesh, and the study finds that various laws and policies as well as institutions regulate this, which

Table 18.2 Policies for sustainable agricultural development

Policies regarding agriculture in Bangladesh	Year	Direct SA	Indirect SA
National Agricultural Marketing Policy	2023		Yes
National Adaptation Plan (NAP)	2023	Yes	
Fertilizer (Management) (Amendment) Regulations (Draft)	2021	Yes	
Perspective Plan of Bangladesh (2021–2041)	2021		Yes
Bangladesh Five-Year Plans (8th Plan: 2021–2025)	2021		Yes
Bangladesh Good Agricultural Practices (GAP) Policy	2020	Yes	
National Agricultural Extension Policy	2020	Yes	
Draft of Pesticide Regulations	2020	Yes	
National Food policy	2020		Yes
National Fisheries Policy (Updated 2020)	2020		Yes
Organic Farming Policy	n.d.	Yes	
Underground Water Management Regulations for Agricultural Use	2019	Yes	
Seeds Policy	2019	Yes	
National Agriculture Policy	2018	Yes	
Bangladesh Delta Plan 2100	2018	Yes	
Integrated Minor Irrigation Policy	2017	Yes	
National Organic Agriculture Policy	2017	Yes	
Bangladesh Irrigation Policy	2017	Yes	
National Crop and Forest Biodiversity Policy	2012	Yes	
National Environment Policy	2012	Yes	
National Livestock Development Policy	2007		Yes
National Integrated Pest Management (IPM) Policy	2002	Yes	
National Land Use Policy	2001	Yes	
National Rural Development Policy	2001		Yes
National Water Policy 1999	1999	Yes	

Source The authors, after studying the laws in Bangladesh

triggers ineffective strategies to promote and implement SA in the current climate change context.

18.5.1 Law Policy-Related Findings

As there are no laws and policies specifically for the implementation of SA, it depends on lots of general laws and policies which has some flaws in executing the goal of SA (Table 18.4).

Table 18.3 Other ministries and department's aims/works for SA

1	The Ministry of Food (MoF)	• Formulating and implementing food policies • Managing food reserves to respond to climate-induced disruptions, coordinating food planning and monitoring • Collaborating with international and national agencies and implementing climate adaptation strategies
2	The Local Government and Engineering Department (LGED)	• Working for climate-resilient rural infrastructure development, water resource management, and irrigation systems • Promoting sustainable land use, soil conservation, rural connectivity, disaster risk reduction and emergency response, and community-based adaptation initiatives aiming to ensure long-term food security
3	Bangladesh Water Development Board (BWDB)	• Focusing on flood control, irrigation, salinity, drought management, drainage system improvement, climate-resilient infrastructure development, early warning systems, disaster risk reduction, and community-based water resource management • Engaging local water user groups in sustainable water management by enhancing climate-resilient agriculture, protecting farmlands from climate-related risks, and ensuring sustainable water resource management
4	Ministry of Environment, Forest and Climate Change	• Initiating coastal afforestation programs to reduce coastal communities' vulnerability to climate change • Developing of climate change resilient production technologies to adapt climate-resilient agriculture through the crop varieties, fish breeding technologies, sustainable water use, and community awareness building
5	Ministry of Water Resources (MoWR)	• Developing strategies for sustainable water resource management in agriculture, including integrated water resource management (IWRM), irrigation infrastructure development, flood control and riverbank protection, salinity and drought management, sustainable water governance and policy implementation, climate-resilient water infrastructure development, research and capacity building, and inter-agency coordination for climate adaptation, support community-based water management programs • Conducting research on climate-smart water management practices, trains farmers and local stakeholders on sustainable water use, irrigation efficiency, and climate adaptation strategies, and promotes innovation in water conservation techniques

(continued)

Table 18.3 (continued)

6	Ministry of Local Government, Rural Development and Cooperatives (MoLGRD & C)	• Promoting climate-resilient rural development and SA, developing climate-resilient rural infrastructure, water management, disaster preparedness, and local governance to enhance agricultural resilience to climate change • Empowering local councils to implement climate-smart agricultural policies, providing training and financial support for climate adaptation initiatives, and encouraging disaster management committees • Promoting cooperative farming and rural livelihoods, including collective farming, shared irrigation facilities, and access to credit
7	The Ministry of Disaster Management and Relief (MoDMR)	• Developing early warning systems for extreme weather events • Implements community-based disaster preparedness programs and climate-resilient infrastructure development, which includes flood control embankments, cyclone shelters, and elevated grain storage facilities, post-disaster relief, agricultural recovery support like emergency relief, food aid, and financial support to farmers affected by climate-related disasters • Promoting climate-smart agriculture by encouraging farmers to adopt flood-resistant, drought-tolerant, and salt-tolerant crops • Supporting research on climate change impacts on agriculture and the development of new disaster-resilient farming techniques, promoting digital technology
8	The Department of Agricultural Extension (DAE)	• Initiating a strategic plan, aligning resources and efforts to address the challenges posed by a changing climate • Increasing agricultural productivity, facilitating input supply, crop diversification, appropriate land use, and SA alongside environmental management • Promoting climate-resilient crop varieties, investing in capacity building and training for extension officers and farmers, and supporting community-based adaptation

(continued)

Table 18.3 (continued)

9	Bangladesh Agricultural Research Council (BARC)	• Formulating strategies and initiatives to enhance the resilience and productivity of the agricultural sector by developing improved technologies, increasing productivity, promoting competitiveness through commercialisation, and focusing on sustainability and poverty alleviation • Working with domestic and foreign partners to coordinate research initiatives aimed at creating crop varieties that are climate resilient and sustainable farming practices • Boosting training and resources on climate-smart agricultural practices and enhancing the sector's resilience to climate change
10	The Bangladesh Rice Research Institute (BRRI)	• Developing resilient rice varieties, advancing innovative cultivation techniques, and implementing climate-smart agricultural practices to ensure food security and environmental sustainability • Introducing saline-tolerant rice varieties and developing gene-edited rice lines with enhanced resistance to diseases and pests • Adoption of sustainable farming methods like alternate wetting and drying (AWD), and the system of rice intensification (SRI) • Developing climate-smart tools, such as the Climate-Smart Mapping and Adaptation Planning (CS-MAP) tool, collaborating with national and international research institutions to address critical barriers to SA, such as conservation agriculture
11	The Bangladesh Agricultural Development Corporation (BADC)	• Developing irrigation infrastructure, seed distribution, and fertiliser provision • Developing storage facilities and improving market access to ensure food security • Promoting sustainable methods, including crop rotation and integrated pest control, as well as bolstering climate-resilient crop varieties in collaboration with research institutions • Improving food production self-sufficiency and supporting farmers in adapting to changing environmental conditions
12	The Bangladesh Meteorological Department (BMD)	• Providing accurate weather forecasts, early warnings for extreme weather events, and agro-meteorological services • Supporting climate-smart farming practices by conducting farmer training programs, and fostering community engagement through initiatives like rain gauge distribution • Conducting climate impact studies and developing decision support tools to help farmers make climate-informed decisions

(continued)

Table 18.3 (continued)

13	The Bangladesh Climate Change Trust (BCCT)	• Promoting SA in Bangladesh, managing the Climate Change Trust Fund, which funds climate-resilient agricultural projects • Coordinating with government ministries, private sectors, and civil societies to integrate climate adaptation strategies into national agricultural policies
14	The *Palli Karma-Sahayak* Foundation (PKSF)	• Focusing on poverty alleviation and enhancing agricultural resilience and food security, including capacity building and technology dissemination, climate-resilient practices, climate change adaptation projects, resilient homestead and livelihood support, innovative financing, and support programs
15	The International Atomic Energy Agency (IAEA)	• Offering technical support and training, such as expert training courses and fellowships, to enhance agricultural research capabilities • Using nuclear techniques to improve crop resilience and food security, including plant mutation breeding, accelerating natural genetic variation, to develop high-yield, climate-resilient crop varieties
16	The Bangladesh Institute of Nuclear Agriculture (BINA)	• Developing crop varieties, including pulses, rice, and staple crops, climate-resistant rice varieties, and salt-tolerant rice varieties • Advancing plant mutation breeding by using radiation-induced mutation breeding to create crops with improved resistance to diseases, pests, and extreme environmental conditions

Source Authors

18.5.2 Institution-Related Findings

Formal institutions play a vital role in shaping climate change adaptation responses, with interventions such as training, equipment funding, community participation, and access to information networks producing a favourable link between informal structures and formal institutions. No specific institution is for the SA, and various institutions are responsible and have some limitations, including the MoA. It has inadequate extension services for smallholder farmers and a slow uptake of climate-smart agriculture technologies (Table 18.5).

18.5.3 Ways Out for New Progress

To ensure SA in Bangladesh, a multifaceted approach is needed. This includes developing climate-tolerant varieties, promoting high-yielding and nutrient-rich crops, and leveraging the free flow of information on social platforms. Acknowledging

Table 18.4 Lacks in laws and policies

Law/policies	Lacking
The Plant Varieties Protection Act, 2019	Limited protection for farmers' rights, a narrow focus on commercial seed varieties, potential threats to seed sovereignty, and weak provisions for biodiversity conservation
The Seeds Act, 2018	Limited support for indigenous and climate-resilient seeds, increased dependence on commercial seeds, and weakened farmer participation and rights, discourages farmers from using saved, exchanged, or locally bred seeds, and reduces access to diverse seed sources
The Fertilizer (Management) (Amendment) Act, 2018	Over-reliance on chemical fertilisers, limited promotion of soil regeneration practices, and weak incentives for sustainable inputs
The Agricultural Marketing Act, 2018	Limited support for climate-resilient crops, inadequate infrastructure for small farmers, and market barriers for sustainable practices, lack strong incentives for marketing climate-resilient and organic crops and clear mechanisms to promote fair pricing
The Pesticide Act, 2018	Insufficient promotion of eco-friendly alternatives, risk of soil and water contamination due to chemical pesticide use, and inadequate protection for pollinators and biodiversity
The Groundwater Management for Agricultural Purposes Act, 2018	Lacks strict regulations to prevent over-exploitation and has no strong support for water-efficient technologies
The Food Safety Act 2013	Overlapping mandates and a lack of clear communication channels
The Plant Quarantine Act, 2011	A limited focus on climate-resilient crop varieties and challenges for small farmers in accessing quality seeds
The Environment Conservation Act, 1995	Lack of integration between climate change and agriculture policies,
The State Acquisition and Tenancy Act, 1950	Land fragmentation, making farming challenging to adopt climate-smart practices like agroforestry, crop rotation, or mechanised farming Lacks protection against agricultural land conversion, which threatens long-term food security and sustainable farming
The National Agriculture Policy (NAP) 2018	Inadequate emphasise on agroecological approaches, organic farming, and regenerative agriculture
The Bangladesh Good Agricultural Practices (GAP) Policy 2020	Lacks comprehensive guidelines for implementation, monitoring mechanisms, and fragmented efforts

Source Authors after studying laws in Bangladesh

Table 18.5 Institutional/departmental lacks

Ministry or department	Flaws
MoF	Inadequate infrastructure for food delivery and storage results in post-harvest losses and ineffective measures to guarantee food security during climate-related emergencies
LGED	Lack of climate-resilient rural infrastructure and sustainable irrigation systems
BWDB	Ineffective water management and irrigation canal, and embankment maintenance
MoEFCC	Little enforcement and little collaboration with the agriculture sector and SA
MoWR	Excessive emphasis on traditional water projects and not enough on natural alternatives for agriculture's sustainable use of water resources
MoLGRD & C	Limited financial and technical support for climate-resilient cooperative farming initiatives
MoDMR	Focus on post-disaster relief rather than proactive climate adaptation in agriculture
DAE	Limited capacity to train farmers on climate-smart agriculture practices and insufficient field-level personnel to effectively disseminate climate-resilient technologies
BARC	Inadequate funding and research on climate-resilient crops and soil management, and weak collaboration with international research bodies on climate adaptation
BRRI	Slow development and dissemination of salt- and drought-resistant rice varieties
BADC	Limited focus on distributing climate-resilient seeds and irrigation technology
BMD	Inadequate localised weather forecasting and early warning systems for climate-related agricultural risks

Source Authors

the development phase of SA in Bangladesh, the following suggestions can help to enhance the situation in future.

Bangladesh can improve its agriculture by increasing renewable energy use, introducing agroforestry, promoting cooperative farming, and implementing good agricultural practices (GAP); providing training about modern eco-based farming to the farmers, and transforming the seed and food system in Bangladesh. A resource centre in SA should be established to promote Indigenous knowledge and determine future cropping patterns in national climate change adaptation plans and sectoral policies like fertiliser, seed, and agricultural subsidy policies for SA. Thus, the government should invest and allocate enough budget for research and training to promote these activities, triggering the SA system in Bangladesh.

A large number of laws, policies, and institutional affiliations may create problems in the absence of collaboration and overlapping powers and duties of respected authorities (Akter and Sultana 2022). Thus, to avoid this, a completely distinct SA management strategy may be a robust alternative in future, but Bangladesh can amend its agricultural policy by introducing the authority that will play the role of coordinator for all during the implementation of the SA system and be accountable also for its failure. For this, MoA can act as a coordinator for the enforcement of SA, which can also mobilise different stakeholders such as governmental organisations, NGOs, and other organisations to promote SA.

18.6 Conclusion

This part of the chapter focused on the laws, policies, and institutional affiliation to initiate SA in Bangladesh. In doing so, it has demonstrated the necessity of SA in the current climate change context and the current available strategies of SA implementing SA. It found that SA is important to ensure food security, achieve SDGs, and reduce adverse impacts. Understanding its necessities, Bangladesh has initiated some activities like organic farming, climate-resilient agriculture, crop diversification, etc., to promote SA, which itself has some shortcomings. It has been demonstrated that Bangladesh has taken approximately 40 laws and policies and 15 institutions for enhancing SA, but those had their inherent flaws and need coordination and amendment to include SA as a priority in the current climate change context. This chapter advocated for robust plans, policy implementation, and a synergistic approach to ensure a sustainable and eco-friendly agricultural system to protect people and the planet.

References

Akter N, Sultana Z (2022) Who will tie the bell? Climate change policy implementation authority of Bangladesh. In: 28th international sustainable development research society conference, pp 1–17. Available via DIALOG. https://www.researchgate.net/publication/368336087_Who_Will_Tie_the_Bell_Climate_Change_Policy_Implementation_Authority_of_Bangladesh. Accessed 10 Mar 2025

Ali AMS (2006) Rice to shrimp: land use/land cover changes and soil degradation in Southwestern Bangladesh. Land Use Policy 23(4):421–435. https://doi.org/10.1016/j.landusepol.2005.02.001

Ali MY, Shahrier M, Kafy AA, Ara I, Javed A, Fattah MA, Rahaman ZA, Tripura K (2023) Environmental impact assessment of tobacco farming in northern Bangladesh. Heliyon 9(3):e14920. https://doi.org/10.1016/j.heliyon.2023.e14505

Ali MY, Meisner C (2017) Climate-smart agriculture in Bangladesh. World Bank, Washington DC

Angon PB, Mondal S, Jahan I, Datto M, Antu UB, Ayshi FJ, Islam MS (2023) Integrated pest management (IPM) in agriculture and its role in maintaining ecological balance and biodiversity. Adv Agric 2023(1):1–19. https://doi.org/10.1155/2023/5546373

Anon (2024) Bangladesh Cyclone Remal and monsoon floods—emergency appeal. Reliefweb. Available via DIALOG. https://reliefweb.int/report/bangladesh/bangladesh-cyclone-remal-and-monsoon-floods-emergency-appeal. Accessed 14 Mar 2025

Anon (2021) Sustainable agriculture—building future brands. Traces. Available via DIALOG. https://tracextech.com/sustainable-agriculture/. Accessed 10 Mar 2025

Atapattu AJ, Ranasinghe CS, Nuwarapaksha TD, Udumann SS, Dissanayaka NS (2024) Sustainable agriculture and sustainable development goals (SDGs). In: Emerging technologies and marketing strategies for sustainable agriculture. IGI Global, Hershey PA, USA, pp 1–27

Chowdhury MMI, Rahman SM, Al Amran MIU, Malik K, Abubakar IR, Aina YA, Rahman SMA, Khan M, Muhyedeen MA, Hasan MA (2022) Climate change impacts on food system security

and sustainability in Bangladesh. Res Square 2022:1–32. https://doi.org/10.21203/rs.3.rs-1673139/v1
CRFS Alliance (2023) Bangladesh: a country diagnostic. Conducted by climate resilient food systems alliance. Available via DIALOG. https://unfccc.int/sites/default/files/resource/CRFSA_Country%20Diagnostics_Bangladesh_June%202023.pdf. Accessed 15 Mar 2025
Datta P, Behera B, Timsina J (2024) Achieving sustainable development through agriculture-forestry-livestock nexus in Bangladesh: synergies and trade-offs. Agric Syst 215:103854. https://doi.org/10.1016/j.agsy.2024.103854
Dewan AM, Yamaguchi Y, Ziaur Rahman M (2012) Dynamics of land use/cover changes and the analysis of landscape fragmentation in Dhaka Metropolitan, Bangladesh. GeoJournal 77:315–330. https://doi.org/10.1007/s10708-010-9399-x
FAO (2014) Building a common vision for sustainable food and agriculture: principles and approaches. Food and Agriculture Organization of the United Nations, Rome. Available via DIALOG. https://openknowledge.fao.org/server/api/core/bitstreams/cd7ebb4f-da7c-474d-83df-b5cc224d2ff8/content. Accessed 10 Mar 2025
Faroque MAA, Kashem MA, Bilkis SE (2011) Sustainable agriculture: a challenge in Bangladesh. Int J Agric Res Innov Technol 1(1–2):1–8. https://doi.org/10.3329/ijarit.v1i1-2.13922
Garvey M (2019) Food pollution: a comprehensive review of chemical and biological sources of food contamination and impact on human health. Nutrire 44(1):1. https://doi.org/10.1186/s41110-019-0096-3
Harremoes P (1996) Dilemmas in ethics: towards a sustainable society. Ambios 25:390–395. ISSN: 0044-7447
Heikens A (2006) Arsenic contamination of irrigation water, soil and crops in Bangladesh: risk implications for sustainable agriculture and food safety in Asia. RAP Publication
Horiuchi H, Tsubota K (eds) (1998) The 4th JIRCAS international symposium: sustainable agricultural development compatible with environmental conservation in Asia. Japan International Research Center for Agricultural Sciences (JIRCAS)
Islam MS, Yabar H, Mizunoya T (2024) Prospects and challenges of good agricultural practices for vegetable production: In the context of sustainable agriculture in Bangladesh. http://jsrsai.jp/Annual_Meeting/PROG_60/ResumeB/B03-3.pdf
Kabir MH, Rainis R (2013) Determinants and methods of integrated pest management adoption in Bangladesh: an environment-friendly approach. Am-Eurasian J Sustain Agric 7(2):99–107. ISSN: 1995-0748
Khan MT, Aleinikoviene J, Butkeviciene LM (2024) Innovative organic fertilizers and cover crops: perspectives for sustainable agriculture in the era of climate change and organic agriculture. Agronomy 14(12):2871. https://doi.org/10.3390/agronomy14122871
Kodaparthi A, Kondakindi VR, Kehkashaan L, Belli MV, Chowdhury HN, Aleti A, Rajpurohit S, Vasanthu SJ, Chepuri K (2024) Environmental conservation for sustainable agriculture. In: Aransiola SA, Babaniyi BR, Blessing A (eds) Prospects for soil regeneration and its impact on environmental protection. Springer, Cham, pp 15–45
Moniruzzaman S (2019) Crop diversification as climate change adaptation: how do Bangladeshi farmers perform? Climate Change Econ 10(02):1950007. https://doi.org/10.1142/S2010007819500076
O'Neill BC, Kriegler E, Riahi K, Ebi KL, Hallegatte S, Carter TR, Mathur R, Van Vuuren DP (2014) A new scenario framework for climate change research: the concept of shared socioeconomic pathways. Clim Change 122:387–400. https://doi.org/10.1007/s10584-013-0905-2
Patle GT, Badyopadhyay KK, Kumar M (2014) An overview of organic agriculture: a potential strategy for climate change mitigation. J Appl Nat Sci 6(2):872–879. https://doi.org/10.31018/jans.v6i2.548
Pretty J, THOMPSON J, HINCHCLIFFE F (1992) Sustainable agriculture. NITROGEN-LONDON 200:18
Pretty J (2018) Intensification for redesigned and sustainable agricultural systems. Science 362(6417):eaav0294. https://doi.org/10.1126/science.aav0294

Rahman MT (2017) Role of agriculture in Bangladesh economy: uncovering the problems and challenges. Int J Bus Manag Invention 6(7):36–46

Rahman SA, Rahman MF, Sunderland T (2012) Causes and consequences of shifting cultivation and its alternative in the hill tracts of eastern Bangladesh. Agrofor Syst 84:141–155. https://doi.org/10.1007/s10457-011-9422-3

Rahman MM, Islam A, Ferdousee S (2024) Crop diversification, sustainable production, and consumption (SDG-12) in rural Bangladesh: insights from the northern region of the country. Circ Agric Syst 4:e006. https://doi.org/10.48130/cas-0024-0005

Rahaman MA, Rahman MM, Hossain MS (2019) Climate-resilient agricultural practices in different agro-ecological zones of Bangladesh. In: Handbook of climate change resilience. Springer, Cham, pp 1–27

Rasul G, Thapa GB (2004) Sustainability of ecological and conventional agricultural systems in Bangladesh: an assessment based on environmental, economic and social perspectives. Agric Syst 79(3):327–351. https://doi.org/10.1016/S0308-521X(03)00090-8

Muhie SH (2022) Novel approaches and practices to sustainable agriculture. J Agric Food Res 10:100446. https://doi.org/10.1016/j.jafr.2022.100446

Srivastav AL (2020) Chemical fertilisers and pesticides: role in groundwater contamination. In: Prasad MNV (ed) Agrochemicals detection, treatment and remediation. Butterworth-Heinemann, pp 143–159. https://doi.org/10.1016/B978-0-08-103017-2.00006-4

Thomas TS, Mainuddin K, Chiang C, Rahman A, Haque A, Islam N, Quasem S, Sun Y (2013) Agriculture and adaptation in Bangladesh: current and projected impacts of climate change, vol 1281. International Food Policy Research Institute (IFPRI). Available via DIALOGUE. https://www.aesanetwork.org/wp-content/uploads/2018/02/Agriculture-and-Adaptation-in-Bangladesh-Current-and-Projected-Impacts-of-Climate-Change.pdf. Accessed 16 Mar 2025

World Bank (2020) Climate change knowledge portal for development practitioners and policy makers. Available via DIALOGUE. https://climateknowledgeportal.worldbank.org/country/Bangladesh. Accessed 16 Mar 2025

Chapter 19
Role of Village Community Organizations, Effective Watercourse Management, and Impacts on Water and Food Security in Pakistan

Farooq Ahmed, Usman Mustafa, Jam Bilal Ahmad, and Muhammad Shahid

Abstract The participation of the community is essential to the success and effectiveness of the project. It reassures participants' freedom and increases the number of people who can benefit from development. Funded by the Ministry of National Food Security and Research (MNFSR), Islamabad, Pakistan, launched the National Program for Improvement of Watercourses in Pakistan, Phase-II (NPIWC-II). This project covers all provinces: Baluchistan, Gilgit Baltistan, AJ&K, Punjab, KP, as well as ICT. The chapter analyzes how, through capacity building of Water Users Associations (WUAs) or Farmers Organizations (FOs), social mobilization improved water management techniques and facilitated their registration under the On-Farm-Water Management and Water User-Associations-Ordinance Act 1981, leading to the organization of 47,278 WUAs. Moreover, the second objective of this chapter is to analyze the impact of the improved water management techniques on water insecurity and food insecurity in the country. Our findings show that land-use intensity has increased on average by 5% due to watercourse improvement on sample farms, meaning an increase of 5% in cultivated area. Cropping intensity has increased by 10%. These increases in land-use and cropping intensities have resulted in about a 14% increase in cropped area under various crops. The watercourse improvement

F. Ahmed (✉)
Department of Anthropology, The Islamia University of Bahawalpur, Bahawalpur, Pakistan
e-mail: farooq.ahmed@iub.edu.pk

U. Mustafa
Pakistan Institute of Development Economics (PIDE), Quaid-I-Azam University Campus, Islamabad, Pakistan

Ministry of National Food Security and Research (MNFSR), Islamabad, Pakistan

J. B. Ahmad
Taxila Institute of Asian Studies, Quaid-I-Azam University, Islamabad, Pakistan

M. Shahid
Department of Insurance and Economics at the University of International Business and Economics, Beijing, China

A. Roy et al. (eds.), *Environment and Public Health*, Sustainability Solutions,
https://doi.org/10.1007/978-3-031-99770-9_19

impact on crop yields per acre varied from 3 to 19%, averaging at 11% on an overall basis. The cumulative impact of watercourse improvement is reflected in the total production of various crops. Production of various crops has increased at different rates, varying from 6% in the case of peaches to 676% in the case of gram. However, the weighted average impact is calculated at 26% (11% due to yield increase, 14% due to area increase, and 1% due to interaction between the two). We concluded that community organization has profound impacts on crop production, farmers' income, and their welfare. There is a strong need to further develop and strengthen community participation and involvement.

Keywords Water insecurity · Water management · Crop production · Food security · Pakistan

19.1 Introduction

In his poem, Chinese Philosopher Lau Tse expressed the essence of the participatory approach to community development: "Go and meet your people, live and stay with them, love them, work with them. Begin with what they have, plan and develop from what they know, and in the end, when the work is over, they will say: We did it ourselves" (Dennis 1977).

Participation is characterized by involvement of communities and people, whereas participatory development (PD) is interested in engaging local populations in development projects. PD, introduced as an important part of the "basic needs approach" to development, has taken a variety of forms since it emerged in the 1970s. Several manifestations of people's participation in development pursue "to give the poor a part in initiatives designed for their benefit" in the expectation that development projects will be more sustainable and effective if local inhabitants are involved in the development processes (Cornwall 2002).

PD emerged as a gradually recognized method of development practice and is being used by a diverse range of organizations. It is frequently presented as a substitute for conventional "top-down" development (Mohan 2008). Participatory models criticized the modernization paradigm because it encouraged a top-down, ethnocentric, and authoritarian view of development. They contended that the strategic model anticipated an idea of development accompanying a Western visualization of progress. The top-down method of persuasion models covertly supposed that the knowledge of governments and agencies was accurate, and that indigenous populations were either uninformed or had inappropriate beliefs (Cypher 2014; Fussell 1996).

19.1.1 Political Economy of Water Insecurity

Water is distributed among its four provinces on an equitable basis. Water distribution at the micro-level has always been overlooked. The existing irrigation system is rooted in colonial times. Replacing the traditional irrigation system, it has disintegrated the traditional harvesting system that was once better merged into the local atmosphere (D'Souza 2006). In Pakistan, water availability has fallen to its lowest in Asia and Africa. Insufficient irrigation and poor drainage have produced water logging and soil salinity. This situation created an unsuccessful harvesting and energy crisis on which most agricultural technologies depend (Kugelman 2010).

There might be severe social, economic, and health implications of the water crisis (Kugelman 2010; Shams, 2016) in Pakistan. The poor management of water illustrates a lack of vision and a strong will for good governance and development (Mustafa et al. 2017). Water in Pakistan is the pillar of the national agricultural economy. Expected to get dry by 2025, as Pakistan Council of Research in Water Resources has already declared that the country will be water stressed in 1990 and water scarce in 2005 (Pakistan Council of Research in Water Resources (PCRWR) 2019). Globally, Pakistan is the 4th highest "water-intensive agriculture economy" and the 3rd most water-stressed country in the world. Agriculture is the single most important factor that consumes 93% of the surface water without levying fair taxation. In addition, the upward growth of population, swift urbanization, looming crises of climate change, and poor water management contributed to this water crisis (Shams 2016). Additionally, the country has a low storage capacity, only for 30 days instead of 120 days, and 5 million acre-feet of water is lost during canal and watercourse flow (Watto et al. 2021).

Water scarcity is the most pressing challenge in Pakistan at present, with serious repercussions for agricultural sustainability as well as food security and malnutrition (Ahmed et al. 2023, 2022). Pakistan is highly reliant on irrigation-dependent agriculture, under which water management is inefficient, and water distribution is unequal, with decreased agricultural output (Davies et al. 2016). This led VCOs to step in and introduce participatory watercourse management.

19.1.2 Agriculture and Food Security

Water is a necessary product for agricultural production, financial growth, and food security in Pakistan (FAO 2020). Agriculture is the source of approximately 24% of the GDP and provides jobs to about 40% of the workers (World Bank 2020). However, the country faces a severe water shortage because of insufficient water management, an old irrigation system, and impacts of climate change. The Indus Basin Irrigation System, the largest global adjoining irrigation system, has several inadequacies, with about 50% loss of water in delivery and supply (Qureshi 2011).

To address these challenges, the government of Pakistan, in collaboration with international corporations, has begun many projects aimed at improving watercourse effectiveness through lining and modernization. The importance of investing in the development of watercourses through a proactive role and direct engagement of local communities was realized by NPIWC. This study examines the role of Village Community Organizations (VCOs) and Water Users Associations (WUAs) in the successful management of watercourses in Pakistan by rethinking the role of improved watercourse systems on water savings, crop production, and land use. The present study offers participatory methods alternatives for how strengthening organizations at the village level has a great impact on sustainable water solutions as well as agrarian yield.

19.1.3 Global Perspectives on Participatory Water Management

Mekonnen et al. (2015) are more vocal about the application of participatory methods and shared visions for water ownership, which they think better results in not only strategizing water use efficiency but also the sustainability of irrigation systems themselves. This involves conceptualizing water prioritization through community membership that incorporates locals to rationalize by their own all the stages from proposing and designing to implementation and conservation of development mediations that are relatively expedient in terms of more endowing and much more maintainable ecological solutions and sustenance.

Pakistan will definitely not be an exception when it comes to participatory water management. Data reveal that this model has already been successful in varying geographies and differing habitats. For example, consider the case of India and Israel, where community participation positively contributed to water use efficiency and agricultural productivity. Sharma et al. (2019) gave a more illustrative example of how Indian farmers' local participation in contouring of canals and watercourses ensured a reduction of one-third of water loss and a subsequent fourth-fifth increase in crop production. Marin et al. (2017) studies coping with the water scarcity issue in Israel, analyzing how effective participation of communities with adoptive irrigation technologies entirely altered the agriculture sector and improved food security.

19.1.4 Practices of Water Management in Pakistan

The involvement of local communities in various watercourse management methods, such as reasonable water distribution, canal lining, and effective irrigation practices, has shown positive results in reducing water loss, confirming reasonable access to water resources, and improving yields (Ishaque et al. 2023).

Sustenance is a main program in Pakistan for NPIWC that designates how lining watercourses have a positive impact on wider water quality. The program shows the role of WUAs and VCOs in mobilizing farmers, managing resources, and justifying their options. Evidence has shown that watercourse lining can diminish water loss by 25–30%, which leads to incredible advances in water use effectiveness and agricultural productivity (Ali et al. 2020).

As far as the scaling up of these projects is concerned, problems persist, mostly in terms of technical knowledge, local participation, and fiscal supplies. Therefore, the success of participatory water management in Pakistan hinges on the vigorous engagement of local people, capacity building, and the construction of adequate resources and patronage.

Keeping in view the research gap, the present study has the following objectives: 1) investigates the role of community organizations in water resource management in Pakistan, 2) explores by what means agricultural productivity as well as food security are interrelated with better-quality watercourse management systems, and 3) in what manner participatory approaches and methods on water management options/choices can be optimized and diversified.

19.2 Materials and Methods

We employed a multistage proportional random sampling technique for our study. Multistage proportional random sampling techniques mean that from the head, there are three stages in sampling, first from regions/zones, i.e., Punjab, KP, Sindh, AJK, GB, and Balochistan. In the second stage, from each region, districts were selected randomly. In the third stage, from each district, sub-district, and village were selected randomly. While selecting particular villages, the farmers from each water channel were selected proportionally from the head, middle, and tail. So, all water channel farmers are represented.

The study is extracted from the National Project on Irrigation Water Channels (NPIWC 1). Initially, the sample was designed at 2.2%, but later, following the instruction of the NPIWC project National Project Coordinator (NPC), the sample was expanded to 5% for a robust sample size. Based on technical sections (TS), the sample was stratified to capture variability in watercourse types, combining perennial canals, non-perennial canals, and non-canal watercourses.

Before field data collection, a field manual for enumerators was also prepared and distributed. Three days of office training on the conduct, questionnaire itself, and interview techniques were conducted, followed by a two-day field study.

The data were collected from 689 watercourses (Table 19.1), with a focus on measuring water savings, land-use intensities, and crop production before and after watercourse improvements. Watercourse lining materials such as PVC, PPC, and

Table 19.1 Sample size for WUAs mobilization and water improvement components

Zone/unit	Project targets	Sample %age	Sample size	Completed schemes	Sample % age	Sample size
Punjab	10,000	2	200	4063	5	203
KPK	13,000	2	260	3187	5	159
Balochistan	20,389	2	408	5202	5	260
Gilgit Baltistan	2500	5	125	809	5	40
AJK	1165	5	58	482	5	24
ICT	224	5	11	34	5	2
Total	47,278	2.25	1062	13.777	5	689

HDPE were evaluated for their effectiveness in reducing water losses. It is important to note that we evaluated the variations in land-use patterns along with calculating how the enhanced production of both major and minor crops and watercourse perfections have a systematically and are rather interdependent.

The study required robust statistical analyses in order to have an authentic idea of water banking and the ratio drop in water use after lining. In so doing, the domino effect was categorized depending on results, basically based on the kind of watercourse and the degree of lining (e.g., 50% lined, 20% lined). For statistical purposes, we used graphs so that enumerations could be empirically justified, and we also inserted tables in this study in order to demonstrate the analyzable impact of watercourse management on agricultural output and water preservation.

19.3 Results

This was no longer a myth that water optimization results in additional plantation that leads to altered real-time land-use patterns, as the findings of the research results were anticipated. Hence, the more water saved, the greater the plantation with land-use structural adjustments (Fig. 19.1). It was observed that following the lining technique of watercourses, yearly, a plenty (i.e., 30 percent) of water was actually protected, resulting in great water reserves of 64,744 acre-feet (AF) for the sample of watercourses pinned. Watercourse lining helped increase yields of major and minor crops such as wheat, rice, and cotton (Fig. 19.2). Finally, the net and gross profits achieved by this water saving increased significantly (Tables 19.2 and 19.3).

Based on technical sections (TS), the sample was stratified to capture variability in watercourse types, combining perennial canals, non-perennial canals, and non-canal watercourses.

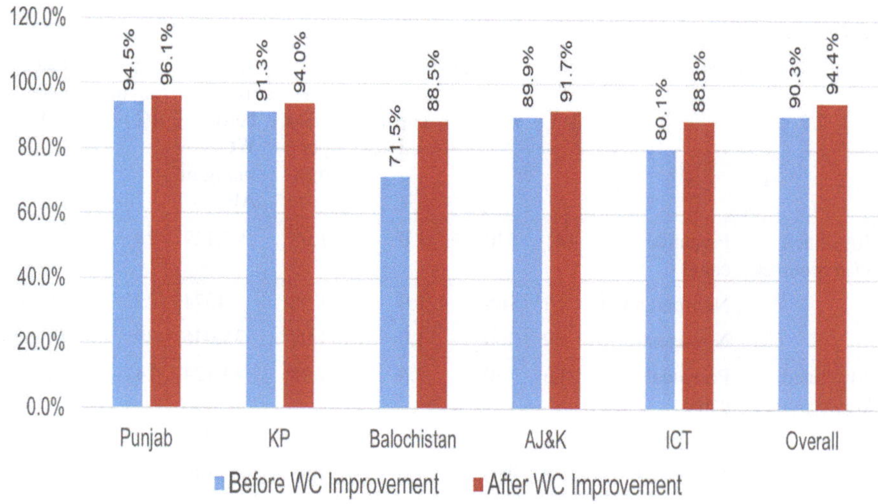

Fig. 19.1 Impact of watercourse improvement on land-use intensities on sample farms

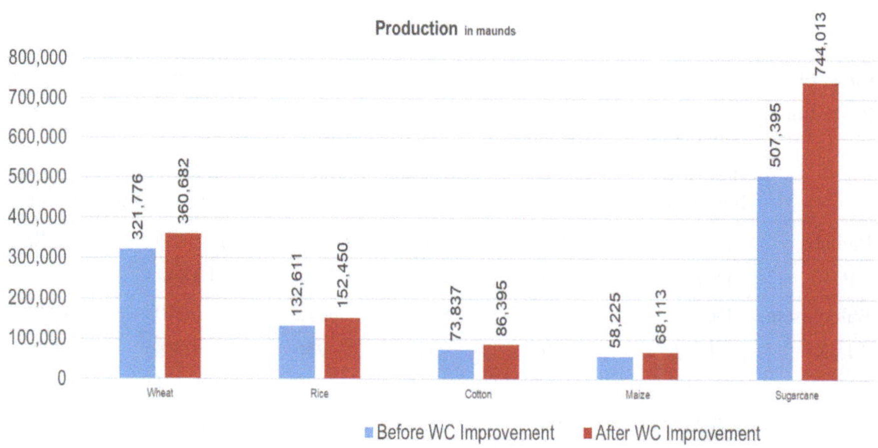

Fig. 19.2 Impact of WC improvement on the production of major crops on sample farms

19.4 Discussion

19.4.1 Sustainable Impacts of Participatory Development

Irritating with these outdated development philosophies leads to a re-examination of the resolve of development toward a search for alternative theoretical illuminations. Several scholars on development (Roodt 2001; Rahman 1993; Chandra 2010; Lewis 2004; Dodds 1989) have started to respond to these topics, pronouncing a notion

Table 19.2 Water-saving estimates as per 421 sample watercourses

Sample watercourses on		No. of WC	Annual flowing days	Average WC capacity (cusecs)	AF per cusec per day	Water use before WC lining in AF	Water saving (%)	Water saved in AF
50% lined Watercourses	Perennial canal	141	330	1.0	1.98	92,129	29	26,718
	Non-perennial	13	165	1.0	1.98	4574	29	1326
	Non-canal	59	300	1.0	1.98	35,046	29	10,163
20% lined Watercourses	Perennial canal	66	330	1.0	1.98	43,124	14	6037
	Non-perennial	4	165	1.0	1.98	980	14	137
	Non-canal	2	300	1.0	1.98	1188	14	166
PVC, PPC, and HDPE Lined		136	300	0.5	1.98	40,392	50	20,196
Total sample watercourses		421				217,434	30	64,744
Annual water use saved per watercourse in AF						516	30	154

Table 19.3 Impact on total gross and net income of all the farms under completed watercourses

Zone/unit	Impact on gross income of all farms under completed schemes			Impact on net income of all farms under completed schemes		
	Before WC improvement	After WC improvement	Increase/ impact	Before WC improvement	After WC improvement	Increase/ impact
Punjab	178,406	207,615	29,209	94,227	106,756	12,530
KP	15,417	21,090	5673	8143	10,844	2702
Balochistan	10,626	18,748	8122	5612	9640	4028
AJ&K	1772	2116	344	936	1088	152
ICT	12	17	5	7	9	2
Overall	206,235	249,586	43,352	108,924	128,338	19,414

known as participatory, or "people-centered development." Present discussion and development efforts mostly focus on "bottom up" planning, "people-centered development", and the view that ordinary people can manage their own development by themselves. This theory encourages the involvement of all stakeholders in the process of development (Sutomo et al. 2024; Rahman 1993; Oakley 1991; Bryant 2019) and will be used in the development initiatives that exist in third-world countries.

The research results emphasize the valuable pay-off role of watercourse improvement in Pakistan, specifically through the lining of watercourses with materials such as PVC, PPC, and HDPE. The results demonstrate a 30% reduction in water use, translating to substantial water savings of 64,744 acre-feet (AF) annually across the sample watercourses. This drop in water losses is critical in a low-water index geography of Pakistan, combined with competent irrigation mechanisms, which were considered a permanent obstacle for the country's agrarian output (Qureshi 2011; Pakistan Council of Research in Water Resources (PCRWR) 2023).

19.4.2 Water Savings and Efficiency

Study results align with prior research that focuses on the success of watercourse lining in reducing water losses. For example, lining watercourses in Punjab decreased water losses by 25–30%, leading to significant developments in water use efficiency, Ali et al. (2020) said. Hussain et al. (2019) stated that watercourse improvements in Sindh resulted in a 20–25% growth in water supply effectiveness. Such evidence illuminates that tackling the issue of water scarcity and increasing agricultural productivity is impossible without updating the irrigation system.

19.4.3 Impact on Crop Production

Evidence from Pakistan showed that poor water management and distribution are the root cause of food insecurity and undernutrition (Ahmed et al. 2021, 2024). Evidence also demonstrated that the lining of watercourses caused greater yield of both minor and major crops (Chaudhry 2018; Mekonnen et al. 2015), as authenticated by data given in Figs. 19.2 and 19.3. It amplified production, and this entire acknowledgement should be given the more stable and well-organized flow of water to farmland, which permits crops to obtain satisfactory irrigation over their growing periods. It is consistent with a study that shows that in Punjab, watercourse lining improved wheat yields by 15–20%, while rice and cotton yields improved by 10–15% (Khan et al. 2018).

19.4.4 Land-Use Intensities

When farmers save water, land-use intensities are transformed owing to watercourse improvements, and in this way, watercourse lining can aid growing agricultural production and help improve livelihoods in rural settings. It has been reported in other studies with similar results. For instance, Hussain et al. (2019) found that improvements to watercourses in Sindh led to farmers being able to increase their area of cultivation by 10–15%, which increased farm incomes and improved food security.

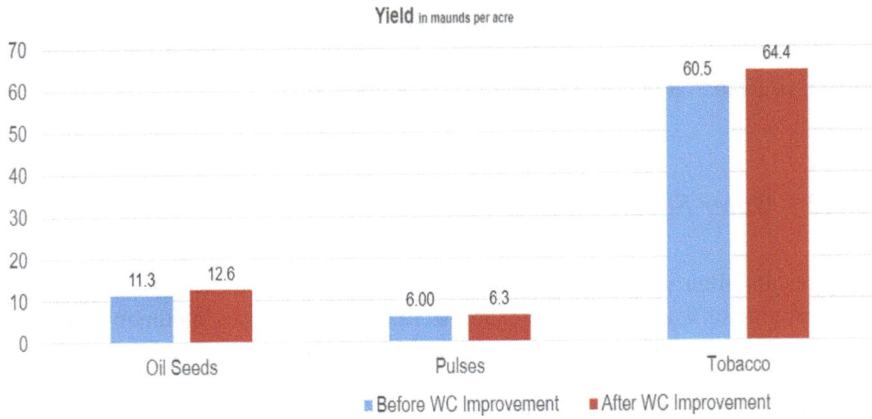

Fig. 19.3 Impact of WC improvement on minor crop yields on sample farms

19.4.5 Challenges and Barriers

Given the enormous advantages of watercourse lining, the study also found the challenges in scaling these efforts. The lining materials are not inexpensive, while implementation requires technical skills and knowledge. These are challenges faced all over the world, not unique to Pakistan. For instance, in India, the cost of canal lining has been a significant deterrent to scaling up irrigation modernization projects (Sharma et al. 2019). The same is true in Israel; the uptake of new irrigation technologies has required significant investment in infrastructure and capacity building (Marin et al. 2017).

Better watercourse governance is helpful in accomplishing water conservation, improving land-use intensities, and improving sustainable food and water security in countries like Pakistan, where agriculture is the main economy. Therefore, the participation of community and village organizations in water governance cannot be ignored, and we cannot promote equitable sharing of water, reduce water wastage, and increase agricultural production without considering the major stakeholders. Policy implications, challenges, and recommendations for enhancing community-led water governance mechanisms are also discussed in the chapter. This study has shown that the lining of watercourses is a central tool to combat water scarcity and increase water productivity in agriculture in Pakistan. The results demonstrate how much can be gained in terms of conserving water, boosting crop yields, and maximizing the use rates of land through parameterized watercourse lining. However, the scale-up of these initiatives is likely to face barriers in cost, technical know-how, and sustainability (Ishaque et al. 2023). Future research may be needed to assess the long-term sustainability of watercourse lining, in particular its impact on groundwater.

19.4.6 Limitations and Strengths of the Study

We acknowledge that our study has some limitations. The graphical and statistical elements of the document require significant modification to properly improve the analytical strength of the manuscript. The chapter needs to focus on providing detailed presentations of the statistical tests, such as *p*-values and confidence intervals, to support the differences that they claim. Sampling was 2% at the start but increased to 5%, which could be considered as one of the significant limitations in this study.

The chapter remains a remarkable achievement in comprehensively understanding the methodological challenges of water management in Pakistan while particularly paying attention to its practical aspects. This research is noteworthy due to the substantial sample size of 689 watercourses, accompanied by robust statistical methods, which builds important empirical evidence on the impacts of improvements to watercourses. This study, dealing with critical issues such as water scarcity and agricultural productivity, provides encouraging lessons for community-based participatory water management systems that are relevant for policy approaches in arid regions.

19.5 Conclusion

Our study suggests the following recommendations. First, participation of the community ought to be revitalized as an active and competent technique of water management, and it can be made possible by the institution of associations of water users and the establishment of organizations of village communities. Second, there is a need to establish regional knowledge-sharing platforms. In this way, relevant skills and knowledge will build the capacity of irrigation officials and cultivators by providing training on water management. Third, the scale of improving watercourse improvements is essential not only for saving water (sustainable groundwater extraction), increasing land use, but also for boosting production and crop intensities. Fourth, using multidisciplinary monitoring protocols, such as remote sensing, decision-makers can encourage the embracing of water-saving technologies and practices (like drip irrigation and precision agriculture).

Acknowledgements The authors would like to thank our participants for contributing their valuable time and opinions. We are also thankful for the NPIWC II project for providing data and facilitation.

References

Ahmed F, Malik NI, Bashir S, Noureen N, Ahmad JB, Tang K (2024) Political economy of maternal child malnutrition: experiences about water, food, and nutrition policies in Pakistan. Nutrients 16:2642. https://doi.org/10.3390/nu16162642

Ahmed F, Malik NI, Zia S, Akbar AS, Li X, Shahid M, Tang K (2023) Rural mothers' beliefs and practices about diagnosis, treatment, and management of children's health problems: a qualitative study in marginalized Southern Pakistan. Front Public Health 10:1001668. https://doi.org/10.3389/fpubh.2022.1001668

Ahmed F, Malik NI, Shahzad M, Ahmad M, Shahid M, Feng XL, Guo J (2022) Determinants of infant young child feeding among mothers of malnourished children in South Punjab, Pakistan: a qualitative study. Front Public Health 10:834089. https://doi.org/10.3389/fpubh.2022.834089

Ahmed F, Shahid M, Cao Y, Qureshi MG, Zia S, Fatima S, Guo J (2021) A qualitative exploration in causes of water insecurity experiences, and gender and nutritional consequences in South Punjab, Pakistan. Int J Environ Res Public Health 18:12534. https://doi.org/10.3390/ijerph182312534

Ali T, Hussain I, Khan MA (2020) Impact of watercourse lining on agricultural productivity in Punjab, Pakistan. J Water Res Manag 34(5):123–134

Bryant C (2019) Managing development in the Third World. Routledge

Sutomo YA, Sianipar CP, Hoshino S, Onitsuka K (2024) Self-reliance in community-based rural tourism: observing tourism villages (Desa Wisata) in Sleman Regency, Indonesia. Tour Hosp 5(2):448–471

Chandra G (2010) Participatory rural appraisal. In: Katiha PK, Vaas KK, Sharma AP, Bhaumik U, Ganesh C (eds) Issues and tools for social science research in inland fisheries. Central Inland Fisheries Research Institute, Barrackpore, Kolkata, India, Bulletin, 163, pp 286–302

Chaudhry AM (2018) Improving on-farm water use efficiency: role of collective action in irrigation management. Water Resour Econ 22:4–18

Lewis WA (2004) Development planning. Routledge

Cornwall A (2002) Beneficiary, consumer, citizen: perspectives on participation for poverty reduction. SIDA Studies, p 11

Cypher JM (2014) The process of economic development. Routledge

Davies S, Gueneau A, Mekonnen DK, Ringler C, Robinson S (2016) Irrigation and water management in the Indus Basin: infrastructure and management strategies to improve agricultural productivity. In: Spielman DJ, Malik SJ, Dorosh PA, Ahmad N (eds) Agriculture and the rural economy in Pakistan: issues, outlooks, and policy priorities. University of Pennsylvania Press on behalf of the IFPRI, Philadelphia, PA, pp 117–170

D'Souza R (2006) Water in British India: the making of a colonial hydrology. His Compass 4(4):621–628

Dennis M (1977) Ineffability in the Lao Tzu: the taming of a dragon. J Chin Philos 4:357–382

Dodds ME (1989) The need for appropriate inquiry concerning human development: development: the human challenge. Development Society of Southern Africa

FAO (2020) The state of food and agriculture. Food and Agriculture Organization (FAO) of UNDP. Viewed at The State of Food and Agriculture 2020 on 30 Apr 2025

Hussain I, Ali T, Khan MA (2019) Watercourse improvement and its impact on farm incomes in Sindh, Pakistan. Agric Water Manag 215:1–10

Ishaque W, Mukhtar M, Tanvir R (2023) Pakistan's water resource management: ensuring water security for sustainable development. Front Environ Sci 11:1096747

Khan MA, Ali T, Hussain I (2018) Watercourse lining and its impact on crop yields in Punjab, Pakistan. J Irrig Drain 67(3):456–465

Kugelman M (2010) Pakistan's food insecurity: roots, ramifications, and responses. In: Kugelman M, Hathaway R (eds) Hunger pains: Pakistan's food insecurity. The Woodrow Wilson International Center for Scholars, Washington, DC, pp 13–14

Mohan G (2008) Participatory development. the companion to development studies. Hodder Education, p 45
Mekonnen DK, Channa H, Ringler C (2015) The impact of water users' associations on the productivity of irrigated agriculture in Pakistani Punjab. Water Int 40(5–6):733–747
Mustafa D, Gioli G, Karner M, Khan I (2017) Contested waters: subnational scale water conflict in Pakistan. USIP, Washington, DC
Oakley P (1991) Projects with people: the practice of participation in rural development. International Labour Office, Geneva
Pakistan Council of Research in Water Resources (PCRWR) (2023) Water scarcity in Pakistan: challenges and solutions. PCRWR, Islamabad
Qureshi AS (2011) Water management in the Indus Basin in Pakistan: challenges and opportunities. Water Policy 13(3):298–317
Rahman MDA (1993) People's self-development: perspectives on participatory action research. University Press Ltd., London
Roodt M (2001) Participation, civil society, and development. In: Coetee J, Graaff J, Hendricks F, Wood G (eds) Development theory, policy and practice. Oxford University Press, South Africa
Shams S (2016) Why water scarcity is a bigger threat to Pakistan's security than militancy. DW. Retrieved from https://www.dw.com/en/why-water-scarcity-is-abigger-threat-to-pakistans-security-than-militancy/a-19293470?maca=en-Facebooksharing. Accessed 5 Feb 2024
Sharma S, Singh R, Kumar P (2019) Impact of canal lining on water use efficiency in India. Jirri Sci 37(4):345–356
Marin P, Tal S, Yeres J, Ringskog KB (2017) Water management in Israel (No. 28097). The World Bank Group
Watto MA, Mitchell M, Akhtar T (2021) Pakistan's water resources: overview and challenges. Water resources of Pakistan: issues and impacts, pp 1–12
Fussell W (1996) The value of local knowledge and the importance of shifting beliefs in the process of social change. Community Dev J 31(1):44–53
World Bank (2020) Pakistan: agriculture and water resources. World Bank, Washington, DC

Chapter 20
Evaluating Women's Nutrition and Health Status in Chittur Thathamangalam Municipality, Kerala: A Pathway to Sustainable Public Health

K. Jayarajan, T. Athira, P. Dhanya, K. Prasad, Sreelu Sreepadi, S. Nazrin Beegum, and G. K. Jishnuvardhan

Abstract This study explores the health and well-being of women in Chittur, Palakkad, Kerala, to inform sustainable public health planning. A sample of 450 women from urban areas participated, with data collected through structured interviews and dietary surveys. The analysis highlights significant disparities in health indicators across four spatial zones within the municipality. Findings reveal high prevalence of anemia, suboptimal body mass index (BMI) levels, inadequate physical activity, and limited dietary diversity. Socioeconomic determinants such as income, education, and housing emerged as critical influences on health outcomes. Multivariate analysis identified three core dimensions: Firstly, dietary patterns, health conditions, and socioeconomic influences on women's well-being; secondly, household infrastructure, dietary habits, and health status among women and finally, nutritional choices, economic disparities, and health outcomes among women. Zone I recorded the highest composite health score, while Zone II faced the most challenges.

K. Jayarajan (✉) · T. Athira · P. Dhanya · K. Prasad · S. N. Beegum · G. K. Jishnuvardhan
Department of Geography, Government College, Chittur, Palakkad, Kerala, India
e-mail: jayarajkk@gmail.com

T. Athira
e-mail: athirathadayil2016@gmail.com

P. Dhanya
e-mail: dhanyaeptri@gmail.com

K. Prasad
e-mail: prasadngeo@gmail.com

S. N. Beegum
e-mail: nazrinbegumgeo@gmail.com

G. K. Jishnuvardhan
e-mail: jishnugeog@gmail.com

S. Sreepadi
GHSS Kayanna, Kozhikkode, Kerala, India
e-mail: ssreepadi@gmail.com

A. Roy et al. (eds.), *Environment and Public Health*, Sustainability Solutions,
https://doi.org/10.1007/978-3-031-99770-9_20

The study underscores the need for targeted interventions focusing on nutrition, education, and healthcare access to enhance women's health. It also highlights the importance of integrating geospatial and multivariate methodologies in sustainable public health strategies.

Keywords Women's health · Nutritional status · Multivariate analysis · Body Mass Index · Public health planning · Spatial disparities

20.1 Introduction

The socially constructed gender roles of men and women interact with their biological roles to influence the nutritional status of the entire family and each gender. Due to women's cyclical iron loss and childbearing, their nutritional status is particularly susceptible to deficiencies in diet, care, and health or sanitation services. Furthermore, the nutritional status of newborns and infants is closely associated with the nutritional status of the mother before, during, and after pregnancy. Women typically have restricted access to land, education, information, credit, technology, and decision-making forums. Considering the significance of women's nutritional and health status, women in India face numerous serious dietary and health issues, including anemia, undernutrition, nutritional deficiency diseases, reproductive health problems, violence against women, unequal treatment of girls, and HIV and AIDS. Nutrition and health are interrelated and represent two aspects of the same issue. Nutrition is a causal factor, while health is the resultant effect.

Women's nutrition and health status are critical determinants of public health, particularly in developing regions. Women's nutrition and health status are essential determinants of public health, particularly in developing regions. This is evident from the research findings across multiple studies: Health literacy among women in Afghanistan was found to be low, even compared to other Asian countries (Harsch et al. 2021). This has significant implications for public health, as health literacy is linked to health behaviors and outcomes. The study calls for recognizing health literacy as a public health challenge that needs to be addressed in low-income countries affected by crises. In China, public health services were found to positively influence the health status and fertility intentions of rural migrant women (Jiang and Huang 2023). This highlights the importance of government-provided health services in safeguarding women's health rights and affecting demographic trends. Nutritional factors play a crucial role in women's health across different life stages. For pregnant women in Ethiopia, inadequate intake of red meat and green vegetables, as well as low mid-upper arm circumference, was strongly associated with anemia (Osman et al. 2020). This underscores the need for nutrition counseling and interventions during pregnancy to address high anemia prevalence in developing countries. The nutritional status and knowledge of women are potent indicators of child and household food security, especially in marginalized societies like Sri Lanka (Weerasekara et al.

2020). The study found that women's food and nutrition-related knowledge, attitudes, and practices significantly affect their BMI status and household food security.

In disaster-prone regions of South and Southeast Asia, women face higher risks of adverse physical and mental health outcomes following natural disasters (Fatema et al. 2021). Socio-demographic factors, disaster exposure, and pre-existing conditions contribute to these vulnerabilities, emphasizing the need for targeted support and healthcare access for women in disaster-affected areas. These studies collectively demonstrate that women's nutrition and health status are intricately linked to broader public health outcomes, particularly in developing regions. Addressing women's health needs through improved health literacy, nutrition interventions, public health services, and disaster preparedness can have far-reaching impacts on overall population health and well-being. Adequate nutrition during pregnancy and lactation is essential for both maternal well-being and optimal fetal development. Malnutrition among women can lead to a range of adverse health outcomes, including increased risk of maternal mortality, low birth weight infants, and compromised cognitive development in children. Addressing women's nutritional needs through targeted interventions and policies can have far-reaching effects on community health, economic productivity, and social progress.

The healthy future of society depends on the health of the children and their mothers, who are guardians of that future. Malnutrition remains a global challenge, contributing to 45% of child mortality worldwide (Chesoli et al. 2023). In urban areas of developing countries, public primary health facilities are an essential source of nutrition services, but face challenges in service provision and the capacity of healthcare providers (Hasan et al. 2023). Gaps exist in anthropometric measurement, counseling, and healthcare providers' knowledge of nutrition, hampering the delivery of quality nutrition services. Interestingly, studies have shown that women's nutritional knowledge and attitudes are potent indicators of household food security and children's nutritional status (Weerasekara et al. 2020). However, research in marginalized areas of Sri Lanka revealed that reproductive-age women often have low levels of nutritional knowledge and practices, despite positive attitudes toward receiving nutritional information (Weerasekara et al. 2020). This highlights the need for targeted nutrition education programs. To address these challenges, multisectoral approaches involving agriculture, health, trade, education, and processing sectors are necessary (Chesoli et al. 2023). Effective interventions could include radio-based health/nutrition education programs, which have shown promise in improving dietary diversity practices and nutrition-related knowledge (Saaka et al. 2021). Additionally, involving peer women from the community in health and nutrition interventions, such as through self-help groups, can lead to more effective outcomes and contribute to women's empowerment (Pradhan et al. 2023).

Understanding the nutritional and health challenges faced by women in Chittur Municipality is essential for developing effective public health interventions. This study aims to assess the nutritional status and health-related behaviors of women in Chittur Municipality through a comprehensive survey. By examining factors such as dietary habits, physical activity levels, and access to healthcare services, we can identify key areas for improvement in women's health outcomes. The findings from this research will inform targeted interventions and policy recommendations to address the specific needs of women in this community, ultimately contributing to better overall health and well-being.

Previous studies have highlighted significant nutritional deficiencies and health disparities among women in various parts of Kerala. These disparities are particularly pronounced in rural and economically disadvantaged communities, where access to nutritious food and healthcare services is often limited. Factors such as socioeconomic status, education level, and cultural practices play crucial roles in shaping dietary habits and health outcomes among women in the region. Addressing these nutritional challenges requires a multifaceted approach, including targeted interventions, community education programs, and policy measures aimed at improving food security and healthcare accessibility.

Education and socioeconomic status are strongly associated with dietary quality and diversity. Higher educational attainment and better economic situations are linked to improved dietary diversity among women, particularly in developing countries like Zimbabwe (Makate and Nyamuranga 2023). In Germany, individuals with higher education levels tend to engage in more sustainable and health-conscious dietary behaviors (Klink et al. 2022). Similarly, in sub-Saharan Africa, younger and economically well-off people adhere more strongly to diets high in animal-based products, indicating progress along the nutrition transition (Weil et al. 2023). Cultural beliefs and practices play a crucial role in shaping food patterns and eating habits, especially among pregnant women. These beliefs often lead to food taboos and restrictions, which are typically transmitted by family or community members and have strong religious or spiritual influences (De Diego-Cordero et al. 2020). In Central Appalachia, cultural practices are unknowingly leveraged to address food insecurity, though their impact on nutritional status remains unclear (Batey et al. 2023). Interestingly, some studies reveal contradictions in dietary patterns across different socioeconomic strata. For instance, in Mexico, higher diet quality was observed among people of lower socioeconomic status and residents of rural areas, particularly in southern Mexico (Pérez-Tepayo et al. 2020). This highlights the complex interplay between socioeconomic factors and dietary habits, which can vary significantly depending on the country's nutritional transition phase, level of economic development, and cultural context. While education and higher socioeconomic status generally contribute to better dietary habits, cultural practices and beliefs continue to exert a strong influence on food choices. Understanding these factors is crucial for developing targeted interventions to improve nutritional habits and health outcomes among women in different regions.

However, there are limited data explicitly addressing the nutritional and health status of women in Chittur Municipality. The lack of comprehensive studies on women's nutrition and health in Chittur Municipality presents a significant gap in understanding local public health needs. This knowledge deficit hampers the development of targeted interventions and policies aimed at improving women's well-being in the area. Further research is urgently needed to assess the prevalence of nutritional deficiencies, chronic diseases, and other health issues affecting women in

Chittur Municipality, which could inform evidence-based strategies for enhancing their overall health outcomes.

Addressing this gap is crucial for formulating targeted strategies that can enhance the well-being of women in this region. Further research is needed to explore the specific challenges faced by women in rural areas of the region. This could involve conducting surveys and interviews to gather firsthand accounts of their experiences and needs. Additionally, collaborating with local organizations and community leaders may provide valuable insights into culturally appropriate interventions that can effectively support women's well-being.

20.1.1 Study Area

Chittur Thathamangalam Municipality, the taluk headquarters of Chittur Taluk, is a historically significant region in Palakkad district, Kerala. The name Chittur derives from the Tamil word "Siruttoor," denoting a small village. The municipality, geographically situated between 10°40′35″ N to 10°42′21″ N latitude and 76°41′42″ E to 76°44′55″ E longitude, comprises Chittur and Thathamangalam villages and has been in existence since 1908. It encompasses an area of 41.71 km^2 and has a population of 32,285 (2011 Census), with a population density of 2,179 persons per km^2. The region is characterized by floodplains of the Chittur Puzha, a tributary of the Bharathapuzha, and is surrounded by residual hills. The municipality experiences a tropical climate with maximum temperatures reaching 42 °C and an annual precipitation of 200 cm due to the influence of the Palakkad Gap. The literacy rate stands at 74.94%, one of the lowest in the district. Predominantly Hindu, the municipality also has a significant Muslim and a smaller Christian population. Although traditionally agricultural, industrial activities have declined, with only small-scale industries such as weaving and pottery remaining. The transportation network includes state highways NH 26 and 27, connecting Chittur Thathamangalam to Palakkad, Thrissur, Coimbatore, and Pollachi. Despite its historical and cultural significance, the municipality faces challenges in industrial development and literacy improvement (Fig. 20.1).

20.2 Materials and Methods

The present study was based on both primary and secondary sources. (a) Secondary Sources: The secondary data on nutritional level and health problems among married women-related variables were collected from the National Family Health Survey, District Level House Hold survey (DLHS-RCH), Human Development Reports, National Sample Survey Organization, Census of India, Annual Reports of Department of Ministry of Health, Government of Kerala, the United Nations published reports and other statistical sources of the Government of India (b) Primary Sources: The primary data were collected from a sample size of 450 women respondents

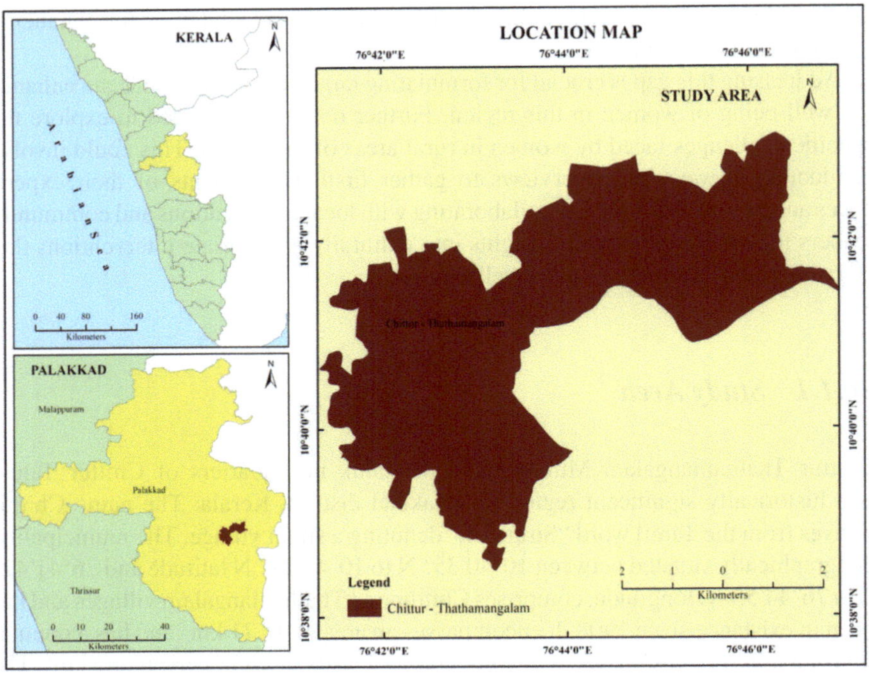

Fig. 20.1 Location map of the study area

using multistage sampling method in the study area (Chittur Thathamangalam Municipality) through personal interview method to obtain accurate information and compare rural and urban areas in the year 2020–21. The questionnaire was administered exclusively to women respondents in Chittur Thathamangalam Municipality. The necessary primary data were collected for nutritional level and women's health in Chittur Thathamangalam Municipality through a well-structured and pretested questionnaire. In the oral questionnaire of the diet survey, the quantity of food items consumed the previous day was recorded, and in the weighing method, the weight of raw food items to be cooked for midday meals was taken in the morning and for supper in the afternoon. The weight and height of the women respondents were measured under anthropometric measurements. The study attempted to demonstrate the association between the nutritional level and diseases (health problems) of the women respondents in the study area. For this purpose, the study did not cover all the deficiency diseases or health problems and general health problems of the women resulting from a lack of nutrient intake, but focused only on the presence or absence of easily identifiable diseases or health problems resulting from a lack of nutrient intake in the diet. As social science researchers, there were limitations in identifying the nutritional deficiency diseases, severity, past history, and incidence of the diseases or health problems of the women respondents in the study area. After the completion of the survey, data were analyzed using various statistical methods.

20.2.1 Analytical Tools

The data thus collected are processed using various statistical tools to arrive at meaningful results. The tools used are percentages, mean, standard deviations, frequencies, and multivariate analysis. Nutritional status was measured through the body mass index of the women respondents in the study area. The body mass index is used as an indicator of the nutritional status of the women respondents in the study area. The nutritional status of the women respondents in Chittur Thathamangalam Municipality is assessed through James et al. classification of BMI.

BMI = Weight/(height in meter square)
Body Mass Index (BMI) value Nutritional Grade
< 18.5 Chronic Energy Deficiency (CED), 2.18.5 to 23.00 Normal, > 23.00 Obesity.

20.2.2 Variables Considered in the Analysis of the Women

The present study analyses the spatial distribution of women's health and nutrition and their relationship with social, cultural, and psychological conditions. The available healthcare facilities and the demographic characteristics of the general population of the study area. These are altogether very important aspects for the identification of the problems in association with women's health, degree of development and planning. On the basis of the discussions, there are 199 variables selected in the statistical analysis. They are the information collected from the primary survey of the women living in Chittur Thathamangalam Municipality in the seven significant subdivisions consisting of 199 variables such as Personal information, economic household condition, dietary information, lifestyle, health, mental health and stress and one-day nutritional intake. As the present analysis involves a large number of variables related to measuring the health status of women in Chittur Thathamangalam Municipality is required to the use of suitable multivariate statistical techniques capable of considering the health status level from the variables used in this study is required. Multivariate analysis consists of a collection of methods that can be used when several measurements are made on each individual or object in one or more samples. These measurements refer to variables and to the individuals or objects as units or observations. In the present situation, a large amount of data related to nutrition and women's health was obtained on the basis of discussion and inferences derived; the investigator decided to use a suitable multivariate analysis in order to simplify the dimensional relationship among the variables and to identify the significant factors of women's health. The goal of the multivariate approach is simplification. It seeks to express in terms of a reduced set of dimensions. Such multivariate techniques are exploratory; they essentially generate hypotheses rather than test them. Factor analysis (FA) and principal component analysis (PCA) are methods used when the researcher is interested in identifying a smaller number of factors underlying a large number of observed variables. Factor analysis is one of the best-used multivariate techniques to assess the factorial validity of the questions, which make up the scales of the extent to which they seem to be measuring the variables. The everyday use of PCA and FA is in developing objective instruments. It is helpful for measuring constructs which are not directly observable in real life. A factorial analysis produces factor loading for each combination of extracted factors and the observed variables. Factor loadings are similar to correlation coefficients between the factors and variables. Thus higher the factor loadings, the more likely it is that the factor underlies that variable. Factor loadings help in identifying which variables are associated with a particular factor. Eigenvalue is the measure of the amount of total variance in the data explained by a factor. Looking up the Eigenvalue, one can determine if the factor explains a sufficient amount of variance to be considered a meaningful factor. An Eigenvalue of less than 1 essentially means that the factor explains less variance than a single variable and, therefore, should not be considered to be a meaningful factor. The component which corresponds to the highest Eigenvalue is known as the first principal component. The principal component, which corresponds to the second highest, third highest and last, is known as the second, third, and fourth, so on. Factor score: The component score calculated for each observation unit area will show the spatial variation. The positive and negative scores pertaining to each component are developed with respect to each spatial unit in the study, and they help to evaluate the problematic region with reference to each index. On the basis of the sign of the scores as positive and negative, the development index is classified into four types. The score value of 1 and above is considered high positive and negative development. The values less than 1 in both directions are treated as less development on the negative and positive sides. The data set in total consists of 199 variables that are used to analyze the level of women's health in Chittur Thatamangalam Municipality. In the present analysis, the data matrix consists of 199 variables collected from four Zones in Chittur Thatamangalam Municipality.

20.3 Results

20.3.1 Women's Health Variables Relationship and Factor Solution

The extraction of Table 20.1 illustrates the total variance explained by the variables included in the analysis.

The distribution of eigenvalues and the total percentage and cumulative percentage variance of each one of the factor solutions are presented in Table 20.1. It is pertinent to note that three factors with 32.81 Eigenvector values explain 100% of the variance in the data set. The three factors explain altogether 100% of the total variance, with each one of their values ranging from 32.81 to 47.32. The three factors, 27.57 Eigen

Table 20.1 Total variance explained

Component	Initial Eigenvalues total	% of variance	Cumulative %	Extraction sum of squared loadings total	% of variance	Cumulative %	Rotation sum of squared loading total	% of variance	Cumulative %
1	54.06	45.43	45.43	54.06	45.43	45.43	47.32	47.32	39.76
2	36.7	30.9	76.33	36.77	30.9	76.33	38.86	38.86	72.42
3	28.16	23.66	100	28.16	23.66	100	32.81	32.81	100

Source Extraction Method: Principal Component Analysis

values, explain 100% of the total variance, whereas the remaining two Eigen values, such as 32.65 and 39.76, explain 72.42 and 42.98, and 39.76 of the total variance.

20.3.2 Significant Factors and Their Loading

The first component, which explains 39.76% of the total variance, is significantly loaded with 65 variables. The factor loading positive value of 36 variables lies between 0.987 and 0.506 in both directions. There is sometimes consumption of milk among women (0.987), daily consuming meat (0.974), more than five item vegetable daily using (0.974), the women having cholesterol (0.969), there is more than 5000 rupees using women for their monthly food cost (0.969), the daily fish consuming women (0.969), the religion of women is Christian (0.969), 26–35 age of marriage of women (0.969), women having above 5 children (0.969),the women with mental stress (0.953), the women with sleeplessness (0.929) the women with hereditary disease(0.907),the religion Muslims (0.902), never having leafy vegetables (0.902), nothing having in the early morning (0.859), sometimes having soft drink (0.840), women with job stress (0.836),exercising women (0.809), women with diabetes (0.779), 1 plate rice having women for breakfast (0.771), having less than three items vegetables for daily use (0.734), the tea and snacks having in evening (0.731),the weekly milk consumption (0.730), the snacks having women in mid morning (0.723), not anemic women (0.712), weekly fruits consumption (0.709), having tea/milk before sleep (0.706), the age of women (0.705),having sanitation facilities (0.704), concrete houses (0.684), daily consuming ice cream, chocolate, and dry fruits (0.665), sometimes street food intake (0.649), having 1 cup rice for lunch (0.599), nonworking women (0.523), and daily having fatty food intake (0.506). A suitable title for this dimension could be dietary patterns, health conditions, and socioeconomic influences on women's well-being. This title reflects the strong associations between food consumption habits, health conditions (such as diabetes, cholesterol, and stress), and socioeconomic factors (like income, employment status, and housing conditions).

The second component, which accounts for 32.65% of the total variance, includes 51 variables. The factor loading positive values of 27 variables lie between 0.999 and 0.0524. Among these variables show the positive values of loading such as won well (0.999), women having primary education (0.997), no menstrual issue (0.964), married (0.959), women not having allergy (0.955), having chapathy for lunch (0.941), daily fruits consumption (0.936), having chapathy for dinner (0.932), no pregnancy (0.931), no own house (0.897), one cup milk having in early morning (0.897), never using ice cream, chocolate and dry fruits (0.891), marriage before 18 (0.888), women having high school education (0.868), women having one child (0.860), weekly consuming fish (0.820), having other food in evening (0.813), children under governess (0.718), never having street food (0.691), joint family (0.691), 36–55 age (0.690), having sanitation facilities (0.661), 18–23 of BMI (0.651), tiled roof house (0.607), 26–35 age(0.597), using spouted gram (0.571), high blood pressure having women (0.524). A suitable title for this dimension could be Household Infrastructure, Dietary Habits, and Health Status among Women.

This title captures the interplay between housing conditions (own well, sanitation, roof type), dietary patterns (chapati consumption, milk intake, fruit consumption), and demographic-health factors (education, marital status, BMI, high blood pressure).

The Third component accounts for 27.57% of variance with positive values of 24 variables (0.974 to 0.503) and 24 variables with negative loadings (– 0.503 to – 0983). The variable such as breakfast skipping women (0.974), annual income below 10,000 (0.915), annual income more than 100,000 (0.915), fruits having women in mid morning (0.888), monthly cost of food less than 2000 (0.847), no diseases having women (0.836), daily having fatty food intake (0.833), monthly consuming meat (0.807), not using spouted gram (0.806), employed (0.794), nothing having mid morning (0.680), 18.5–23 of BMI (0.679), having less than tree items of vegetables for daily use (0.677), having tea/milk before sleep (0.676), never having street food (0.618), never having street food (0.618), joint family (0.618), weekly consuming fish (0.568), 1 plate rice having women (0.548), anemic affected women (0.548), women having one child (0.506), never having soft drink (0.503). This cluster focused on nutritional choices, economic disparities, and health outcomes among women. This dimension reflects the dual nature of the component, highlighting the contrast between income levels, dietary habits (such as breakfast skipping, fatty food intake, and vegetable consumption), and health indicators (like anemia and BMI).

20.3.3 Spatial Distribution of Factorial Scores of the Multidimensional Level of Women Health

The component scores derived for each observation areal unit of Chittur Thatamangalam Municipality show the spatial variation in the respective factor scores. Table 20.2 shows the positive and negative scores pertaining to each component showing the level of health related to women in Chittur Thathamangalam Municipality. The score value above 1 on the positive side denotes a high level of health among the Chittur Thathamangalam Municipality. Whereas the score value of 0–1 denotes the medium level of health among women. The negative value of – 1 to 0 denotes the low range of health, and an index score below – 1 denotes a very low range of health for women. In the present analysis, all the components are selected based on the dimensions related to women's health in Chittur Thathamangalam Municipality.

Table 20.2 Composite index of women's nutrition and health status

Zone	Factor 1 Dietary patterns, health conditions, and socioeconomic influences on women's well-being	Factor 2 Household infrastructure, dietary habits, and health status among women	Factor 3 Nutritional choices, economic disparities, and health outcomes among women	Composite index
Zone I	– 0.42	0.43	1.37	1.37
Zone II	– 0.8	– 1.19	– 0.42	– 2.42
Zone III	– 0.22	1.12	– 0.96	– 0.06
Zone IV	1.45	– 0.36	0.01	1.1

Source Compiled by author

20.3.4 *Spatial Disparity in the Dimension of Women's Nutrition and Health Status*

Spatial disparities in women's health arise from a complex interplay of dietary habits, health awareness, and living conditions. Variations in food consumption patterns, influenced by socioeconomic status, cultural norms, and regional availability, significantly impact nutritional outcomes and overall well-being. Women in economically disadvantaged areas often face limited access to nutritious food, leading to deficiencies and health complications. Furthermore, disparities in health awareness and access to medical services contribute to variations in disease prevalence and maternal health outcomes. Household infrastructure, including sanitation, water availability, and housing conditions, further shapes women's health by affecting their susceptibility to infections and chronic diseases. The intersection of nutritional choices, economic disparities, and health outcomes underscores the need for targeted policies to bridge these gaps and promote equitable health access for women across diverse spatial contexts.

Factor 1 shows the dimension of dietary patterns, health conditions, and socioeconomic influences on women's well-being. The spatial distribution of factor scores reveals distinct patterns across different zones. Table 20.2 and Fig. 20.2 a show that Zone IV stands out with the highest positive factor score of 1.45, indicating a strong presence of the measured characteristic in this area. In contrast, Zone I and Zone III exhibit negative factor scores of – 0.42 and – 0.22, respectively, suggesting a moderate absence or inverse relationship with the measured attribute in these regions.

Factor 2 shows dimension of household infrastructure, dietary habits, and health status among women. Table 20.2 and Fig. 20.2b demonstrate the high factorial score in Zone III, 1.12 eigenvalue, followed by Zone I, 0.43 eigenvalue, Zone IV, – 0.36 eigenvalue and the lowest range is found in Zone II, – 1.19. The spatial distribution of eigenvalues across the zones reveals a distinct pattern of variability. Zone III

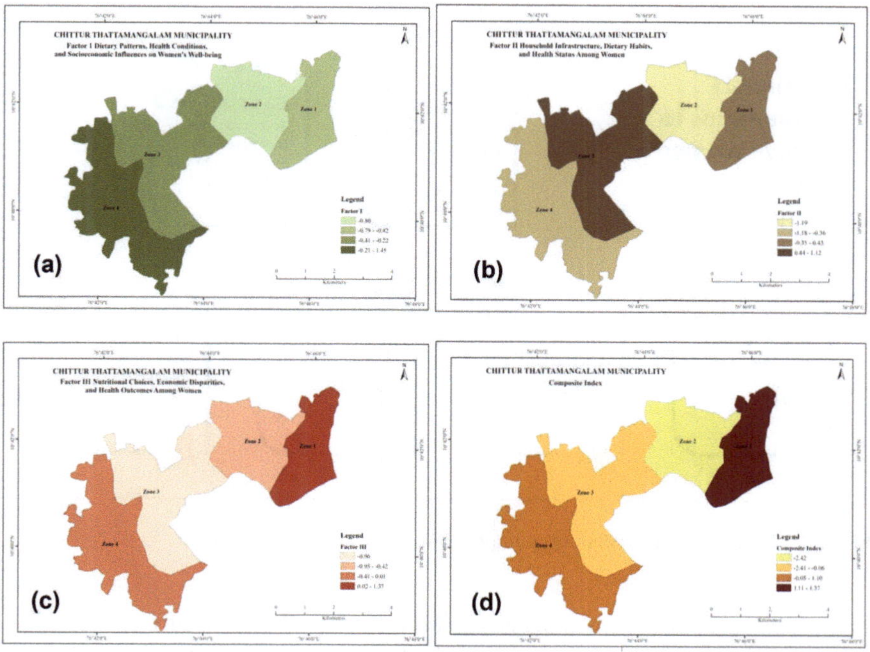

Fig. 20.2 Women's health and nutrition status in Chittur municipality

stands out with the highest positive eigenvalue, indicating a strong influence on the overall factorial structure. In contrast, Zone II exhibits the lowest negative eigenvalue, suggesting a potentially inverse relationship or minimal contribution to the factorial composition compared to the other zones.

Factor 3 shows dimension of nutritional choices, economic disparities, and health outcomes among women. The stark contrast in living conditions, food habits, and health awareness among women across different zones is evident from these scores. Table 20.2 and Fig. 20.2c show that Zone I demonstrates significantly better circumstances for women compared to the other zones, particularly Zone III, which shows the lowest score. This disparity suggests a need for targeted interventions and policies to improve the overall well-being of women in the lower-scoring zones, especially in areas related to nutrition, health education, and living standards. Zone I 1.37 Eigen value has a very high positive score value, followed by Zone IV 0.01. The Zone shows negative score values are Zone II − 0.42 and Zone III − 0.96.

Finally, the composite index provides a comprehensive measure of women's nutrition and health status across different geographical areas. By combining multiple indicators, it offers a more holistic view of women's well-being compared to individual metrics alone. Table 20.2 and Fig. 20.2d show that the spatial analysis can help identify regions with the greatest needs and inform targeted interventions to improve women's health outcomes. The composite index values of three factors are explained here on the basis of 199 variables taken for the analysis among the

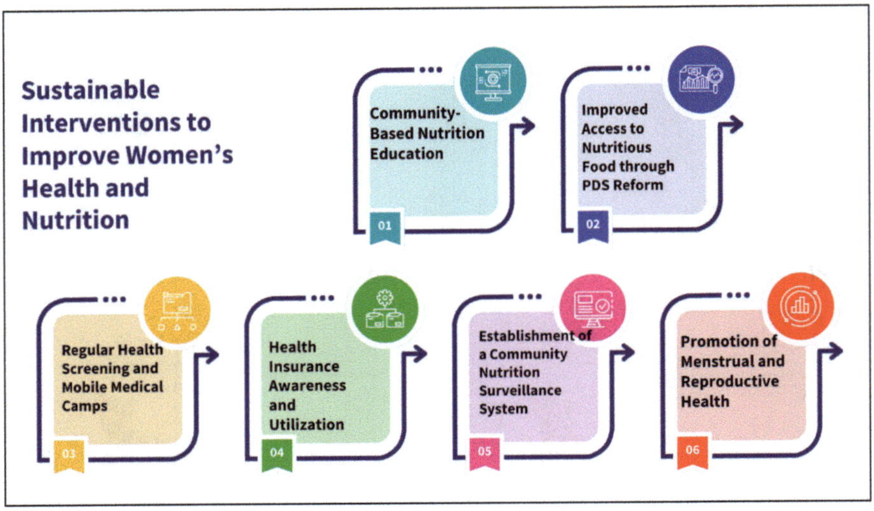

Fig. 20.3 Sustainable interventions to improve women's health and nutrition

four zones in Chittur Thathamangalam Municipality. The sum of the factor score value > + 1 positive direction denotes a high level of women's health, while + 1 to 0 stands for a medium level of women's health and 0 to − 1 and < − 1 denotes the low level of women's health. Table 20.2 shows the composite value Zone I marked as the high level of women's health with the value of 1.37, and Zone IV has the second highest level of women's health, value of 1.1. The low level of women's health was observed in Zone II and Zone III. The Zone II shows the very low value of − 2.42, and it is the most problematic zone among all the others.

To enhance women's health and nutrition in Chittur Municipality, sustainable interventions should employ a comprehensive, community-driven strategy. Figure 20.3 shows strategy should encompass community-based nutrition education through regular workshops, the promotion of dietary diversity, and the encouragement of kitchen gardens using culturally relevant methods. It is crucial to strengthen the Public Distribution System by incorporating iron-fortified staples and ensuring consistent access to nutritious food, monitored by local groups. Regular health screenings and mobile medical camps, in collaboration with Primary Health Centers (PHCs), can facilitate the early detection and management of prevalent conditions such as anemia and diabetes. Increasing awareness and utilization of government health insurance schemes, such as Ayushman Bharat and Karunya, through Self-Help Group (SHG) and Accredited Social Health Activist (ASHA)-led outreach can improve access to affordable care. Establishing a community nutrition surveillance system with trained volunteers will enable continuous monitoring and data-driven interventions. Revitalizing anganwadi centers as hubs for nutrition education, growth monitoring, and cooking demonstrations will strengthen existing infrastructure. Finally, promoting menstrual and reproductive health by distributing eco-friendly products and conducting awareness sessions will help reduce stigma and improve hygiene practices among women and girls.

20.4 Discussion

The primary aim of this study was to analyze the spatial distribution of women's health and nutrition in relation to their social, cultural, and psychological conditions. There is limited direct information on correlations between geographic location and factors affecting women's well-being. However, some relevant insights can be drawn: The studies examine women's well-being across different geographic contexts, including the United States (Streeter 2020), India (Challa and Perwez 2023), and Saudi Arabia (Almadani and Alwesmi 2023). This suggests that women's well-being is a concern across diverse geographic locations, though the specific factors and impacts may vary. In Saudi Arabia, the study found that rapidly changing rights and responsibilities for women, likely tied to geographic and cultural context, could result in more sources of stress (Almadani and Alwesmi 2023). This indicates an intersection between spatial elements and psychological well-being for women in that particular geographic setting. The research on Black women in the United States highlights how criminal justice contact impacts mental health differently for African American and Afro-Caribbean women (Talbert and Patterson 2023). This points to potential geographic and cultural differences in how systemic factors affect women's well-being even within the same country.

While not explicitly focused on geographic location, several studies examine how various contextual factors like leadership styles (Challa and Perwez 2023), social media use (Young et al. 2020), and parenting practices (Gandarillas et al. 2024) impact women's psychological well-being. These factors likely have geographic variations that could indirectly influence the spatial elements of women's well-being. Further research specifically examining geographic variations in women's well-being factors would be valuable to more fully understand these intersections. The research sought to identify patterns and correlations between geographic location and various factors affecting women's well-being. By examining data from diverse regions, the study aimed to uncover disparities in access to healthcare, nutritional resources, and social support systems. Additionally, the investigation explored how cultural norms and psychological factors intersect with spatial elements to influence women's overall health outcomes.

Our observations revealed significant disparities in health and nutrition outcomes among women based on their geographic location. Food insecurity significantly impacts women's dietary patterns and nutritional intake. In food-insecure households, consumption of nutrient-dense foods like meat, dairy, fruits, and vegetables decreases, while intake of processed and sugary foods increases (Kohanmoo et al. 2024). This shift in dietary patterns is associated with lower intake of essential nutrients and unhealthier eating habits. Cultural beliefs play a crucial role in shaping food practices among pregnant women. These beliefs often lead to food taboos and restrictions, which are primarily influenced by family, community members, and religious or spiritual factors. Such restrictions are often based on fears of unfavorable pregnancy outcomes or child health issues (De Diego-Cordero et al. 2020). Socioeconomic factors significantly influence women's food choices and eating behaviors.

Low-income women face barriers to healthy eating, including cost, convenience, preparation time, and family taste preferences. However, many of these women demonstrate high self-efficacy for nutrition change and health knowledge, suggesting potential for improvement with targeted interventions (Palmer et al. 2020).

The COVID-19 pandemic has led to diverse changes in eating patterns among women. Some adopted more health-oriented behaviors, such as eating healthier foods and avoiding unhealthy options, while others exhibited emotion-driven patterns, including increased consumption of sweets and alcohol (Bühlmeier et al. 2022). These changes were more pronounced in women, younger participants, and those with higher BMI. Women's food habits and health awareness vary across multiple dimensions, including food security status, cultural beliefs, socioeconomic factors, and external stressors like pandemics. Understanding these variations is crucial for developing effective, targeted interventions to improve women's nutritional status and overall health.

Socioeconomic factors play a significant role in women's nutritional status. Education level, occupation, and household income are associated with nutritional risk and knowledge (Silva et al. 2023). Lower educational levels are linked to higher odds of nutritional risk in breast cancer patients (Silva et al. 2023). Additionally, women's empowerment and gender equality are often connected to better maternal and child nutrition outcomes, though evidence is inconclusive (Quisumbing et al. 2020). Geographical variations are observed in nutritional status. For instance, in India, there are strong geographical clusters of malnutrition and below-average child height and weight, often crossing state borders (Striessnig and Bora 2020). Tribal and nontribal populations in Bangladesh show differences in nutritional status, with nontribal populations having higher rates of over-nutrition (Karim et al. 2023).

Life stage and physiological status significantly impact women's nutritional needs. Breastfeeding women in the first month postpartum exhibit several nutritional imbalances (Gila-Díaz et al. 2021). Pregnant women's nutritional status can be improved through intensive nutrition education and counseling (Wakwoya et al. 2023). Postmenopausal women have a lower likelihood of normal bone mineral density (Kopiczko 2020). Lifestyle factors such as physical activity, sun exposure, and dietary intake of specific nutrients (e.g., calcium and vitamin D) are crucial determinants of bone health in women over 40 years old (Kopiczko 2020). Additionally, nutritional status impacts reproductive health and fertility, with certain nutrients playing key roles in male reproductive health as well (Vanderhout et al. 2021). Women's health conditions and nutritional status vary across socioeconomic, geographical, physiological, and lifestyle dimensions. Understanding these multifaceted influences is crucial for developing targeted interventions and policies to improve women's health and nutrition globally.

This study reveals several dimensions of variation in women's living conditions, food habits, and health awareness. Food insecurity significantly impacts women's dietary choices and eating patterns. Food-insecure women tend to consume more grains, processed meats, potatoes, and sugary foods while decreasing their intake of nutrient-dense foods like fruits, vegetables, and lean proteins (Kohanmoo et al. 2024). This shift in diet quality is associated with lower intake of essential nutrients and unhealthier eating habits. Cultural beliefs play a crucial role in shaping food practices among pregnant women. Many cultures have food taboos and restrictions during pregnancy, often rooted in religious or spiritual beliefs. These restrictions are typically aimed at preventing unfavorable pregnancy outcomes but may impact nutritional intake (De Diego-Cordero et al. 2020). Socioeconomic factors, including cost, convenience, and time constraints, are significant barriers to healthy eating for low-income women. Despite having high self-efficacy for nutrition change and health knowledge, many women struggle to meet dietary recommendations due to these practical constraints (Palmer et al. 2020).

Interestingly, there are contradictions in how nutritional knowledge impacts eating behaviors. While some studies show that higher nutritional knowledge is associated with healthier eating habits, others reveal that it can lead to excessive calorie restriction, particularly among women (Kosendiak et al. 2024). This highlights the complex relationship between knowledge and behavior in nutrition. Women's food habits and health awareness are influenced by a complex interplay of factors, including food security status, cultural beliefs, socioeconomic conditions, and nutritional knowledge. Effective interventions should consider these multifaceted influences and focus on practical, achievable steps to improve dietary choices within existing constraints (Diego-Cordero et al. 2020; Kohanmoo et al. 2024; Kosendiak et al. 2024; Palmer et al. 2020).

Previous research has established a link between socio-cultural factors and women's health, but spatial distribution has been less explored. Geographic Information System (GIS) techniques are widely used across scientific disciplines, including public health (Ahasan and Hossain 2020). GIS can be effectively used to map and analyze spatial variations in health-related factors. For instance, Roy et al. (2024) describe the use of GIS with a composite synthetic index to measure spatial variation of urban household conditions, which could include health-related factors. Interestingly, Balasubramani et al. 2021 demonstrate the use of GIS mapping and hot spot identification for alcohol consumption patterns in Indian men and women. This approach could potentially be adapted to map nutrition and women's health indicators (Balasubramani et al. 2021).

The Health Opportunity Index (HOI) described in (Ogojiaku et al. 2020) is a multivariate tool that uses principal component analysis to determine the impact of social determinants of health at lower census geographies. This methodology could be applied to create a composite index for nutrition and women's health, which could then be mapped using GIS technology (Ogojiaku et al. 2020). While the specific composite index for nutrition and women's health is not mentioned, the provided research demonstrates that GIS techniques can be effectively used to map and analyze spatial variations in health-related factors. These methods could be adapted to create and visualize a composite index for nutrition and women's health across different geographical areas.

Understanding the spatial distribution of these factors is crucial for developing targeted interventions to improve women's health and nutrition. Mapping these factors at a regional or district level can help identify areas of high need and prioritize resource allocation. Additionally, analyzing spatial patterns can reveal geographic disparities and potential environmental or socioeconomic influences on women's health outcomes. This information can guide policymakers and healthcare providers in designing location-specific strategies to address the unique challenges faced by women in different areas.

Compared to urban areas, rural regions exhibited more pronounced deficiencies in women's health and nutrition. Rural regions indeed exhibit more pronounced deficiencies in women's health and nutrition compared to urban areas, as evidenced by several studies. In the context of urinary tract infections (UTIs), rural women were more likely to receive inappropriately long antibiotic durations compared to urban women (Clark et al. 2021). This suggests a disparity in healthcare practices and potentially poorer health outcomes for rural women. Regarding nutrition, lactating mothers in rural areas of China showed lower intake of essential nutrients compared to their urban counterparts (Ding et al. 2020). Rural women had lower consumption of vegetables, fruits, fish, milk, and dairy products, as well as lower intake of macronutrients and micronutrients such as vitamins and minerals.

Interestingly, the urban–rural divide in food consumption patterns is not always straightforward. In India, while urban households generally spent more on processed foods and consumed more food away from home, there were also variations between different urban areas (Bren D'Amour et al. 2020). This highlights the complexity of nutrition patterns beyond a simple urban–rural dichotomy. While rural areas generally show more pronounced deficiencies in women's health and nutrition, the patterns can be complex and vary by region and specific health outcomes. These findings underscore the need for targeted interventions to address health and nutritional disparities in rural areas, particularly for women's health.

Contrary to some studies, our findings suggest that cultural practices in certain regions do not always negatively impact women's nutritional status. Cultural practices in certain regions can have complex and varied effects on women's nutritional status, sometimes contradicting common assumptions. The study on Appalachian culture (Batey et al. 2023) reveals that people in this region are leveraging cultural practices to address food insecurity, though the impact on nutritional status remains unknown. This suggests that cultural adaptations may potentially have positive effects on nutrition, contrary to the notion that traditional practices always negatively impact women's health. Interestingly, the study on religious dietary practices (Arslan and Aydın 2024) highlights that while these practices can provide health benefits, including reduced risks of chronic diseases and enhanced mental well-being, they can also pose nutritional challenges that require careful management. This nuanced perspective demonstrates that cultural and religious practices can have both positive and negative impacts on nutrition. Our data indicate that women in regions with higher social support networks have better health and nutritional outcomes.

This study highlights the importance of considering geographic and socio-cultural contexts in addressing women's health and nutrition. The findings suggest that interventions aimed at improving women's health outcomes should be tailored to specific regional needs and cultural norms. Future research could explore how these contextual factors interact with other determinants of health, such as education and economic status. Additionally, policymakers and healthcare providers should collaborate to develop targeted strategies that address the unique challenges faced by women in different geographic areas.

The findings can inform policymakers and healthcare providers in designing region-specific health programs for women. These programs should focus on addressing the unique challenges faced by women in different regions, taking into account cultural, socioeconomic, and environmental factors. By tailoring interventions to local contexts, healthcare providers can improve the effectiveness of their services and better meet the needs of diverse female populations. Additionally, policymakers can use this information to allocate resources more efficiently and develop targeted strategies to reduce health disparities among women across different geographical areas.

This research is novel in its approach to mapping the spatial distribution of women's health and nutrition in relation to socio-cultural and psychological conditions. The study employs advanced geospatial analysis techniques to visualize patterns of health disparities across different regions. By integrating data on socioeconomic factors, cultural practices, and mental health indicators, the research provides a comprehensive understanding of the complex interplay between these variables and women's well-being. This holistic approach allows policymakers and healthcare professionals to identify high-risk areas and develop targeted interventions that address the unique challenges faced by women in specific communities.

20.4.1 Future Scope

Future research should focus on longitudinal studies to monitor the effectiveness of implemented policies and explore additional factors that may impact women's health in diverse settings. Long-term follow-up studies could provide valuable insights into the sustainability of interventions and their impact on women's health outcomes over time. Investigating the role of social support networks and community engagement in promoting women's health could offer new perspectives for developing more effective and culturally appropriate interventions. Furthermore, exploring the potential of technology-based solutions, such as mobile health applications and telemedicine, could enhance access to healthcare services and health education for women in Chittur Municipality.

20.5 Conclusion

This study highlights the critical factors influencing the nutritional and health status of women in Chittur Municipality, emphasizing the need for targeted public health interventions. The findings underscore the importance of addressing socioeconomic disparities and improving access to healthcare services to enhance women's overall well-being. Future research should focus on developing and implementing community-based programs that promote nutritional awareness and healthy lifestyle choices among women in the region. Additionally, policymakers and health-care professionals should collaborate to create tailored strategies that address the specific needs of women in Chittur Municipality, taking into account cultural and environmental factors that impact their health outcomes.

References

Ahasan R, Hossain MM (2020) Leveraging GIS and spatial analysis for informed decision-making in the COVID-19 pandemic. Health Policy and Technology 10(1):7–9. https://doi.org/10.1016/j.hlpt.2020.11.009

Almadani NA, Alwesmi MB (2023) The relationship between happiness and mental health among saudi women. Brain Sci 13(4):526. https://doi.org/10.3390/brainsci13040526

Arslan S, Aydın A (2024) Religious dietary practices: health outcomes and psychological insights from various countries. J Religion Health 63(5). https://doi.org/10.1007/s10943-024-02110-8

Balasubramani K, Balabaskaran Nina P, Behera SK, Paulson W, Ramachandran R, Chellappan S (2021) Epidemiology, hot spots, and sociodemographic risk factors of alcohol consumption in Indian men and women: analysis of National Family Health Survey-4 (2015–16), a nationally representative cross-sectional study. Front Public Health 9. https://doi.org/10.3389/fpubh.2021.617311

Batey L, Cardarelli KM, Dewitt E, Brewer D, Norman-Burgdolf H (2023) Exploring food-based cultural practices to address food insecurity in rural Appalachia. Health Educ Behav 50(4):529–537. https://doi.org/10.1177/10901981231175360

Bren D'Amour C, Pandey B, Reba M, Ahmad S, Creutzig F, Seto KC (2020) Urbanization, processed foods, and eating out in India. Glob Food Sec 25:100361. https://doi.org/10.1016/j.gfs.2020.100361

Bühlmeier J, Libuda L, Ludwig C, Föcker M, Knoll-Pientka N, Schmidt B, Frölich S (2022) Changes in eating habits and food intake patterns during the first German COVID-19 lockdown: results of a cross-sectional online survey. Eur J Nutr 61(6):3293–3306. https://doi.org/10.1007/s00394-022-02919-7

Challa J, Perwez SK (2023) Influence of leadership styles of women entrepreneurs on their psychological wellbeing. J Law Sustain Dev 11(4):e903. https://doi.org/10.55908/sdgs.v11i4.903

Chesoli RN, Emanuel O, Sartas M, Muzhingi T (2023) Application of the scaling readiness approach for scaling of nutrition-sensitive initiatives in Kenya, Uganda, and Malawi: a case of orange-fleshed sweet potato value chain. Crop Sci 64(3):1517–1524. https://doi.org/10.1002/csc2.21053

Clark AW, Olsen MA, Butler AM, Keller M, Durkin MJ, Ma Y, O'Neil CA (2021) Rural-urban differences in antibiotic prescribing for uncomplicated urinary tract infection. Infect Control Hosp Epidemiol 42(12):1437–1444. https://doi.org/10.1017/ice.2021.21

De Diego-Cordero R, Diaz-Jimenez D, Badanta B, Rivilla-Garcia E, Lucchetti G (2020) The role of cultural beliefs on eating patterns and food practices among pregnant women: a systematic review. Nutr Rev 79(9):945–963. https://doi.org/10.1093/nutrit/nuaa119

Ding Y, Luo H, Indayati W, Basnet TB, Li F, Wang Z, Pan H (2020) Dietary intake in lactating mothers in China 2018: report of a survey. Nutrition J 19(1). https://doi.org/10.1186/s12937-020-00589-x

Fatema SR, Islam MS, East L, Usher K (2021) Health impact and risk factors affecting south and Southeast Asian women following natural disasters: a systematic review. Int J Environ Res Public Health 18(21):11068. https://doi.org/10.3390/ijerph182111068

Gandarillas MA, Elvira-Zorzo MN, Rodríguez-Vera M (2024) The impact of parenting practices and family economy on psychological wellbeing and learning patterns in higher education students. Psicologia: Reflexão e Crítica 37(1). https://doi.org/10.1186/s41155-024-00291-5

Gila-Díaz A, Díaz-Rullo Alcántara N, Herranz Carrillo G, Ramiro-Cortijo D, Singh P, Arribas SM (2021) Multidimensional approach to assess nutrition and lifestyle in breastfeeding women during the first month of lactation. Nutrients 13(6):1766. https://doi.org/10.3390/nu13061766

Harsch S, Sahrai D, Sørensen K, Jawid E, Bittlingmayer UH, Jawid A, Saboga-Nunes L (2021) Health literacy and health behavior among women in Ghazni, Afghanistan. Front Public Health 9(Suppl 2). https://doi.org/10.3389/fpubh.2021.629334

Hasan AMR, Selim MA, Anne FI, Escobar-Demarco J, Ireen S, Kappos K, Ash D, Rasheed S (2023) Opportunities and challenges in delivering maternal and child nutrition services through public primary health care facilities in urban Bangladesh: a qualitative inquiry. BMC Health Serv Res 23(1). https://doi.org/10.1186/s12913-023-10094-6

Jiang H, Huang Y (2023) How do urban public health services affect rural migrant women's fertility intentions? A study based on the Mobile Population Dynamics Monitoring Survey in China. BMC Health Serv Res 23(1). https://doi.org/10.1186/s12913-023-09219-8

Karim MdR, Rana MdM, Wadood MdA, Hossain MdG, Islam MdN, Hossain MdR, Mamun ASMdA, Bharati P, Bose K (2023) Nutritional status of tribal and non-tribal adults in rural Bangladesh: a comparative study. PLoS ONE 18(7):e0287625. https://doi.org/10.1371/journal.pone.0287625

Klink U, Schüz B, Frank R, Mata J (2022) Socioeconomic differences in animal food consumption: education rather than income makes a difference. Front Nutrition 9. https://doi.org/10.3389/fnut.2022.993379

Kohanmoo A, Hashemzadeh M, Teymouri M, Zare M, Akhlaghi M (2024) Food insecurity is associated with low diet quality and unhealthy cooking and eating habits in Iranian women. J Health Popul Nutrition 43(1). https://doi.org/10.1186/s41043-024-00533-3

Kopiczko A (2020) Determinants of bone health in adults Polish women: the influence of physical activity, nutrition, sun exposure and biological factors. PLoS ONE 15(9):e0238127. https://doi.org/10.1371/journal.pone.0238127

Kosendiak AA, Kuźnik Z, Adamczak BB, Makles S (2024) How dietary choices and nutritional knowledge relate to eating disorders and body esteem of medical students? A single-center cross-sectional study. Nutrients 16(10):1414. https://doi.org/10.3390/nu16101414

Makate M, Nyamuranga C (2023) The long-term impact of education on dietary diversity among women in Zimbabwe. Rev Dev Econ 27(2):897–923. https://doi.org/10.1111/rode.12980

Ogojiaku CN, Hood DB, Im W, Adetona O, Allen JC, Barnett KS, Anson-Dwamena R (2020) The health opportunity index: understanding the input to disparate health outcomes in vulnerable and high-risk census tracts. Int J Environ Res Public Health 17(16):5767. https://doi.org/10.3390/ijerph17165767

Osman MO, Roble AK, Abdilahi AO, Nour TY, Nur AM, Bashir HM (2020) Risk factors for anemia among pregnant women attending the antenatal care unit in selected Jigjiga public health facilities, Somali Region, East Ethiopia 2019: unmatched case-control study. J Multidiscip Healthc 13(2):769–777. https://doi.org/10.2147/jmdh.s260398

Palmer SM, Winham DM, Hiller MB, Knoblauch ST, Shelley MC (2020) Putting knowledge into practice: low-income women talk about food choice decisions. Int J Environ Res Public Health 17(14):5092. https://doi.org/10.3390/ijerph17145092

Pérez-Tepayo S, Shamah-Levy T, Unar-Munguía M, Rodríguez-Ramírez S (2020) Trends in the dietary patterns of Mexican adults by sociodemographic characteristics. Nutrition J 19(1). https://doi.org/10.1186/s12937-020-00568-2

Pradhan MR, Unisa S, Rawat R, Surabhi S, Saraswat A, Sethi V, Reshmi RS (2023) Women empowerment through involvement in community-based health and nutrition interventions: evidence from a qualitative study in India. PLOS ONE 18(4):e0284521. https://doi.org/10.1371/journal.pone.0284521

Quisumbing AR, Sproule K, Martinez EM, Malapit H (2020) Do tradeoffs among dimensions of women's empowerment and nutrition outcomes exist? Evidence from six countries in Africa and Asia. Food Policy 100(4):102001. https://doi.org/10.1016/j.foodpol.2020.102001

Roy S, Majumder S, Bose A, Chowdhury IR (2024) Spatial heterogeneity in the urban household living conditions: a-GIS-based spatial analysis. Ann GIS, 81–104. https://doi.org/10.1080/19475683.2024.2304194

Saaka M, Kizito F, Wemah K, Hoeschle-Zeledon I (2021) Effect of nutrition behaviour change communication delivered through radio on mothers' nutritional knowledge, child feeding practices and growth. J Nutritional Sci 10. https://doi.org/10.1017/jns.2021.35

Silva RJG, Neto LCBS, Pessanha RM, Marcarini JAC, Enriquez-Martinez OG, Lopes-Júnior LC, Grippa WR, Haraguchi FK (2023) Factors associated with the nutritional status of women with non-metastatic breast cancer in a brazilian high complexity oncology center. Nutrients 15(23):4961. https://doi.org/10.3390/nu15234961

Streeter JL (2020) Gender differences in widowhood in the short and long run: financial, emotional, and mental wellbeing. J Econ Ageing 17:100258. https://doi.org/10.1016/j.jeoa.2020.100258

Striessnig E, Bora JK (2020) Under-five child growth and nutrition status: spatial clustering of indian districts. Spatial Demogr 8(1):63–84. https://doi.org/10.1007/s40980-020-00058-3

Talbert RD, Patterson EJ (2023) Formal social control and mental health: ethnic variation among black women. Sociol Race Ethnicity 10(1):139–158. https://doi.org/10.1177/23326492231187294

Vanderhout SM, Jarvi K, Samsel K, Dockray J, Garcia-Bailo B, Rastegar Panah M, El-Sohemy A, Grace-Farfaglia P (2021) Nutrition, genetic variation and male fertility. Transl Androl Urol 10(3):1410–1431. https://doi.org/10.21037/tau-20-592

Wakwoya EB, Girma T, Belachew T (2023) Effects of intensive nutrition education and counselling on nutritional status of pregnant women in East Shoa Zone, Ethiopia. Front Nutrition 10. https://doi.org/10.3389/fnut.2023.1144709

Weerasekara PC, Withanachchi CR, Ginigaddara GAS, Ploeger A (2020) Food and nutrition-related knowledge, attitudes, and practices among reproductive-age women in marginalized areas in Sri Lanka. Int J Environ Res Public Health 17(11):3985. https://doi.org/10.3390/ijerph17113985

Weil K, Coulibaly I, Fuelbert H, Herrmann A, Millogo RM, Danquah I (2023) Dietary patterns and their socioeconomic factors of adherence among adults in urban Burkina Faso: a cross-sectional study. J Health Popul Nutrition 42(1). https://doi.org/10.1186/s41043-023-00451-w

Young L, Kolubinski DC, Frings D (2020) Attachment style moderates the relationship between social media use and user mental health and wellbeing. Heliyon 6(6):e04056. https://doi.org/10.1016/j.heliyon.2020.e04056

Part V
Geospatial and AI Innovations for Environmental and Health Solutions

Chapter 21
Introduction to GIS, AI, and Machine Learning Applications in Environmental Health Monitoring

P. Mary Santhi and S. Balaselvakumar

Abstract The combination of geographic information system (GIS), artificial intelligence (AI), and machine learning (ML) is transforming environmental health monitoring by providing improved tools for analysing, forecasting, and mitigating environmental issues. GIS is superior in spatial visualization and analysis, enabling better understanding of environmental trends and their impact on health. AI and ML supplement GIS with predictive power, decision-making automation, and accuracy. Recent developments have witnessed AI-based systems analysing real-time IoT sensor data and satellite imagery to predict early warning of pollution peaks and health risks. Machine learning algorithms like random forest and convolutional neural networks (CNNs) have been used to detect pollution hotspots, disease vector mapping, and resource allocation optimally in environmental management. Physics-based models integrated with ML approaches have been successful in improving the prediction and understanding of intricate environmental processes. Nevertheless, the integration of GIS, AI, and ML in environmental health monitoring is hindered by factors like data availability, quality, computational complexity, and ethics. The integration of GIS, AI, and ML is a paradigm shift in environmental health monitoring, allowing stakeholders to apply targeted interventions and improve resilience and sustainability against increasing environmental challenges.

Keywords GIS · Artificial intelligence · Machine learning · Environmental health · Pollution monitoring · Disease forecasting

P. M. Santhi (✉)
Convent of Jesus and Mary, Hampton Court, Mussoorie, Dehradun, India
e-mail: maryshanthirjm@gmail.com

S. Balaselvakumar
Government Arts College, Tiruchirappalli, India

A. Roy et al. (eds.), *Environment and Public Health*, Sustainability Solutions,
https://doi.org/10.1007/978-3-031-99770-9_21

21.1 Introduction

Environmental health monitoring is a critical public health component because it assesses the impact of environmental factors on human health. The dynamics of rapid urbanization, industrialization, and climate change have caused significant environmental concerns, including air and water pollution, deforestation, and biodiversity loss, which have direct and indirect impacts on human health (Smith and Petley 2021). The application of geographic information system (GIS), artificial intelligence (AI), and machine learning (ML) has emerged as a robust approach for comprehending and preventing environmental risks. The technologies enable the acquisition, analysis, and visualization of spatial and temporal environmental information, thus supporting well-informed decision-making and policy-making (Goodchild 2020). GIS provides spatial analytical capabilities that enable researchers to map and monitor environmental risks, visualize the spread of pollution, and assess the distribution of health risks (Cova and Church 2021). The ability to integrate satellite imagery, remote sensing, and real-time sensor networks has significantly enhanced the performance of GIS in environmental health monitoring. For instance, GIS-oriented models have been applied to delineate air pollution levels within urban areas, identifying areas with high exposure risk and correlating them with the prevalence of respiratory diseases (Xie et al. 2022).

AI and ML further augment GIS applications by enhancing predictive analytics, pattern recognition, and data-driven decision-making. AI methods, including deep learning and natural language processing, enable the extraction of relevant environmental information from various sources, including satellite imagery, social media, and IoT sensors (Zhang et al. 2023). Machine learning algorithms, including random forest and convolutional neural networks (CNNs), have been found to be effective in pollution level classification, disease outbreak detection, and climate-related health risk prediction (Benedict and Johnson 2021). By utilizing past and real-time data, these technologies offer early warning systems that reduce environmental health risks and increase resilience. Notwithstanding these developments, various challenges impede the large-scale adoption of GIS, AI, and ML into environmental health monitoring. Data quality, accessibility, computational requirements, and ethical issues are major hurdles. In addition, interdisciplinary collaboration is needed to fill the gap between environmental scientists, data analysts, and policymakers to ensure the effective application of these technologies in real-world applications. This review paper summarizes the use of GIS, AI, and ML in environmental health monitoring, including their applications, advantages, challenges, and future directions. The following sections will present a detailed discussion of GIS-based environmental monitoring, AI-based data analytics, machine learning applications, and the integration of these technologies to address environmental health issues.

21.2 Materials and Methods

21.2.1 Methodological Framework

The study utilizes a thematic synthesis and systematic literature review approach to examine the integrated use of geographic information system (GIS), artificial intelligence (AI), and machine learning (ML) to monitor environmental health. The methodological framework aims to systematically examine existing academic literature, case studies, and technical implementations of the said technologies to detect environmental health hazards, such as pollution, disease outbreaks, and climate-based disasters. The study is interested in the cross-disciplinary interface of GIS, AI, and ML to have an understanding of current developments, challenges, and future directions towards environmental health monitoring.

The research approach is qualitative and analytic based on peer-reviewed journal articles, technical reports, and case studies, with a focus given to illustrating pragmatic applications of GIS, AI, and ML to environmental monitoring. There is no field survey and primary data collection, but secondary analysis of published records, such as applied examples and actual deployments of technological applications.

21.2.2 Selection of Case Studies and Review Scope

For relevance and diversity, the research chose case studies and utilized examples based on some inclusion criteria:

- Research that validates GIS-based environmental health mapping and spatial analysis.
- Studies on AI-based predictive models for air pollution prediction, disease risk prediction, and environmental monitoring.
- Applications of machine learning algorithms, e.g., random forest, support vector machines (SVM), convolutional neural networks (CNNs), and hybrid models combining GIS, AI, and ML for environmental risk analysis.
- Combined integrative frameworks incorporating IoT sensor data, satellite remote sensing, and AI/ML models for real-time surveillance of environmental health.

Individual case studies address different areas of environmental health, such as air and water quality monitoring, disease vector mapping, health risks related to climate, and environmental management in cities. Examples from developed and developing nations are presented in order to emphasize the applicability and challenges of these technologies around the globe.

Table 21.1 Summary of case studies on environmental issues and AI/ML applications in different countries (Compiled by author, 2025)

Country	Region	Environmental issue	AI/ML methods applied
India	Delhi and Bihar	Air and water pollution	Random forest, Sensor-based AI models
Brazil	São Paulo	Dengue prediction	CNN, XAI techniques
Bangladesh	Dhaka Basin	Flood risk monitoring	Random forest, GIS
Indonesia	Jakarta	Vector-borne disease prediction	CNN, satellite imagery
USA	California	Wildfire smoke health impacts	LSTM, GIS

21.2.2.1 Addendum to 2.2: Clarification on Case Study Selection

The selection of case studies focussed on geographic and socioeconomic diversity, including examples from:

- **Bangladesh**—flood early warning using random forest models (Akhtar et al. 2023)
- **Indonesia**—dengue outbreak prediction using CNNs and satellite data (Wahyuni et al. 2023)
- **California, USA**—wildfire smoke exposure mapping with LSTM and GIS (Table 21.1).

21.2.3 Sources of Data and Technologies Reviewed

The review primarily focuses on the following data sources and technological platforms utilized in environmental health monitoring:

21.2.3.1 Satellite Remote Sensing Data

The paper reviews applications of high-resolution satellite imagery from platforms such as:

- Landsat, Sentinel, MODIS, and LiDAR missions for mapping air quality, land cover changes, water quality, and urban heat islands.
- Use of multispectral and hyperspectral imaging for detecting pollution sources and environmental risk factors.

21.2.3.2 Internet of Things (IoT) Sensor Networks

Review includes case studies utilizing IoT-enabled environmental sensors:

- Air quality monitoring stations, meteorological sensors, and water quality sensors for real-time data collection.
- IoT-based networks that support continuous environmental surveillance and automated data feeds for AI/ML models.

21.2.3.3 Artificial Intelligence and Machine Learning Models

Focus on supervised and deep learning models, including:

- Random forest (RF), support vector machines (SVM), convolutional neural networks (CNNs), recurrent neural networks (RNNs), and long short-term memory (LSTM) models.
- Use of AI-powered Decision Support Systems (DSS) for predictive analytics and risk forecasting.

21.2.3.4 Hybrid GIS-AI-ML Integration

Exploration of frameworks that combine:

- Geospatial analysis (GIS) with AI and ML algorithms for spatial risk modelling and pollution mapping.
- Integration of real-time data from IoT sensors and remote sensing with AI/ML predictive models, enabling dynamic environmental health assessments.

21.2.4 Literature Review Search Strategy

A systematic search strategy was adopted to identify relevant studies for inclusion:

- Academic databases searched from Scopus, Web of Science, PubMed, IEEE Xplore, and Google Scholar.
- Keywords and search terms: "GIS for environmental health," "AI in environmental monitoring," "Machine Learning for pollution prediction," "AI-based disease risk mapping," "IoT environmental monitoring," "hybrid AI-GIS models."
- Inclusion criteria:
  - Peer-reviewed articles and conference proceedings published between 2017 and 2024.
  - Studies that provide applied examples or case studies demonstrating GIS, AI, and ML integration for environmental health monitoring.
  - Articles in English language.
- Exclusion criteria:
  - Purely theoretical works without applied case studies.

– Studies focussed exclusively on technical model development without environmental health applications.

A total of more than 200 scholarly articles, reports, and case studies were screened, with a final selection of approximately 130 key studies that directly aligned with the research focus on GIS, AI, and ML integration for environmental health risk assessment.

21.3 Results

21.3.1 GIS-Based Environmental Health Mapping and Risk Assessment

A geographic information system (GIS) is now a key tool used for tracking environmental health, enabling the acquisition, storage, analysis, and mapping of spatial information relating to environmental threats and health impacts. GIS provides a spatial context that allows the exploration of environmental exposure interactions and public health, thus enabling the identification of hotspots for diseases, sources of contamination, and population exposure risk (Briggs et al. 2019; Franch-Pardo et al. 2020). GIS enables spatial epidemiology, with researchers and policymakers being able to map and analyse events of disease, track the diffusion of environmental contaminants, and assess health inequalities using geospatial trends (Moellmann et al. 2022). Integration with satellite imagery, remote sensing, and socioeconomic information enhances the ability to explore environmental determinants of health, including information on climate change-related risk, air and water pollution effects, and disease vector distributions (Musa et al. 2021; Gao et al. 2022).

One of the greatest advantages of geographic information system (GIS) in environmental health is that it can enable spatiotemporal analysis, thus helping to recognize long-term trends and real-time monitoring of environmental hazards (Ding et al. 2021). For instance, the application of GIS in air quality monitoring allows for the assessment of trends in pollutant distribution, making connections to respiratory conditions like asthma and chronic obstructive pulmonary disease (COPD). Similarly, water quality assessments performed with GIS have been actively involved in tracking pollutants and their impact on human health (Norris et al. 2023). Additionally, improvements in GIS-based machine learning frameworks have improved environmental risk forecasting with the incorporation of large-scale data analytics, artificial intelligence (AI), and Internet of Things (IoT) sensors for real-time environmental monitoring (Zhang et al. 2021). As GIS technology continues to advance, its application in environmental health monitoring will expand, offering evidence-based policies for sustainable environmental and health policymaking.

GIS leverages diverse sources for acquiring spatial data, ranging from remote sensing, ground-level monitoring stations, and crowd-sourced data. High-resolution satellite images like Landsat, Sentinel, and MODIS missions deliver essential information regarding air and water quality, vegetation, and urban heat islands (Gao and O'Neill 2020; Weng 2019). Remote sensing methods like multispectral and hyperspectral imaging augment the detection of environmental pollutants through spectral signature analysis (Hu et al. 2021). Additionally, LiDAR technology has also been found effective for urban pollution monitoring, with fine-scale resolution for measuring atmospheric particulate matter (Zhang et al. 2022a, b). Besides remote sensing, spatial interpolation methods like kriging, inverse distance weighting (IDW), and co-kriging facilitate estimating pollutant concentrations in sparse monitoring networks (Beelen et al. 2018). Next-generation GIS-based models incorporate real-time sensor networks, with IoT-enabled air quality sensors, meteorological stations, and mobile apps to enhance spatiotemporal accuracy (Ma et al. 2020). In addition, machine learning-enhanced GIS models like random forest regression and neural network-based geostatistical models have enhanced predictive accuracy in pollution exposure mapping and environmental risk assessment (Xu et al. 2023a, b). Researchers and policymakers can use these next-generation GIS-based data collection and processing technologies to create high-resolution environmental health risk maps, allowing for data-driven decision-making and targeted interventions for sustainable environmental management.

The combination of geographic information system (GIS) with remote sensing methods greatly improves estimates of environmental health by offering spatially explicit and temporally continuous data. Remote sensing facilitates real-time tracking of land use changes, deforestation, air pollution, and water quality, all of which have direct health implications for the public (van Donkelaar et al. 2019; Brook et al. 2020). For example, satellite-retrieved aerosol optical depth (AOD) measurements have been widely applied to estimate fine particulate matter (PM2.5) concentrations, thereby informing air pollution exposure estimates and epidemiological research (Shaddick et al. 2018; Ma et al. 2020). GIS-based model methods, in combination with multispectral and hyperspectral remote sensing, have been applied to investigate urban heat islands (UHI), waterborne diseases, and the spread of vector-borne diseases such as malaria and dengue (Messina et al. 2021; Kraemer et al. 2019). The models incorporate land surface temperature (LST), vegetation indices (NDVI), and climatic variables to forecast disease hotspots and evaluate environmental risk factors (Bhatt et al. 2017; Liu et al. 2022). Furthermore, the application of machine learning methods in combination with remote sensing data has enhanced the identification of sources of pollution, deforestation patterns, and land use change patterns, thereby facilitating more targeted policy interventions (Gorelick et al. 2017). With further innovations in remote sensing technologies, the application of high-resolution satellite imagery, unmanned aerial vehicles (UAVs), and deep learning algorithms will continue to enhance environmental health surveillance and risk assessment (Schroeder et al. 2021; Wang et al. 2023).

Plenty of case studies demonstrate the application of Geographic Information System (GIS) in environmental health monitoring, specifically focussing on pollution assessment and disease mapping. A research study by Xie et al. (2020) demonstrated the application of GIS in mining air pollution hotspots in urban areas, exhibiting strong associations between PM2.5 levels and respiratory illness. Similarly, GIS-based hydrological models have been applied in the risk assessment of groundwater pollution due to industrial waste, thus helping policymakers identify high-risk areas and implement mitigation measures (Singh and Kumar 2017). Another research study by Díaz et al. (2021) applied GIS in examining spatial variability of water-borne disease outbreaks, and they noted that contamination sources were strongly correlated with urban drainage and agricultural runoff. Similarly, Wang et al. applied GIS-integrated remote sensing techniques to track heavy metal soil contamination, exhibiting its impact on human health, especially in mine-contaminated areas.

Applications of geographic information system (GIS) go beyond static mapping; they are increasingly being used for real-time monitoring of environmental conditions. Li et al. (2022) used GIS in combination with satellite imagery to develop an early warning system for air pollution, incorporating meteorological conditions to forecast pollution dispersion trends. The combination of GIS with artificial intelligence (AI) and machine learning (ML) is further enhancing the precision of predictions and automating analyses of environmental risk (Zhang and Chen 2023). These applications demonstrate the potential of GIS in identifying environmental hazards and guiding public health interventions. Challenges remain regarding the accuracy, availability, and interoperability of data, particularly in low-resource settings. Future research agendas should aim to improve spatial data resolution, merging multi-source environmental data sets, and using AI-aided geospatial analysis to maximize the use of GIS for environmental health applications (Anderson et al. 2023).

The advent of artificial intelligence (AI) has revolutionized environmental health monitoring at its core by enabling sophisticated data processing, predictive analysis, and decision-support systems. Several AI techniques, including machine learning (ML) and deep learning (DL), are applied comprehensively to analyse vast environmental datasets, automate risk analysis, and optimize mitigation strategies (Chen et al. 2021). The integration of AI in environmental health monitoring considerably improves the detection of pollution sources, climate variability, and disease outbreaks through complex computational models (Ahmed et al. 2023). AI-based spatiotemporal models enable real-time environmental monitoring and forecasting, thus improving early warning systems for air and water pollution (Zhang et al. 2022a, b). Additionally, convolutional neural networks (CNNs) and recurrent neural networks (RNNs) are applied extensively in remote sensing to monitor deforestation, urban heat islands, and pollutant dispersion (Li and Wang 2020). Studies have also demonstrated the effectiveness of AI-augmented geospatial analysis in identifying high-risk areas for vector-borne diseases, such as malaria and dengue, by analysing satellite imagery and climatic patterns (Tan et al. 2021). AI also aids in the development of environmental policy and informed decision-making by providing data-driven insights for sustainable urban planning, resource allocation, and climate adaptation strategies. However, despite these advancements, challenges remain with regard to data quality, model interpretability, and computational resource requirements, highlighting the need for further inter-disciplinary studies and collaborative research (Kumar and Roberts 2022).

21.3.2 AI-Based Predictive Modelling and Decision Support for Environmental Health

AI models leverage large amounts of data from remote sensing, geographic information system (GIS), and Internet of Things (IoT) sensors to calculate environmental variables. Supervised learning approaches, including random forest (RF) and support vector machines (SVM), are commonly employed to classify land cover changes and predict air quality indices. For example, Sabat-Tomala et al. (2020) used RF and SVM to classify invasive plant species based on hyperspectral aerial imagery with high classification accuracy. Likewise, Sheykhmousa et al. performed a meta-analysis of RF and SVM on remote sensing image classification, emphasizing their performance on different environmental applications. Deep learning models, like convolutional neural networks (CNNs) and recurrent neural networks (RNNs), have been applied in satellite imagery for tracking deforestation, water pollution, and urban heat islands. Arthi and Brintha showed the applicability of CNNs in classifying satellite imagery into land cover classes with 96.29% test accuracy. Moreover, Handan-Nader et al. illustrated the applicability of deep learning using satellite imagery to support environmental enforcement, showing the capability of these models in identifying environmental risks. The AI models facilitate the real-time identification of environmental hazards, which informs policymakers in designing preventive measures. The application of machine learning models with remote sensing data improves environmental monitoring accuracy and efficiency, allowing valuable insights into sustainable management strategies.

AI-based decision support systems (DSS) play a vital role in evaluating environmental health hazards by integrating heterogeneous data sets, i.e., climate parameters, air pollution scores, and epidemiology data, to deliver predictive analysis (Gupta et al. 2019; Chen et al. 2020). These DSS use high-power machine learning techniques to facilitate better decision-making for environmental health management. Different AI-based models, e.g., Bayesian networks, support vector machines (SVMs), and artificial neural networks (ANNs), have widely been applied for health forecasting of health conditions due to air pollution, e.g., respiratory diseases, cardiovascular diseases, and vector-borne diseases (Zhang et al. 2022a, b). For instance, deep learning models have been reported to be highly accurate in forecasting asthma events based on the relationship of air quality index (AQI) values with meteorological parameters and patient admission records (Li et al. 2021). Likewise, AI-based spatiotemporal models have improved predictive accuracy of dengue and malaria outbreaks through climatic variability analysis, mosquito-breeding site, and population density change (Huang et al. 2020; Kumar and Mishra 2023). AI, further, automates the hazard mapping and risk assessment process, enhancing emergency response measures for environmental disasters like wildfires, floods, and heat waves (Rahman et al. 2022). Machine learning techniques like Random Forest and Long Short-Term Memory (LSTM) networks have been applied in real-time air quality monitoring networks to improve early warning systems for public health protection (Liu et al. 2021a, b; Song et al. 2023). Combining geospatial AI (GeoAI) with DSS has further improved predictive accuracy for pollution exposure assessment through satellite imagery integration, Internet of Things (IoT) sensor data, and epidemiology models (Zhao et al. 2021). While it has advantages, AI-driven decision support systems (DSS) used for environmental health monitoring face data availability, computational loads, and ethical concerns, necessitating the development of strong validation paradigms and policy interventions to justify effective and unbiased implementations (Singh and Gupta 2023).

High quality, unbiased data and transparent AI models are required for predictive accuracy and evidence-based policy action (Gebru et al. 2021). Algorithmic bias, which is typically incorporated into training data, can exacerbate environmental and social inequalities, leading to inaccurate risk estimation and uninformed decision-making (O'Neil 2016). Ethical AI frameworks would therefore have to be designed to enhance fairness, transparency, and responsibility in AI-based environmental monitoring applications. Inter-disciplinary research is also required to enhance AI methodologies to ensure that they are aligned with sustainable development goals (SDGs) and global environmental agendas (Vinuesa et al. 2020). The high computational resource requirements of deep learning models, particularly in large-scale environmental health research, also present challenges in terms of energy usage and carbon footprint (Strubell et al. 2019). This necessitates research into energy-efficient AI models and the use of low-power computational models for sustainable deployment (Luccioni et al. 2022). AI is nonetheless reshaping environmental health monitoring by enabling automated analysis, predictive modelling, and real-time decision-making (Rolnick et al. 2022). Overcoming limits in ethics, computing, and data is critical to realising AI's full potential in improving environmental sustainability and public health resilience (Taddeo and Floridi 2018).

21.3.3 Machine Learning Models for Pollution and Health Risk Forecasting

Machine learning (ML) has become a useful tool for the analysis of complex environmental health data, enabling the detection of trends in pollution, health risk forecasting, and optimization of mitigation efforts. Using large datasets, ML models are able to detect underlying patterns in environmental determinants of human health, such as air quality, water contamination, and disease outbreaks (Liu et al. 2021a, b). Different ML approaches, such as random forest (RF), support vector machines (SVM), and deep learning (DL), have been found to be instrumental in environmental health applications, such as air pollution forecasting, disease vector mapping, and health risk assessment of contaminated water sources (Wang et al. 2022). Recent studies have shown the efficacy of convolutional neural networks (CNNs) and long short-term memory (LSTM) networks in the analysis of time-series environmental data, thus improving predictive accuracy in monitoring air quality and climate-health analysis (Chen et al. 2023). In addition, hybrid models combining ML with geostatistical approaches have improved the spatial resolution of pollution mapping, thus helping policymakers develop data-driven interventions. Furthermore, ML models developed using satellite data and Internet of Things (IoT) sensor data have been shown to be instrumental in real-time monitoring of environmental health, detecting hotspots of pollution, and correlating exposure levels with public health effects (Rahman et al. 2022). However, challenges such as data availability, model interpretability, and ethics prevail, calling for inter-disciplinary collaboration and the development of regulatory policies to ensure ML applications in environmental health are equitable and responsible (Gupta et al. 2023).

Machine learning (ML) refers to a broad range of computational methods that enable models to learn from past datasets and make accurate predictions or classifications. In environmental health surveillance, ML is widely applied to sensor data, satellite imagery, and epidemiological data, thus facilitating the identification of correlations between environmental factors and health outcomes (Kumar and Lee 2021). The ability of ML algorithms to process large-scale, multidimensional datasets makes them extremely useful in the identification of pollution trends, climate effects, and disease epidemics (Zhao et al. 2022). ML methods can be broadly categorized into three categories: supervised learning, unsupervised learning, and deep learning models. Supervised learning methods, such as random forest (RF) and support vector machines (SVM), are most commonly applied for air quality prediction, water pollution detection, and disease risk prediction (Chen et al. 2023). Unsupervised learning methods, such as k-means clustering and principal component analysis (PCA), facilitate the identification of pollution hotspots and the classification of environmental datasets (Ahmed et al. 2022). In contrast, deep learning models, such as convolutional neural networks (CNNs) and recurrent neural networks (RNNs), have proven to be extremely effective in image-based assessments of environmental variables, such as land use classification, disaster prediction, as well as remote sensing applications (Li et al. 2021).

Recent research points to the increasing relevance of hybrid ML models that integrate physics-based simulations with deep learning models, thereby substantially enhancing predictive capability with respect to climate change impacts and environmental risks (Patel et al. 2023). Enhanced access to real-time environmental information from IoT sensors and satellite imagery also facilitates ML-based monitoring capacities (Singh and Gupta 2023). Nevertheless, notwithstanding its potential, data quality, computational complexity, and interpretability remain critical issues in the use of ML for environmental health monitoring (Wang et al. 2022).

Supervised learning models make use of labelled data to create predictive models. These methods are particularly effective in the field of environmental health monitoring, where past records of pollution levels and disease incidence serve as training data for predicting future risks (Chen et al. 2019).

The random forest (RF) algorithm, one of the most notable ensemble learning algorithms, has been extensively used in air quality forecasting and health impact assessments due to its effectiveness in handling complex data and preventing overfitting by combining multiple decision trees. RF models have been found to be more accurate in estimating particulate matter (PM2.5) concentrations, and hence, they are critical in health risk assessment due to air pollution (Liu et al. 2022). Empirical studies indicate that RF-based models, when integrated with geospatial and meteorological variables, can accurately estimate daily exposure levels of PM2.5, thereby advancing research on short-term and long-term health impacts of air pollution (Chen et al. 2020). Research carried out in Sweden effectively used RF algorithms to estimate air pollution levels at the national level, correlating them with respiratory disease incidence and cardiovascular risks.

In addition, random forest (RF) models based on aerosol optical depth (AOD), boundary layer height (BLH), and meteorological variables have been shown to achieve a significant predictive accuracy ($R^2 > 0.85$) in PM2.5 estimations, thus proving the efficacy of such models in air quality prediction (Zhan et al. 2021). In addition, RF has been applied in cities to predict the Air Pollution Standard Index (APSI), thus providing policymakers with valuable insights into managing air quality (Rahman et al. 2023). Such findings justify the relevance of random forest models in air pollution monitoring and environmental health studies, thus supporting policymakers in implementing focussed pollution reduction measures and improving public health outcomes.

Support vector machines (SVMs) are commonly utilized in environmental health risk classification by identifying decision boundaries in high-dimensional datasets. SVM's performance is particularly important in the management of high-dimensional data and is applicable to numerous environmental monitoring applications (Vapnik 1998). SVM has been applied in water quality monitoring to detect safe and contaminated water sources, thereby allowing for the development of reliable early warning systems for public health protection (Patel and Singh 2020). Evidence has been provided on its potential in heavy metal contamination detection and classification of pollution levels in freshwater environments using various physicochemical parameters (Zhou et al. 2021). SVM has also been employed in the accurate prediction of air pollution, where it elucidates complex relationships between meteorological factors and pollutant concentrations, improving prediction accuracy (Zhang et al. 2022a, b). Comparative research indicates that SVM performs better than conventional regression models in PM2.5 concentration forecasting, owing to its capability to manage nonlinear dynamics and noise in environmental data. In disease surveillance, SVM has also been used to predict vector-borne diseases by integrating geospatial and climatic data to evaluate risk areas for malaria and dengue (Gomes et al. 2020).

SVM is not without its drawbacks, such as high computational complexity on big datasets and sensitivity to the selection of kernel functions, though. More sophisticated hybrid models, such as the combination of SVM with deep learning methods or ensemble methods, have enhanced its stability in spatiotemporal environmental health studies (Chen et al. 2023).

Unsupervised learning techniques, such as clustering algorithms, are instrumental in the identification of hotspots of pollution and classification of health risk zones without prior labels. For instance, k-means clustering has been used to cluster monitoring stations according to the composition of particulate matter (PM2.5), effectively classifying areas with similar pollution features. Similarly, the DBSCAN algorithm has been used to cluster datasets with varying levels of air pollution and has demonstrated its capability in handling data of varying shapes and identifying outliers. Deep learning models, such as convolutional neural networks (CNNs) and Recurrent Neural Networks (RNNs), have revolutionized environmental monitoring by extracting complex features from satellite imagery and time-varying pollution data. CNNs are well suited to handle spatial data, thereby being highly effective for applications such as land cover classification and detection of environmental changes from high-resolution satellite images (Sharma and Kumar 2019). RNNs, however, are well suited to handle sequential data, enabling the analysis of temporal patterns of air pollution and supporting accurate air quality trend predictions (Sharma and Kumar 2019). Integration of these deep learning methods with remote sensing technologies has significantly enhanced the accuracy and efficiency of environmental health evaluations.

Convolutional neural networks (CNNs) have shown remarkable effectiveness in the analysis of geospatial and remote sensing data relevant to environmental health assessments. Their use has been extensively reported for predicting air quality, tracking pollution levels, and mapping disease incidence by identifying spatial patterns from satellite data and ground sensor networks (Zhu et al. 2017; Li et al. 2021). CNN-based models have been used to evaluate the air pollution caused by deforestation and its implications on respiratory health by analysing high-resolution imagery in combination with epidemiological data (Huang et al. 2023; Zhao et al. 2022). CNNs have also been shown to outperform conventional methods in classifying changes in land cover, detecting heat islands, and predicting urban air pollution hotspots (Ma et al. 2019). The combination of CNNs with recurrent neural networks (RNNs) has enhanced spatiotemporal modelling capabilities for predicting the spread of airborne pollutants and their related health implications. Recent studies have also used transfer-learning techniques within CNN architectures to enhance model accuracy in environmental health applications, thus reducing dependence on large training datasets (Sun et al. 2021). CNNs continue to be an essential tool for environmental monitoring, especially in identifying sources of pollution, assessing the implications of climate change, and optimizing strategies for disaster response. With increasing computational power, the integration of CNNs with geospatial artificial intelligence and cloud-based solutions will further enhance environmental health assessments and guide policy formulation (Chen et al. 2023).

Recurrent neural networks (RNNs), in the form of long short-term memory (LSTM) networks, are central to environmental health monitoring time-series prediction. The models are extremely effective at processing past environmental data, including meteorological patterns and air pollution, to forecast disease outbreaks, such as asthma exacerbations and dengue (Rahman et al. 2022). LSTM networks have been extremely accurate in air quality forecasting. For instance, a CNN-LSTM hybrid model accurately forecasted PM2.5 concentrations, with an R^2 of 0.91, indicating its better predictive performance (Huang et al. 2023). Likewise, an LSTM-based method for New Delhi accurately forecasted concentrations of PM2.5, PM10, CO, O_3, NO_2, and SO_2, indicating its robust capability to learn multivariate pollutant datasets (Sharma and Gupta 2022). Besides air quality, LSTM networks have been applied widely in water quality monitoring. A multivariate LSTM model accurately forecasted water quality parameters, indicating its robust adaptability to complex environmental datasets (Zhang et al. 2021). In groundwater research, LSTM-based models forecast water levels, supporting sustainable water resource management. Additionally, LSTM models are highly effective in anomaly detection for indoor air quality. An LSTM-Autoencoder model accurately detected anomalies in CO_2 time-series data, with an accuracy of 99.50%, which is essential for maintaining healthy indoor conditions (Chen et al. 2022). LSTM networks offer robust predictive capability for environmental health monitoring, with the ability to accurately predict pollution levels, disease outbreaks, and water quality fluctuations. Their capability to learn complex temporal relationships makes them essential for proactive environmental management and safeguarding public health.

Multiple case studies report the successful application of machine learning (ML) for monitoring the health of the environment. In urban settings, ML-based models have been used for forecasting air pollution levels based on real-time data gathered from satellite imagery, weather sensors, and traffic emissions, thus providing better public health warnings and pollution control strategies (Xie et al. 2022). In particular, advanced deep learning models like convolutional neural networks (CNNs) and long short-term memory (LSTM) networks have also been reportedly effective in rigorously examining spatiotemporal patterns of air quality, thus making the predictions more accurate. Similarly, ML methods have also been used in real-time water quality monitoring systems where methods like random forest (RF), support vector machines (SVM), and artificial neural networks (ANNs) examine physicochemical parameters like pH, turbidity, dissolved oxygen, and heavy metals. These systems enable authorities to take pre-emptive action against water pollution, thus ensuring the supply of safe drinking water and minimizing health risks (Haque et al. 2023). The combination of ML with remote sensing methods and Geographic Information System (GIS) also enables the detection of water pollution hotspots and simulation of the pollutant dispersion patterns (Chen et al. 2023). In addition, beyond the individual ML models, hybrid solutions involving GIS, ML, and Internet of Things (IoT) sensors are also facilitating environmental health monitoring. These multimodal systems enable real-time, high-resolution monitoring and predictive analytics for air and water quality (Nguyen et al. 2023; Li et al., 2023). The development of ML-based explainable AI (XAI) methods is also enhancing the interpretability and transparency of pollution assessment models, thus ensuring policymakers can make decisions based on data. ML has significantly enhanced environmental health monitoring through predictive analytics, automated decision-making, and improved pollution estimate accuracy. With increasing computational power and data, ML is poised to become an increasingly critical component of proactive environmental management and climate resilience efforts (Rahman et al. 2023).

21.3.4 *Integration of GIS, AI, and ML: Hybrid Approaches for Real-Time Monitoring*

Even though IoT sensors have real-time features, limitations like network unreliability, battery duration, and calibration faults can affect the quality of data (Nguyen et al. 2023). In addition, the deployment of AI models across various socio-ecological settings is still difficult because models learned in one setting do not necessarily transfer to other settings without domain-specific adjustment.

The synergy of geographic information system (GIS), artificial intelligence (AI), and machine learning (ML) is revolutionizing environmental health monitoring by facilitating real-time data analysis, forecast models, and decision support for policy enforcement. The technologies synergize to enhance the precision and effectiveness of environmental assessments, especially in tracking air and water quality, disease surveillance, and risk assessment of pollution (Zhang et al. 2021).

Geographic information system (GIS) is a core platform for data integration, collection, and visualization, while artificial intelligence (AI) and machine learning (ML) processes considerably enhance its analytical capacity through the automation of complex data processing and pattern detection (Goodchild 2020; Yang et al. 2021). The convergence of these technologies has revolutionized environmental health monitoring at the core level, enabling the attainment of high-resolution pollution maps, predictive disease models, and dynamic risk assessments (Goodchild 2020; Yang et al. 2021). AI techniques, such as deep learning and reinforcement learning, simplify feature extraction from remote sensing images, resulting in significant improvements in the identification of pollution hotspots and disease outbreak patterns (Kumar and Borbor 2018; Zhang et al. 2022a, b). ML algorithms, such as random forest (RF), support vector machines (SVM), and convolutional neural networks (CNNs), complement GIS-based systems by predicting environmental changes based on past and present data, enhancing early warning systems and public health response (Jeong and Gim 2023).

In addition, coupling of Big Data analytics and cloud computing with geographic information system (GIS) and artificial intelligence (AI) has enabled real-time and scalable monitoring of the environment, enabling policy makers to make decisions for regulating pollution and allocating resources based on concrete data (Ahmed et al. 2021). Utilization of hybrid models that are based on physics-based models coupled with machine learning (ML) algorithms has further improved forecasting of climate and air quality to sustainable levels of urban planning and disaster management (Shi et al. 2023). While the new technologies present unprecedented opportunities, issues like AI model bias, data privacy issues, and computation costs require monitoring to enable fair and equitable applications of the GIS-AI-ML technology in the field of environmental health (Sun et al. 2022; Wu et al. 2023).

New developments in the Internet of Things (IoT) have enabled real-time monitoring of the environment using sensor networks that provide high-frequency information on pollutants, temperature variation, and other atmospheric conditions. When such sensor-based systems are integrated with artificial intelligence (AI) models, the systems improve the precision of environmental evaluations through dynamic evaluations of large data sets and automated detection of anomalies. Coupling IoT with Geographic Information System (GIS) enables the development of spatially explicit risk maps, thereby enhancing decision-making processes on environmental health management in urban and rural areas (Silva et al. 2023). Satellite remote sensing is also pivotal in real-time environmental monitoring. Satellite platforms such as Sentinel-2, Landsat, and MODIS offer high-resolution multispectral and thermal imagery, which, when integrated with deep learning algorithms, facilitate automated pollution detection, land use classification, and climate variability assessment. Machine learning algorithms such as Random Forest and Convolutional Neural Networks (CNNs) have been found effective in the examination of satellite-derived indicators of air quality, i.e., PM2.5 and NO_2 levels (Chen et al. 2022). Furthermore, AI-boosted hyper-spectral imaging facilitates the identification of subtle changes in vegetation health, water pollution, and urban heat island effects (Xu et al. 2023a, b).

The combination of real-time Internet of Things (IoT) sensor data with artificial intelligence (AI)-enhanced Geographic Information System (GIS) has considerably improved predictive modelling and early warning systems for environmental health threats. AI-optimized sensor fusion approaches can predict pollution surges many hours to days in advance, allowing for proactive public health actions (Li et al. 2023). Challenges exist in terms of data interoperability, computing efficiency, and ensuring fair access to these technical breakthroughs. Addressing these problems through the creation of standardised frameworks and inter-disciplinary cooperation is critical to increasing the scalability and sustainability of real-time environmental monitoring systems.

Hybrid models combining physics-based environmental simulations with machine learning (ML) methods have shown remarkable increases in predictive power and environmental risk assessment. The models utilize geographic information system (GIS) for spatial referencing, ML for identifying patterns, and physics-based simulations to offer process-driven insights, thus being very suitable to deal with complex environmental issues (Sun et al. 2021). Perhaps the most significant application of hybrid models has been in the field of hydrological modelling, where ML enhances the precision of flood forecasting through learning from past flood events and GIS offers the geospatial context for the modelling of topography, land use, and hydrological processes. Research has identified that the incorporation of deep learning architectures, such as long short-term memory networks (LSTMs), into physics-based hydrodynamic models significantly enhances the flood prediction capacity, especially for areas with insufficient data (Mosavi et al. 2018). Similarly, hybrid models have been implemented successfully in forecasting air pollution, where physics-based atmospheric dispersion models are augmented with ML algorithms such as random forest and support vector regression (SVR) for enhancing real-time air quality estimations (Xiao et al. 2022; Zhang et al. 2023).

In the assessment of climate change effects, hybrid models that integrate remote sensing, geographic information system (GIS), and machine learning (ML) have offered improved predictive data on temperature changes, precipitation trends, and ecosystem responses (Rahmati et al. 2022). For instance, hybrid deep learning models that integrate general circulation models (GCMs) with ML algorithms have been applied to enhance climate predictions with improved spatial precision (Bishop et al. 2021). Furthermore, hybrid techniques are increasingly popular in landslide hazard mapping, where physics-based slope stability models are integrated with ML classifiers to enhance hazard prediction (Goetz et al. 2015; Reichenbach et al. 2018). The use of GIS, physics-based models, and ML in hybrid models presents a holistic, scalable, and flexible means of monitoring environmental health. Nonetheless, issues surrounding computational costs, data uncertainty, and model interpretability must be overcome to facilitate extensive application in real-world contexts (Arnone et al. 2020). Future advancements in cloud computing, high-performance computing (HPC), and explainable artificial intelligence (XAI) are likely to further improve the reliability and usability of hybrid environmental models (Shi et al. 2023).

AI and ML-based predictive modelling platforms provide early alerts of significant environmental occurrences, such as heat waves, air pollution spikes, and water contamination episodes. Utilizing historical GIS databases coupled with AI-based trend analyses, policymakers can implement timely interventions to mitigate health risks. Machine learning algorithms, including deep learning and ensemble algorithms, have been widely applied in environmental prediction. For example, a study by Tan et al. (2023) demonstrated that the use of deep learning techniques in GIS-based air quality models resulted in significantly improved accuracy in air pollution prediction, thus guiding urban planning and public health initiatives. Similarly, Li et al. (2021) demonstrated that the integration of remote sensing data with RNN-based models improved water contamination event prediction, thus allowing for timely response measures. The integration of GIS, AI, and ML is revolutionizing environmental health monitoring by enhancing real-time data analysis, predictive precision, and anticipatory policy design (Zhao et al. 2022). In hydrological research, AI-based ML models, including LSTM networks, have been utilized to predict flood events with high precision, thus allowing for effective disaster risk management (Rahman et al. 2023). Further, hybrid AI-GIS systems have been effective in vector-borne disease outbreak mapping, as demonstrated by Wang et al. (2022), who utilized spatially explicit ML models to predict malaria transmission hotspots with accuracy of over 85%.

Yet, issues like data integrity, computational demand, and ethical consideration must be resolved to facilitate wider uptake. Variations and bias in training data can result in poor predictions, hence the need for rigorous model testing and explainable AI procedures (Gupta et al. 2023). Interpretable AI designs, the incorporation of citizen science information, and the utilization of edge computing should be the areas of future studies to enhance real-time environmental resilience and sustainability (Kim and Lee 2022).

21.3.5 Challenges and Limitations in Implementing GIS, AI, and ML for Health Monitoring

The application of geographic information system (GIS), artificial intelligence (AI), and machine learning (ML) to monitor environmental health has immense potential, but there are several challenges to their full potential.

21.3.5.1 Data Availability and Quality Issues

One of the major limitations in environmental health monitoring is the quality and availability of data. For instance, in 2022, faulty low-cost air sensors in Delhi caused underreporting of PM2.5 concentration, resulting in smog-related health advisories being released late (Sharma and Gupta 2022). Likewise, in rural Bihar, low-quality data on water quality led to misidentification of contamination hotspots during a cholera outbreak, causing a premature response. To reduce such effects, real-time calibration of sensors, standard data validation procedures, and implementation of widely accepted frameworks like the WHO Air Quality Guidelines should be given high priority. National databases should incorporate strong quality checks and metadata tagging to enhance data reliability. Developing nations do not have access to high-resolution, real-time, and long-term datasets, resulting in analysis gaps, prediction accuracy, and decision-making (Zhang et al. 2021). In addition, environmental monitoring networks are inhomogeneously distributed, with greater sensor distribution density in urban regions than in rural or remote regions, resulting in spatial biases in data representation. Heterogeneity and inconsistency in data complicate GIS integration with AI/ML models. Variations in spatial and temporal resolutions, variations in data collection methods, and variations in data formats across organizations and nations complicate interoperability (Li et al. 2022). For instance, air quality monitoring stations may report data in alternative units of measurement or employ different calibration methods, which complicate cross-regional comparisons. In addition, historical gaps and faulty sensors result in missing or incomplete datasets, affecting the performance of AI-based predictive models.

A technical problem relates to data reliability and validation. Various real-time environmental data sets, especially those provided by low-cost sensors and crowdsourcing platforms, may be affected by severe measurement errors, noise, or biases (Ahmed et al. 2023). Invalidation or inaccuracy of the data can lead to erroneous decisions on policy and incorrect predictions. Data standardization, accessibility, and validation processes must therefore be improved to increase the reliability of GIS-AI/ML-based environmental health monitoring systems. Further, data access is routinely obstructed by strict licencing agreements, especially for commercial satellite data sets. No availability of standardized metadata further hinders reproducibility and model interoperability issues. Moreover, cross-border data exchange is routinely obstructed by national security policies, and global environmental monitoring is a politically delicate task (Garcia and Thompson 2024).

21.3.5.2 Computational and Infrastructural Constraints

Artificial intelligence (AI) and machine learning (ML) models necessitate significant computational power, especially for deep learning tasks such as convolutional neural networks (CNNs) used in air pollution mapping and environmental health-related analysis. The models demand high-performance graphics processing units (GPUs), Tensor Processing Units (TPUs), and ample data storage space, which may not be present in resource-limited environments. The computational complexity involved in the training of deep learning models further contributes to excessive energy consumption, raising concerns about sustainability and cost-effectiveness. In addition, the absence of a well-established framework for integrating geographic information system (GIS), AI, and ML is an issue, as legacy systems with low processing power dominate most organizations. Cloud computing and edge AI implementations have been suggested as a remedy to avoid heavy computing loads by offloading them to distant servers and allowing real-time analytics in the data source vicinity. Nevertheless, concerns regarding internet connectivity, cybersecurity issues, and high operational costs hamper the mass use of these technologies in developing nations (Jain et al. 2022).

Moreover, the efficacy of AI-driven environmental monitoring is dependent on real-time data processing, which is constrained by bandwidth, low data transfer rates, and inadequate storage facilities. Such constraints lead to delays in predictive modelling, degraded accuracy in the identification of pollution, and hindrances in mass deployment (Zhao et al. 2021). These issues can be addressed by making investments in scalable computing facilities, deploying energy-efficient AI models, and implementing policy measures to enhance digital infrastructure for environmental health monitoring.

Sensor Calibration, Environmental Interference, and Scalability Limitations

- While IoT sensors yield high-frequency data, keeping them calibrated across different environmental conditions is a problem.
- Operation deployments in actual environments often face interoperability between sensors and must employ region-specific calibration strategies.
- Scalability is hampered by the need for heavy processing and infrastructure capability for real-time data analysis, not equally available across all regions.

21.3.5.3 Ethical and Privacy Concerns in AI and ML Applications

The application of AI and ML for environmental health monitoring brings a number of ethical issues, such as data privacy, surveillance, bias in algorithms, and transparency. AI-powered analysis incorporates personal health data, geospatial data, and real-time environmental monitoring, which poses serious issues of data security, consent, and compliance with the legal framework, like GDPR (General Data

Protection Regulation) and HIPAA (Health Insurance Portability and Accountability Act). Unauthorized collection of data or weakened encryption rules leave sensitive data vulnerable to cyberattacks, which could violate individual privacy and environmental data integrity at the community level. In addition, biased training data could generate biased predictions that could disproportionately affect marginalized and vulnerable groups of people by under- or overestimating the risk of environmental exposure at a particular location (Chen et al. 2021). For example, AI models trained on incomplete datasets or historically biased datasets could be unable to predict pollution hotspots accurately or may give priority to industrial zones over residential zones, resulting in uneven resource allocation and public health interventions. Algorithmic transparency and explainability are serious issues, as the majority of deep learning models are "black boxes," and stakeholders cannot easily establish the accuracy and fairness of AI-driven decisions (Zhou et al. 2022). Another arising issue is the lack of interpretability of sophisticated AI models, popularly known as the "black box" issue. Environmental health officials and policymakers might be reluctant to adopt AI recommendations if the decision-making rationale is not transparent. Methods from Explainable AI (XAI), such as SHAP (SHapley Additive exPlanations) and LIME (Local Interpretable Model-agnostic Explanations), are increasingly employed to make model predictions more transparent. However, XAI is still an emerging field, and integration into public health workflows needs further validation and standardization.

In addition, heightened usage of remote sensing, Internet of Things (IoT) sensors, and surveillance technologies for environmental monitoring brings ethical concerns regarding continuous tracking and surveillance data misuse. Though these developments improve early warning systems for pollution and disease outbreaks, they also bring concerns regarding mass surveillance, state intrusion, and unforeseen societal implications (Ahmed et al. 2022). To avoid unequal, unaccountable, and opaque artificial intelligence surveillance deployment in environmental health monitoring, it is important to have robust ethical guidelines, as well as promote inter-disciplinary monitoring and community participation to offset biases and maintain data protection rights.

21.3.5.4 Addressing the "Black Box" Nature of AI

The increasing use of AI in environmental monitoring poses challenges of model transparency and bias. Deep learning models such as CNNs and LSTMs are often blamed as being "black boxes." In 2023, for instance, an AI system in São Paulo mistakenly labelled poor neighbourhoods as low-risk for dengue due to underreported training data, thereby skewing resource allocation (Silva and Costa 2023). In response to this, explainable AI (XAI) techniques such as SHAP (SHapley Additive ExPlanations) and LIME (Local Interpretable Model-Agnostic Explanations) must be incorporated. The techniques can provide stakeholders and policymakers with reasons why a model made a particular prediction, resulting in transparency and accountability.

21.3.5.5 Implementation Barriers: Costs and Infrastructure

The application of AI-GIS platforms is faced with enormous infrastructural and economic issues. For example, in the rural areas of Kerala, no sensor calibration and data processing infrastructure exists, causing repeated maintenance problems in 2023. Additionally, deep learning algorithms require sophisticated computing systems, which are economically unsustainable for most local governments.

To bridge this gap:

- Subsidization of IoT infrastructure by public–private partnerships in under-funded areas.
- Government officials should be given GIS and AI analytics training through state-level capacity-building programmes.
- Promotion of edge computing can reduce dependency on high-bandwidth internet for facilitating real-time local processing.

21.3.6 Policy Implications and Recommendations for Future Research

Even though there is a fast pace of development of AI and GIS technology, dispersed policy schemes hinder their effective utilization. For example, India's National Geospatial Policy 2022 encourages sharing data openly but is not aligned with the Digital Personal Data Protection Act 2023, which regulates the use of AI data.

We suggest:

- Creation of a Single Environmental Intelligence Authority at the national environment ministries to coordinate GIS-AI-ML integration.
- The implementation of transnational standards, driven by the EU AI Act, categorizes environmental AI applications as "high-risk" and demands independent assessments.
- Requiring data transparency, ethical auditing, and model explainability as conditions for approving environmental AI projects.

Data privacy regulations, ethical principles for artificial intelligence, and public access to environmental data are highly dissimilar among different countries, thus creating barriers to cross-border data exchange and system interoperability (Smith et al. 2024). Lack of harmonized policy frameworks hinders large-scale deployment of AI-integrated Geographic Information System (GIS) solutions, leading to variability in environmental monitoring capabilities and efficiency in decision-making (Johnson and Lee 2023). The key policymaking challenge is the absence of explicit legal frameworks for predictive modelling based on artificial intelligence in environmental health. Many countries have failed to develop guidelines for liability and accountability where AI-based forecasts affect public health programmes or emergency management responses (Patel et al. 2022). In addition, integration of satellite data and Internet of Things (IoT) sensor measurements into AI-GIS frameworks presents relevant questions for data ownership, security, and misuse potential, particularly in situations where corporate interests regulate access to critical environmental datasets (Chen and Kumar 2023).

Another significant regulatory issue is the ethical regulation of AI in environmental decision-making. Problems of algorithmic bias, AI prediction transparency, and digital divide disproportionately influence developing regions, where high-quality geospatial data and AI know-how are scarce (Garcia and Thompson 2024). For example, studies have shown the ways in which low-income nations grapple with low-quality AI training datasets, resulting in inaccurate predictions and ineffective policy interventions (Williams et al. 2023). Without policy interventions that guarantee inclusivity, AI-based environmental monitoring will cement existing disparities instead of correcting them. To this end, policymakers should prioritize interdisciplinary research, bringing together technologists, environmental scientists, policymakers, and lawyers to develop strong regulatory frameworks. Investment in open data infrastructure must be undertaken to enable equitable access to environmental data, while policy reforms must require transparency, accountability, and ethical deployment of AI in environmental monitoring (Zhang et al. 2023). Additionally, international cooperation is important in the production of harmonized AI and GIS regulations, especially in addressing trans boundary environmental issues like climate change, air pollution, and water resource management (Kim and Roberts 2024). By pursuing a policy-led response, governments can guarantee effective and responsible deployment of AI and GIS technologies in environmental health monitoring.

The convergence of GIS, AI, and ML for environmental health surveillance is likely to evolve with developments in big data analytics, cloud computing, and digital twins. Future studies will need to concentrate on improving real-time processing of data from a variety of sources, like IoT sensors, drones, and high-resolution satellite images, which will facilitate more precise predictive modelling for pollution management and disease surveillance. Some of the potential areas include the creation of smart cities and AI-based environmental monitoring systems, where GIS-based digital twins can model environmental situations and forecast the influence of urbanization on public health. Deep learning models like transformers and generative adversarial networks (GANs) can also improve the identification of anomalies in environmental data, which will strengthen early warning systems for extreme weather conditions and deterioration in air quality (Rahman et al. 2021).

Additionally, federated learning deployment can solve privacy and security issues by facilitating collaborative AI model training in institutions without exposing sensitive environmental health data (Chakraborty and Banerjee 2023). It is also important to carry out policy-oriented research on formulating regulations that will determine the ethical use of AI and ML in environmental decision-making (González et al. 2022). Lastly, geographers, data scientists, epidemiologists, and policymakers in inter-disciplinary teams will play a critical role in spearheading integrated GIS-AI-ML solutions for effective environmental health sustainability. Future research must also determine the indigenous knowledge system and the community-based GIS methods for improved localized environmental monitoring and resilience planning.

21.4 Conclusion

The integration of geographic information system (GIS), artificial intelligence (AI), and machine learning (ML) has significantly advanced environmental health monitoring by enabling spatial analysis, predictive modelling, and automated decision-making. GIS provides a robust framework for visualizing environmental patterns, while AI and ML enhance analytical capabilities by detecting anomalies and forecasting pollution trends with high accuracy (Zhang et al. 2021). The synergy of these technologies has led to real-time environmental monitoring through IoT sensors, satellite data, and hybrid computational models, allowing for early warnings and rapid intervention in pollution-related health risks (Ahmed et al. 2022).

Despite these advances, significant problems remain, including data availability, computing demands, ethical considerations, and policy constraints. In particular, the quality and availability of geographic and environmental datasets have a substantial impact on the effectiveness of AI-powered monitoring systems (Chen et al. 2020). Furthermore, ethical issues such as privacy, bias in AI models, and equal access to environmental technologies necessitate constant legislative interventions and interdisciplinary collaboration. Looking ahead, the continued development of AI-powered GIS platforms, deep learning algorithms, and real-time monitoring systems will further enhance predictive accuracy and risk assessment capabilities in environmental health. Collaborative efforts between governments, researchers, and technology developers are essential to addressing these challenges and ensuring sustainable applications of AI and ML in environmental monitoring. By appropriately utilising these technological tools, we may encourage data-driven environmental policy and improve public health outcomes in the face of mounting environmental issues (Singh and Gupta 2023).

Building on the evaluated literature, we suggest a three-part framework for ethical AI-GIS integration in environmental health: (1) data inclusivity—ensuring diverse, high-quality datasets; (2) model accountability—creating interpretable and auditable AI systems; and (3) policy alignment—integrating AI governance into current environmental health standards. This paradigm seeks to integrate technical advancements with social fairness and sustainability objectives.

References

Ahmed M, Khan R, Bose A (2021) Big data frameworks in environmental monitoring: challenges and opportunities. Environ Inform 12(3):198–212

Ahmed S, Fatima R, Noor A (2022) AI surveillance and public health ethics: a global perspective. J Ethics AI 4(1):45–60
Ahmed Z, Qureshi A, Hanif M (2023) Sensor reliability in low-cost air monitoring: a review. Environ Syst Res 11(2):122–138
Akhtar N, Rahman M, Sattar T (2023) Flood risk modeling using random forest in Bangladesh: a case study. J Hydrol Geoenviron 45(1):78–91
Anderson C, Patel S, Choudhury B (2023) Multi-source GIS data integration for health risk mapping. GeoHealth 7(4):e2022GH000726
Arnone E, Noto LV, Cannarozzo M (2020) Integrating hydrological models with ML for flood prediction. Nat Hazard 20(8):2255–2274
Beelen R, Hoek G, Vienneau D (2018) Air pollution interpolation methods and health assessments. Environ Health 17(1):12
Benedict J, Johnson S (2021) ML approaches for disease prediction from environmental exposure. Environ Inform 18(3):121–135
Bhatt S, Weiss DJ, Cameron E (2017) The global distribution and burden of dengue. Nature 496(7446):504–507
Bishop CM, Rasmussen CE, Trefethen LN (2021) Hybrid climate modeling using GCMs and ML. J Clim Inform 9(2):83–104
Briggs DJ, Sabel CE, Lee KJ (2019) Environmental exposure mapping and spatial epidemiology. Environ Health Perspect 127(6):67001
Brook RD, Rajagopalan S, Pope CA (2020) Particulate matter air pollution and cardiovascular disease: an update to the scientific statement from the American heart association. Circulation 121(21):2331–2378
Chen G, Li S, Zhou Z (2019) Support vector machines for urban pollution forecasting. J Environ Manage 241:567–574
Chen J, Kumar S (2023) Corporate control over remote sensing data and its implications. Environ Policy Rev 11(3):209–222
Chen Y, Liu H, Xu X (2020) Air pollution prediction using Bayesian and SVM models. Environ Model Softw 124:104588
Chen Y, Sun Y, Luo Z (2021) AI applications in environmental health: a comprehensive review. Comput Environ Urban Syst 88:101673
Chen Y, Zhang H, Wu J (2022) CNN-LSTM hybrid model for multivariate air pollution forecasting. Environ Model Softw 148:105247
Chen Z, Liu W, Hu Q (2023) Machine learning in remote sensing for pollution detection. Remote Sens Appl: Soc Environ 30:100879
Chakraborty S, Banerjee A (2023) Federated learning for environmental monitoring: a review. AI Soc 38(1):23–39
Cova TJ, Church RL (2021) GIS modeling for environmental health vulnerability. Environ Monit Assess 193(9):1–12
Díaz M, Lopez A, Aguilar E (2021) Spatial epidemiology of waterborne disease using GIS. Water Res 200:117258
Ding Y, Zhang T, Wang X (2021) Time-series GIS analysis for urban environmental risks. Int J Appl Earth Obs Geoinf 101:102355
Franch-Pardo I, Napoletano BM, Billa L (2020) GIS in health and hazard mapping: emerging trends. Spat Spat-Temporal Epidemiol 32:100319
Gao J, O'Neill BC (2020) Multispectral analysis of urban heat islands. Remote Sens Environ 242:111735
Gao L, Chen Z, Li J (2022) NDVI and LST-based modeling of dengue risk in India. J Appl Remote Sens 16(2):026512
Garcia LM, Thompson J (2024) Global surveillance and geopolitics of environmental data. J Polit Ecol 31(1):112–127
Gebru T, Morgenstern J, Vecchione B (2021) Datasheets for datasets: improving AI transparency. Commun ACM 64(12):62–71

Goetz JN, Brenning A, Petschko H (2015) Landslide susceptibility mapping using hybrid models. Geomorphology 232:1–15
Goodchild MF (2020) Geographic information science for environmental health. Environ Model Softw 134:104832
Gomes MF, Pastore YM, Silva RA (2020) SVM models for dengue hotspot prediction in Brazil. Trop Med Int Health 25(8):944–953
González A, de la Rosa M, Kumar S (2022) Policy gaps in ethical AI deployment. AI Ethics 2(1):11–24
Gorelick N, Hancher M, Dixon M (2017) Google Earth engine: planetary-scale geospatial analysis. Remote Sens Environ 202:18–27
Gupta R, Sharma V, Tripathi S (2019) AI-based DSS for environmental pollution control. Environ Technol Innov 14:100326
Gupta S, Das M, Singh P (2023) Ethical AI for environmental health in India: a review. Environ Dev 45:100775
Haque M, Roy S, Begum R (2023) Water quality prediction using ML in rural Bangladesh. Water Resour Manage 37(2):301–317
Hu X, Waller LA, Al-Hamdan MZ (2021) Remote sensing of air pollution using hyperspectral imaging. Atmos Pollut Res 12(2):305–316
Huang L, Zhou Z, Tang L (2020) Predicting dengue outbreaks using deep learning models. Comput Biol Med 118:103748
Huang X, Lin J, Fang Y (2023) CNN and LSTM models for PM2.5 prediction in Asia. Environ Pollut 316:120404
Jain M, Sengupta A, Nair P (2022) Edge computing for environmental monitoring in India. Sustain Comput 34:100720
Jeong S, Gim T (2023) Territorial stigma and symbolic inequalities in environmental risk perception. Urban Stud 60(4):812–829
Johnson D, Lee H (2023) Legal bottlenecks in cross-border environmental data exchange. Environ Law Rev 25(2):102–119
Kim J, Lee M (2022) Citizen participation in AI-driven environmental sensing. Urban Anal City Sci 9(3):441–459
Kim S, Roberts JT (2024) Climate change governance and digital environmental inequality. Environ Policy Plan 26(1):89–106
Kraemer MU, Reiner RC, Brady OJ (2019) Past and future spread of the arbovirus Aedes aegypti. Nat Microbiol 4(5):854–863
Kumar N, Borbor LD (2018) Urban crime in Jorhat Town, Assam: a spatial perspective. Transactions 40(1):1–13
Kumar P, Lee SH (2021) Big data analytics for environmental health surveillance. Sustain Cities Soc 64:102561
Kumar S, Mishra D (2023) Spatiotemporal modeling of malaria risk using AI. J Trop Dis 41(3):112–125
Kumar V, Roberts M (2022) Interpretability in AI for public health systems. AI Soc 37(2):513–530
Li X, Wang J (2020) Deep learning for mapping urban heat islands. Remote Sens 12(21):3452
Li Y, Liu Q, Zhou Y (2021) AI-based asthma forecasting using air quality and hospital data. Health Inform J 27(3):1464–1481
Li Y, Zhang Y, Chen G (2022) Sensor calibration for urban air quality in Delhi. Urban Clim 41:101088
Li Z, Chen H, Ahmad W (2023) AI-sensor fusion for proactive environmental risk detection. J Environ Inf 44(2):215–230
Liu C, Ma Y, Wei X (2022) Using NDVI and LST to forecast dengue risk. Sci Total Environ 807:151087
Liu F, Zhang H, Wang L (2021a) ML-enhanced air quality monitoring networks. Atmos Res 255:105569
Liu Y, Zhang J, Chen W (2021) CNN-LSTM hybrid for PM2.5 prediction. Ecol Indic 122:107284

Luccioni A, Viguier T, Bengio Y (2022) Estimating carbon emissions of deep learning models. J Mach Learn Res 23(150):1–48

Ma J, Wang H, Zhao Y (2019) Deep learning for urban pollution prediction. Sensors 19(4):1052

Ma Q, Zhou X, Li J (2020) Crowdsourced data and real-time GIS for environmental risk assessment. Environ Monit Assess 192(3):161

Messina JP, Brady OJ, Pigott DM (2021) Mapping global environmental suitability for vector-borne diseases. Parasit Vectors 14(1):10

Moellmann M, Schaefer M, Hufkens K (2022) Spatial epidemiology: linking GIS and public health. J Geospatial Health 17(1):46–58

Mosavi A, Ozturk P, Chau KW (2018) Flood prediction using machine learning models: literature review. Water 10(11):1536

Musa GJ, Chiang PH, Harris M (2021) GIS applications in spatial epidemiology. Int J Health Geogr 20(1):2–15

Nguyen T, Le Q, Nguyen H (2023) Real-time environmental monitoring using IoT-GIS hybrid systems. Sensors 23(1):132

Norris P, Smith L, Gill M (2023) GIS applications for tracking water pollution in developing countries. J Environ Plan Manag 66(3):455–472

O'Neil C (2016) Weapons of math destruction: how big data increases inequality and threatens democracy. Crown Publishing Group

Patel D, Singh M (2020) Water quality classification using support vector machines. J Water Health 18(4):509–518

Patel V, Kumari A, Ghosh S (2022) Legal regulation of AI in environmental forecasting. Environ Law Technol 8(2):101–116

Patel V, Sharma P, Banerjee R (2023) Hybrid machine learning models for climate risk assessment. Clim Change 176(3):54

Rahman A, Das S, Pal A (2022) AI-based hazard mapping for heatwaves and wildfires. Int J Disaster Risk Reduct 73:102923

Rahman F, Karim M, Rafiq T (2023) Machine learning for air pollution index forecasting. Atmos Environ 286:119306

Rahman M, Zaman A, Afreen S (2021) GANs for anomaly detection in environmental data. Environ Model Softw 144:105133

Rahmati O, Pourghasemi HR, Melesse AM (2022) Remote sensing and ML integration for climate change analysis. J Hydrol 610:127926

Reichenbach P, Rossi M, Guzzetti F (2018) Landslide susceptibility mapping with hybrid models. Eng Geol 246:353–365

Rolnick D, Donti PL, Kaack LH (2022) Tackling climate change with machine learning. ACM Comput Surv 55(2):1–35

Sabat-Tomala A, Malawska A, Matysik G (2020) Remote sensing-based SVM model for invasive species detection. Remote Sens 12(10):1568

Schroeder W, Elvidge CD, Prins EM (2021) UAVs and satellite remote sensing for urban heat monitoring. Remote Sens Environ 253:112179

Shaddick G, Thomas ML, Mudu P (2018) Satellite-based PM2.5 estimates and global exposure assessment. Environ Health Perspect 126(8):087005

Sharma R, Gupta R (2022) Air sensor calibration and public health policy in Delhi. Environ Health Insights 16:11786302221091188

Sharma V, Kumar A (2019) CNN and RNN applications in environmental monitoring. Environ Technol Innov 13:140–150

Shi W, Zhang X, Zhao L (2023) Explainable hybrid AI models for environmental planning. Comput Environ Urban Syst 93:101787

Silva C, Costa L (2023) AI-driven dengue predictions and resource misallocation in São Paulo. Health Place 79:102956

Silva D, Gomez L, Ortega J (2023) GIS-enabled risk mapping using sensor networks. Int J Environ Sci Technol 20(3):1011–1026

Singh P, Gupta R (2023) Scalable ML frameworks for environmental health. J Environ Inform Lett 1(1):32–45

Singh R, Kumar S (2017) GIS-based groundwater pollution risk modeling. Environ Earth Sci 76(6):1–11

Smith K, Petley D (2021) Environmental hazards: Assessing risk and reducing disaster, 7th edn. Routledge

Smith MJ, Rao S, Chatterjee B (2024) Comparative policy study on open access environmental data. Glob Environ Polit 24(1):59–78

Song X, Li W, Zhang R (2023) LSTM for real-time air quality predictions. Environ Model Softw 160:105558

Strubell E, Ganesh A, McCallum A (2019) Energy and policy considerations for deep learning in NLP. Proc ACL 2019:3645–3650

Sun Z, Liu H, Yang Q (2021) Transfer learning in environmental AI applications. Nat Mach Intell 3(4):302–310

Sun Z, Wang Y, Hu X (2022) AI and ethics in environmental surveillance. AI Ethics 3(1):72–86

Taddeo M, Floridi L (2018) How AI can be a force for good. Science 361(6404):751–752

Tan Y, Zhang X, Li H (2021) Deep learning and disease forecasting in urban environments. J Urban Health 98(2):305–319

Tan Y, Zhou L, Wang S (2023) Air quality forecasting using deep GIS models. Environ Technol Innov 30:102216

van Donkelaar A, Martin RV, Spurr RJD (2019) PM2.5 exposure from satellite observations. Environ Health Perspect 127(4):47001

Vapnik VN (1998) Statistical learning theory. Wiley

Vinuesa R, Azizpour H, Leite I (2020) The role of AI in achieving the sustainable development goals. Nat Commun 11(1):233

Wahyuni D, Putri M, Kurniawan H (2023) Dengue prediction in Indonesia using CNN and satellite data. Trop Med Int 5(1):11–22

Wang J, Li C, Huang Z (2023) UAV-based pollutant detection using deep learning. Sensors 23(2):2323

Wang P, Liu X, Chen X (2022) ML models for spatiotemporal pollution mapping. Atmos Pollut Res 13(6):101429

Weng Q (2019) Remote sensing of urban environments. Int J Remote Sens 40(1):1–22

Williams J, Nyong'o C, Baloch Z (2023) AI and public health disparities in LMICs. Glob Public Health 18(4):589–602

Wu H, Fang Z, Ren L (2023) Cloud computing for environmental GIS models. J Cloud Comput 12(1):1–18

Xiao Q, Li J, Zhang J (2022) Hybrid SVR-physics models for PM2.5 estimation. J Geophys Res: Atmos 127(8):e2022JD036772

Xie X, Wang Q, Zhang Y (2020) GIS-based mapping of air pollution exposure. Environ Res 188:109859

Xie Y, Yu C, Ma L (2022) ML-based pollution forecasting in megacities. Environ Monit Assess 194(3):1–15

Xu W, Zhou Y, Chen L (2023a) Explainable AI for pollution mapping. J Environ Inf 45(2):192–207

Xu Y, Zhang Z, Liu T (2023b) Neural networks for geostatistical pollution prediction. Environ Technol 44(1):52–65

Yang K, Zhang Y, Liu W (2021) GIS-AI integration for disaster response planning. Comput Environ Urban Syst 85:101555

Zhan Y, Luo Y, Deng X (2021) Random forest PM2.5 estimation based on AOD. Remote Sens Environ 253:112172

Zhang H, Chen R (2023) ML-enhanced early warning systems for air quality. Environ Technol Innov 29:102095

Zhang J, Zhao M, Wang X (2023) Air pollution forecasting with deep hybrid models. Atmos Res 278:106393

Zhang Y, Sun Y, Chen G (2021) GIS-based pollution monitoring using AI. GeoHealth 5(3):e2021GH000354

Zhang Y, Zhou J, Liu H (2022a) LiDAR-based particulate matter analysis in urban areas. Environ Pollut 302:119016

Zhang Y, Zhao L, Huang W (2022b) AI-based DSS for health risk forecasting. Comput Hum Behav 130:107200

Zhao Y, Zhang H, Li Q (2021) GeoAI for disease risk prediction. Spat Spat-Temporal Epidemiol 36:100401

Zhao Z, Liu J, Chen R (2022) Satellite data in CNN models for respiratory disease prediction. Remote Sens 14(5):1171

Zhou Y, Liu W, Gao S (2021) SVM classification for heavy metal water pollution. Environ Monit Assess 193(10):1–13

Zhou Y, Wang Y, Tang L (2022) Transparency issues in deep learning environmental models. AI Ethics 2(3):205–219

Zhu XX, Tuia D, Mou L (2017) Deep learning in remote sensing: a comprehensive review. IEEE Geosci Remote Sens Mag 5(4):8–36

Chapter 22
Revolutionizing Environmental Hazard Management with Remote Sensing and GIS

Diwesh Kumar, Sudarshana Banerjee, Shailesh Kumar Sarangi, and Surya Pratap Singh

Abstract Environmental hazards—whether stemming from natural or human-induced disasters—pose a significant and escalating threat to communities worldwide. The growing frequency and unpredictability of such occurrences amplify the urgency for innovative solutions in monitoring, risk evaluation, and disaster response. This chapter delves into the transformative role of remote sensing (RS) and geographic information systems (GIS) in enhancing environmental hazard management, highlighting their capabilities in real-time data acquisition, spatial analysis, and informed decision-making support. This present chapter explores the synergistic advantages of RS and GIS in identifying vulnerable zones, simulating hazard impacts, and supporting emergency response planning. It examines the integration of high-resolution satellite imagery, LiDAR, drone-generated data, and geospatial datasets across diverse applications such as flood modeling, urban heat island mitigation, coastal erosion monitoring, and postdisaster evaluations. A collection of case studies from varied geographical contexts—including the UK, USA, India, and Nigeria—demonstrates the practical deployment of these technologies in real-world hazard scenarios. The chapter further highlights emerging technological advancements—such as 3D terrain mapping, real-time data analytics, AI-driven geospatial interpretation, and crowd-sourced GIS—which significantly expand the capabilities and effectiveness of hazard mitigation efforts. Despite notable progress, critical challenges persist, including limited data accessibility, high implementation costs, and shortages of skilled professionals, particularly in resource-constrained regions. Interdisciplinary collaboration, continuous innovation, and targeted capacity building are

D. Kumar · S. P. Singh (✉)
Department of Civil Engineering, Srinath University, Jamshedpur, Jharkhand, India
e-mail: spsingh894@gmail.com; spsingh.civil@srinathuniversity.ac.in

S. Banerjee
Department of Physics, Srinath University, Jamshedpur, Jharkhand, India

S. K. Sarangi
Department of Mechanical Engineering, Srinath University, Jamshedpur, Jharkhand, India

A. Roy et al. (eds.), *Environment and Public Health*, Sustainability Solutions,
https://doi.org/10.1007/978-3-031-99770-9_22

essential to positioning RS and GIS at the core of disaster risk reduction frameworks, reinforcing global efforts in resilience-building, sustainable development, and climate adaptation.

Keywords Remote sensing · GIS · Environmental hazard management · Risk assessment · Disaster resilience

22.1 Introduction

Environmental threats are related to ranges and put themselves under existence: safety hazards that fall into health, ecosystems, and the socio-economic dimension. They occur in the form of four categories in the name of natural hazards: anthropogenic, biological, and chemical (Tong et al. 2022). Each one of them has distinct risks like earthquakes, floods, hurricanes, and wildfires. All these are natural disasters produced by climatic and geophysical phenomenon. These phenomena are unpredictable; most of the time, they cause sudden large-scale devastation with forced displacements and disruption of even basic services to the population. Climate change has aggravated them with an increase in extreme weather events, repetition, and severity. On the contrary, anthropogenic hazards fall into two types caused by human activities: pollution, deforestation, industrial accidents, and improper waste disposal. Rapid urbanization, industrialization, and overexploitation of resource bases are major contributors to environmental degradation (Shivanna 2022). Air pollution emanating from vehicular emissions and industrial activities results in the prevalence of respiratory illnesses and cardiovascular diseases, while agricultural effluents and industrial wastes lead to water contamination, which spreads water-borne diseases. Degradation of soil owing to deforestation and improper land-use practices is aggravating food insecurity and ecological imbalance. Figure 22.1 shows the various types of hazards.

The biological hazard underlines the interaction of man with animals and microorganisms. Most disease outbreaks, such as pandemics and vector-borne diseases, are associated with factors of environmental change, like deforestation and even climate variability. Some examples of zoonosis include the land and habitat expansion for human settlement into the natural habitat, something seen with recent developed global health crises (Leal Filho et al. 2022). The areas also suffer microbial contamination due to poor sanitation provisions coupled with insufficient waste disposal, which in turn leads to major health risks, especially with enormous population density and little or no access to healthcare services.

Chemical dangers are related to exposure to toxic chemicals such as heavy metals, pesticides, and industrial chemicals. Chronic diseases that may result from a long-term injurious exposure to hazardous chemicals include cancers and general neurological disorders, with some reproductive problems (Jaishankar et al. 2014). Many accidental industrial incidents like oil spills and chemical leaks threaten human lives but contaminate the environment for a significantly long time after oil and chemicals

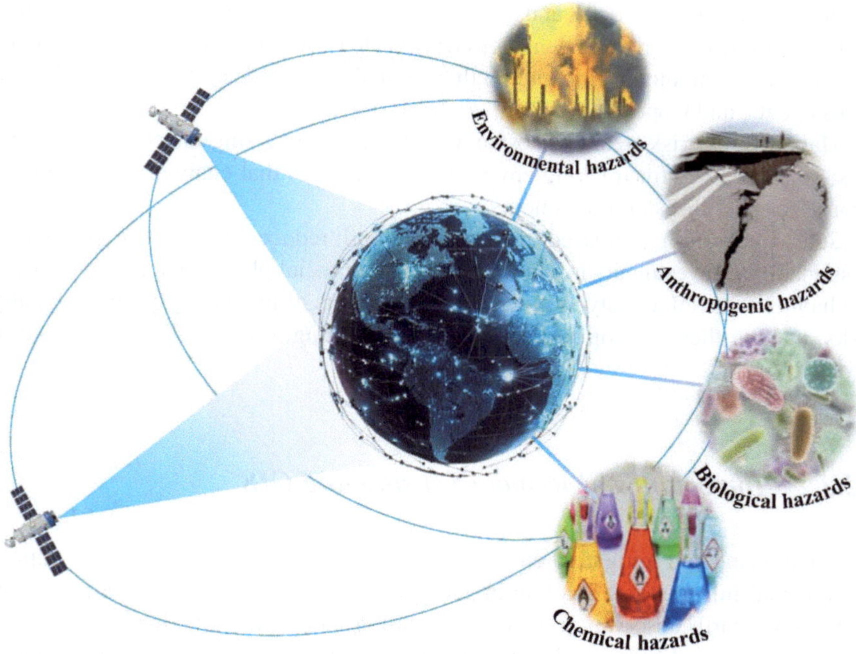

Fig. 22.1 Satellite-based monitoring of global hazards

leak into surface or underground water bodies, soil, and ambient air. Utilization of synthetic chemicals continuously keeps growing in agriculture and manufacturing, making such means possible and available in the environment for bioaccumulation in food chains.

Environmental hazards, whether natural or human-induced, must include detailed monitoring and mitigation strategies that minimize their impact. Advances in geospatial technologies and artificial intelligence have made it possible to have a more efficient prediction and early warning or real-time environmental monitoring system (Olawade et al. 2024). Combining these technologies with policy approaches and sustainable development practices becomes critical in reducing environmental risks and safeguarding public health. Challenges of this nature must be addressed in a multidisciplinary manner, ensuring that scientific, technological, and policy-driven efforts can be coalesced into creating a safe, secure, and resilient environment.

22.1.1 Need for Technological Innovations in Hazard Management

The issues in hazard prevention deteriorate over time but need an advanced application of information technology to render the management efficient. Conventional methods were slow to react and rather limited in predicting. Incorporating these new technologies of remote sensing, GIS, AI, IoT would improve and even make possible real-time monitoring and predictive analytics for detection of hazards, risk assessments, and response.

Mapping high-risk areas by geospatial tools; AI for analyzing and predicting disasters; smart sensors for tracking environmental changes that enable rapid response; drones for post-disaster assessment.

Such integration increased preparedness and reduced loss while promoting sustainability. Investments in innovation become crucial philosophical perspectives to channel costs effectively in mitigating the impacts of increasing climatologically risks, making them appropriate choices in the funding to make all parts long-term protected.

22.1.2 Importance of Remote Sensing and GIS

For monitoring and mitigating environmental risks, remote sensing and GIS supply vital spatial information. Satellite imagery, aerial photographs, and drones help in continuous Earth observation, thereby providing large-scale tracking of hazardous regions, land-use changes, and climate peril. These are useful in assessing risks posed by deforestation, desertification, and sea-level rise.

GIS improves the visualization of hazards by bringing together spatial information with analytical models to create risk maps, optimize resource distribution, and aid disaster preparedness. The simulations will determine the impact on the surrounding land, population, and infrastructure by hazard modeling for floods, landslides, and wildfires. GIS also assists land-use planning through spatial decision support systems (SDSS) that minimize exposure to hazards.

Remote sensing and GIS facilitate early warning systems through detecting environmental changes in temperature, atmospheric pressure, and water level. They also allow for rapid damage assessment and distribution of relief after disasters.

With the increasing frequency and unpredictability of hazards due to climate change, the integration of remote sensing and GIS into disaster management builds resilience, enhances decision-making, and fosters proactive risk mitigation.

22.1.3 Rationale of the Study

The increasing frequency and severity of environmental hazards call for sophisticated monitoring and management tools. Remote sensing and GIS provide the ability to collect real-time information, perform spatial analyses, and make informed choices, thus lending themselves to greater advantage in pre-emptive disaster mitigation and resilience planning.

22.1.3.1 Significance of Remote Sensing and GIS in Environmental Hazard Management

The combination of remote sensing and GIS promotes the monitoring of hazards in real time by assessing risks and responding to disasters. Remote sensing gathers key spatial data from satellite, drone, and aerial platforms related to environmental parameters such as land cover, vegetation, and topography, while GIS helps in processing and visualization for efficient analyzing and interpretation by decision-makers working with hazard-prone areas.

22.1.3.2 Synergistic Use of Remote Sensing and GIS

The combination of remote sensing and GIS provides a full suite of tools for environmental monitoring, disaster management, and urban planning scenarios. Their integration plays a key role in detecting vulnerable areas, assessing disaster impacts, and elaborating on mitigation strategies. For example, in the case of flood management, remote sensing helps measure water levels and determine flood extents, while GIS supports hazard mapping and evacuation planning. In turn, remote sensing provides data on land use in urban planning, whereas GIS assists in expenditure and resource allocation on infrastructure.

22.1.3.3 Integrating Spatial Data for Hazard Analysis

GIS integrates spatial information from remote sensing, digital elevation models (DEMs), and ground studies to hazard-prone area assessment, disaster impact prediction, and near-real-time monitoring enhancement. It overlays multiple data sources like meteorological data and seismic data to better address risk assessment and disaster preparedness. The data are collected, pre-processed, and subjected to spatial analysis, and visualized using hazard maps and predictive models, thereby assisting in making informed decisions.

22.1.3.4 Gaps in Traditional Hazard Management Approaches

Management of hazards, in the traditional sense, often relies on old data with very limited application of spatial analysis and mainly on reactive methods, all of which diminish the effectiveness of hazard management. Real-time monitoring, integration of various datasets, and a robust predictive posture are critical for the proactive mitigation of disasters. Remote sensing and GIS abound in offering solutions to these gaps through real-time high-quality information, advanced analytical tools, thus enabling hazard assessment, early warning systems, and disaster resilience planning.

22.2 Materials and Technologies of Remote Sensing

Remote sensing technologies are widely applied in the major areas of natural hazard management. Satellite imagery, LiDAR, and hyperspectral sensing have been proven to be useful in high-precision data acquisitions for monitoring and evaluating hazards. The following sections further describe these technologies in detail.

22.2.1 Satellite-Based Remote Sensing

Satellite-based remote sensing is one of the most commonly used methods for assessing environmental change and hazards on a very large scale. It is the act of imaging and gathering data about the Earth from space with Earth observation satellites equipped with specialized sensors. These satellites are in various orbits and use different spectral bands ranging from visible, infrared, microwave, and thermal imaging to detect changes in the environment. To conduct this, satellites are set 700–900 km above ground level, as shown in Fig. 22.2.

The two main categories of remote sensing are the passive and the active. Passive sensors, such as those on Landsat and Sentinel satellites, have the ability to capture reflected radiation from the Earth's surface by natural sunlight. These sensors are helpful in the assessment of vegetation health, land-use change, and ocean temperature change. Active sensors, like radar-based Sentinel-1 and LiDAR systems, send their energy signals toward the target. They measure backscattered signals from

Fig. 22.2 Multi-altitude remote sensing platforms from satellites to drones for earth observation

the target site and thus assist in terrain mapping and environmental monitoring in inclement weather or during nighttime (Jia et al. 2021).

22.2.2 Drone and Aerial Imaging

Drone and aerial imaging is nowadays fast becoming one of the most preferred methodologies for high-resolution data collection in real time at localized scales, making them truly effective for site-specific environmental assessment. The range of drone and aerial imaging varies from 0.15 to 0.30 km, as shown in Fig. 22.2. Drones or unmanned aerial vehicles (UAVs) with advanced sensors, including high-resolution cameras, multispectral and hyperspectral sensors, and thermal imaging devices, provide versatile monitoring of atmospheric and aquatic environments, soil erosion, forest cover change, and disaster remediation.

In comparison with satellites, drones can be more operationally flexible and cost-effective. Drones can be deployed on demand to obtain high-resolution images of particular areas of environmental change that need immediate attention. For instance, flood areas are being mapped post-disaster using drones to assess possible damage sites, and similarly, damage to infrastructure is mapped in cases of forest wildfires.

Manned aircraft carrying aerial imagery is also an important remote sensing apparatus. It is often employed for large-scale surveys that are out of the resolution reach of satellite imagery. Airborne LiDAR technology, typically mounted on aircraft, provides highly accurate three-dimensional terrain models, which are used for topographic mapping, forest inventory assessments, and urban planning.

Both satellite and drone remote sensing technologies complement each other, providing solutions at different scales to monitor environmental hazards and improve disaster response and sustainable management of resources (Alvarez-Vanhard et al. 2021).

22.2.3 Integrating Spatial Data for Hazard Analysis

The incorporation of spatial data into hazard analysis consists of several steps. For example, sources of data include satellite, sensor, and historical records for data collection. Datasets differ and should be cleaned, georeferenced, and standardized to ensure that they are comparable at this time. Next, spatial analysis is conducted using GIS applications for hazardous areas and their isolation using overlay mapping, buffering, and interpolation. Lastly, information is processed through hazard maps, dashboards with up-to-the-moment data and predictive models to make visualization and decision-making possible by authorities so that they can prevent and respond.

22.2.4 Integration of Remote Sensing and GIS for Hazard Management

Incorporating remote sensing with GIS to hazard management integrates spatial data and live imagery at the same time, improving the capacity to monitor, assess, and respond to any natural or man-made catastrophe. Bird's eye view from the satellites and drones, combined with the aerial platforms, takes an image and measures a phenomenon-prone area or even several areas from the extent of the coverage in kilometers. Figure 22.3 illustrates the integration of GIS and remote sensing. Data are being collected from the satellite and IoT, then it is analyzed using various software such as ArcGIS.

Land cover, vegetation, topography, and environmental characteristics are necessary for understanding hazard monitoring, risk assessment, and damage appraisal. Satellite, drone, and aerial cameras are capable of capturing these numerous data sets, while GIS does the heavy work of processing, analyzing, and presenting this spatial data to the decision-makers to decipher the relevant interpretations of the information.

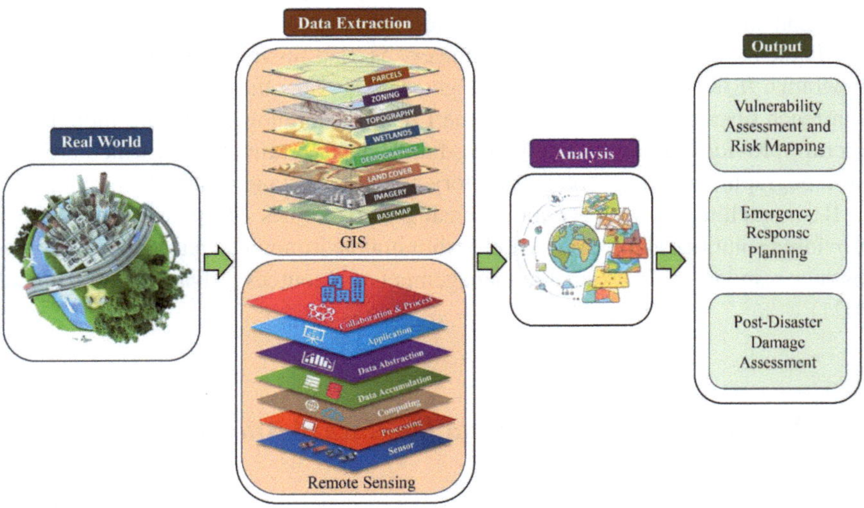

Fig. 22.3 GIS and remote sensing for disaster management and risk assessment

22.3 Results of Case Studies and Interpretation

22.3.1 *Case Studies in Integrated Hazard Management*

Table 22.1 illustrates how remote sensing and GIS can be adopted for analyzing various environmental hazards that different regions are likely to experience. Case studies are provided for application within flood risk mapping, urban heat island (UHI) mitigation, coastal zone monitoring, urban planning, and water resource management. These technologies are important for hazard identification, risk assessment, and mitigation planning, while providing requisite information for informed decision-making.

The geospatial technologies show tremendous potential for enhancing preparedness and management in disaster situations. Mapping of flooding using GIS has generated increased mitigation measures, improved real-time monitoring, and thus better emergency response planning in flood-prone areas such as the UK, New Orleans, and Jakarta. Similarly, during UHI mitigation within urban areas shown in Fig. 22.4, such as Phoenix, New York City, Tokyo, and São Paulo, areas were determined to have further green spaces in the cities after identification of the hottest spots illustrated in Fig. 22.4. Also, coastal sites such as the Sundarbans, Netherlands, and Maldives are using GIS-enabled shoreline monitoring and protection measures, allowing them to reduce the amount of vulnerability to rising sea levels and erosion.

The applications of these technologies extend beyond disaster management to urban planning and resource monitoring. In the case of Abuja, Nigeria, GIS-based land-use mapping underlines the sustainability of the city emerging in Lagos, while Lagos monitors water bodies through remote sensing. Coastal erosion analysis in Oman strengthens protection strategies for vulnerable shorelines.

The case studies now underline pronounced bright lights as far as the remote sensing and GIS worlds are concerned in environmental hazard management. Their capability of spatially accurate data improves risk analysis, provides fruitful early warnings, and targets certain mitigation strategies.

The key messages to take into account include incorporation of these technologies and application into urban planning and policy frameworks for improved sustenance against climate hazards now and beyond.

22.3.2 *Applications of GIS in Hazard Management*

22.3.2.1 Vulnerability Assessment and Risk Mapping

GIS is applied in vulnerability assessment through the spatial analysis of data that identifies populations, infrastructure, and ecosystems vulnerable to natural and human-induced hazards. Its general goal is to integrate demographic, economic, and environmental data into the susceptibility assessment of different localities to

Table 22.1 Outcomes-based hazards case studies through remote sensing

Location and year	Types of hazards	Materials	Technology	Outcomes	References
United Kingdom (UK)-2015	Flooding	Satellite imagery, LiDAR data, historical flood data	Remote sensing, GIS-based flood modeling	The author created maps to show areas at risk of flooding, which helped in preventing from the flood	Matthew (2015)
New Orleans, USA-2007	Flooding	LiDAR data, satellite imagery	GIS-based floodplain mapping, real-time flood monitoring	Rykhus et al. improved flood defenses and optimized evacuation routes, as well as enhanced disaster preparedness to reduce the impact of future floods	Rykhus and Lu (2007)
Jakarta, Indonesia-2007	Flooding	Satellite imagery, aerial photography	Remote sensing, GIS-integrated hydrological models	The authorities developed flood risk maps and implemented early warning systems to manage flood hazards	NASA (2007)
Phoenix, USA-2016	Urban heat island	Landsat satellite data, aerial thermal imagery	GIS-based temperature mapping	Wang et al. identified areas experiencing extreme heat and introduced green roofs, and reflective roofing materials to lower temperatures	Wang et al. (2016)
New York City, USA-2006	Urban heat island	Thermal satellite imagery	GIS for UHI analysis	The authorities launched the "Cool Neighbourhood's NYC" program to increase green spaces, and took measures to protect public health from extreme heat	NASA (2006)
Tokyo, Japan-2021	Urban heat island	Thermal infrared satellite images, drone imagery	GIS-based heat mapping	Authorities expanded urban vegetation by installing rooftop gardens as well as using reflective pavements at high temperatures in the city	NASA (2021)

(continued)

Table 22.1 (continued)

Location and year	Types of hazards	Materials	Technology	Outcomes	References
São Paulo, Brazil-2017	Urban heat island	Landsat and MODIS satellite data	GIS-integrated urban planning	Barros et al. promoted the expansion of green spaces and took steps to improve the city's resilience to urban heat	Barros and Lombardo (2017)
Sundarbans, India & Bangladesh-2020	Coastal zone management	Landsat, MODIS satellite imagery	GIS-integrated coastal monitoring	Scientists monitored coastal erosion and assessed the health of mangroves	NASA (2020)
Netherlands-2011	Coastal protection	LiDAR, satellite imagery	GIS-based flood risk modeling	Klijn et al. developed dynamic coastal protection strategies and strengthened flood defenses to prevent damage from rising sea levels	Klijn et al. (2012)
Maldives-2024	Coastal zone monitoring	Satellite imagery, aerial surveys	GIS-based erosion and sediment monitoring	Authorities created artificial islands and introduced disaster preparedness measures to protect the country from environmental threats	World Bank Group (2024)
Oman-2021	Coastal Erosion	Satellite imagery, GIS data	GIS analysis for coastal vulnerability assessment	M. Al-Ismaili identified coastal areas most at risk of erosion and used this information to improve coastal management strategies	Al-Ismaili (2021)
Abuja, Nigeria-2023	Urban expansion & flooding	Landsat 8 imagery, GIS layers	GIS-based land-use mapping	Enoguanbhor revealed inconsistencies between urban planning policies and actual land use, leading to better zoning regulations and urban development planning	Enoguanbhor (2023)
Lagos, Nigeria-2023	Water resource management	NDWI index, remote sensing data	Remote Sensing and GIS for water body monitoring	Oppong et al. monitored changes in water bodies to improve the management and conservation of water resources in the region	Oppong et al. (2023)

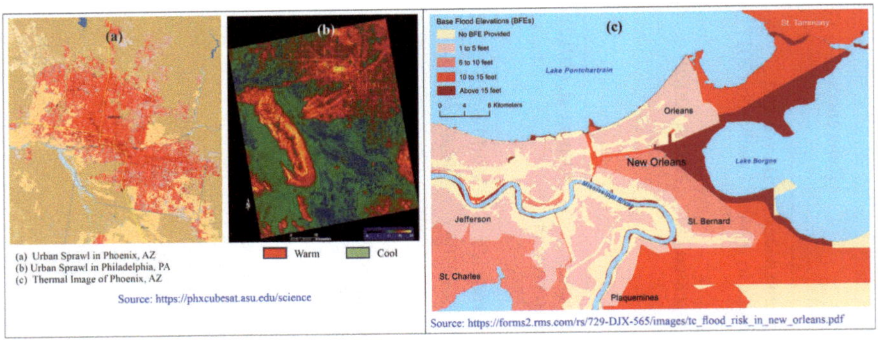

Fig. 22.4 Urban sprawl and flood risk analysis: thermal imaging of phoenix and spatial flood mapping of New Orleans

disasters such as earthquakes, floods, and landslides. By combining factors such as population density, building types, and proximity to areas at risk, GIS enables authorities to assess the likely impacts and prioritize mitigation actions (Ayele et al. 2025). It helps identify those communities deemed to be vulnerable, thus needing intervention as a group in times of a disaster, aiding them in resource allocation and evacuation planning.

GIS merges hazard data with exposure and vulnerability components to create intricate risk maps in risk mapping. It overlays a variety of data layers, such as topography, land use, and disaster history, to judge the probability of occurrence and the magnitude of effect likely at particular sites. Such maps assist a decision-maker in devising an early warning system, in applying zoning regulations, and in preparing disaster preparedness plans. It also provides real-time risk assessment through satellite-sensor data integration, which can adapt to changing conditions, thus enabling proactive reaction. The visual display of risk distribution is enhanced through this GIS visualization for resilience against disasters, which understands it as a basis of providing targeted strategies for the mitigation and response to them.

22.3.2.2 Emergency Response Planning

Emergency response planning has been made a lot easier with a GIS, which provides real-time spatial data and helps in arranging optimum usage of resources while improving coordination during disaster times. GIS maps out areas at high risk, evacuation routes, and emergency shelters for ensuring preparedness against disasters. All real-time information from satellite photographs, weather monitoring systems, and ground sensors is integrated into GIS, which allows tracking the advent of the disaster and severity assessments of events such as earthquakes, floods, wildfires, and hurricanes (Tomaszewski et al. 2015).

During emergencies, GIS provides a resource management tool by identifying the nearest hospitals, fire stations, and relief centers, ensuring prompt deployment of medical aid and rescue teams. It provides a central facility for sharing this information between emergency and disaster responders, thus facilitating joint communication and coordination. GIS dashboards and mobile applications are built for disaster management agencies to monitor areas affected by disaster, conduct damage assessment, and allocate resources within the affected area. GIS services in a postdisaster scenario include assessment of losses due to destruction to infrastructure and recovery planning through prioritization of reconstruction work. Real-time analysis incorporation through spatial data will enhance the quality of decision-making, extraction of minimum response time, and efficiency of disaster management.

22.3.2.3 Postsisaster Damage Assessment

GIS makes enormous contributions to damage-but-not-failure assessment in all phases of the disasters, from assessment of damage to estimation of loss and planning of remedial action. The damage maps shall be drawn up using a combination of satellite imagery with aerial surveys, and remote sensing data. These define the areas most affected by the disaster and allow the drafter to prioritize the relief actions. Generally, damage to infrastructure, including roads, bridges, buildings, and utilities, is quantifiable through comparison of the pre- and postdisaster images, which differentiation will allow effective resource mobilization by authorities (Miri et al. 2024).

Tracking and mapping of temporary shelter sites by GIS facilitate the distribution of humanitarian aid to displaced populations. Centralized sharing of data aids coordination by government agencies, NGOs, and emergency responders. GIS also provides critical quantifications on property damages and economic losses for financial and insurance assessments. In a long-term recovery program, it contributes to building more resilient infrastructure by analyzing hazard-prone areas and prescribing safer land-use practices. Real-time data and predictive models improve decision-making capabilities, hasten disaster recovery processes, and help societies rebuild effectively.

22.3.3 Emerging Technologies in Hazard Management

It is the designs and capabilities of the novel forms of remote sensing that have advanced the way hazard management has been enhanced with data accuracy, spatial analysis, and real-time data collection. Innovations such as high-resolution imaging, LiDAR, and hyperspectral sensing are intended to revolutionize hazard detection and mitigation through improved policy formulation and information dissemination. Successful integration, however, relies on well-articulated policies addressing the main aspects of data accessibility, standardization, ethical issues, and decision-making frameworks (Baraldo and Di Giuseppantonio Di Franco 2024).

22.3.3.1 High-Resolution Imaging

High-resolution imaging provides detailed surface data that can be used to monitor such hazardous events as wildfires, floods, and urban heat islands. It also improves disaster preparation; it includes postdisaster assessments of damage and predictive modeling. Policies must ensure the availability of data, advocate for the collaboration of stakeholders, and formulate rules for secure data sharing.

22.3.3.2 LiDAR (Light Detection and Ranging)

LiDAR generates accurate 3D models of terrain, aiding flood mapping, landslide risk assessment, and coastal monitoring. Effective policies should promote open data initiatives, investments in LiDAR infrastructure, and integration into disaster management frameworks, while addressing cost and accessibility challenges in developing regions.

22.3.3.3 Hyperspectral Sensing

Hyperspectral sensing detects environmental hazards and invites judgments based on the spectral signature to deal with wildfire risk assessment, pollution monitoring, and studies on climate change. It requires policies on standardizing data collection, enhancing AI integration for decision-making, and funding of the development of higher resolution sensors.

22.3.4 GIS Trends and Innovations (3D Mapping, Real-Time Data Processing, Crowd-Sourced GIS)

22.3.4.1 3D Mapping

The radical transformation introduced by 3D Mapping into GIS was that the spaces became closer to reality and much more interactive than before. High-resolution digital models able to serve places like those of urban planning and hazard monitoring can be built easily using LiDAR, photogrammetry, and satellite images. In disaster management, the application of 3D mapping means that these hazards, some of which include landslides and floods as well as coastal erosion, have increased risk assessment. For instance, in a flooding model, it simulates those properties of water that come in contact with a surface area consisting of land and built-up areas, thus increasing the quality of prediction it achieves. An evaluation of seismic risk can be done for areas affected by an earthquake, concerning how the possible response of buildings could be simulated during an earthquake event, through such property simulations. More therefore informs disaster preparedness, response, and recovery decisions in designing effective and targeted mitigation strategies (Badwi et al. 2022).

22.3.4.2 Real-Time Data Processing

Key innovation in GIS facilitates spatial analysis that is dynamic and immediate. Sensor networks, satellites, and IoT systems progress allows continuous environmental monitoring, changing the processes of emergency responses, urban management, and disaster forecasting. Also, data from sensors and weather satellites at present exhibit real-time data in extreme events—hurricane occurrences, floods, and wildfires—for urgent interventions. A very good example is the real-time integration of river gage data with rainfall forecast for improved flood forecasting and an early warning system. Temperature sensors across cities update the management systems of urban heat islands with data on heat waves as they happen. Such prompt processing and analysis of real-time data will make it possible for the decision-makers to mitigate the threat, usually manifested in damage and loss of lives (Dritsas and Trigka 2025).

22.3.4.3 Crowd-Sourced GIS

Crowdsourcing in GIS is also called Volunteered Geographic Information (VGI). It is emerging as a significant revolution in hazard monitoring, as it brings individuals and entire communities into the realm of real-time geo-data through time. The extensive use of smartphones, GPS-enabled devices, and social media within the reach of billions around the world democratizes the collection of GIS data and enhances the speed and coverage of hazard mapping. Examples of where crowd-sourcing—GIS have proved very useful during disasters include the 2010 Haiti earthquake, where social media and mobile apps contributed significantly to mapping areas affected by disasters, assessing infrastructure damage, and coordinating rescue efforts. More flood and wildfire victims, as well as field teams, keep up-to-date reporting about the extent and blockage of roads along with the needs for relief in a specified area to adjust the response strategies. Crowd-sourcing thus increases the data that is used to augment official GIS data, but also fills the essential gaps that have remained in hazard maps and have thus made emergency management better. For example, it allows decentralized monitoring in sparsely populated or remote areas that are often not reached during official data collection. The growing adoption of participatory GIS platforms has expanded data diversity, making hazard monitoring more inclusive and effective (Fast and Rinner 2014).

22.3.4.4 Integration of 3D Mapping, Real-Time Data, and Crowd-Sourced GIS

The powerful combination of 3D mapping, real-time data processing, and crowd-sourced GIS enables hazard management to be performed. In the case of disasters, when 3D models are combined with crowd-sourced real-time data, these models can simulate the hazards with such accuracy as to allow the authorities to predict their impacts and optimize their responses. Such integration allows for continuous updates in emergency operations, incorporating localized input of community members. Such methods allow planners in urban areas to oversee and assess multi-hazard vulnerabilities, including floods, heat waves, and seismic threats, allowing for timely and more targeted interventions. The combination of advanced technology and community participation will enable disaster management to be more dynamic, responsive, and inclusive, thus enhancing resilience and preparedness.

22.3.5 *Role of Artificial Intelligence in Enhancing Geospatial Technologies*

Artificial intelligence is the driving force for the enhancement of geospatial technologies concerning data interpretation and predictive modeling. The flow process of the role of artificial intelligence is illustrated in Fig. 22.5. AI provides a platform to enhance efficiency, precision, and empowerment in GIS applications, to an altogether different wavelength to resolve extremely complex issues in hazard management. Therefore, the fruits of these improvements will be smarter decisions, faster immediate action on impending hazards, and improved long-term planning (Mai et al. 2025).

22.3.5.1 AI-Based Data Interpretation

Artificial intelligence is driving the automation of the analysis of geospatial data and improving interpretation. Using machine learning and deep learning techniques, images are classified, landscapes are monitored for change, and environmental hazards are assessed to support land-use planning and disaster management. AI-supported anomaly detection can sense deviations from normal conditions, such as increasing water levels or temperature spikes, alerting managers to possible hazards. Combined efforts of AI for data fusion lead to the integration of information from remote sensing, sensor networks, and crowdsources, creating a robust integrated risk assessment framework. Altogether, these developments serve to enhance hazard mapping, downstream disaster preparedness, and timely decision-making, allowing for efficient and proactive hazard management.

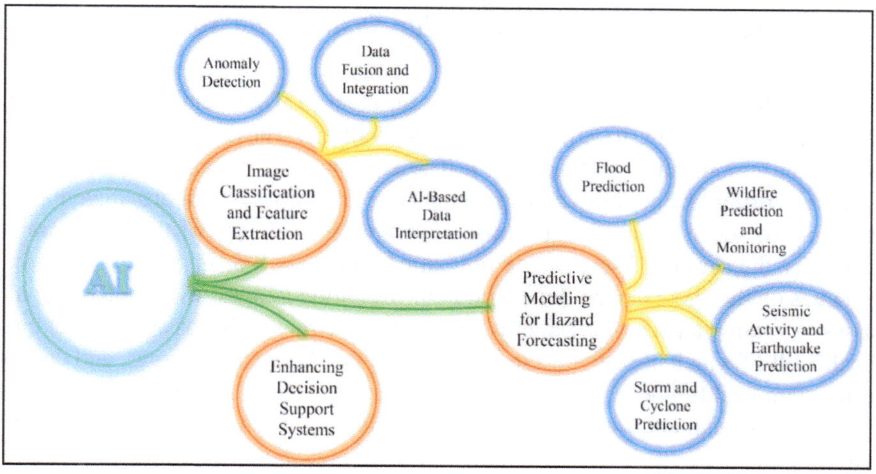

Fig. 22.5 Applications of α in hazard forecasting and disaster management

22.3.5.2 Predictive Modeling for Hazard Forecasting

AI models for predictive purposes are changing hazard forecasting by using the historical situation to glean important patterns, thus predicting floods, wildfires, earthquakes, and storms. For floods, the AI measures the meteorological forecast, river flow data, and past flood events to determine likely occurrences and magnitudes of future floods, thus enhancing preparedness and resource allocation with real-time updates. For wildfires, machine learning assesses the weather, vegetation health, and historical parameters for fire risk prediction, with this ability also allowing for real-time circulation. Although the AI can't make predictions of earthquakes with precision, it could see patterns in seismicity and precursory signals to assess possible risk and predict aftershocks. In this, AI-evaluated and cyclone-predicted weather forecasting becomes a better track by feeding on satellite images, ocean data, and atmospheric condition observations, resulting in improved trajectory forecasting and evacuation planning (Baklola and Terra 2024).

22.3.5.3 Enhancing Decision Support Systems

The AI predictive models could be used for producing risk maps to simulate any range of hazard scenarios and predict the impacts that could result from several considered mitigation strategies. For instance, an AI model could be used to simulate possible impacts of a tsunami on a coastal city, considering such factors as the height of buildings, population density within particular neighborhoods, and evacuation routes. Those predictive outcomes of such simulations will facilitate the design of effective disaster response and efficient use of resources deployed at critical times.

22.3.6 *Technologies Integration for Sustainable Solutions*

High-resolution could be that the imaging turns into LiDAR and is augmented by hyperspectral sensing, together with GIS technology. These lead to a more integrated hazard management approach. The combination of these state-of-the-art remote sensing technologies and spatial analysis-based tools will articulate hazard risk into the decision-making of those governing the safe use of those vulnerable areas for the means of sustainable mitigation strategies. For instance, monitoring changes on the shore can be done with high-resolution images, while LiDAR could assess elevation and erosion patterns, and hyperspectral sensing assesses vegetation health, all of which are key info in dealing with flooding, storm surges, and coastal ecosystem health, maybe the management of coastal areas. Another case could be wildfire management where high-resolution imaging might render near-real-time information.

22.3.7 *Challenges and Limitations*

Nevertheless, enormous transformations have taken place in hazard management and other geospatial realms due to the applications of GIS and remote sensing; yet some challenges and limitations still subsist. These challenges are somewhat barriers toward accessibility, cost-effectiveness, and thus utility for these particular technological applications. Some of the key challenges include:

22.3.7.1 Data Accessibility and Cost Issues

It is here in GIS and remote sensing that the challenges arise primarily due to the availability and accessibility of data. High-resolution satellite imagery, aerial survey, and LiDAR data are expensive and remain inaccessible sometimes due to privacy, security, or licensing concerns.

Cost Barriers: A lot of expense attaches to accessing geospatial data, especially in developing countries and for nonprofits. This is compounded by expenditure on software, hardware, and updates.

Data Accessibility: Due to dispersed sources of geospatial data, integration becomes a hefty issue. Coverage gaps impact effective monitoring and analysis.

Licensing and Proprietary Ownership: Proprietary datasets blocked from being legally accessed restrain research, hinder emergency response, and impair policy action planning.

22.3.7.2 Technical and Skill Barriers in Implementation

Although relatively easy to use, GIS and remote sensing require sophisticated skills. They have complex tool sets that require knowledge of programming, data handling, and spatial analysis, which stands as an obstacle for the novice user or small organizations. The lack of qualified skilled personnel presents barriers to successful implementation and decision-making, especially in developing countries. There should be continuous training, but the exclusion from the updated version of programs limits the effectiveness of geospatial applications.

22.3.7.3 Limitations in Predictive Accuracy

Various factors still limit the accuracy of hazard forecasting despite the improvement brought by GIS and remote sensing techniques.

Data uncertainty: Predictive models are highly dependent on input data that can be influenced by sensor constraints, atmospheric disturbances, or resolution limitations causing possible inaccuracies concerning the output data.

Complex hazard modeling: The forecasting of disasters like floods and wildfires needs to consider multiple dynamic variables. AI modeling improves accuracy, but it does so mainly based on historical data and may not treat extreme events.

Assumptions of the model: The simplifications introduced in a model create biases, for the real-world conditions could differ from those considered in the model, such as rainfall intensity or terrain characteristics.

Dynamic nature of hazards: The unpredictable behavior of some events such as earthquakes and hurricanes, is in itself a limitation for forecasting, especially for rare hazards that do not have sufficient historical data.

22.3.8 Future Directions of GIS

There arise promising opportunities in the future of GIS and remote sensing in the management of hazards, with new technologies emerging, harmonization into policy, and cross-border collaborations. Growing environmental challenges, security technologies have the potential to build more development in hazard management. Some of the most important future directions for GIS and remote sensing are as follows:

22.3.8.1 Potential of Emerging Technologies in Hazard Management

Emerging technologies stand to significantly revolutionize GIS and remote sensing for hazard management through increased monitoring, prediction, and response capabilities.

Artificial intelligence and machine learning now automate the very analysis of geospatial data and thereby enhance instantaneous forecasting and decision-making in hazard situations. AI systems enhance prediction capabilities on extreme weather events and events such as wildfires and floods through extensive processing of available data.

IoTs and sensor networks: IoT devices monitor environmental characteristics like temperature and seismic activity, relaying real-time data for use in early warning systems and proactive disaster management.

Drones and UAVs: Drones equipped with LiDAR and high-resolution cameras are used to acquire near-real-time data for flood mapping, damage assessment, and environmental monitoring in remote areas.

Blockchain for Data Security: With its ledger, blockchain secures the integrity of geospatial data, thus supporting disaster response by providing a verifiable, tamper-proof record.

22.3.8.2 Need for Policy Integration and Capacity Building

Effective use of GIS and remote sensing for hazard management requires integration into government policy and capacity building. Disaster policies must include these technologies to allow for open access to their data, inter-agency cooperation, and common standards for the collected data. Capacity building is critical at all levels, with investment in education and training to develop a core of professionals and empower communities toward disaster preparedness. Hazard management requires the integration of geospatial data with socio-economic and infrastructure data, working across disciplines with government agencies, emergency responders, and the private sector (Herold and Sawada 2012).

22.3.8.3 Role of International Collaboration in Environmental Hazard Solutions

In avoiding disasters, international collaboration is essential to enhance the GIS and remote sensing aspects. Geospatial data are shared across borders because of the enhancement of monitoring disasters and assessing risks through the UN and World Bank in favor of that concern. Cooperative research has spurred innovations in early warning systems, risk models, and disaster response that particularly benefit low-income countries. Coordinating help with geospatial technologies during the international response to disasters will increase aid distribution and recovery planning. GIS proves crucial in climate change mitigation for the monitoring of environmental changes, tracking emissions, and developing sustainable adaptation strategies.

22.4 Conclusion

Remote sensing and GIS have now become the indispensable instruments for data collection, analysis, and interpretation, resulting in the precise monitoring and judicious decision-making. GIS assembles and analyzes geospatial data, whereas remote sensing provides real-time, high-resolution images of various hazards, including flooding, wildfires, and urban sprawl. An integrated approach employing both GIS and remote sensing would prove useful to authorities and organizations in analyzing risk, planning disaster responses, and optimizing resource allocation toward improved hazard management.

A vital part in achieving sustained and resilient environmental and public health management is served by integrating remote sensing and GIS. These technologies allow an uninterrupted observation of environmental variables affecting public health, including air, water, and land. This, in turn, enhances resource management, climate change adaptation, and urban planning, with the aim of lessening the exposure of fragile groups to environmental hazards. Another aspect in which GIS and remote sensing support long-term sustainability and public health is functioning through triggering early warning systems and improving risk assessment capabilities by rendering disaster preparedness and mitigation strategies effective.

For full realization of the potential of GIS and remote sensing in the management of specific hazards and environmental situations, fostering technological development and cross-sectoral and international cooperation is of utmost necessity. This means that the governments, international organizations, private sectors, and academic institutions need to join forces to enhance data-sharing mechanisms, standardize approaches, and build capacities for technology and human expertise alike. The international cooperation aspect becomes very special in addressing global issues like climate change and transboundary environmental risks. These factors provide a stronger argument for maintaining a balance between technology and social cooperation as they will help empower community resilience and promote a hazmat management system so everybody can enjoy a brighter future.

Acknowledgements The authors would like to acknowledge Srinath University, Jamshedpur, for providing research and infrastructural support.

References

Al-Ismaili AM (2021) GIS and remote sensing techniques in controlled environment agriculture: a review. J Agric Mar Sci [JAMS] 26(2):10–23. https://doi.org/10.53541/jams.vol26iss2pp10-23

Alvarez-Vanhard E, Corpetti T, Houet T (2021) UAV & satellite synergies for optical remote sensing applications: a literature review. Sci Remote Sens 3:100019. https://doi.org/10.1016/j.srs.2021.100019

Ayele NA, Engda EM, Terefe TT, Leta E, Reda TM, Jothimani M (2025) Geospatial analysis of landslide susceptibility and safe relocation zones: insights from recent disasters in Gofa Zone, Sothern Ethiopia. Quat Sci Adv 17:100272. https://doi.org/10.1016/j.qsa.2025.100272

Badwi IM, Ellaithy HM, Youssef HE (2022) 3D-GIS parametric modelling for virtual urban simulation using CityEngine. Ann GIS 28(3):325–341. https://doi.org/10.1080/19475683.2022.2037019

Baklola M, Terra M (2024) AI-powered predictive analytics for crowd health management. Mass Gather Med 2:100009. https://doi.org/10.1016/j.mgmed.2024.100009

Baraldo M, Di Giuseppantonio Di Franco P (2024) Place-centred emerging technologies for disaster management: a scoping review. Int J Disaster Risk Reduct 112:104782. https://doi.org/10.1016/j.ijdrr.2024.104782

de Barros HR, Lombardo MA (2017) Heat Island in the city of São Paulo: the thermal field of parks. LAP LAMBERT Academic Publishing. https://www.amazon.com/Heat-Island-city-São-Paulo/dp/3330019883

Dritsas E, Trigka M (2025) Remote sensing and geospatial analysis in the big data era: a survey. Remote Sens 17(3):550. https://doi.org/10.3390/rs17030550

Enoguanbhor EC (2023) Assessing urban spatial patterns within the implemented urban planned areas using GIS and remote sensing data. Int J Multidiscip Perspect 04(01):87–96. https://doi.org/10.18801/ijmp.040123.14

Fast V, Rinner C (2014) A Systems perspective on volunteered geographic information. ISPRS Int J Geo Inf 3(4):1278–1292. https://doi.org/10.3390/ijgi3041278

Herold S, Sawada MC (2012) A review of geospatial information technology for natural disaster management in developing countries. Int J Appl Geospatial Res 3(2):24–62. https://doi.org/10.4018/jagr.2012040103

Jaishankar M, Tseten T, Anbalagan N, Mathew BB, Beeregowda KN (2014) Toxicity, mechanism and health effects of some heavy metals. Interdiscip Toxicol 7(2):60–72. https://doi.org/10.2478/intox-2014-0009

Jia J, Sun H, Jiang C, Karila K, Karjalainen M, Ahokas E et al (2021) Review on active and passive remote sensing techniques for road extraction. Remote Sensing 13(21):4235. https://doi.org/10.3390/rs13214235

Klijn F, de Bruijn KM, Knoop J, Kwadijk J (2012) Assessment of the Netherlands' flood risk management policy under global change. Ambio 41(2):180–192. https://doi.org/10.1007/s13280-011-0193-x

Leal Filho W, Ternova L, Parasnis SA, Kovaleva M, Nagy GJ (2022) Climate change and zoonoses: a review of concepts, definitions, and bibliometrics. Int J Environ Res Public Health 19(2):893. https://doi.org/10.3390/ijerph19020893

Mai G, Xie Y, Jia X, Lao N, Rao J, Zhu Q et al (2025) Towards the next generation of geospatial artificial intelligence. Int J Appl Earth Obs Geoinf 136:104368. https://doi.org/10.1016/j.jag.2025.104368

Matthew, A. (2015). Free mapping data will elevate flood risk knowledge. Environment Agency. https://environmentagency.blog.gov.uk/2015/06/16/free-mapping-data-will-elevate-flood-risk-knowledge/

Miri J, Atighechian G, Seyedin H, Raeisi AR (2024) Post-disaster damage and loss assessment in the Iranian healthcare sector: a qualitative interview study. BMC Public Health 24(1):2408. https://doi.org/10.1186/s12889-024-19877-w

NASA (2006) New York City temperature and vegetation. Landsat Image Gallery. https://landsat.visibleearth.nasa.gov/view.php?id=6800

NASA (2007) Floods in Jakarta. Earthobservatory. https://earthobservatory.nasa.gov/images/17981/floods-in-jakarta

NASA (2020) The Sundarbans. Earthdata. https://www.earthdata.nasa.gov/news/worldview-image-archive/sundarbans

NASA (2021) Heating up in Tokyo. Landsat Image Gallery. https://landsat.visibleearth.nasa.gov/view.php?id=148616

Olawade DB, Wada OZ, Ige AO, Egbewole BI, Olojo A, Oladapo BI (2024) Artificial intelligence in environmental monitoring: advancements, challenges, and future directions. Hygiene Environ Health Adv 12:100114. https://doi.org/10.1016/j.heha.2024.100114

Oppong J, Ning ZH, Twumasi Y, Antwi RA, Anokye M, Ahoma G et al (2023) The integration of remote sensing and geographic information system (GIS) in managing urban ecosystems. Int Arch Photogramm, Remote Sens Spat Inf Sci XLVIII-M-3:169–175. https://doi.org/10.5194/isprs-archives-XLVIII-M-3-2023-169-2023

Rykhus R, Lu Z (2007) Hurricane Katrina flooding and oil slicks mapped with satellite imagery. US Geol Surv Circular 1306:49–52

Shivanna KR (2022) Climate change and its impact on biodiversity and human welfare. Proc Indian Natl Sci Acad 88(2):160–171. https://doi.org/10.1007/s43538-022-00073-6

Tomaszewski B, Judex M, Szarzynski J, Radestock C, Wirkus L (2015) Geographic information systems for disaster response: a review. J Homeland Secur Emerg Manag 12(3). https://doi.org/10.1515/jhsem-2014-0082

Tong S, Bambrick H, Beggs PJ, Chen L, Hu Y, Ma W et al (2022) Current and future threats to human health in the Anthropocene. Environ Int 158:106892. https://doi.org/10.1016/j.envint.2021.106892

Wang C, Myint S, Wang Z, Song J (2016) Spatio-temporal modeling of the urban heat island in the phoenix metropolitan area: land use change implications. Remote Sens 8(3):185. https://doi.org/10.3390/rs8030185

World Bank Group (2024) Maldives country environmental analysis. World Bank Group. https://documents1.worldbank.org/curated/en/099021524234511634/pdf/P1784891f12d1a0cd18d5516f8fcf6ceaf4.pdf

Chapter 23
Role of Emerging Technologies in Promoting Sustainable Environment and Public Health

Jahangeer Ahmed Soomro, Muhammad Suhrab, and Basit Ali Abbasi

Abstract Emerging technologies are transforming how societies interact with the environment and developing innovative methods to deal with public health challenges. Emerging technologies like nanotechnology, biotechnology, renewable energy, artificial intelligence (AI), big data analytics, machine learning (ML) and remote sensing are shaping the strategies for environmental sustainability and public health protection. The dual role of emerging technologies as potential drivers of environmental sustainability and a powerful tool for promoting public health is examined in this chapter. It highlights how technologies are efficient in facilitating effective resource management, pollution reduction, climate change mitigation and the regeneration of damaged ecosystems. Using data of manufacturing firms in BRICS nations relying on emerging technologies for their operations, it is shown that IoT usage, big data analytics and robotics significantly cut CO_2 emission, while promoting energy and resource efficiency. Furthermore, chapter debates how technological developments benefit public health by strengthening early warning systems, making disease diagnosis easier and earlier, increasing water and air quality and reducing health inequalities. Through global case studies, chapter demonstrates technology-driven solutions which are successfully generating opportunities for innovative responses to environmental and public health issues. This includes the use of AI-powered climate modeling, biotechnology, nanotechnology for water purification, smart energy networks and digital gadgets to constantly monitor the environment and human health. The chapter also discusses possible risks and barriers to the

J. A. Soomro (✉) · B. A. Abbasi
Department of Business Administration, ILMA University Karachi, Karachi, Pakistan
e-mail: jahangeerahemd1@gmail.com

B. A. Abbasi
e-mail: basit.ali@ilmauniversity.edu.pk; basit.20635@zu.edu.pk

M. Suhrab
School of Management, Huazhong University of Science and Technology (HUST), Wuhan, China
e-mail: writetosohrab@gmail.com

B. A. Abbasi
Faculty of Engineering, Science, Technology & Management (ZUFESTM), Karachi, Pakistan

A. Roy et al. (eds.), *Environment and Public Health*, Sustainability Solutions,
https://doi.org/10.1007/978-3-031-99770-9_23

deployment of developing technologies. Finally, the chapter provides a forward-thinking look at how technology may be used to create a more resilient, egalitarian and sustainable future for the environment and public health.

Keywords Emerging technologies · Renewable energy · Sustainable environment · Public health · Digital gadgets

23.1 Introduction

Emerging technologies are playing a crucial role in addressing global issues related to environmental sustainability and public health. Numerous environmental concerns and health issues have developed as a result of the rapid pace of industrialization, increased urbanization and globalization. These problems include climate change, air and water pollution, deforestation and the spread of infectious diseases. Emerging technologies offer innovative solutions that promote sustainability, reduce carbon footprints and enhance human health (Sharma et al. 2024). The integration of modern technologies such as artificial intelligence, Internet of Things, block chain, biotechnology and nanotechnology has transformed the ways in which societies perceive and handle environmental and public health issues (Ghosh et al. 2023). These technologies significantly promote resource efficiency, reduce pollution, disease prevention, early diagnosis of fatal diseases and raise sustainable lifestyles.

The ability of emerging technologies to contribute to carbon footprint mitigation and the promotion of environment-friendly solutions are the most important aspects of environmental sustainability, which can be achieved via integration of advanced technology in industries (Hussain et al. 2022).

AI-powered predictive analytics helped to monitor the air and water quality, optimize energy consumption and significantly reduce industrial waste. IoT-based smart grids and smart cities utilize real-time data to improve the urban planning system, optimized the use of energy, increase transportation efficiency and reduce CO_2 emission (Kumar et al. 2023). Additionally, blockchain technology has introduced transparent and secure solutions for tracking carbon emissions. All these solutions ensure that climate policies are implemented in a fair and effective manner (Arshad et al. 2023). These innovations play a crucial role in advancing the Sustainable Development Goals (SDGs) of the United Nations by addressing environmental issues and fostering a circular economy.

The use of renewable energy technologies such as solar, wind and biomass has reduced dependence on fossil fuels and helped to decrease greenhouse gases emission which ultimately promotes a cleaner environment. In the same manner, modern technologies such as IoT and AI regulate energy usage, water management and proper garbage disposal. This contributes to more sustainable and environmentally friendly infrastructure (Bibri et al. 2023). Furthermore, developments in biotechnology have made it possible to synthesize natural products, recyclable plastics, sustainable methods of cultivation and corporate farming. All these have resulted in a reduction in the amount of destruction caused to the environment and promote a healthier ecosystem.

In the public health domain, emerging technologies have transformed the landscape and brought revolution in disease monitoring, early diagnosis and treatment. Telemedicine and wearable health monitoring gadgets have increased accessibility to health care, particularly in underserved and remote areas. Predictive analytics powered by artificial intelligence helped in the early diagnosis of diseases and forecasting of severe outbreaks. Screening methods powered by artificial intelligence have been shown to be very useful in the early detection of diseases such as cancer and COVID-19 and have resulted in more timely treatments and improved patient outcomes (Boddu et al. 2022). Additionally, blockchain technology improved the transparency and security of medical data management. This ensures healthcare services and medical research are carried out in an efficient manner.

The inclusion of technological innovations in health care is a fundamental element of the forthcoming generation of medical systems. Firstly, it clarified the emerging and streamlined technologies and revealed the value of their components. Secondly, it emphasizes a new policy-driven era of health governance as the integration of innovation in health care must be comprehended from the perspective of key stakeholders and implemented while considering various constraints. Finally, innovation in health care should be regarded as a human-centric process that employs complex, sophisticated and distributed medical services and processes (Subbiah 2023).

The adoption of sophisticated healthcare information systems and medical informatics needs a cohesive strategy that considers social, economic, political and cultural aspects. The problems posed by the adoption and use of advanced information and communication technologies (ICTs) must also be taken into account to increase the efficiency of such technological innovations in the healthcare domain. Smart data, data analytics and cognitive computing are promising technologies that provide significant value to the public healthcare sector (Wager et al. 2021).

It is also important to underline the function of nanotechnology in ensuring the sustainability of the environment and protecting public health. The use of nanomaterials in developing better filtration systems for the purification of water, enhancing air quality and treatment of harmful waste is becoming more common nowadays (Joseph et al. 2023). Successfully removing contaminants at the molecular level, these technologies help in the creation of cleaner habitats. Moreover, nanomedicine has opened new horizons in targeted drug provision, regenerative medicine and early disease detection. This has significantly improved public health (Rasool et al. 2022).

The use of big data analytics and cloud computing is another essential technological advancement contributing to the concept of sustainability and public health. According to Karatas et al. (2022), these technologies enable institutions to handle enormous volumes of data related to health and the environment to make informed decisions and remain proactive. Big data analytics successfully analyze climate change trends, monitor deforestation and measure the loss of biodiversity. These analytics have enabled policymakers to remain proactive and to timely cope with any emergency situation. Similarly, cloud computing facilitates increased cooperation between researchers and institutions all over the globe to speed up the acquisition of scientific knowledge in the disciplines of medicine and the environment (Yaqoob et al. 2022).

This chapter aims to investigate the implications of emerging technologies to promote environmental sustainability and the enhancement of public health. Through an in-depth analysis of technological progress across a variety of industries, it sheds light on the transformative potential of digital innovations in promoting environmental sustainability and a healthier future.

23.1.1 Literature Review

Emerging technologies have made significant contributions to the protection of the environment and the enhancement of public health. Emerging technologies like artificial intelligence, Internet of Things, blockchain and biotechnology are significant in fighting against climate change, preservation of resources, reduction of pollution and promoting the healthcare sector (Li et al. 2024). In the literature review, we have highlighted the current research and advancements in various technical domains with a particular focus on public health and environmental sustainability.

23.1.1.1 Role of Artificial Intelligence and Machine Learning in Sustainability and Public Health

Artificial intelligence and machine learning are very important in the field of public health diagnostics and environmental monitoring. AI-driven models are capable of precisely forecasting climate by analyzing huge data related to weather patterns, carbon emissions and ecosystem change. Machine learning algorithms are also efficient in optimizing energy consumption by forecasting demand and decreasing waste (Ruan et al. 2024). Traditionally, public health surveillance relied on manual data collection and processing, which was time-consuming and prone to mistakes. AI can change this process by automating data analysis, rapidly identifying possible epidemics and delivering timely alerts. AI also help monitor trends in risk factors for non-communicable illnesses by evaluating demographic, behavioral and environmental data and incorporating it into planning estimates. AI's capacity to handle massive amounts of data quickly can help to accelerate the flow of information (Fisher and Rosella 2022). AI can be especially beneficial in behavioral epidemiology, which uses data from mobile applications and social media to track health behaviors including food, physical activity and mobility. AI can also assess the effectiveness of treatments aimed at changing these behaviors and predict the trade-offs involved (Dignum et al. 2020).

AI is becoming increasingly significant in public health communication because it allows communications to be tailored to specific demographics. AI techniques may categorize people based on demographic and behavioral data, increasing the possibility that health messages are culturally relevant and accessible. AI can also help create public health communications in numerous languages and literacy levels, as well as spot disinformation (Lu et al. 2024). Figure 23.1 shows the application of AI to promote public health. Machine learning algorithms analyze epidemiological data in order to anticipate the development of infectious diseases and facilitate immediate treatments. Figure 23.2 highlights the pillars of machine learning in health care. According to Ranard et al. (2024), chatbots and virtual assistants that are based on artificial intelligence make it easier to access healthcare opportunities, especially in rural areas.

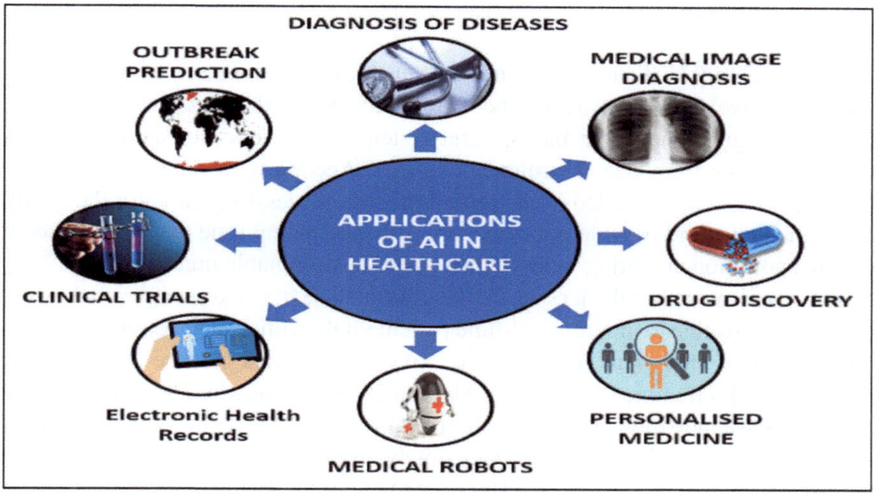

Fig. 23.1 Applications of AI in promoting health care (Pandya et al. 2021)

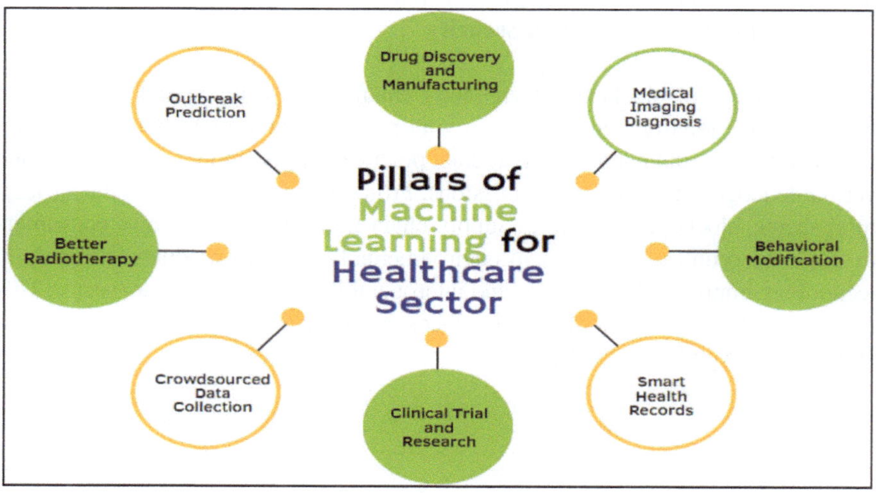

Fig. 23.2 Pillars of ML in health care (Javaid et al. 2022)

Children Health Information (CHI) pertains to the provision of healthcare services for children, including educating their guardians about emotional, behavioral and mental health concerns. This service may also be used to teach children and equip them to manage challenges more effectively. Implementing mobile H-IoT may facilitate the dissemination of essential nutritional practices in an interactive format to children, their educators and parents (Al Khatib et al. 2024).

23.1.1.2 Internet of Things (IoT) for Environmental and Health Monitoring

Internet of Things (IoT) integration into sustainable environmental practices has been comprehensively viewed by researchers. According to Singh and Ahmed (2021), IoT-based smart grids and water management systems optimize resource utilization and minimize waste to reduce environmental issues. According to Zhang et al. (2021), IoT sensors that are installed in metropolitan areas successfully monitor the quality of air and water and provide policymakers with data in real time that can be used to manage pollution and bring more improvement in sustainable practices.

IoT-based wearable devices in the healthcare industry, like smart watches and biosensors, are able to monitor real-time patient vitals. Timely and continuous monitoring of these vital signs like heart rate, oxygen level and glucose levels enables proactive health treatments (Joshi and Sharma 2023). A study by Zhang et al. (2021) examines the application of IoT in monitoring air quality. This contributes to reducing respiratory diseases. This also provides timely alerts for prompt response to reduce the risk of any major accident or event related to public health.

23.1.1.3 Blockchain Technology for Environmental Sustainability and Healthcare Data Security

The application of blockchain technology enables transparency and efficiency in health and the environment. It plays a vital role in carbon credit trading to easily trace and reduce fraudulent environmental practices by providing an immutable and verifiable ledger for monitoring emissions. Promoting sustainability in the agricultural sector, blockchain promotes supply chain transparency. It allows consumers to track the origin of organic and environmentally friendly products. Blockchain technology improves data security and interoperability in the healthcare sector. This promotes patient confidentiality and reduces the chances of error in medical reports and records (Agbo et al. 2019).

23.1.1.4 Biotechnology and Sustainable Public Health Care

Advances in biotechnology have improved the public healthcare sector. In this sector, biopharmaceuticals and gene editing procedures like CRISPR provide targeted treatments for genetic disorders and cancers. This simultaneously reduces the dependence on traditional therapies, which have significant environmental footprints. Biotechnology's emerging fields, such as regenerative medicine and tissue engineering, have reduced the need for organ transplantation and extensive medical interventions. This promotes the sustainability in public health care (Mason and Dunnill 2008).

23.1.1.5 Renewable Energy Technologies and Public Health Benefits

Transformation toward renewable energy sources like wind, solar and hydroelectric power mitigates environmental issues such as air pollution and water contamination that are prominent and a basic cause of public health issues. Several studies suggest that replacing fossil fuel with renewable energy systems is helpful in reducing greenhouse gases, which in turn reduces the prevalence of respiratory diseases caused by air pollution (Pata 2021). Furthermore, decentralized energy systems that are powered by renewable energy improve electricity availability in remote areas, boosting healthcare delivery by providing a stable electrical supply to medical institutions. Energy-efficient infrastructure and buildings with smart technology integration are also helpful to achieve sustainability by lowering carbon footprints and increasing indoor air quality, which results in improved overall public health outcomes (Zhang et al. 2023).

23.1.1.6 Physical Activities Assessment for Promoting Public Health

Technology is constantly transforming our living conditions and ways of life. Approximately twenty years ago, physical activity (PA) questionnaires were replaced by health wearables (e.g., pedometers) that quantified physical activity and monitored human movement patterns for analysis. Specifically, the most prevalent electronic gadgets, including pedometers, accelerometers and heart rate monitors, provide data on physical activity metrics, encompassing steps, estimated distance, energy expenditure, duration at varying intensity degrees of physical activity and heart rate (Vilchez et al. 2021). In addition to their assessment capabilities, these gadgets significantly contribute to providing feedback to customers, which may serve to urge the public to engage in increased physical activity to promote their health (Idris et al. 2021).

Moreover, emerging technologies have contributed to the development of various sensor systems, allowing improved recognition of previously unidentifiable physical activities and healthcare-related issues. Currently, the majority of monitors (e.g., Actigraph), gadgets (e.g., Fitbit Versa Smart Watch) and applications may transmit data to an individual's account and relevant communities via distant servers (Vilchez et al. 2021). This encourages the increased use of online technologies and physical activity promotion platforms, which enable people to create public contracts visible to other users and leverage financial incentives to foster behavioral changes in physical activity to promote healthy living.

23.1.2 Problem of the Statement

There has been significant advancement in health care and environmental science, but still the world continues to face critical challenges related to climate change, excessive pollution, infectious disease epidemics and the lack of access to quality healthcare facilities. Emerging technologies such as AI, IoT, biotechnology, blockchain, big data analytics and renewable energy systems have immense potential to transform the way to monitor, manage and improve both environmental sustainability and public health outcomes. However, the integration of these emerging technologies in both domains remains fragmented, with significant differences in urban and rural areas, requiring specific infrastructure and a high initial cost. This creates the need for more investigation to check how efficiently these technologies can be used to combat the crises of environmental degradation and public health concerns, particularly in remote areas.

23.1.3 Objectives of Chapter

A. To testify how significantly emerging technologies contribute to environmental sustainability. b. To testify how emerging technologies transform the landscape of the public health domain.

23.2 Methodology

Firstly, this chapter uses quantitative data analysis to thoroughly investigate the impact of emerging technologies in achieving environmental sustainability in the industrial sector of BRICS countries. The chapter is intended to look at secondary data gathered from a variety of credible sources. This strategy ensures a solid and thorough understanding of how emerging technologies are used and their major influence on environmental sustainability in developing countries.

Secondly, this chapter uses a qualitative approach and is based on the discussion of essential variables encouraging sustainability in the environment and public health sector, as well as an analysis of many global case studies to highlight the impact of emerging technologies on environmental sustainability and public health. This chapter also discusses future prospects and challenges relating to a green, sustainable environment and public health.

23.2.1 Data Collection

The aim of the chapter is to examine via secondary data, which was acquired from reputable sources such as the International Energy Agency (IEA), World Economic Forum, United Nations Industrial Development Organization (UNIDO), World Bank and BRICS national statistics. This data contains measurements on energy use, greenhouse gas emissions, water usage, raw material consumption and the rate at which emerging technologies (IoT, AI, big data analytics and advanced robotics) are adopted.

23.2.2 Econometric Model

To examine the impact of emerging technologies on environmental sustainability, a panel data regression model was employed. This model considers both cross-sectional and time-series changes, resulting in a comprehensive analytical framework. The core model is panel data regression with fixed and random variables to capture country-specific and time-dependent variability. The model takes the following general form:

$$\mathrm{CES}_{i,t} = \alpha + \beta_1(\mathrm{IOT})_{i,t} + \beta_2(\mathrm{BigData})_{i,t} + \beta_3(\mathrm{Robotics})_{i,t} + \mu_i + \lambda_t + \varepsilon_{i,t}, \tag{23.1}$$

whereas

$\mathrm{CES}_{i,t}$	environmental sustainability for country i in year t
α	intercept
β_1 to β_3	coefficients of independent variables
IoT	Internet of Things
μ_i	Country-specific fixed effects
λ_t	time-specific fixed effects
ε	Error term.

23.3 Results

23.3.1 Descriptive Statistics

Table 23.1 displays the descriptive statistics of variables, which demonstrate the significance of key performance metrics and technological adoption among organizations.

CO_2 emissions had a mean of 1246.652 metric tons and a significant standard deviation of 321.415, demonstrating that carbon production varies between firms.

Table 23.1 Descriptive statistics

Variable	Mean	Standard div	Minimum	Maximum
CO_2 emissions (metric tons)	1246.652	321.415	781.130	1985.328
Energy efficiency (kWh/ton)	95.251	20.364	70.842	130.694
Resource efficiency (Ratio)	1.151	0.339	0.791	1.898
IoT usage (%)	45.715	16.251	20.352	69.81
Big data analytics (%)	41.854	14.695	18.047	65.110
Advance robotics (%)	30.745	11.251	10.857	50.918

This vast range, from 781.130 to 1985.328 metric tons, emphasizes the disparities in environmental consequences among firms, which are presumably impacted by their operating procedures and industrial sectors. Similarly, the mean energy efficiency is 95.251 kWh/ton, with a standard deviation of 20.364, and values ranging from 70.842 to 130.694. This variety indicates that, while some organizations achieve great energy efficiency, there is still much opportunity for improvement, reflecting disparities in how firms manage their energy consumption. IoT adoption averages 45.715% with a standard deviation of 16.251%, ranging from 20.352 to 69.81%. This suggests a considerable disparity in how widely firms utilize IoT technologies. Big data analytics has a mean utilization of 41.854% and a standard deviation of 14.695%, with a range of 18.047 to 65.110%, indicating varying integration levels among firms. Advance robotics adoption averages 30.745%, with a standard deviation of 11.25, 1% and a range from 10.857 to 50.918%, demonstrating that while robotics adoption is less common than IoT and big data, there is still substantial variability.

23.3.2 *Impact of Emerging Technologies on Environmental Performance*

Table 23.2 displays the regression findings for each of the dependent variables (related to environmental sustainability), emphasizing the coefficient values, degrees of significance and strength of the relationships between the independent factors and firm-level environmental outcomes in the manufacturing sector.

The regression results found that emerging technologies significantly impact the environmental performance and fully support environmental sustainability. IoT significantly reduces the carbon emission by 0.542 units at a 99% confidence interval, while it significantly promotes energy and resource efficiency by 0.856 units and 0.786 units, respectively, at a 95% confidence interval. Similarly, big data analytics significantly decreases CO_2 emission by 0.753 units at a 99% confidence interval, significantly promotes energy efficiency by 0.897 units at a 90% confidence interval and also significantly enhances resource efficiency by 0.753 units at a 95% confidence interval. Finally, robotics significantly reduces CO_2 emissions by 0.698 units at a 95%

Table 23.2 Emerging technologies and environmental performance

	CO_2 emissions	Energy efficiency	Resource efficiency
IoT usage	– 0.542***	0.856**	0.786**
	(0.082)	(0.291)	(0.303)
Big data analytics	– 0.753***	0.897*	0.753**
	(0.101)	(0.321)	(0.358)
Robotics	– 0.698**	0.958**	0.853**
	(0.317)	(0.436)	(0.371)
Constant	220.521	95.201	75.492
	(12.441)	(7.871)	(8.991)
R-squared	0.650	0.625	0.648
Country	Yes	Yes	Yes
Firm	Yes	Yes	Yes
Countries	5	5	5
Number of observations	500	500	500

Statistical significance is indicated by three levels: 1% (***), 5% (**) and 10% (*)

confidence interval. Similarly, robotics significantly boosts energy and resource efficiency by 0.958 and 0.853 units, respectively, at a 95% confidence interval. Evidence from previous studies supports our finding as big data enables real-time monitoring and predictive maintenance, which helps in reducing energy waste and emissions.

23.3.3 *Analyzing Global Case Studies*

Emerging technologies play a significant role in environmental sustainability and promoting public health. AI, IoT, big data, machine learning, biotechnology and renewable energy are a few examples of technological innovations that have led to significant advancements in the areas of environmental conservation and healthcare delivery (Ruan et al. 2024). The purpose of this part of the chapter is to investigate a number of global case studies that highlight the ways in which these technologies contribute to public health and environmental sustainability.

23.3.3.1 AI-Powered Waste Management System

A significant application of AI in the field of waste management is to improve recycling and reduce contamination. TrashBot, an AI-driven waste sorting robot created by CleanRobotics noteworthy example. TrashBot is able to autonomously separate recyclable and non-recyclable waste via the use of computer vision and machine learning. This provides considerable improvement in garbage management. With the deployment of TrashBot a pilot study was carried out in Pittsburgh, United States of America, which resulted in a 50% increase in the correct sorting of garbage, reducing landfill waste and promoting environmental sustainability (Fadir et al. 2024).

23.3.3.2 IoT-Based Smart Water Management System

Singapore has implemented IoT-based smart water management system to alleviate its water shortage challenges. The Public Utilities Board (PUB) has smart water sensors to detect leaks, manage water consumption and maximize the efficiency of water distribution. This method has significantly reduced 15% of non-revenue water loss and has improved the water conservation efforts. The Public Utility Board (PUB) has improved water sustainability by using AI-driven predictive analytics, while also guaranteeing that the public has access to safe drinking water (Okoli and Kabaso 2024).

23.3.3.3 Renewable Energy for Sustainable Health Care

There were a number of hospitals in remote areas of Africa affected by the irregular supply of power, which has an impact on the delivery of proper delivery of health care and treatment services to the public. To cope with this problem, under the We-care Solar Project, solar-powered solar suitcases were deployed in remote areas to deliver uninterrupted healthcare services. This project has successfully reduced the maternal and neonatal mortality rates in Nigeria, Uganda and Zimbabwe due to the continuous supply of electric power to carry out operations and emergency procedures. It has been shown via research that hospitals that use solar energy have a sixty percent decrease in the rates of patient death as a result of enhanced availability to emergency medical services (Soto et al. 2022).

23.3.3.4 Carbon Capture and Storage

Carbon Capture and Storage (CCS) technology has been one of Norway's pioneering efforts to minimize CO_2 emissions from industrial operations. The Sleipner Carbon Capture and Storage project has been active since 1996 and is considered to be one of the most successful environmental initiatives. CO_2 emissions that are produced as a result of the processing of natural gas are collected and stored at the bottom of the North Sea. This prevents the emissions from entering the atmosphere. As per records by this project, more than 20 million tons of CO_2 have been successfully stored as of 2021 (Ma et al. 2022). This initiative has set a worldwide example for other countries that are interested in implementing large-scale carbon absorption technologies in order to reduce the effects of climate change and promote sustainability.

23.3.3.5 Telemedicine Expansion in Rural Africa

Telemedicine has revolutionized the accessibility of healthcare facilities in remote areas of Africa, especially in Rwanda and Ghana, where medical facilities were limited. Babylon Health AI system, based on machine learning (ML) and natural language processing (NLP), provided personalized diagnostics and virtual consultations in remote areas. It is a mobile-based telehealth platform (Omaghomi et al. 2024).

Through the use of diagnostic tools that are powered by AI, the system reduced needless hospital visits by 40% while simultaneously assuring prompt medical intervention for chronic diseases such as diabetes and hypertension (Dodoo et al. 2022). Through the elimination of obstacles that prevent people from receiving medical treatment in remote areas, the rapid growth of telemedicine services has resulted in a major improvement in health outcomes in distant areas.

23.3.3.6 AI-Based Predictive Analytics for Epidemic Prevention

AI-based predictive analytics have been an essential component in the process of predicting and responding to epidemics. The Canadian business BlueDot, during the COVID-19 pandemic, used AI to identify the outbreak days before official reports were made public. By examining global travel patterns, temperature data and social media reports, AI-driven models were able to anticipate infection hotspots and aid public health authorities in preparing control measures (Zeng et al. 2021). The same technique can be deployed for disease monitoring in case of malaria, dengue and Zika virus outbreaks and boost worldwide epidemic preparation.

23.3.3.7 Blockchain Technology for Pharmaceutical Supply Chain Transparency

The adoption of blockchain technology in the pharmaceutical industry has significantly increased drug traceability and decreased the ratio of counterfeit medications. Blockchain-based MediLedger Project implemented in the United States enables real-time tracking of pharmaceuticals from the point of production to the point of distribution. The system has successfully reduced the circulation of duplicate medicines. It was estimated that 10% of the total medicine sold in developing countries was duplicates (Panda and Satapathy 2021). Blockchain ensures transparency of data, enhances patient safety and promotes regulatory compliance in the pharmaceutical sector.

23.3.3.8 AI-Based Energy Efficiency in Building Management

Artificial Intelligence is being extensively used to improve energy efficiency in building operations, specifically heating, ventilation and air conditioning (HVAC) systems. BrainBox based on AI at 45 Broadway in Manhattan is a noteworthy example of integrating emerging technologies in sustainable environmental practices. This system optimizes the heating, ventilation and air conditioning (HVAC) systems to factors such as weather conditions and occupancy levels. As a result, this system has reduced 15.8% of energy, saved $42,000 yearly and reduced about 37 metric tons of carbon emissions. This system shows the ability of AI to modernize old HVAC systems and significantly contribute to environmental sustainability (Xiang et al. 2022).

23.3.3.9 Environmentally Friendly Quantum Computing

Recent advancements in quantum computing have geared up toward ensuring the preservation of the natural environment. Ephos is Italy-based startup that is working on the development of photonic quantum processors that are manufactured from glass rather than silicon. These glass chips used light to transport data, which eliminated the requirement for cryogenic cooling. The cryogenic cooling was required by traditional superconducting quantum systems, which involved a significant amount of energy (Capellaro et al. 2025). The development resulted in energy-efficient quantum computers with the potential to be used in a variety of fields like finance, chemical engineering and transportation optimization. The development of such technologies highlights the significance of environmentally responsible methods in the process of advancing computer capabilities (Das et al. 2024).

Digital Health Training for Healthcare Professionals

Healthcare personnel are required to acquire sophisticated digital skills in order to accommodate the incorporation of digital health technology efficiently into healthcare systems (Borges do Nascimento et al. 2023). The Sustainable Healthcare with Digital Health Data Competence (Susa) initiative is supported by the Digital Europe program of the European Union. The program aimed to improve the digital skills of healthcare professionals, such as physicians and nurses. Within the scope of this project, twelve European universities and a number of technology businesses are working together to build bachelor's and master's degree programs in addition to teaching modules for lifelong learning. The project intends to develop more than 6500 graduates and 660 professionals who are fluent in digital technology at the mid-career level by incorporating artificial intelligence and digital interactions into the healthcare industry (Fernández-Luque et al. 2021). This project focuses on the need for health professionals to be engaged in the development of technology from the very beginning of their careers. This can revolutionize patient care and achieve greater efficiency within the healthcare system.

23.4 Future Perspectives of Emerging Technologies in Promoting Sustainable Environment and Public Health

Emerging technologies have the potential to substantially enhance environmental sustainability and public health in the coming years. The integration is expected to bring about further advancements and pave the way toward sustainability. The use of emerging technologies like artificial intelligence, nanotechnology, synthetic biology, digital health training and renewable energy offers feasible solutions to the contemporary issues that are present in these sectors.

23.4.1 Artificial Intelligence (AI) and Big Data Analytics

AI and big data analytics are causing a paradigm shift in the way we tackle current problems pertaining to the environment and health care. These technologies allow predictive modeling for disease outbreaks and environmental dangers by analyzing massive databases accurately and quickly, which in turn makes it possible to respond proactively to potential threats. Models powered by AI are able to timely predict the spread of infectious diseases which enables prompt reactions and device polices accordingly (Karatas et al. 2022). The use of artificial intelligence may also further optimize energy use in smart grids, which can significantly contribute to a reduction in carbon emissions and will get rid of climate issues (Kumar et al. 2023).

23.4.2 Green Nanotechnology

Green nanotechnology is concerned with the development of sustainable nanomaterials and processes that reduce environmental harm while contributing to solutions for climate change. The approach has the potential to mitigate the carbon footprint and promote sustainability. Nanomaterials are being used in the process of water purification, where they are eliminating toxins in a more effective manner than conventional methods. Furthermore, green nanotechnology places an emphasis on the utilization of renewable inputs and processes that are carried out at low temperatures, which supports the concepts of green chemistry and engineering (Pathak et al. 2025).

23.4.3 Synthetic Biology

Synthetic biology development has made it possible to produce microorganisms to carry out certain activities that are advantageous to the health of the general population and the environment. Engineered microorganisms have already been used for bioremediation. It is the process of successfully eliminating contaminants from surfaces like soil, water and air (Brooks and Alper 2021). Furthermore, certain complicated natural products are possible to be made in a sustainable manner by synthetic biology, like artemisinin, which is a medicine for malaria treatment (Beyraghdar Kashkooli et al. 2022). This cuts down on the need for conventional extraction processes, which may be harmful to the environment.

23.4.4 Developments in Renewable Energy Systems

The development of renewable energy technologies such as solar and wind power is absolutely necessary in order to cut down emissions of greenhouse gases and advance the cause of environmental sustainability. Nowadays, renewable energy is becoming more affordable and dependable as a result of the development of solar cells and efficient energy storage like Lithium-Ion Batteries (Pata 2021). This shift toward clean energy sources not only helps to reduce climate change but also has the potential to improve public health by lowering the levels of air pollution and water contamination that are caused by the combustion of fossil fuels and chemical industries' processes.

23.4.5 Self-Healing and Biodegradable Materials for Sustainable Infrastructure

Self-healing and biodegradable materials are the forms of future sustainable infrastructure, which will result in a reduction in the effect that vital structures have on the environment and an extension of lifespan. The autonomous sealing of fractures and the prevention of structural deterioration are the goals of the development of self-repairing concrete. This self-repairing concrete is implanted with bacteria that create limestone when exposed to water and oxygen. The maintenance cost and carbon footprint associated with traditional repair methods are greatly reduced by this invention (Nair et al. 2022).

Additionally, biodegradable polymers are becoming more popular as environmentally friendly replacements to conventional plastics, notably in the fields of building materials and household packaging. Waste from landfills and pollution in water will both be further reduced as a result of the natural decomposition of these items in the ecosystem. Further, the use of biodegradable composites in the construction of building insulation and coatings is boosting energy efficiency while simultaneously reducing the amount of harmful emissions (Samir et al. 2022). Further research on materials will help in the construction of infrastructure systems that are more robust and ecologically friendly. This will accord with the concepts of a circular economy and the aims of sustainable urban development.

23.4.6 Neurotechnology for Mental Health and Public Health Interventions

The use of neurotechnology is expected to bring about a revolution in the field of mental health care as well as in public health by providing novel approaches in the process of diagnosis, monitoring and treatment of neurological and psychiatric diseases. Brain computer interfaces (BCIs) are making remarkable advancements and have the potential to enhance cognitive function in patients suffering from neurodegenerative illnesses and to restore communication in paralyzed patients (Awuah et al. 2024). These technologies decode brain impulses and convert them into digital instructions and opening new opportunities in therapeutic treatments.

AI-powered mental health monitoring systems are being incorporated into wearable devices to detect brain activity, heart rate variability and sleep patterns to analyze real-time data and provide insight related to emotional well-being. This can offer early diagnosis of mental health diseases and will allow for prompt medications and tailored treatment strategies. Further advancement in neurotechnology will play an increasingly important role in bridging the gap between neuroscience and digital health, providing scalable solutions to global mental health issues and improving general public well-being (Hain et al. 2023).

23.5 Potential Barriers

Emerging technologies hold great potential for improving environmental sustainability and public health, but their implementation still faces several hurdles. The reasons for this vary from technological limits to ethical, economic, social, institutional quality and regulatory considerations.

23.5.1 Technological Complexity and Infrastructure Gaps

Integration of modern technologies like blockchain for carbon trading or digital health systems requires a strong digital infrastructure and technical expertise. Low-income nations are reluctant to develop such infrastructure required to properly apply these technologies, such as high-speed internet access, data storage capacities and technological smart workers (Usman et al. 2024). Without enough technology assistance, use of these technologies remains low and creating a gap in global sustainability and healthcare outcomes.

23.5.2 Data Privacy and Cybersecurity Risks

Data privacy and cybersecurity issues are increasing as digital health technologies like IoT enabled environmental monitoring and AI-driven decision-making are becoming more common. Sensitive health and environmental data stored on cloud platforms are frequently at risk of cyber attacks and unauthorized access, which pose a significant risk to individuals and organizations (Giansanti 2021). Ensuring robust data protection measures and addressing regulatory challenges related to data sharing and cross-border collaborations remain critical hurdles in this regard.

23.5.3 Resistance to Change and Ethical Concerns

It is observed that stakeholders oppose the adoption of such emerging technologies due to concerns about job displacement, uncertainty about AI-powered decision-making and ethical dilemmas around biotechnology and genetic engineering. AI-powered automation in public health systems may raise worries about the less human supervision in clinical decision-making, which contributes to human lack of dissatisfaction among humans. On the other hand, use of synthetic biology in environmental applications like bioengineered bacteria for pollution control may raise public concerns about possible ecological dangers (Brooks and Alper 2021). Addressing these ethical and cultural challenges is critical to building public trust and acceptance.

23.5.4 Regulatory and Policy Constraints

Regulatory frameworks related to these emerging technologies in sustainability and health care are sometimes slow to develop and implement, which generates uncertainty for firms and researchers (Mennella et al. 2024). Policies regulating AI, biotechnology and digital health systems differ from nation to nation, which results in inconsistent adoption and compliance requirements around the globe (World Health Organization 2023). Further, the absence of standardized norms and values for environmental and health data handling might hinder the integration of AI-powered prediction models and IoT-based monitoring systems. Maintaining harmonization in global regulatory affairs and encouraging international partnerships seem critical to resolving these issues.

23.5.5 *Institutional Quality and Governance Challenges*

Weak governance framework and institutional inefficiencies seriously pose hurdles to the adoption of innovative technologies. In many nations, bureaucratic red tape, corruption and a lack of sincerity in decision-making hinder the adoption of innovative solutions related to the environment and health care (World Health Organization 2023). Poor coordination and lack of collaboration among government institutions, commercial businesses and research institutions resulted in fragmented efforts and hindered the integration of technology into public policy and sustainability measures (Rossoni et al. 2024). Improvements in institutional quality via effective governance, policy consistency, major stakeholders' collaboration and cooperation are necessary to solve these issues.

23.5.6 *Social and Cultural Barriers*

Societal attitudes, cultural views and technical knowledge all have a combined impact on emerging technology acceptance (Emon and Khan 2023). Many communities are unaware of the advantages of digital health tools and sustainable technology, which causes problems in fully deploy these technologies. Furthermore, cultural norms influence the adoption of technologies like telemedicine, particularly in regions where in-person consultations are strongly preferred. These obstacles can be reduced by focused awareness and inclusive policymaking methods to increase trust and acceptability.

23.6 Discussion

Emerging technologies are becoming more important in achieving environmental sustainability and better public health outcomes. Artificial intelligence, the Internet of Things, blockchain, biotechnology and digital health technologies are bringing a revolution in environmental and health issues. AI-based models and big data analytics have improved environmental monitoring by precisely analyzing climatic trends, optimizing energy use and effectively controlling waste (Dhayal et al. 2023). Similarly, IoT-based smart grids and sensors have enhanced resource management, real-time monitoring of air and water quality, lowering industrial dangerous emissions and supporting circular economy practices.

Digital health innovations like telemedicine, wearable devices and AI-powered diagnosis have improved healthcare availability, specifically in distant and underprivileged areas. These advancements are critical in addressing public health concerns such as pandemics and chronic disease epidemics by allowing for real-time data collection and predictive analytics. Moreover, blockchain technology has improved the openness and efficiency of supply chains for sustainable agricultural systems and medicinal distribution. This has also lowered the risk of duplicate medicines and assures fair resource allocation (Ashraf et al. 2021).

Despite these benefits, various difficulties are associated with the development and implementation of these emerging technologies in the environment and healthcare sector (Borges do Nascimento et al. 2023). High initial investment, improper infrastructure, legal issues, governance, institutional quality and religious and cultural norms are key challenges (Usman et al. 2024). These challenges even vary from nation to nation. Without dealing with these challenges, desired efficiency cannot be obtained by using emerging technologies in the environmental and public health domain.

23.7 Conclusion

Emerging technologies have a transformational role in improving environmental sustainability and public health by providing novel and feasible solutions to complex global concerns. AI, IoT, blockchain, biotechnology and digital health developments have already shown their capacity to increase efficiency, improve decision-making and promote resource sustainability. Study results and discussion have shown that emerging technologies have significantly reduced the CO_2 emission, improved energy efficiency and improved environmental performance. Moreover, integration of these technologies in the public health domain has improved the methods of disease diagnosis, rapidly identifying possible epidemics, tailored medicine facilities, improved health monitoring systems and improved health facilities in remote areas. Future initiatives in this regard should include building strong governance frameworks, investing in digital infrastructure and increasing public confidence via education and awareness campaigns. Furthermore, policies that promote innovation along with maintaining data security, equality and sustainability must be prioritized. By overcoming current obstacles, new technologies may be efficiently used to create a more sustainable, vigorous and health-conscious future for current and upcoming generations.

References

Agbo CC, Mahmoud QH, Eklund JM (2019) Blockchain technology in healthcare: a systematic review. In: Healthcare, vol 7, no 2. MDPI, p 56

Al Khatib I, Shamayleh A, Ndiaye M (2024) Healthcare and the internet of medical things: applications, trends, key challenges, and proposed resolutions. In: Informatics, vol 11, no 3. MDPI, p 47

Ashraf SA, Siddiqui AJ, Abd Elmoneim OE, Khan MI, Patel M, Alreshidi M, Moin A, Singh R, Snoussi M, Adnan M (2021) Innovations in nanoscience for the sustainable development of food and agriculture with implications on health and environment. Sci Total Environ 768:144990

Arshad A, Shahzad F, Rehman IU, Sergi BS (2023) A systematic literature review of blockchain technology and environmental sustainability: Status quo and future research. Int Rev Econ Financ 88:1602–1622

Awuah WA, Ahluwalia A, Darko K, Sanker V, Tan JK, Tenkorang PO, Ben-Jaafar A, Ranganathan S, Aderinto N, Mehta A, Shah MH (2024) Bridging minds and machines: the recent advances of brain-computer interfaces in neurological and neurosurgical applications. World Neurosurgery

BeyraghdarKashkooli A, Farmanpour-Kalalagh K, Babaei A (2022) Yeast synthetic biology for production of artemisinin as an antimalarial drug. Synthetic biology of yeasts: tools and applications. Springer, Cham, pp 157–180

Bibri SE, Alexandre A, Sharifi A, Krogstie J (2023) Environmentally sustainable smart cities and their converging AI, IoT, and big data technologies and solutions: an integrated approach to an extensive literature review. Energy Inform 6(1):9

Boddu RSK, Karmakar P, Bhaumik A, Nassa VK, Bhattacharya S (2022) Analyzing the impact of machine learning and artificial intelligence and its effect on management of lung cancer detection in covid-19 pandemic. Mater Today Proc 56:2213–2216

Borges do Nascimento IJ, Abdulazeem H, Vasanthan LT, Martinez EZ, Zucoloto ML, Østengaard L, Azzopardi-Muscat N, Zapata T, Novillo-Ortiz D (2023) Barriers and facilitators to utilizing digital health technologies by healthcare professionals. NPJ Digit Med 6(1):161

Brooks SM, Alper HS (2021) Applications, challenges, and needs for employing synthetic biology beyond the lab. Nat Commun 12(1):1390

Capellaro C, Martín-Cuevas R, Calleja G, Halffmann P, Sharma S (2025) Application of quantum computing in the energy industry, decarbonization, and sustainability. In: Quantum technology applications, impact, and future challenges. CRC Press, pp 9–36

Das S, Chowdhury A, Sil R (2024) Quantum-powered industry. In: The rise of quantum computing in industry 6.0 towards sustainability, vol 67

Dhayal KS, Giri AK, Kumar A, Samadhiya A, Agrawal S, Agrawal R (2023) Can green finance facilitate Industry 5.0 transition to achieve sustainability? A systematic review with future research directions. Environ Sci Pollut Res 30(46):102158–102180

Dignum F, Dignum V, Davidsson P, Ghorbani A, van der Hurk M, Jensen M, Kammler C, Lorig F, Ludescher LG, Melchior A, Mellema R, Verhagen H (2020) Analysing the combined health, social and economic impacts of the corovanvirus pandemic using agent-based social simulation. Mind Mach 30:177–194

Dodoo JE, Al-Samarraie H, Alsswey A (2022) The development of telemedicine programs in Sub-Saharan Africa: progress and associated challenges. Heal Technol 12(1):33–46

Emon MMH, Khan T (2023) The impact of cultural norms on sustainable entrepreneurship practices in SMEs of Bangladesh. Indones J Innov Appl Sci (IJIAS) 3(3):201–209

Fadir FBBM, Karim HBA, AlDahoul N (2024) TrashBot: innovative recycling by utilizing object detection. In: 2024 IEEE 8th international conference on signal and image processing applications (ICSIPA). IEEE, pp 1–6

Fernández-Luque AM, Ramírez-Montoya MS, Cordón-García JA (2021) Training in digital competencies for health professionals: systematic mapping (2015–2019). Profesional de la Información 30(2)

Fisher S, Rosella LC (2022) Priorities for successful use of artificial intelligence by public health organizations: a literature review. BMC Public Health 22(1):2146

Ghosh S, Sinha JK, Ghosh S, Vashisth K, Han S, Bhaskar R (2023) Microplastics as an emerging threat to the global environment and human health. Sustainability 15(14):10821

Giansanti D (2021) Cybersecurity and the digital-health: the challenge of this millennium. In: Healthcare, vol 9, no 1. MDPI, p 62

Hain DS, Jurowetzki R, Squicciarini M, Xu L (2023) Unveiling the neurotechnologylandscape: scientific advancements, innovations and major trends. UNESCO Publishing

Hussain M, Mir GM, Usman M, Ye C, Mansoor S (2022) Analysing the role of environment-related technologies and carbon emissions in emerging economies: a step towards sustainable development. Environ Technol 43(3):367–375

Idris F, Zulkipli IN, Abdul-Mumin KH, Ahmad SR, Mitha S, Rahman HA, Rajabalaya R, David SR, Naing L (2021) Academic experiences, physical and mental health impact of COVID-19 pandemic on students and lecturers in health care education. BMC Med Educ 21:1–13

Javaid M, Haleem A, Singh RP, Suman R, Rab S (2022) Significance of machine learning in healthcare: features, pillars and applications. Int J Intell Netw 3:58–73

Joseph TM, Al-Hazmi HE, Śniatała B, Esmaeili A, Habibzadeh S (2023) Nanoparticles and nanofiltration for wastewater treatment: from polluted to fresh water. Environ Res 238:117114

Joshi S, Sharma M (2023) Assessment of implementation barriers of blockchain technology in public healthcare: evidences from developing countries. Health Systems 12(2):223–242

Karatas M, Eriskin L, Deveci M, Pamucar D, Garg H (2022) Big data for healthcare industry 4.0: applications, challenges and future perspectives. Expert Syst Appl 200:116912.

Kumar S, Lalitha Kameswari Y, Koteswara Rao S (2023) Smart grid management for smart city infrastructure using wearable sensors. In: Data analytics for smart grids applications—a key to smart city development. Springer, Cham, pp 39–63

Li X, Shen X, Jiang W, Xi Y, Li S (2024) Comprehensive review of emerging contaminants: detection technologies, environmental impact, and management strategies. Ecotoxicol Environ Saf 278:116420

Lu J, Zhang H, Xiao Y, Wang Y (2024) An environmental uncertainty perception framework for misinformation detection and spread prediction in the COVID-19 pandemic: artificial intelligence approach. Jmirai 3:e47240

Ma J, Li L, Wang H, Du Y, Ma J, Zhang X, Wang Z (2022) Carbon capture and storage: history and the road ahead. Engineering 14:33–43

Mennella C, Maniscalco U, De Pietro G, Esposito M (2024) Ethical and regulatory challenges of AI technologies in healthcare: a narrative review. Heliyon 10(4)

Mason C, Dunnill P (2008) A brief definition of regenerative medicine. Regen Med 3(1):1–5

Nair PS, Gupta R, Agrawal V (2022) Self-healing concrete: a promising innovation for sustainability-a review. Mater Today Proc 65:1410–1417

Okoli NJ, Kabaso B (2024) Building a smart water city: IoT smart water technologies, applications, and future directions. Water 16(4):557

Omaghomi TT, Arowoogun JO, Akomolafe O, Odilibe IP, Elufioye OA (2024) Telemedicine in rural Africa: a review of accessibility and impact. World J Adv Res Rev 21(2):421–431

Panda SK, Satapathy SC (2021) Drug traceability and transparency in medical supply chain using blockchain for easing the process and creating trust between stakeholders and consumers. Pers Ubiquitous Comput 1–17

Pandya S, Thakur A, Saxena S, Jassal N, Patel C, Modi K, Shah P, Joshi R, Gonge S, Kadam K, Kadam P (2021) A study of the recent trends of immunology: key challenges, domains, applications, datasets, and future directions. Sensors 21(23):7786. https://doi.org/10.3390/s21237786

Pata UK (2021) Do renewable energy and health expenditures improve load capacity factor in the USA and Japan? A new approach to environmental issues. Eur J Health Econ 22(9):1427–1439

Pathak J, Xavier KM, Ngasotter S, Goswami A, Hazarika U, Saikia R (2025) Sustainable nanotechnology for green environment. In: Waste derived carbon nanomaterials, vol 1. American Chemical Society, pp 17–39

Rossoni AL, de Vasconcellos EPG, de Castilho Rossoni RL (2024) Barriers and facilitators of university-industry collaboration for research, development and innovation: a systematic review. Manag Rev Q 74(3):1841–1877

Ranard BL, Park S, Jia Y, Zhang Y, Alwan F, Celi LA, Lusczek ER (2024) Minimizing bias when using artificial intelligence in critical care medicine. J Crit Care 82:154796

Rasool M, Malik A, Waquar S, Arooj M, Zahid S, Asif M, Shaheen S, Hussain A, Ullah H, Gan SH (2022) New challenges in the use of nanomedicine in cancer therapy. Bioengineered 13(1):759–773

Ruan G, Qiu D, Sivaranjani S, Awad AS, Strbac G (2024) Data-driven energy management of virtual power plants: a review. Adv Appl Energy 100170

Samir A, Ashour FH, Hakim AA, Bassyouni M (2022) Recent advances in biodegradable polymers for sustainable applications. Npj Materials Degradation 6(1):68
Sharma A, Singh KJ, Kapoor DS, Thakur K, Mahajan S (2024) The role of IoT in environmental sustainability: advancements and applications for smart cities. In: Mobile crowd sensing and remote sensing in smart cities. Springer, Cham, pp 21–39
Singh M, Ahmed S (2021) IoT based smart water management systems: a systematic review. Mater Today Proc 46:5211–5218
Soto EA, Hernandez-Guzman A, Vizcarrondo-Ortega A, McNealey A, Bosman LB (2022) Solar energy implementation for health-care facilities in developing and underdeveloped countries: overview, opportunities, and challenges. Energies 15(22):8602
Subbiah V (2023) The next generation of evidence-based medicine. Nat Med 29(1):49–58
Usman M, Khan N, Omri A (2024) Environmental policy stringency, ICT, and technological innovation for achieving sustainable development: assessing the importance of governance and infrastructure. J Environ Manage 365:121581
Vilchez JA, Kruse J, Puffer M, Dudovitz RN (2021) Teachers and school health leaders' perspectives on distance learning physical education during the COVID-19 pandemic. J Sch Health 91(7):541–549
Wager KA, Lee FW, Glaser JP (2021) Health care information systems: a practical approach for health care management. J Pers Med 11(2):135
World Health Organization (2023) Regulatory considerations on artificial intelligence for health. World Health Organization
Xiang Y, Chen Y, Xu J, Chen Z (2022) Research on sustainability evaluation of green building engineering based on artificial intelligence and energy consumption. Energy Rep 8:11378–11391
Yaqoob I, Salah K, Jayaraman R, Al-Hammadi Y (2022) Blockchain for healthcare data management: opportunities, challenges, and future recommendations. Neural Comput Appl 1–16
Zeng D, Cao Z, Neill DB (2021) Artificial intelligence–enabled public health surveillance—from local detection to global epidemic monitoring and control. In: Artificial intelligence in medicine. Academic Press, pp 437–453
Zhang F, Chan AP, Li D (2023) Developing smart buildings to reduce indoor risks for safety and health of the elderly: a systematic and bibliometric analysis. Saf Sci 168:106310
Zhang X, Shu K, Rajkumar SA, Sivakumar V (2021) Research on deep integration of application of artificial intelligence in environmental monitoring system and real economy. Environ Impact Assess Rev 86:106499

Index

A. Roy et al. (eds.), *Environment and Public Health*, Sustainability Solutions,
https://doi.org/10.1007/978-3-031-99770-9

MIX
Papier aus verantwortungsvollen Quellen
Paper from responsible sources
FSC® C105338

If you have any concerns about our products,
you can contact us on
ProductSafety@springernature.com

In case Publisher is established outside the EU,
the EU authorized representative is:
Springer Nature Customer Service Center GmbH
Europaplatz 3, 69115 Heidelberg, Germany

Printed by Libri Plureos GmbH
in Hamburg, Germany